SCIENCE AND CULTURE

BOSTON STUDIES IN THE PHILOSOPHY OF SCIENCE

VOLUME 231

SCIENCE AND CULTURE

By

JOSEPH AGASSI
Tel-Aviv University and York University, Toronto

KLUWER ACADEMIC PUBLISHERS
DORDRECHT / BOSTON / LONDON

A C.I.P Catalogue record for this book is available from the Library of Congress.

ISBN 1-4020-1156-3

Published by Kluwer Academic Publishers
PO Box 17, 3300 AA Dordrecht, The Netherlands

Sold and distributed in North, Central and South America
by Kluwer Academic Publishers,
101 Philip Drive, Norwell, MA 02061, USA

In all other countries, sold and distributed
by Kluwer Academic Publishers,
PO Box 322, 3300 AH Dordrecht, The Netherlands

Printed on acid-free paper

Printed and bound in Great Britain by MPG Books, Bodmin, Cornwall.

For Mario Bunge:
Grand philosopher,
Steadfast friend.

… all we know is infinitely less
than all that still remains unknown …
(William Harvey, *Disquisition*, Dedication)

The most beautiful thing we can experience is the
mysterious. It is the source of all true art and science.
(Albert Einstein, *What I believe*)

… the world is not rational, but …
we [should] submit or subject it to reason,
as far as possible.
(Karl Popper, *The Open Society and Its Enemies*)

ABSTRACT

Two different false philosophies divorce science from culture in general and from tradition in particular: scientism and relativism.

Scientism is the view of science as certain (or almost certain) plus the demand to endorse and act on scientific ideas. It is the view that science and science alone embodies rationality. It thus negates or ignores the rest of culture. It seemingly supports science, but it does not. By recommending the authority of science in preference to the autonomy of the individual, it is authoritarian at heart. As autonomy is essential for rationality, as rationality is essential for research, and research is essential to science, scientism is opposed to science. (It views science as given, not as a growing concern.)

Relativism shares with scientism all its defects and not all of its advantage. Relativism limits the promise of scientism to the scientific culture. It is the view that every culture embodies its own rationality, and it is then in force only locally. It seemingly supports science locally but its aim is to support the ignorance of science that is shared by many cultures, including the parts of the academy where familiarity with science is rare. Relativists boast a broad familiarity with diverse cultures but they boast scientific illiteracy. They approve of any tradition, and so it is authoritarian in the name of science or in the name of some local tradition. It is seemingly tolerant in limiting each authority to its own domain, but only at the cost of separating the different domains and making each of them the local unrestricted authority. It is authoritarian and irrationalist.

Science is an integral part of culture. All traditional claims for scientific certitude (or for certitude surrogates) are false and obstacles to its freedom and integration in the general culture. It is the most rational intellectual activity, but not the only one. It is special in its systematic effort to comprehend, and its persistent critical attitude towards its own products. This way it plays diverse roles within human culture, and its contribution to it deserves distinct study.

Today science bestows high status on its practitioners and associates. They tend to behave as guardians of the interests of science. They often expect philosophers and historians of science to advocate scientism. It is useless to attack these guardians of the interest of science. Their hostility is the mirror image of that of relativists, who are often ignorant of science. Both scientism and relativism ignore the cultural value of science. They share the coarse view that science is only for scientists. Popular science is vital for culture at large but also for science. We must regularly seek newer, friendlier ways to make scientific ideas popular, to apply them to broadening areas, and to use them to widen horizons and rationalize life. Popular science and technology, the history of science and of technology, applied science – especially social science – they all break the isolation of science from the rest of culture.

Scientism imposes the edicts of scientific reason; relativism limits this imposition to the scientific culture; irrationalism opposes it, if not altogether, then at least where things matter. They are all in error: the brain is a powerful organ and its use invites no justification and no imposition and no outside constraints. We may learn to use it with no deep problem, the way we use any other organ.

The book is in five parts, dealing with autonomy, tolerance, reason, philosophy, and responsibility. For better or worse, these are the right topics to put on top of the agenda of democratic philosophy today. They share the idea of critical rationalism in a simple manner. It is useless to identify reason with certitude. Doing so has led to scientism and to the relativism that is an effort to limit it without challenging the quest for certitude, as this quest is for an escape from autonomy and from responsibility. Either scientism or relativism leads inexorably to intolerance and to irrationalism. When people say, they cannot decide as they have insufficient information, they say, they wish the fact and the rules of rationality to decide for them. Yet, decisions require deliberations, and the information that may influence the deliberations is necessary means for making decisions responsible. Some philosophers advocate the irrationalism that decries deliberations by appeal to gut feelings; others advocate the relativism that decries deliberations by appeal to established norms; and still others advocate the scientism that decries deliberations by appeal to the authority of science. All these dodge the obvious: we should make decisions as responsibly and reasonably as possible and then hope for the best.

TABLE OF CONTENTS

3. REASON

4. PHILOSOPHY

5. RESPONSIBILITY

FOREWORD

Thirty years ago, Wartofsky and I introduced Joseph Agassi's *Science in Flux* with appreciative flattery. We wrote that he is a critic, a gadfly, a debunker and deflater, a fun-loving Peck's bad boy, sharp and pungent, puzzling, perverse, and provocative. He pursues a constructive path in his books and essays, but rarely in a straight line. Digression is heart-warming means for him to bring a point home, and here in these newly gathered studies of *Science and Culture* the reader will be cajoled, tempted, fascinated, and apparently misled along many garden paths. But we said then, and I say it all the more again now: stay with him, for Agassi means exactly what he says.

Thus, in his deep critique of Arthur Koestler's book on the scientific revolution, *The Sleepwalkers*, Agassi takes as his first motto the bold and dreamy words of the great Kepler: "If a science has to be supported by fraudulent means, let it perish". Would that it were so plain, we hear Agassi mumbling as he brings to bear his insightful meditations on the social, historical, cultural contexts of the sciences. A bit later in his second collection for us, his *Science and Society*, he seizes upon Max Weber's classic lecture on 'science as a vocation', translates it to Weber's 'scientific religion', locates Weber's ultimate irrationalism, and headlines for us, in emotional neon lights, Weber's "Picture of the Scientist as a Maniac".

What sort of philosopher of science will hold such a point of view, or such a style of argument? In a brief sharp contribution to a volume honoring Marx Wartofsky, Agassi put the matter lucidly. How clear it seems: optimism and pessimism play cultural roles, for they are myths. Agassi neatly admits how crude they are in his typically condensed way: "since the world is a set of lost opportunities, as any side-glance at the past indicates, it is really impossible to say which of the two is nearer the mark". But myths carry us, and he cites the ominous words of Levi-Strauss: *myths think for us*. How to understand myths in cultures, optimisms and pessimisms, and all the partial in-betweens, in different cultures, and thereby see the optimistic rationalism of science in our culture too? So Agassi will have you wrestle with the putative myth of science, but with his critical momentum always directing our way, ever the falsificationist way, we can avoid both relativism and self-assured scientism. This volume, *Science and Culture,* carries forward his philosophy of science, science in relations to contexts, embedded within culture, science teasing, leading, at times drenching culture. But Agassi is clear: there is no boundary between science and the rest of culture.

R.S. Cohen
Boston
March 2003

REFERENCES

Joseph Agassi, *Science in Flux* (Boston Studies 28, 1975).
Joseph Agassi, *Science and Society* (Boston Studies 65, 1981).
Joseph Agassi, 'The Philosophy of Optimism and Pessimism', in Carol C. Gould and R.S. Cohen (eds), *Artifacts, Representations and Social Practice* (Boston Studies 154, 1994).

PREFACE

It is a great pleasure to see this new selection of essays of mine, *Science and Culture*, appear in the highly esteemed *Boston Studies in the Philosophy of Science* series. It is a sequel to earlier selections of my essays there, *Science in Flux* (1975) and *Science and Society* (1981). My collections of essays (among them my *The Gentle Art of Philosophical Polemics,* 1988), include chapters that have appeared earlier in learned periodicals, but are hardly accessible. They all appear here in revised versions to suit general audiences as well. Reading this book requires no special prior knowledge. Let me state now all the philosophical information that I expect my readers to know. Here it is.

Philosophy comprises attitudes, theories, and their criticism. The basic attitude is of curiosity, detached but friendly. The theories answer basic questions. Among their developments are science and democracy. Science appears in different guises. Here are the four best-known portraits of it.

1. The oldest view of science is intellectualism: science rests on purely rational foundations, with no need to appeal to experience. Foremost thinkers from Plato through Galileo and Descartes to Kant have advocated it. Champions of the dominant view of science malign it. I will reluctantly ignore it here.

Intellectualism fails. Each guarantee for correctness invites a guarantee for its own correctness. This is the argument from infinite regress.

2. The dominant view of science is empiricism – science rests on experience – and inductivism (of Francis Bacon, the father of modernism). It is the view of science as error avoidance: science disregards all tradition and relies on facts alone to guarantee the truth of its theories (or at least their high probability).

Inductivism fails too, and for the same reason for which intellectualism does.

3. Consequently, instrumentalism (of Pierre Duhem) suggests that science is but a mathematical tool-kit. Depriving scientific theories of informative content, instrumentalists remain with no view of the world. They endorse some traditional or arbitrary views of the world. Viewing science as merely practical (as applied mathematics) they cannot possible explain its practical success.

4. The only serious alternative to these theories is the critical view (Karl Popper) that takes scientific theories at face value as true or false and research as the process of explanatory conjectures and their tests – their attempted refutations. It is an endless process of error elimination. Technology is a social affair: to explain it we need theories from the social sciences to bridge between the

theories that we apply and their successful applications. The idea that applications of applicable technology are unproblematic is obviously false.

This is all the required background information. I will explain it repeatedly. My main concern here is with social and political affairs with an accent on individual autonomy, liberalism, pluralism, and democracy. Science is in no need for defense. At times I describe the evil of some philosophical arguments – as more anti-democratic than as anti-scientific – because democracy always needs boosting and science does not. But this is hardly a challenge: in the context of boosting democracy, only pluralism is somewhat challenging, since tradition regards science as a set of views that command unanimity. This situation invites criticism in the form of discussions of liberalism and pluralism in manners agreeable to the scientific culture. This is a challenge I try to meet.

Sociological tradition separates the two concepts, of science and of culture. Around 1900 Ferdinand Tönnies made a distinction between community and society as well as between culture and civilization. His aim was to extol what they call today family values and to put down science as sheer gadgetry of no moral worth. When C. P. Snow spoke of the two cultures, the arts and the sciences, he brought about a protest from F. R. Leavis, the then leading professor of literature. I share Snow's view of the scientific culture as parallel to the artistic one, and the religious and the political ones besides. Unlike Snow, I find no boundary between them, and the contrary view as an excuse for ignorance. All ignorance, however, is poverty: it is no crime, and so it needs no excuse.

What flatters me most about my public image is the article on me in the Russian dictionary of philosophy. It reports on my view of science as an integral part of culture. Many works of the cultural background to science are exciting. It is my conceit to suppose that I have anchored them in the philosophy that I have acquired from Popper. For this he has my constant and profound gratitude. (My deviations from his views come to better accommodate this central idea that all reasonable people share.)

Science is even more exciting than its background. Despite all its defects, the academy that houses it is just about the best part of modern society. It is perhaps the only modern institution with relatively free entry, and one that punishes no one for excellence (unless one insists). Comparing the commonwealth of learning to the market or a university to a business concern, my assessment stands: scholarship is not quite open, but it is more open than business. Though the academy is relatively open, the world of learning should remain more open. At least academics should always remember that one might be a scholar without being a member of an institution; and *vice versa*. The worst thing about the commonwealth of learning today is its current tendency to identify the scholar with the academic. The commonwealth of learning is wider and more open than the academy. The difference between the two will diminish when the academy will be more open and less of an exclusive club. We should do all we can to make it so.

Herzlia, Israel, Summer, 2002.
WebPages: http://www.tau.ac.il/~agass/

ACKNOWLEDGMENTS

Earlier versions of the following chapters in this volume have appeared previously. I am grateful to editors and publishers for permissions to republish. The following list of sources indicates my indebtedness.

1.1 "Science Fiction: This Message is for You. Maybe" (originally titled "This Message is for You. Maybe") previously appeared in *Philosophy and Literature*, 7, 1983, 95-8.

1.2 "The Consolations of Science" previously appeared in *American Philosophical Quarterly*, 23, 1986, 129-41.

1.3 "The Moral Base of Science, or, The Architectonic of Open-ended Reason" previously appeared in *Iyyun, The Jerusalem Philosophical Quarterly*, 48, 1999, 93-110.

1.4 "The Theory and Practice of Critical Rationalism", previously appeared in J. Misiek, editor, *Rationality: On the Problem of Rationality of Science and Its Philosophy. Popper vs. Polanyi. Boston Studies in the Philosophy of Science*, 160, 1995, 1-18.

1.5 "Science and the Philosophy of Autonomy" (originally titled "Autonomy and the Philosopher") previously appeared in *Methodology and Science*, 25, 1992, 1-10.

1.6 "Science and the Call of the Wild" (originally titled "The Place of Sparks in the World of Blah") previously appeared in *Inquiry* (www.tandf.no/~inquiry), 24, 1980, 445-69, by permission of Taylor & Francis AS.

2.1 "Faith in the Open Society: the End of Hermeneutics" previously appeared in *Methodology and Science*, 22, 1989, 183-200.

2.2 "The Functions of Intellectual Rubbish" previously appeared in *Research in the Sociology of Knowledge, Science, and Art*, 2, 1979, 209-27.

2.3 "Science and the Interpersonal" previously appeared in *Studies in Science and Culture*, 2, 1986, 30-46.

2.4 "Science and Pluralism" (originally titled "Pluralism in Science") previously appeared in *Methodology and Science*, 24, 1991, 99-119.

2.5 "Wild Goose Chase" (originally titled "Towards a Rational Theory of Superstition") previously appeared in *Zetetic Scholar*, 107-29, 1979.

2.6 "Science and its Public Relations" (originally titled "Towards An Honest Public Relations of Science") previously appeared in *The Significance of Popper's Thought, Poznań Studies in the Philosophy of the Sciences and the Humanities*, 49, 1996, 39-57.

2.7 "Irrationalism Today" previously appeared in *Dialectica*, 36, 1982, 465-80.

3.1 "Minimal Criteria for Intellectual Progress" previously appeared in *Iyyun, The Jerusalem Philosophical Quarterly*, 43, 1994, 61-83.

3.2 "Bloodletting" (originally titled "The Concept of Scientific Theory as Illustrated by the Practice of Bloodletting") previously appeared in *Medical Opinion and Review*, 5, 1969, 156-69.

3.3 "The Critique of Linearity" (originally titled "The End of Linearity: The Science-Technology Interaction from a Philosophical Perspective"), previously appeared as a Working Paper, Department of the History of Science and Technology, Royal Institute of Technology, Stockholm, 1996.

3.4 "Science, Politics, and Objectivity" (originally titled "Objectivity in Science and Politics") previously appeared in *Philosophy and Social Action*, 10, 1984, 191-6.

3.5 "Science and the Detective Novel" (originally titled "The Detective Novel and Scientific Method") previously appeared in *Poetics Today*, 3, 1982, 99-108.

3.7 "Science and Technology" (originally titled "How Technology Aids and Impedes the Growth of Science") previously appeared in P. D. Asquith and T. Nickles, editors, *Philosophy of Science Association, Proceedings, 1982/83*, Vol. 2, 1983, 585-597.

4.1 "Science and Philosophy" (originally titled "The Philosophy and the Sciences of Man") previously appeared in *Specificité des sciences humaines et tant des sciences, The International Academy for the Philosophy of Science*, Brussels, 1979. Also published in *Epistemologia*, 2, 1979, Special Issue, 155-66.

4.2 "Progress in Science and in Art" (originally titled "Progress in the Arts and the Sciences") previously appeared in *Sonus, A Journal of Investigations into Global Musical Possibilities*, 18, 1998, 57-68.

4.3 "Technology as both Science and Art" (originally titled "Technology as Both Art and Science") previously appeared in *Research in Philosophy and Technology*, 6, 1983, 55-63.

4.4 "Artificial Intelligence" previously appeared in *SIGart Newsletter,* 105, July 1988, pp. 15-22.

4.6 "Science and Art" (originally titled "Art and Science") previously appeared in *Scientia*, 73, 1979, 127-140.

5.1 "Validation" previously appeared in *Iyyun, The Jerusalem Philosophical Quarterly*, 47, 1998, 57-76.

5.2 "The Politics of Science" previously appeared in *Journal of Applied Philosophy*, 3, 1986, pp. 35-48.

5.5 "Rationalizing Politics" previously appeared in N. Oren, editor, *Intellectuals in Politics*, The Magnes Press, the Hebrew University, 1984, 82-100.

In addition to publishers, editors and referees, many friends and colleagues have my gratitude for their repeated help. They are too many to list. For help in preparing this volume I am grateful to Nili Alon, Nimrod Bar-Am, and particularly Chen Yehezkeli, all of Tel-Aviv University, as well as to Stefano Gattei of the University of Milan, Robert S. Cohen of Boston University and to I. C. Jarvie of York University, Toronto.

FIRST INTRODUCTION: SCIENTISM AS A SETBACK

The thesis of this volume is trite, but some corollaries from it are not. This should correct some widespread misconceptions that set back even the most advanced fields of study. Here is my thesis. Science is an integral part of the modern culture. For some purposes, it is reasonable to view it in isolation. But to ignore that isolation of images distorts them is to invite confusion. The distorted image is then misleading. It leads to scientism, to the view that science is, or can be, the sole part of our culture that matters intellectually. If so, then the unscientific in our culture is a throwback that we should make as marginal as possible. Scientism is a dangerous setback. It has harmful consequences even for science. But it is also appealing. This appeal merits better study – more appreciative and more critical – than it has thus far gained.

Scientism is appealing in its projection of an extremely unproblematic, simple image of reason. It does so with immense authority. It solves many troublesome problems and it prevents others from appearing. It is a version of quietism. It does not look this way, because it includes the exhortation to invest in science as much of one's energies as possible. But this is the exhortation to overlook the social and individual context of the students of science. Hence, it is the exhortation to behave as scientists within its context and to behave like the lilies in the fields otherwise. Scientism is also a false prophecy. It does not look this way because it seems to be the refusal to assert anything other than what science already asserts, and with sufficient authority. Yet, scientism allows, nay, encourages, its adherents to conform to their social milieu in matters cultural, social, political, and religious. And, indeed, surprisingly, they are exceptionally willing to conform. They do not look that way, because they conform to the norms of the social sub-culture that their society respects most. But this is nonetheless the fact. It is evident from the speed with which they can change from the conservative to the radical, when the norm of their sub-culture turns radical, and back again when fashion alters.

Scientism is dangerous because it projects an extremely poor image of humanity and it does so with authority. We see this when we take in isolation any other aspects of humanity, any other valuable aspects of human life. This way the image of humanity becomes too fragmented, and thus too distorted. When we ignore the interaction between the different aspects of life, they all lose their meanings. We cannot possibly hope to achieve an integrated image of humanity by simply putting its diverse aspects together without catering for their interactions. Perhaps it is impossible to achieve this integration anyway. But we can try, and we can hope for a better outcome if we seek the necessary modifications in order to weld the pieces together. This search is fascinating. In particular, it involves questions about the role that each aspect of human culture – here I center on science – plays in some other part of human culture – the diverse arts, religion and theology, politics, interpersonal relations, and even superstition. And we may examine the interplay between these diverse aspects of culture. We may find such questions intriguing, and studying them profitable. It is trite to say that every item of culture exhibits some characteristics that are more specific to other items. Things may go in many strange directions. We may consider characteristics that are specific to one field of culture, be it art or science, and ask what role these characteristics play in the other field, be it science or art. Einstein touched a chord when he said that Mozart influenced him and that he found Bohr's model of the atom harmonic. Was this a mere analogy? Or was there a theory behind it? This depends not only on what we think art is, or science, but also on what we think analogies are, and what is their value. If it sounds as if this may be a wild goose chase, then perhaps we should add that Einstein did not take such things lightly. This need not be convincing, but it is intriguing enough to merit some initial investigation. More I do not claim to be trying to do.

Scientism is the traditional view of science. It is taking science in isolation as a matter of course – quite implicitly but quite clearly all the same. The implicit manner in which it enters our discourses makes its message particularly powerful. And the implicit view that so many serious thinkers find so self-evident that it needs no statement and no examination is that science is perfection itself. This view is also license for those ignorant of science to ignore it. This is obvious from its obviously erroneous corollary that many cultural historians take for granted and use as their guide: they deem science an exception, as it has no need for any background. They take it for granted then that cultural history can do without taking notice of science. (They do pay much attention to fruits of scientific technology, though.) Similarly, historians of science concluded that the context of science is irrelevant to the understanding of its inner logic. This is pure internalism. It is too limiting, though it is infinitely better than pure externalism, the novel idea that the cultural context of science fully determines it.

Scientism is a purely internalist view of science. It ignores some obvious facts of the influence of science on culture such as the Copernican Revolution and the Darwinian Revolution that have changed our view of our place in God's world. And it ignores the absence of science from most places, an absence that makes obvious the scarcity of the preconditions for the very possibility of its presence. It also ignores the strong influence of scientific technology on other aspects of culture – for better or for worse. These are serious oversights. They are legitimate for some studies, since every study proceeds in disregard for much that is significant and interesting. It is the systematic oversight that is at fault. Adherents of scientism deny that they overlook all this. Rather, they demand that science itself should mend this oversight. They therefore advocate scientific research, such as the study of the sociology of societies that endorse scientific ideas. This attitude breaks down in its application to influences in the opposite direction. Social conditions challenge technology and technology may challenge science. Scientism allows for the fact that science accepts challenges and attacks given problems. It ignores the context in which these problems arise. Again, adherents to scientism may delegate such questions to different fields of scientific research. And, of course, adherents to scientism may endorse all that we may wish to say about science and culture, as long as what we say will be scientific. And so, to endorse the scientific attitude without the wish to ignore any aspect of life, leads to the claim that all study of culture should be scientific. What, however, is the status of culture as such, of art or religion as experience, not as subject to research? What is the status of science itself as experience, not as subject to scientific study? Is the life devoted to science always the best? If not, under what conditions is it? We do not know, and we should try to find out. Scientism is in the way, however. We will do better without it.

Perhaps the culturally most relevant aspect of science is its philosophy. And the culturally most relevant aspect of its philosophy is the place of internalism in culture as such and particularly in the scientific culture. As long as the view of science as certain knowledge prevails, it may serve as support of the view of it as free of the constraints that contexts place on human conduct. The argument that science has a context is ineffective as long as certainty is assured. For, certainty amounts to having taken into account all relevant aspects of the context in which it obtains. Scientific proof is the proof of the power of humans to achieve perfect intellectual independence. (It does not matter here what scientific proof is; what matters here is the question, does it exist?) So, scientific proof can serve as proof of the ability of humans to develop and to do so with no support from any church. Science has offered the faith in the siblinghood of humanity. Under the impact of the tremendous revolutions in physics in the beginning of the twentieth century, we had to give up the view of science as certain (or very probable) knowledge. And then, rather than view its value despite its fallibility – not to say because of it – the view was proposed, and became popular, that science is no knowledge at all, but a mere instrument, a mere technology. The recognition of the fallibility of science has led to the despair in reason. This is a mistake, since some mistakes, especially some great scientific theories, are very interesting and exciting. Otherwise, science would not be so fruitful and productive despite its imperfection. This is the basic thought that guides this volume. I have learned it from the writings of Einstein and more so from the teaching of Popper. His views of science, too, were a wonderful mistake. In particular, he still tried his best to be right and to struggle with every item of criticism that anyone offered him, especially some of his former students, myself included. This was valiant, but not always successful. What I value most in what he has taught me is his constant emphasis on the fact that our heritage is a series of errors that should not be forgotten, some of which are admirable and some of which are even noble. This is the idea that I hope runs through this volume.

Adherents of scientism may easily endorse the idea of the sociology of science as its pioneer, Robert K. Merton, has developed it in the first half of the twentieth century. He wrote after the crisis in physics, yet he trusted scientific proof. He took as his starting point the sociology of knowledge, the study of popular views about the world, in disregard for the fact that experts deem them false. It rests on the (true) idea that popular views play important roles in society. Merton developed the sociology of science on the assumption that what holds for false or possibly false ideas *a fortiori* holds also for true ones, namely for science. He suggested that the scientific revolution owed its very existence to the rise of Puritanism. And he developed a hypothesis about the code of scientific research in action. Decades later, a few empirical studies about laboratory life have appeared. Their authors offered refutations of Merton's hypothesis about the code of scientific research in action. They report observations about violations of that code. The situation is not clear, as it is not clear what exactly his hypothesis says, and what is the status of the empirical findings that seem to belie it. Nevertheless, something important is lurking here. If science is perfect, then at least we have trouble saying what part of our scientific culture has reached the lofty status of science. Not the raw data that it includes, for example. But then, all work is incomplete. So, we need not try to sift the perfect from the imperfect. Science is noble, warts and all.

People find it amazing that mistakes can be noble, useful, or interesting. This is amazing. For, we can all remember examples of such mistakes – from history, from stories, and from our own limited experiences. Otherwise, Popper could not possibly have thought out his view of science as a series of conjectures and refutations.

The leading question within the philosophy of science today concerns rational belief: what gives credibility? This question is intriguing and paradoxical, since scientific theories usually meet with great incredulity. True, researchers often find that the incredulity melts away; often rejection and seldom endorsement replace it; researchers then may swear by new ideas that seemed to them absurd only recently. How come? More generally, what makes them prefer to believe in one theory rather than in another? Everyone takes it for granted that somehow this relates to experience. Observations back up theories and render them credible and then the credence in them, of experts and then of the population at large.

Though the factual part of the previous paragraphs is fairly accurate, the ideas in it are false, and more so the ideas behind it. It alludes to many of the social and cultural aspects of science. The philosophical literature takes them for granted as if they were utterly unproblematic. There is the hint that the way researchers behave is the right way, and that the task of philosophers is merely to justify it. There is the supposition that others do and should take what researchers tell them upon trust. This is somewhat idealized. Should we not try to work without this idealization? No, suggests the standard literature on the philosophy of science: let us go on justifying science by-and-large until we succeed in this, and only then, we will attend to fine-tuning. But perhaps this is the wrong approach. Popper suggested that knowledge of when science goes off the road is informative about the road. And at least we should ask, when does science go off the road?

Science is important for different reasons, all valid. It is tremendously useful. It is tremendously interesting. It helps rethink many things that we have wrongly taken for granted. No part of our culture is free from a strong interaction with science, not even its most stable parts, such as religion. It is not easy to be a fanatic in the face of facts. And science discovers new facts in efforts to be critical. Critics need not force their findings on the public. It is enough that the facts are on public display. One may avert one's gaze from them. This has only a temporary effect. Science keeps culture changing and this enriches life in many ways, among them making things more interesting. It helps develop autonomy, democracy, and peaceful coexistence with as many cultures. This deeply involves the imperfection of science. The claim that science is perfect in any sense is a residue of the idea that only divine revelation is valuable, that everything human is ephemeral and so worthless. This conflicts with the humanism that honors humanity despite its imperfection.

Scientism is parasitic on science. It is useful, however, as a support for science. In traditional societies it is potent, as it may catapult individuals from superstition to science. But their sticking to scientism is their uncritical absorption of the culture and values of current scientific society, making them strangers at home. I will not discuss this ailment in the present book. You may find more about it in my *Technology: Philosophical and Social Aspects.*

Relativism is parasitic on science too, but only after denuding it of all cultural value. Its isolationism thus allows technology transfer, but not the transfer of liberalism and of democracy. The exception to this is Marxism. If you hear a rustle now, then it is Marx turning in his grave, muttering, "I am not a Marxist", which, it is well known, he said when his fans embarrassed him. I will not discuss this ailment either. You may find more about it in the ever-fresh classic of Karl Popper, *The Open Society and Its Enemies.*

SECOND INTRODUCTION: SCIENCE AND POPULAR CULTURE

Let me sketch a popular image of science and of culture as my parents saw them. For, I found almost no description that fits it in the many elevated circles that I was fortunate to roam and in the varied literatures that came my way through the decades of my happy academic career. This is strange, since, before the twentieth century, science and culture grew in the community at large.

Currently, experts view popular culture as low and thus as poor. They often deride it with expressions of indulgence, perhaps laced with traces of compassion. The obvious refutation of their view is their own appraisal of medieval popular culture. It was low and yet admirable. But let this ride. Their view, the quasi-official view of the academy, is inapplicable to the world in which my parents spent their lives, where culture (including science), like everything else there, appeared as a mix of excitement and crushing heartache. They comprised for them a scarce and vital commodity that helped lessen the pain of their everyday existence.

My mother had no schooling at all, since orthodox Jews – even scholars like her father – limit education to religious studies and bar women from them. Orthodox Jewish women often dedicated their lives to providing for their husbands, who dedicated their lives to poring over sacred books. These women had usually no opportunity to acquire even some vocational training. My mother shared the fate of her sisters in this backward culture. In line with ancient tradition, she was married off early in life to an equally young man whom she first met under their wedding canopy. At the demand of tradition, she had to submit to an immediate consummation of the marriage. She referred to that day as to her burial day, as other of her sisters in that culture did. She was barely literate. She could read the newspapers, at least the headlines, and write a few words, which she did in the margins of her children's letters to her kin. Yet, she had contacts with science and with culture at large. Growing up late in nineteenth-century in a far corner of the Ottoman Empire, in isolated, neglected Jerusalem, she learned about hygiene and nutrition and world affairs. This is amazing. She was fluent in the vernacular Arabic in addition to Yiddish, her mother tongue. Zionism changed all that. Her children had to speak Hebrew. She helped her children acquire it as their mother tongue. She taught us proper Hebrew, though I suppose she never saw a grammar book. My parents, both Yiddish native speakers, switched to speaking only Hebrew. The little Yiddish that I have, I picked up from aunts and other relatives who failed to acquire the new language in adult life.

My father came from a Hungarian-speaking small town in a far corner of the Austro-Hungarian Empire. In his youth he showed talent and his community sent him to the big city to study Jewish Law. Aspiring to the life of an intellectual, he lost favor with his rather backward supporters. He could not support himself. He migrated to Palestine. He never integrated into either section of the Jewish society there. The old sector was too strict for him; the new was too lax. He functioned as a schoolteacher for a while. He was soon rudely disqualified for teaching and thus for a proper career. With no contacts, he was helpless against the Zionist establishment. Bitterness consumed him. He wanted his children to succeed where he had failed. What I most value of what he gave me is what he shared with my mother – the love of learning.

My parents often took me to public lectures – separately; they hardly ever went out together. These were the obvious sources of my mother's astonishingly broad, up-to-date knowledge. I remember that I learned from some of these public lectures and from her comments on them more than from my proper schooling. Adult education remained always very dear to me, and I always had very warm feelings towards researchers who paid special attention to it – Galileo, Boyle, Euler, Faraday, Maxwell, Boltzmann, and Einstein – not to mention philosophers – Russell, Buber, Popper – and artists. Schoenberg and Sessions, Casals and Menuhin, and Daumier and Chagall stand out, but there were many others. During my scientific training, a philosophical problem took me in. It was the problem of demarcation of science, namely, the question, what is peculiar to science? But I felt no kinship with almost any of the philosophers whose works on the matter came to my notice. They hardly guarded themselves against the idea that science is an exclusive club and that a proper solution to the problem of the demarcation of science should help the bouncers of the club. Some of them made arrogant virtue of the sad need for specialization. I met this idea repeatedly during my scientific training and throughout my career. Its classical popular expression is the works of Thomas S. Kuhn that I always found unpleasant.

My teacher, Karl Popper, who always spoke of popular science admiringly and of éltism scathingly, did not sufficiently prevent the reading of his own philosophy of science as élitist. He spoke of science as an *élite* club of ideas, not of people, yet this is sufficiently unpleasant: surely, egalitarianism is no less noble for its not belonging to science. Nevertheless, his demarcation of scientific theories as refutable, of science as fallible, fascinated me, partly as it links science with other human endeavors but mostly because of my view of honesty and responsibility as rooted in fallibility. This idea, as the faith in the siblinghood of humanity, was always with me.

My training in physics was a disaster, since I could not understand the first steps of the quantum revolution. It took me years before I had the opportunity to study it properly. I wrote a book on it (*Radiation Theory and the Quantum Revolution*, 1993) of which I am proud, as it addresses audiences both inexpert and expert (physicists and historians of physics). This made it barely publishable. Thirty years passed before I found a publisher for it. I want more people to write books like it, and make science more accessible. If this story elicits a complaint, then the complaint should address the present ethos of science. So, let me discuss it.

The present ethos of science sways uncomfortably between the radicalism of the Enlightenment Movement and the Romantic Reaction to the inflated admiration for science. Auguste Comte presented science (approvingly) as a religion. He called it "positivism". It considers science the sole arbiter in all matters as far as is at all possible. Positivists wanted all practices, all traditions, all customs, and all institutions, to undergo scientific analysis and join science unconditionally or die. This all-or-nothing attitude is intolerant and thus intolerable. The Romantic Reaction could not dismiss science as easily as positivism rejected traditions, so it presented science as one of the components of the output of the human spirit: they placed science, culture, religion, and politics, side by side.

Culture is whatever undergoes cultivation. Archeologists often have to make do with things that dead cultures have left behind. They call them "material culture". Sociologists, by contrast, or rather some of them, speak of our material culture as a civilization, in contradistinction with ethereal culture – the colors of paintings, the sound of music, the charm of words in love poems. They have to view some buildings as works of art and so as vessels of culture, especially medieval cathedrals and castles in Spain. But they value the great buildings themselves no more than the coarse canvasses that display magnificent paintings. These sociologists smuggle the material as vessels of the spiritual, and usefulness and comforts as accompaniments of spiritual pleasures. The value of science than becomes that of technology; science and technology become devoid of the spiritual value that works of art and religion possess. These sociologists are clandestine, dyed-in-the-wool reactionaries.

Today the university is the center of science. To accommodate for this innovation, the university has totally remodeled itself a few times. Yet, at heart it still is the medieval institution of sinecure that caters for education only as a sideline. It has absorbed as many aspects of modern culture as it can, including all the new sciences, all sorts of art schools, all sorts of religious institutions, and all sorts of technologies, physical, biological, social, and communication. Members of the academy who are ignorant of science are often defensive and hostile to it. This began as a discussion of the imperialism of science: its received theories being most rational plus the desirability of maximal rationality, lead to the preference for science over all else. All disputes thus become superfluous and irrational and obscurantist. Abiding by the verdict of science is then obligatory. This is barely tolerable, philosophically, morally, and politically. It denies serious opponents the respect they deserve. It conflicts with the democratic freedom of belief and with the democratic system of standard and constant parliamentary disputes. All this comes in the guise of the search for the justification for the demand to believe in the right theory. It is much more important and interesting to replace this search with a search for the demarcation between rational disputes, as most disputes between scientists are, and irrational ones, as many disputes between politicians are – such as ones about racism and sexism: one can try to educate racists and sexists, if they are willing, but one can hardly argue with them rationally as equals, in a two-way street, as their views are utterly indefensible.

The debunking of science targets the harsh unanimity that it allegedly imposes. To soften this harshness, some philosophers empty science of its theoretical content altogether and ascribe to it mere computational powers, some magic that leads to useful quantitative results. This too is debunking, yet lovers of science undertake it in good faith as it boosts unanimity. They demote science to the level of applied mathematics, and they demote engineering to the same level. Science, applied science, and engineering, each has its own dignity. Their fusion harms them all.

Bertrand Russell tells a remarkable story that may serve as an example of the unintentional debunking of science. His *History of Western Philosophy* describes William James's pragmatism as the theory that the truth-value of an opinion is its cash value. He was surprised to meet with the accusation that this way he had rendered the theory vulgar. He responded by observing that what he had said was a quotation or a near-quotation. James was a lover of science and of religion, and he was tolerant. He spread pragmatism in an effort to break the grip of intolerance on science and on religion. He failed, and looking at the failure, we are liable to forget his initial, noble intent.

Not only intolerance but also refutation raised problems. The supposition was that science includes only demonstrated truths. The revolution in physics showed that whatever the classical theories were, they were not absolutely true. So, the suggestion became popular that though they are not true, at least they are not false either. If they are neither true nor false, then they are not theories proper, not informative assertions. The suggestion then was that they are mere instruments for technology. This is demeaning: some false descriptions are noble and intellectually valuable – more so than bags of tools that work for mysterious reasons. Instrumentalism demeans technology too, by making its successes inexplicable, since it allows for no scientific explanations. This will not do.

We do not know what science is. We do not know why technology works. As Russell has noted (*Human Knowledge: its Scope and Limits*) if what science tells us is any near the truth, then most of the universe is uninhabitable. Our very existence is precarious. Surely, this must have a tremendous impact on our culture and on its mood, said Jean-François Lyotard. Science-based technology increases the dangers of our impending extinction. This contributes severely to the somber mood that has engulfed our culture ever since we have the ability to destroy life on Earth. So, we need new attitudes towards the new problems of our survival. Nothing has influenced post-World-War-II culture more than the dreadful events of that war, especially Auschwitz and Hiroshima, and their aftermath. This is no place to discuss global politics; I have devoted another volume to it (*Technology: Philosophical and Social Aspects*). Here my point is that this is tremendously significant for the current ethos.

Most cultures have always viewed the human race as evil, and the Moderns viewed it as good. Both these views are extremist and false. This may be the most important single characteristic that depicts us uniquely as a new (global) culture, barely half-a-century old. We should take what is good in our different traditions and amend them to suit our tastes, especially our rejection of the view of the human race as good and the view of it as evil. We should not decide whether we are good or evil; rather, we should decide whether we wish to relapse or to improve, and how exactly we should go about it. The idea that we should strive for improvement runs through most of the important items of the verbal popular arts and all the important popularizations of science – though some items of art, popular and high, and some items of science, popular and expert, are bestial and seriously worrying. We have to try to improve. Common people understand this, and many are willing to act whenever they can see how. We do not know how to improve, but we can try. We can express our autonomy this way, and now that philosophy gave up hope to acquire certitude, autonomy is perhaps the only bedrock that philosophy can rest on. At least this is a lovely idea, and an expression of the blessed desire for freedom and for the improvement of our common lot. It is worth a serious try.

THIRD INTRODUCTION: SCIENTIFIC PHILOSOPHY AND ITS CULTURE

Unlike scientific knowledge, scientific philosophy is not a part of science. A scientific philosophy is any philosophy that considers science culturally valuable. The label 'scientific philosophy' is old-fashioned. The use of the label 'the philosophy of science' as its modern equivalent, however, is confusing. It applies traditionally only to views about it of philosophers who value science. Courses about the philosophy of science overlook all irrationalist philosophy of science. This is unfortunate, since irrationalism is very popular and very important, and so is its attitude to science, that ascribes to science no cultural value.

Scientific philosophy concerns not only science: it is the philosophy of those who deem science central to the life of reason and reason the highest human faculty. Of course, some advocates of scientific philosophy say hardly anything on any topic that does not directly concern science. Some leading philosophers of science of the twentieth century published almost exclusively on logic and on the language and methods of science. This reflects a narrow attitude to science that W. V. Quine, for one, expressed openly: on empirical matters, he said, he simply trusted science. This is regrettable, since philosophy traditionally included critical assessments of science and of its ideas. One of the central questions in traditional philosophy is central to metaphysics: what is there? The study of this question, ontology, is still very much alive and interesting. Quine found no need to discuss it since, he said, science answers it. This is an error. Science is ontological, but as yet it has not offered a comprehensive, unproblematic ontology. Science is more a set of problems than solutions.

Traditional scientific philosophy centered on proofs of scientific truths. Modern philosophy of science followed suit. Modern scientific philosophy has begun a discussion of errors. Some errors are objectionable; others are admirable. How?

What makes an error objectionable? In a free country anyone can assert every idea. Likewise, anyone can declare any idea an objectionable error. When an objectionable error may lead to sanctions, the allegation that some assertion was an objectionable error might invite investigation; and judgment on the matter demands validation. What examination leads to validation? What validates the allegation that a given assertion was an objectionable error? Practically all philosophers of science agree that good reasons – strong arguments – must back up any opinion, and thus render it valid. This holds also for the allegation that that some assertion was an objectionable error. Question: What is this validity? What criteria must it answer? How do they make an argument acceptable?

The literature on the philosophy of science centers almost exclusively on this question. Most philosophers of science agree: outside logic and mathematics no idea is valid *a priori* (namely, independently of experience), and to make it *a posteriori* valid (namely base it on experience) we need an *a priori* valid criterion for *a posteriori* validity, so that we are back where we started. Hence, it is impossible to find arguments that validate claims for acceptability. This is Hume's critique; no one has found an answer to it. Philosophers of science go on looking for an answer, since life without validation is life without discrimination between ideas – it is absurd. The task of any fallibilist is to allow for such discrimination while doing justice to Hume's critique. Most philosophers of science ignore unacceptability, as they care only about acceptability, and they assume that whatever is not required is objectionable, and they identify the acceptable with the required and the rest as objectionable. They view science as black-and-white. Fallibilists recommend – in the wake of Socrates – to begin with the objectionable: what opinion is objectionable? After handling the objectionable, we offer as a desideratum from any proposed criterion for acceptability that it should not allow the acceptance of the objectionable. This demand amounts to the rejection of the relativist thesis that any criterion is as good as any other: it amounts to the admission that any criterion can serve as a starting point, plus the recommendation of some current legal system as the most reasonable starting point available. Current legal systems are obviously imperfect, since they diverge. But (contrary to relativism), they are open to criticism and improvements, usually through the democratic process. Legal systems are inherently fallible, and the democratic ones among them pay much attention to the question, what makes an error objectionable. Also, these legal systems employ diverse criteria for what is objectionable. In particular, different criteria apply to ordinary citizens, to government officials, to the press, to the police, to the district attorney and to the law courts. They are all open to critical assessment and correction. So much for the unacceptable.

What then is acceptable within science? All reports of observations are acceptable, even obligatory, but only after their repetition. Philosophers of science often violate this rule as they discuss the problem of acceptability of generalizations of unique observations. The philosophical literature deals with acceptability in the abstract. What statement is acceptable? When is it right to assert it? When does it merit assertion? To use current jargon, what warrants assertability or credibility of a sentence? As philosophers of science ask this question in the abstract, with no context, it puzzles non-philosophers. Philosophers influenced by Ludwig Wittgenstein demand that, in order to avoid confusion, abstract questions should appear in context. Nevertheless, most of them insist on discussing this question in the abstract. It is difficult to know when one feels sufficiently at home in philosophy to meet this abstract question as an old acquaintance without batting an eye. To be on sufficiently friendly terms with this abstract question is to suffer from hopeless confusion. Admittedly, some famous philosophers today write on it; some of them are famous as paragons of clarity. Nevertheless, they are confused. No public opinion to the contrary will straighten things up.

Back to our initial question, then: what makes some error admirable? Of course, the idea that science is admirable makes this problem include as a special case the problem of demarcation of science, which has fascinated many great modern philosophers, Bacon, Descartes, Whewell, and Popper. Faraday, Peirce, and Popper declared science not above error. The problem posed here addresses fallibilists. Faraday did not address it. Peirce did, and his answer is not fallibilist. Popper did, as he offered a criterion: ideas open to criticism are scientific, and the more open the better. This is great progress though it does not differentiate between a great breakthrough and a routine scientific advance. Also, it intentionally does not apply to metaphysics. And unintentionally it does not differentiate between science and technology, as severe tests are common in both. Those in technology are important but not necessarily interesting. Later in his career, Popper spoke of criteria for the adequacy of metaphysics. We want more: we want criteria of adequacy of any intellectual activity. A theory has to serve as an answer to a question, and so its adequacy is that of an answer to a given question. Also, the adequacy of a question must be open to questioning.

This has led Ian Jarvie and myself to view intellectual activity as a special case of rational activity, since the measure of the rationality of all action is the measure of adequacy to end. Technology is not autonomous: it receives its ends from its environment. Science is different: it has its own end: it aims at true comprehension. The end of tests, says Popper, in science, in technology or anywhere else, is to find some faults. This sounds wrong, but it is sheer common sense. Finding faults is not an autonomous activity: it borrows its intent from its environment. For example, diagnosticians seek defects in the state of health of patients. To overlook some defect, then, will be an error. Whatever are our opinions and expectations, much as we do not want bad things to happen, as they do happen, we want to hear about it. Nor do we always want to be right. Alberto Coffa has noticed that when he discussed the philosophy of science of the celebrated Viennese philosopher Moritz Schlick. Opposing Schlick, Coffa said, we do not always want to be right: we want our fears refuted – say, the fear that we suffer from a terminal illness.

This is odd: surely, Schlick knew that we want our fears refuted and our hopes confirmed. He did not think about them in the context of his philosophy of science: he fixed his mind on the desire of researchers to confirm their hunches about things abstract, but he did not specify the context and so Coffa could refute his philosophy, as he did, by reference to a simple, familiar fact.

The end of scientific research is to trace God's blueprint of the universe, to use Einstein's metaphor. Empiricism and intellectualism compete by offering *a posteriori* and *a priori* research methods – empiricism, and intellectualism. Now that empiricism is fashionable, it is interesting to see the comparative value of intellectualism: it offers comprehensiveness. Why then not use both methods? From the traditional viewpoint the answer is, the two methods may clash. This was unthinkable as long as philosophers viewed science as authoritative and infallible. The fallibilist viewpoint differs: we may endorse both and play them against each other. In politics the unreliability of an institution leads to the effort to limit it by another institution in a system of checks and balances. The same may hold for the foundations of science: as offering foundations, the two methods fail miserably; but as limitations on each other they may succeed. The hostility to metaphysics so characteristic of classical scientific philosophy related to the popularity of Aristotelian metaphysics. Today things are different. Today crude versions of scientific metaphysics, of mechanism or of materialism, are popular in circles that exhibit some familiarity with science. And as the metaphysics of these circles appears in the guise of science, they hold on to it not less dogmatically than the Aristotelians. The danger of crude pretense to scientific authority invites efforts, like Popper's, to demarcate science. What else can minimize it is hard to say, though playing one metaphysical system against the other is the technique that Kant uses in his dialectic of pure reason. Yet he deemed this dialectic futile, and he declared its aim to be but the prevention of damage. It is possible to have a dialogue between reason and experience as Popper suggested, and to declare all ideas open to this dialectic. It is likewise possible to have a dialogue between metaphysical and scientific theories: they may conflict, and efforts to settle one conflict may open the road to another– until we find a scientific metaphysics. This, then, is the chief message of this volume: science need not fear metaphysics; the enmity of science to any competitor is lamentable and expensive. As science is in no risk from the alternatives, and as criticism should be beneficial all round, science must remain as friendly to all alternative systems as it is possible, and explain its hostility to systems that it deems imperative to combat. This will render the combat more efficient, hopefully end it rapidly, and science will then be able to return to its traditional task.

1. AUTONOMY

Chapter 1.1

SCIENCE FICTION: THIS MESSAGE IS FOR YOU. MAYBE

There is a mood often enough conjured in science-fiction literature to be familiar to fans, the mood of seemingly intentional yet probably random contact between individuals across immense space-time expanses. After a long, complicated chase story, the chased person has long ago lost contact with the mother planet and drifted into a strange planet. And there, right now, just walking across the plaza, the drifter is casually glancing at a huge poster displaying a piece of commercial advertisement or some similar message of purely local import. On the margins of the poster runs a puzzling arabesque that quickens the drifter's pulse. It intrigues and fascinates – until it hits instantaneously in full force: the arabesque is a message in the drifter's long-neglected native language. It conveys instructions as to how to reestablish contact.

The story, as I say, has many variants. And it has, it seems, a widespread appeal. It appears in terms of adventures stories, but its mood is not that of adventure. The mood is more of the one associated with the idea of or the hope for the successful termination of an adventure, the return home. It is the eternal return, the idea of timelessness or of time standing still (and space vanishing). Western philosophies usually associate this idea with Nicolas, the Renaissance mystic bishop of Cusa, though it is recurrent in all cultures. It is the idea of the identity of vanishingly small and the infinitely large, of an eye blink and all eternity. Indeed, the meeting of the infinite line and the point of vanishing length symbolizes the random encounter of the message with its intended recipient.

Whence this universality? I suppose it is biological. Think of the scent that a living thing, be it a fish or an antelope, leaves behind to contact a possible mate. It is a few complex molecules in a sea of molecules of astronomical numbers: just this room contains more molecules than you can imagine. And another animal follows this scant clue along a tortuous path to find its mate.

William Craig and Konrad Lorenz have suggested this hypothesis. An animal does not have the slightest inkling of what it is after. Even while building a nest, it has no complete picture of its task. In different stages the animal goes after diverse kinds of things, yet it knows not which things, let alone their inborn, rigidly fixed sequence that gives them their meanings. Yet the animal in charge of a task, of finding a mate or of finding a twig for a nest, will be restless, say Craig and Lorenz, and move about, and try to read the message in any piece of arabesque, until one day something will make sense and enable its finder to complete the next step in its mysterious complex task and open up the search for the next message, the next step further along the unfolding series of pre-ordained ordeals that life is strewn with.

Consider writing proper, like my writing this message for you right now. We do not know the background to literacy, much less its actual origins, except that these were far from unique. Though many cultures are illiterate or have acquired literacy from more advanced neighbors, literacy evolved independently on many places on earth. Yet, even here we see a sequence. Many independent ideographic scripts, some of them evolving into syllabic scripts and, two of these, hieroglyphs and cuneiforms, evolving into alphabets proper – perhaps simultaneously, perhaps not quite independently – the hieroglyphs and the cuneiform turning into the two Hebrew alphabets (the one, the Sinaic that has survived to this day, and the other that is the Ugarith that is ancient history). How and why, we do not know.

There is room for speculation, though. It is hard to imagine how the very first writer conceived of the very first message and wrote it without consulting any recipient first. This is the way Rudyard Kipling has imagined it, and the confusion that the message must have caused. Other options suggest themselves. First, that writer and recipient are one: writing was a form of memorizing, or of consigning matters to a memory more stable than that of the human brain – the granite rock, for instance. The ancient Egyptian god of literacy was the god of memory. Illiterate shepherds use knots in their ropes and cuts on their sticks as *aides mémoire*, as bookkeeping of sorts. Some scholars consider these the origins of writings. The evidence for it is the runic writings of the Norse, whose letters are images of sticks with cuts on them, Yet, Runic writing is alphabetic and thus the product of mating two levels of culture. Many facts support the view of writing as originally mnemonic. Most extant ancient texts are, (a) sacred texts, (b) inventories, and (c) monuments.

Suppose, then, that originally writing was mnemonic. Writing messages was then a further invention. How did this come about? Perhaps it happened when someone used a record as a letter proper. The Bible *(Genesis*, 38) tells an interesting story of that sort. It concerns the patriarch Judah, son of Jacob and ancient ancestor of the tribe of Judah and of legendary King David. He had a daughter-in-law who became a widow with no offspring. So, she was obliged to have offspring from her deceased husband's kin. This is the law of levirate that, records of social anthropology inform us, survive in many primitive societies. Judah considered the lady, Tamar by name, a bad omen, as she already lost two husbands, both his sons, and he withheld his third son, intending to keep her waiting quite indefinitely. Donning a veil, she played prostitute and lured Judah himself in an awkward moment, when he could not pay for her services. He left his strings, sticks, and signet-ring with her as collateral. She conceived and he ordered her burned for not waiting. She sent him his records and ring with the (oral?) message saying, I am pregnant to the owner of these. Here was a bit of tact that could have cost the lady her life had Judah not been a gentleman. She deliberately allowed him to feign ignorance. Perhaps she thought, if he is no gentleman, then I had rather die at the stake than have him. Was this involved story a dim memory of the cryptic invention of cryptic writing?

There is more evidence that writing evolved deviously. Messengers had to prove genuineness of messages. As proof of genuineness, they had rings or seals – signet rings – that prospective senders had received from prospective recipients. Alternatively (and presumably later), they replaced the signet ring as means for preventing fake messages with half of a simple coin to match the other half – two individuals would split a coin, hoping to hear from each other. Putting the two pieces together was proof of genuineness. To put together is, in ancient Greek, to symbolize (from *syn*, together, and *ballein*, to throw). To date, the written word still is a symbol.

Messages to oneself, or from a person to a friend who expects it, obviously belong to a different dimension from monuments open boldly to public gaze. To this day, I echo the excitement I felt as a youth, when I first read the contents of the famous Thermopylae inscription. Wayfarer, it addressed its anonymous reader, when you reach town, tell them you saw us lying here, we who fell for its defense. I am not translating literally; the document includes the idea that the motherland instructs soldiers to die for her, and this is both terrible and stupid, as well as far from exciting – at least, so I found it in my youth – I am trying to capture the positive content of the message, its impact. Wayfarer, it addressed its anonymous reader, when you reach town, tell them you saw us lying here, we who fell for its defense. On a dusty road, the long dead heroes address an unknown passer by who may or may not pass through town, requesting to pass a message to just about anybody upon arrival.

The message is uninformative bunk, stuff and nonsense, or worse, the kind of intoxicant that patriotism and religion often serve. It is strong but not nutritious. It is intriguing all the same, even if it is but a seeming message. The fake indicates familiarity with the real thing: con artists know what resonates in their audience's ears as the true stuff. The more elemental, the easier it is to fake and raise powerful echoes. Any person with even a modicum of literary competence can write a fake correspondence between a widow and her soldiering son who is wasting away at the front – a correspondence that expresses the mere longing to reunite – as so many wartime newspapers and dime novels illustrate. We are all open to persuasion when stuff of this kind plays on our feelings.

The mood of an intended yet random message through vast expenses of space and of time, the sentiment that hits you right in your solar plexus, is common to the animal conduct that fits the Craig-Lorenz hypothesis, to religion, primitive and sophisticated; and to its modern substitutes, patriotism and science fiction.

Not so letter writing. There is a middle ground between a private letter and a monumental message to the world. Touching as this can at times be, due to the accidental nature of its contact with the wayfarer or with the archeologist of some distant future, a message is touching more powerfully when it is a message proper yet to an unknown recipient, particularly when the unknown recipient is in all likelihood yet unborn. In Isaac Asimov's *Foundation* trilogy, the hero, Hari Seldon, is the father of psychohistory, and a poetic caricature of Karl Marx – except that unlike Marx he is a real scientist. His knowledge rests on statistics and other branches of high-powered mathematics and all that, and, nevertheless – or therefore – he does not consider his knowledge of future history as infallible as Marx considered his. Seldon uses a sophisticated tool to pass to his posterity some sort of an office memo. He places holograms of himself in a time vault to appear in the safe haven that he has devised for his followers in a series of future points in time. He then appears to the public and speaks. He always ends his message by expressing doubt and by reminding his audience that he is long dead. And, indeed, on one occasion, Asimov makes him blunder, and the inappropriateness of his message leaves his audience in a state of queer disappointment that is hard to describe yet is so familiar it barely invites a description. It is the disappointment of one who finally arrives home only to find it in shambles.

When I think of Benedict Spinoza, I think of him in this vein, and perhaps more so I think of his predecessor Uriel d'Acosta. The Jewish community maligned and ignored Spinoza, but at least they left him in peace. He cherished his peace. He had friends outside the community and they listened to him. Some of them were politically influential and together they planned a better society. They failed. A hired mob lynched his friends. But he survived and lived to write a better study of politics, with a better idea for a future and more civilized society. He did not finish writing it, as illness took him young. His predecessor in the same stern community, d'Acosta, wrote his autobiography to nobody living, and then committed suicide. His loneliness was too painful for him to allow him to go on breathing the fresh air that was free for all, yet the style of his autobiography is most relaxed, conveying a pleasant atmosphere. You can hear the fellow talking to you, a distant-future, unknown, intimate friend. It is a piece of fantasy essentially akin to that of a teenager dreaming of a powerful sexual adventure – the need for friends is no less potent than the need for sexmates – and this, I suppose, is how an animal, a fish or an antelope, feels when leaving scent in the vicinity for a possible mate or partner to follow. There is a sense of oppressive, intense loneliness here, mingled with an exciting anticipation of the real, overpowering joy of an expected encounter. The encounter may be accidental, but the fit of the two meeting mates or friends is not. Or, is it? It gives the impression that the two halves of a symbol are reuniting when possibly they merely meet by a freak accident. Our lives are rich with accidents and it is by accident that we fall into place. The point and the infinite line unite.

Perhaps the best science fiction literature addresses readers in the distant future – in the hope that humanity will survive that long.

Chapter 1.2

THE CONSOLATIONS OF SCIENCE

Science offers consolation. This merits attention in view of the topical debate concerning the aim of science: does it seek the truth or only utility? Pierre Duhem was the greatest advocate of the view that the aim of science is only utility (since the truth is unattainable). On top of utility, science offers terrific consolation in its very intellectual and aesthetic value, he added. Being a religious Catholic, he found consolation mainly in his faith. Yet, the consolation of science mattered to him. What is this consolation is not clear. And since this consolation is the object of attention here, it is not clear from the start what sort of consolation is of concern here. We may, of course, postpone this question for a while, and ask, what consolation do some individuals find in science? This may hopefully offer a clue to the more objective question, what consolation does science offer?

For the first and obvious answer to the question, we may go back to *The Consolations of Philosophy* of Boethius. He, and generations of his readers, viewed philosophy, or rather religion, as today many people view science. It was a source of understanding and reliable knowledge that sustains hopes and allays fears. Many writers have advocated this view and many readers accepted it. It has lost much of its convincing power and it should lose it completely. For, no guarantee is possible that our fears are groundless. Of the reasons for this, some pertain to specific fears, and other pertain to fear as such. Philosophers often identify this fear as philosophical skepticism. It is problematic, gnawing doubt. The means with which to overcome it are such things as love, friendship, and faith, not science. This raises the strange question, is science at all compatible with love and friendship? Or does the scientific examination of love and friendship kill them?

The examination of the first answer, then, not only fails to tell us how and why science consoles, but even raises the possibility that science unnerves or upsets rather than consoles. We should discuss this possibility seriously. We will then discover that we seek the manner in which science is compatible with love and friendship – and this is of concern to all of us, religious or not. Possibly, consequently, science will be compatible with religion too. Hence, the consolation of science cannot be a substitute for the consolation of religion. This should rid us, once and for all, of the first and obvious answer. Whatever the consolation of science is, it is no competitor to religion, it is no assurance that our understanding is solid knowledge, it is no support for our hopes. Perhaps, then, science offers us no consolation at all. Possibly it does, and then what it offers is nothing more than an illusion, a mere psychological fact that the scientific activity is a kind of tonic for tormented spirits. Possibly, however, it is genuine. But then, to offer consolation without competing with religion, science need not allay fears. And then, whatever else we propose that the consolation of science may be, we must propose it with no proof whatsoever, as proof will lead us back to the first and rejected answer. Whatever the proposal is, then, it comes in the spirit of skepticism. We thus abandon the demand for proof that puts a wedge between religion and science, between faith and reason. The conflict disappears then without any attempt at reconciliation! Skepticism allows one to endorse both some religion and science in some interpretation. Moreover, its best expression may be the view of science as a religion of sorts – as faith in reason. Is science compatible then with love and faith? This question vanishes: it is faith of sorts.

Both views are inadequate – the view that science rests on proof (scientism) and the view that science rests on faith (fideism). This clears the way to the introduction of a non-justificationist view of science (to use the expression of W. W. Bartley III). Non-justificationism is both rationalistic and skeptical, and as such, it may be unsatisfactory and uncomfortable: it may block all consolations of science. Whether this is a satisfactory view of science or not, it is the only reasonable view of the consolation of science as the worship of free people, to use the expression of Bertrand Russell. Science is not necessarily opposed to religion even as a consoler, since, having no foundations, it offers the consolation to free spirits that seek no assurance but make efforts best to understand and find in such efforts the highest consolation possible.

THE LIMITS OF SCIENCE

I

Boethius' *Consolations of Philosophy* is a book acclaimed as one of the most popular in the Middle Ages. (1) It is hard to say whether philosophy or religion consoled Boethius and his readers: it was a religious philosophy. Perhaps it was just this mixture of faith and reason that consoled them. This possibility, however, is not obvious: what was it that consoled the poor suffering prisoner that the author was, and that his readers saw themselves at least allegorically if not practically? The traditional answer is that faith represents yearning, reason represents a promise, and the consolation is the supposed agreement between yearning and promise. Yet, this agreement can only be a part of the yearning not of any promise; otherwise, the promise becomes highly questionable as mere wishful thinking.

Scientific reason may establish cosmic harmony or religious grace may unlock personal salvation – and heal a suffering individual of tormenting doubts. Skeptics have questioned whether science can establish universal law and order. (2) When rationalists declared faith irrational, they wished science to undertake also the burden of offering some personal salvation. The skeptics declared this impossible. Religious critics of rationalism did so too. How can rationalists respond? (3)

The religious critics of rationalism concede much to skepticism. They declare only an appeal to faith can dispel the skeptic criticism. Faith and reason cease to cooperate and to complement each other, then, as they become competitors for the position of buffers against doubt, for the position of providers of consolation to suffering humanity. In this contest, rationalism has to show its strength as a response to skepticism, as the ability to present scientific proof. The alternative to rationalism, to the demand for proof, was fideism, the demand for faith. (4)

Science has won general recognition. Some deem its chief value its intellectual worth, its enlightenment. Others value only its usefulness. Galileo's mighty opponent, St. Roberto, Cardinal Bellarmino, rejected Copernicanism as a theory yet he instituted it as useful tool in the hands of astronomers. (5) What he feared was that science would prove to be a serious competitor to the Church of Rome, the way Protestantism proved to be. (He acknowledged this when he mentioned to Galileo the decisions of the Council of Trent that came to counter the Protestant Reformation.) Bellarmino was instrumental in helping his fear realize itself: he forced scientists into a corner by demanding of them to provide proofs before allowing them to make any intellectual claims, thus coercing science to make claims for proofs and thus to make science a competitor to established religion, since with proof they offered a new kind of consolation. Proof assured science its practical value and thus a better future for humanity. And this may console a prisoner, such as, say, Condorcet (see below), no less than the hope for personal salvation in afterlife may console a prisoner such as Boethius.

Science plays regularly the role of a consoler within the rationalist literature. Quite a few writers have noted this. Prominent among them is Sir Leslie Stephen. (6) On the whole, it has suffered neglect, both because of the desire to conceal the tension between science and religion, and because the consolation of science may be secondary and obvious, since allegedly science offers proofs before it offers consolation. Proofs then also validate the consolation that accompanies them. The consolation of science does away with wishful thinking, then, with mere yearning and vain promises. The classical variant of this appears in the powerful title of a section of a book by Friedrich Engels that became a pamphlet with instant success. It is, *The Development of Socialism from Utopia to Science*. (7) False prophecy dies hard. It was for long, and perhaps it still is, the consensus that the future is going to be rosy – even though today only a few agree with Engels that it will be socialist.

Nineteenth century optimism is past history: few among us today are willing to believe that science breeds utopia. But this may make the consolations of science all the more desirable. We do not view ourselves so much in the medieval fashion as prisoners in dungeons. But we do feel adrift on Spaceship Earth. We want science to navigate us towards our chosen destiny. It is hardly possible to specify our destiny as clearly as the classical utopians and socialists did. Yet we do have a general idea, and some specific suggestions. What matters, however, is not the specific or the general idea, but a simple fact: the consolation that science still gives to many of its fans is a false sense of security.

II

An example will illustrate this false sense of security and the strong hold it still has. The following instance of faith in science is an error that persists in the literature. This persistence shows that the false sense of security is as strong as ever, that science still plays the role of a promise and of a proof to back this promise with. The example concerns the question, is the solar system stable? Towards the end of the eighteenth century, Kant gave a proof for the negative answer. Soon, Laplace gave two proofs for the affirmative answer.

Laplace's first proof is very lengthy and complex and depends on the doubtful supposition that the solar system is strictly isolated. It rests on the tautology that periodical motion is stable. He calculated the seemingly a-periodical perturbations of Newtonian mechanics as periodical, and in accord with the observed ones (of Saturn and Jupiter). The proof remained unscathed to the end of the Newtonian era. Since it is eminently Newtonian, it has lost its claim for validity after Eddington's solar eclipse observation of 1919, which decided in favor of Einstein as against Newton. A few decades earlier, the mathematician Jacques Hadamard showed that the proof is defective. Despite Hadamard's criticism, it refuses to go. (8)

Laplace also offered a much simpler proof of the much simpler claim that the sun will rise tomorrow. It is his rule of succession. (9) It says, if all past n trials have succeeded, then there is a chance of $n / (n + 1)$ that the next trial will also succeed. Records of the sunrise exist for three thousand years. So $n = 365 \cdot 3{,}000$. Geological records suggest that we have been observing the sunrise for a million years or more. This makes a virtual certainty the prediction that the sun will rise, not only tomorrow, but even for the next couple of thousand years.

Kant's forecast (10) to the contrary is familiar to modern science fiction readers since H. G. Wells published his famous *Time Machine*: the solar system is cooling off. The cooling off is in part due to the friction between the earth and the waves of the tides that are due to the interaction with the sun. This friction turns into heat that dissipates. The generation of heat is at the cost of some kinetic energy, namely, the reduction of the earth's motion. A more modern prediction is that the sun will probably become a nova – explode like a huge hydrogen bomb – since it is a hydrogen furnace. A still more modern prediction is, the sun will become a red giant; it will expand and swallow its planets. Hardly any text, scientific or historical, ascribes to Kant the idea that the solar system will not stay stable forever. The publicity given to the questionable proof that Laplace offered to the claim that the sun will rise tomorrow and the day after tomorrow, is in strong contrast with the absence of publicity given to the much simpler and still valid considerations of Kant to the limitation of this proof. This illustrates the positive character of science; it also shows that this positive character is science fiction, wishful thinking. What kind of consolation is this? Is it characteristic? At times science fiction proper is more realistic than science proper. How can that be?

III

Science fiction authors offer another kind of consolation, which is utopian, of course, and thus quite different from the one that Laplace has offered and that science textbooks propagate. People often prefer the consolation of science because a proof supports it. As the proof is sham, however, we have here, in lieu of a proof, the mere hope, the faith in the possibility that a proof exists.(11)

The idea that science is rational and rationality is proof is peculiar to western scientific culture. It is our inheritance from the ancient Greek philosophers. They offered not proof but faith in proof – faith in the proposition that proof is around the corner. Proof of the truth of a proposition is peculiar to Greek philosophy and its derivatives. Martin Buber made this last claim as he noted the existence of *Two Types of Faith*. (12) There is faith in the Lord, or trust in Him, he said, and there is faith in the truth of this or that doctrine, especially the doctrine that He exists and is omnipotent *etc.* Buber declares the one (the hope to be redeemed) to be Hebraic and the other (the hope that one's view is true) Greek. The detail of the historical illustration of that contention in that book is too much of a distortion. But it is not hard to see Buber's point nonetheless: the allegedly Hebraic type is a grim gut response of a person who never questions God's existence but does question – in a gutsy way, to be sure – God's good will and concern for me, and finally comes down on the affirmative, of course. The allegedly Greek type is concerned with the truth of a statement, i.e., with its correspondence with the fact it purports to capture, namely, with a detached objective reality. It is, in short, cerebral. Now, why does all this matter? What good is it to trust a Being whose very existence is doubtful? Buber found this question the epitome of the scientific attitude. He labels it the I-it relation: deny all that you possibly can, and doubt all that is capable of being thrown into doubt – and you end up denying that all others exist, other humans and God. This way rationalism leads to utter skepticism and from there on to solipsism. It is then a cold world of reason, dominated by deep loneliness unredeemed. The I-thou world axiomatically takes the other as given: as Buber puts it, there are two different languages, the I-thou and the I-it, and the two types of faith are merely one religion expressed in these two languages. (13)

Buber's philosophy was a denial that we can make do with one of the two languages – we need both the I-thou and the I-it language, he said: one expresses intellectual, scientific, cerebral responses and the other expresses emotional, intuitive, gut responses. His view was that we must synthesize the Greek and the Hebraic in our heritage. Clearly, his *Two types of Faith*, though an exposition of an idea already presented by Georg Simmel, (14) is in Buber's treatment a vehicle for restating his own philosophy. Within it, Boethius' consolation appears with a new significance. From the depth of the desolation and discomfort of the dungeons, he yearned and hoped for comfort: he developed the trust that the Lord will, in the end, truly console and comfort and compensate him. And he held the doctrine firm that an all-powerful, all benevolent, all knowing deity does indeed exist. He thus solved a difficulty. The claim for knowledge makes hope unnecessary, and hope alone is no consolation in the face of fear and despair. The message of Boethius is moving because in the depth of the dungeons he sees his hope shaken by fear and even despair and his reader identifies with him and feels this intensely. How, then, did he resolve the clash of hope with despair? If by knowledge, then both hope and fear vanish, as Epicurus and all his followers propose. If not by knowledge, then by what? By a mixture of trust and hope, says Buber in effect.

Like most everybody else, Buber deemed the approach of Descartes a paradigm of the purely cerebral approach, and he found that approach wanting. Descartes taught that God exists. But already Pascal dismissed Descartes' view of God. In a famous remark in his *Pensée*, (15) he said, "I cannot pardon Descartes". He explained his complaint thus. Descartes kept God outside Nature, unable to participate in Nature's process after His "first kick," i.e., after He granted Nature Her initial quantity of motion that Her Laws forced Her to conserve and thus to keep Him out.

In matters scientific, it is well known, Pascal was a follower of Descartes. It is debatable whether in his religious work he repudiated science. It is possible to endorse Descartes' view of universal harmony yet wish to supplement it to satisfy a need for personal divine consolation. Possibly this is the opinion of Pascal and obviously it is the opinion of Buber, who saw room and need for a combination of cerebral and gut responses.

IV

Perhaps the rescue of the faith in the existence of God has no need for the new device of the I-thou language, one may argue, that pitches trust against critical thinking, gut response against the brains: there is no need to resolve the conflict between the brains and the heart by viewing their responses as complementary. They may cooperate within the I-it language. Descartes supposed that God does not intervene in Nature's working; Newton disagreed. Descartes said God keeps His hands off Nature, and Newton said He keeps correcting repeatedly Her small deviations. (16) Oddly, both hypotheses are within the I-it language, despite their reference to the Eternal Thou. Hence, perhaps there is no need for the I-thou language.

This argument is valid, but it is not relevant to Buber's contribution. For, it is an objection to the introduction of the I-thou language, but in no way is it an objection to the introduction of the I-thou relationship. The I-thou relationship is the recognition of the other, the relationsehip of love and friendship. The I-thou language is the language of expression and communication of the I-thou relationship, perhaps the language in which we describe it. It is possible because love and friendship exist as do their recognition, their communication, and their description. The admission of the I-thou language as a separate language amounts to an admission of personal relation – as something extra-rational. This is Buber's point. Is it a challenge? Is it also a novelty? To see that it is, we may wish to compare the views of Buber and of Russell.

Russell found place for love and friendship, but not for the complementation or supplementation of science or of its language by religion or by its language. Russell discussed this matter in his *Religion and Science* where he responded to new critics of rationalism. Of course, science is not everything and so other matters must supplement it, he observed. Love and friendship, for example. (17) The new critics of rationalism were saying something different. A world dominated by strict rationality, they were saying, has no room for love and friendship; hence, rationality is limited: we put reason in its place in order to allow for love and friendship. Does "Cartesian", "scientific", or "pure" rationality – whatever its maxim exactly is – does it preclude love and friendship? Is Buber right in limiting all science to the I-it language and all that science can study to a cold, loveless, and friendless world? Can one at all defend the view that science allows and even encourages warm feelings, love and friendship, hopes and consolations? It seems difficult to defend this idea, but it is not. And if we are to pay the modern irrationalists their due and explain their intellectual hold on some of the brightest minds around, then we must take this disagreement very seriously.

V

The question raised here is not merely, can I trust my friend and remain a scientist. Doubt goes further and raises the question, can science justify my trust in my friend? Can or should science explain or justify such things as love and friendship? These questions are not the abstract metaphysical ones raised, historically, under the influence of Descartes. He would ask, do my friends exist? That is to say, do I now meet a friend or do I encounter a mere specter or a mere robot? He asked, do I exist? That is to say, am I mere machine? (Am I a part of somebody else's dream? Lewis Carroll has suggested this possibility in his *Through the Looking Glass*. Jorge Luis Borges developed this idea in his terrifying story, "The Circular Ruin".) Let us put these Cartesian questions aside as metaphysical, hold firm to the common sense of our common existence, and accept as scientifically given that in some sense – in the common sense, indeed – my friends and I do exist. This will not console us and will not alleviate our loneliness as it does not yet make science compatible with the love and trust that comprise friendship. Russell conceded this, and with ease: we need our science and we also need our friends. Is this so simple? Is this at least traditional? It is not. Russell, like Buber, altered the scene as the following discussion may illustrate.

Jeremy Bentham considered himself the Newton of the social sciences, because he based them on the law of social attraction: people need people for friends. (18) No one can deny that science may examine friendship the way it examines, say, food. Now, why does the need for food not make Russell say, we need both science and food, yet the need for friends make him concede that we need both science and friendship? Because, I propose, we can examine our food to our hearts' content but it is risky business to examine our friendship over much. Science puts all things into question, analyzes them, and examines them in the cold light of reason. (John Donne: "The new Philosophy calls all in doubt".) And the analyst's lancet is deadly. It is commonplace that all I have to do to kill any friendship and any love is to cut small chunks off it and examine them. This way I come to suspect my friends' fidelity on the minutest trivia, and provoke them to respond in kind. As Russell puts it, the idea that friendship is not a matter for science is quite obvious. The idea that Buber proposed is much stronger. It is that science and friendship exclude each other: looking at one through the prism of the other annuls it. Niels Bohr generalized this idea in his attempt to generalize his principle of complementarity to go beyond the domain of atomic physics. (19) Thus, he said, quite generally, science and life exclude each other since the biologist's lancet kills the living things it dissects. Justice and mercy are likewise mutually exclusive, he added. Whether Buber and Bohr are right or not, however, does not depend on whether scientific rationality is proof, or whether scientific proof is ever valid. We may replace the idea of scientific rationality as proof with the weakest variant of it known. Popper has proposed such an idea: rationality is disproof or criticism, he said. Let us agree with him that science is merely the testing of our opinions. This agrees with the conclusion from the views of Buber and of Bohr: friendship rests on trust; science, reason, rests on examination. The two differ and we must take them as complementary.

This conclusion leads to two important corollaries, one concrete and valid, one abstract and fallacious. First, do not be too rational about your friends and lovers. Second, supplement reason with emotion. The second corollary is false and dangerous. Yet, all falsehoods include some truths. The truth in the false and dangerous corollary is that reason is limited. Russell erroneously proposed that we do not need to limit science for the sake of love and friendship. After all, some medical research is forbidden on moral – not scientific – grounds. This limitation suffices to show that we do limit science quite regularly.

The importance of this lies in the classical understanding (of Condorcet or Laplace) of the consolation of science as the claim that reason suffers no limitation. The hope to gratify the need for consolation rests on the promise of reason, and this promise rests on the assertion that reason suffers no limitation. This promise fails. And then irrationalists propose to recognize the limits of reason. Seemingly, this enables them to propose some other promise to complement or supplement the promise of reason. The new promise comes from faith. And this means that the promise of consolation is no promise but plainly faith. It is the idea that the need (for consolation) is the promise (that the need will be met). What other promise then is at all possible to insure consolation? Perhaps there is none. Faith in either religion or science is an effort to transcend skepticism. The promise of consolation amounts to a promise to transcend skepticism. To no avail: skepticism undermines the promise.

THE LIMITS OF FAITH

VI

Many rationalists, including many scientists, are sincerely religious. Their religion appears as a complement to science / rationality, not as a competitor: if the two compete, then endorsing both is impossible. There is an ambiguity here, however, of the word "impossible". Endorsing inconsistencies is impossible, yet it happens regularly. Are all religious rationalists or religious scientists of necessity inconsistent? Arthur Koestler said, yes. Many people, he suggested, (20) including religious scientists, suffer from a chronic inconsistency and learn to live in the state of controlled schizophrenia. Religious scientists hold one thesis in the church and its contrary in the laboratory.

Perhaps not. Possibly, the proper place for science is the laboratory and the Church is the proper place for something else, for something extra-scientific. And then there is no need for controlled schizophrenia in order to contain the two. This may be so even though it is possible to choose to ignore altogether the church or the lab – or both. One can ignore science /faith even without deserting the lab /church. To do so, one has to declare science / religion empty, devoid of all philosophic import, and view it as no more important and/or interesting than one's workshop for producing tools. The label for this view is "instrumentalism". It is irrationalist: it denies science / religion its autonomous value. Viewing science / religion as a mere instrument is decadent. (Many important and able rationalists have endorsed instrumentalism regarding one or the other. Applying this reasoning to only one of the two is an inconsistent favoritism, though cogent means to resolve inconsistencies are easily available.)

Some say, the lab engages the brains and the church warms the heart. Ignoring then the lab / church is dull / cold. Some reject this parallel. Some places have no labs, but all places have churches. This is so, say some, because faith does not depend on reason, but there is no reason without faith, no lab without church. The advocates of this view are fideists. (21) In religious circles, almost all intellectual or scientific leaders, big and small, are fideists, at least in miniature. (22) They advocate a simple thesis. Science and scientific thinking are excellent for ever so many purposes, but not for all purposes. Contrary to militant rationalist propaganda, fideists do not oppose science and reason; they recommend their exploitation to the utmost. This utmost exploitation, they add, is not enough. They say, rationalism fails to satisfy both the religious and intellectual needs. It cannot confute skepticism as it should. And so we must supplement science by faith, they say – for the sake of science too.

The anthropologists and cultural historians among the fideists mention as support for their view the fact that ancient religions combine faith and reason. They usually are primitive, of course, but as the Pythagoreans count among them, this is not always so. Koestler saw in the Pythagorean lore an ideal of an integration of faith and reason.

In all these deliberations, skeptics appear as the enemy from within, the great demoralizers. Opposite to skepticism stands scientism, the authority of reason, of current science, or else fideism, the authority of faith, of established religion. The authority of religion may oppose science altogether (and view science as of mere practical value). Otherwise, the authority of religion may offer proper foundation for current science and thus legitimize it. These are two different religious views of science, instrumentalism and fideism. Instrumentalism opposes the claim of science to comprise theoretical information and thus food for thought. Fideism endorses this claim of science, and grounds it in faith.

Alternatively, the authority of science may oppose religion (and view religion as stories – as myths and superstitions and folk-tales – or, alternatively, as morality – as stories with a moral – or both). Otherwise, the authority of science may offer proper foundations for religion and thus legitimize it. Of course, religion can legitimize science only after it trims its anti-religious claims. Similarly, science can only legitimize religion within reason alone. This allows parallel kinds of legitimation – fideism (faith is at the foundations of science) and scientism (all and only scientific opinions are admissible). Instrumentalism rests on the despair of legitimation – of faith or of science – and consequently it empties faith or science of content.

The symmetry between the two kinds of legitimation is not perfect. Whereas fideism supports science, scientism allows religion in a condescending, gallant generosity to a losing opponent. (23) As science can overcome skepticism, religion is hardly necessary. It is no authority; it depends on the source of all authority, which, according to scientism, is reason, namely science, and it alone. Consolation, then, may belong to either contestant, but not to both. Not in the strict sense, that is.

The reason for this is in the impossibility to will to believe, to use the famous expression of William James. One may have faith, and it may console. But the wish to have faith and its consolation is just a wish, a yearning, not a faith.

VII

The heart of the matter is the claim that skepticism is the great demoralizer and that both science and religion compete as the means to overcome it. These claims are open to empirical tests. One may make a list of religious and scientific leaders (they need not be top leaders) and press on some members of each list some very standard skeptical points. One will thus find that members of both parties will try first to resist, and then to brush off the skeptical assault in the name of common sense. They do so while expressing some (justifiable) annoyance at it as sheer distraction, and finally they will dismiss the assault as arch-skeptical and so, if at all valid, valid as means with which to destroy almost any proposition. I have witnessed religious and scientific leaders, well known and less well known, let me report, respond in this manner to skeptical assaults. And I read many such responses. If the skeptical argument is valid, they say, then it forces one to the absurd choice between total skepticism and the other party – fideism or scientism, as the case may be. It is surprising empirical information that both parties use the fact that skepticism pushes one to the other party as the final – transcendental – proof that their position is right and the other wrong. They will ignore the commonsense recognition of friendship and other consolations as extra-scientific factors. These, then, rest on sheer faith, on religious belief of sorts, or else they rest on science, on scientific proof of sorts. The very threat of a rejection of both amounts to skepticism. This renders mandatory the choice between the two for all who view them as the only alternatives to the skepticism that they abhor.

The fear of skepticism that advocates of fideism and of scientism exhibit deserves study, and so does the corollary that both opponents have endorsed, namely that they could, and their opponents could not, take care of love and friendship and generally of consolation. The expression of the fear of skepticism is so often all too abstract. Consider, for example, the following argument: if you accept the kind of argument that the skeptics advance (namely that you should not assume this or that thesis, as no proof for it is available), then you can just as well doubt the existence of all things material. This is the argument, to report, that I heard repeatedly since adolescence – as non-religious scientists and philosophically inclined rabbis alike repeatedly uttered – with variations such as, you will have to doubt also that people are not robots. Of course, this addendum contradicts, not supplements, the original claim. For, if you deny that material things exist, then you cannot possibly also assert that your fellow-humans are mere robots. But this is a minor inconsistency that we can remove by viewing it as an expression of a mere doubt.

At times, the argument against skepticism can become more concrete and thereby also more telling. If you are a skeptic, the scientist and the rabbi might say, can you trust that the brakes on your motor-car will function next time you come up against an unexpected pedestrian on the road? Or that the ground on which you stand is solid? Now we come close to a source of the anxiety. Though it has many examples, there is one outstanding representative, a symbolic expression, for the class of examples for such underlying anxieties – a paradigm case. It is this. If you are a skeptic, why do you assure us that the sun will rise tomorrow?

Skepticism begins in the abstract: it applies to any idea, to today's science or to today's religious faith. It raises doubt about it on the most general grounds. It ends up forcing us to doubt that we walk on firm ground. All of a sudden, Laplace starts to make sense; Kant does not. We can put his considerations aside; Laplace's proof may be defective too, but if it is good enough for the next million years, then it is satisfactory: it does offer some consolation and it warms our hearts. So we remember it.

VIII

The image of science as a comforter, as a guarantee that the sun will rise tomorrow, is one that legitimately comforts us, as many a commentator on science assure us. This lends authority to the view of science as authoritative, as Science with a capital S (as Popper has called it), to scientism – the vulgar view of science as a substitute religion: it is the better comforter. What cases does science take care of? Will religion take care of the rest? Briefly, the answer that the same commentators often offer is, yes: science comforts us on matters earthly; religion comforts our souls.

From this we may conclude that we have now transcended both scientism and fideism: rather than having religion under the auspices of science (scientism) or science under the auspices of religion (fideism), we have a division of labor. They need not integrate in harmony; they display a Buberian or a Bohrian division of labor, and are thus totally separate. Whether integration or separation, it is the end of centuries if not millennia of competition. Complementarity replaces competition. This is an innovation, and it rests on one and only one argument: religion cannot compete with science on the facts of nature, whereas science cannot compete with religion on the rest: science has the natural for its domain, religion has the supernatural; the twain never meet. But then Pascal is right: the result is that Earth has no room for religion and Heaven has no room for science, and as the debate takes place on Earth, not in Heaven, religion loses.

This is where Russell and Buber come in, we remember. Each of them offers a division more acceptable to Pascal than Descartes did. Love and friendship are of no concern for science, they both agree. Some affairs are rational and they submit to cold calculation; other affairs are emotional and they submit to gut responses. Russell has no rule for the right attitude of science towards friendship. Buber considers trust axiomatically an attitude that goes beyond science. His argument is, only the axiom that trust is extra-scientific insures its possibility. Buber's reasoning here is straightforward. If you do not assume interpersonal relations axiomatically, you must conclude that they do not exist.

This reading of Buber may be simplistic. There are many commentators who find it barely possible to grapple with the depth of his ideas. To them the interpretation offered here is so simplistic as to be offensive. I claim no authority and no expertise on Buber's writings, but I report that he was never happy with interpreters who sought depth in his views and he expressed to everyone who wished to hear, myself included, the hope that future interpreters would manage to put his views more simply, not more profoundly. (24)

All disagreements and differences between Buber and Russell should not cloud the basic agreement between them: that both opposed scientism as well as fideism. Is there a viable third option? Can there be a viable third option? Can we shed more light on the matter? We had rationalists appeal to reason, and irrationalists appeal to feelings. Intuitively, one might wish to explore the view that perhaps we are partly rationalists and partly irrationalists. (It is an observed fact: humans are partly rational, partly not. Irrationalists use this as an argument. It is valid only against some versions of rationalism.) It is not clear how to approach matters this way, since, obviously, mixing a weak and a strong claim only weakens one's case, and mixing two weak claims is no better. On what ground can a mixture ever be better than its ingredients?

The question looks water-tight, but it is not well-put. It is not the mixture that matters but the rejection of both scientism and fideism. Anyone whose claim is a mixture obviously rejects both views in their pure form, or both purist views. Both Russell and Buber reject both purist views, yet they separate the rational and the irrational, thus maintaining all-round rationalism (Russell) and place the separate rational alongside with separate irrational (Buber). It seems advisable to try a less neat approach. After all, the question was not, what is the best blend of the rational with the irrational. It was, what is the right rationalist attitude towards matters beyond the sphere of reason (love and friendship)?

SKEPTICISM REHABILITATED

IX

The way to approach a situation where all alternatives are wanting is to examine the presupposition common to all of them. If there is no presupposition and no alternative overlooked then there must be one alternative that is true. Clearly, fideism and scientism are the only alternatives traditionally explored; hence the supposition they share is the cause of our displeasure with both. What was it? What was the function either doctrine came to fulfill? It was to offer us consolation by combating fear and doubt – by standing against skepticism. Is there perhaps no consolation anyway? Is it possible that we are better off not combating skepticism?

The traditional answer to this is the shared supposition. Both parties declare that skepticism is poison. Hence, they conclude, it must be confuted, and successfully. And success in vanquishing skepticism opens the way to consolation, to the hope and faith and knowledge that all is well that ends well.

What ends well? Life on Earth? This may be so, but it all too often goes, alas! the other way. Life on Earth, we all think, will end in disaster sooner or later. We have no idea what the end of human life is in general, whether humanity will or will not survive its lost planet. And today we even wonder whether humanity will succeed in its effort to avoid destroying Earth. What is the consolation required? It is the consolation of certitude as such, says the tradition, and the tradition is clearly in error. Seemingly, we want assurance. But we want more: a rosy future. Assurance of doom is no consolation.

It is no good saying, then, never mind skepticism, let us have our consolation. This is mere wishful thinking. It is no good saying, likewise, never mind consolations, let us combat skepticism even if the certitude we will achieve is of the grim knowledge of a bitter fate. Consolation is for both science and religion the very success of the attainment of human knowledge – whether through scientific method or through divine revelation. Knowledge, within tradition, implies optimism; one might say, however, reject optimism and seek knowledge anyway. Once we have knowledge, one might propose, we can take it from there. Not so: knowledge is a good thing even if it is of things most unpleasant: not knowing them will not improve the situation, and knowing them is good even if it is of no other use.

Why combat skepticism? In order to find consolation, of course – even if knowledge is of unpleasant truths. But what assures us that even this meager consolation is at all possible? Tradition claims that skepticism is destructive and demoralizing. Yet, all our doubts have still not destroyed us and they have still not demoralized us. On the contrary, dogmatism may cause destruction, and doubt has repeatedly stood in the way and saved the day. (25)

Tradition will not conclude that since we withstood doubt, it is perhaps not destructive. Rather, tradition claims, since we withstood doubt, we do have certainty. Do we trust science or faith? Is this claim an argument in favor of scientism or of fideism? It is a transcendental proof – the most general form of the most defunct philosophical argument. Transcendental arguments have this form: if x were not the case, then y would be impossible; yet y exists; hence, x is the case. Transcendentalists invite us to check their claim that y exists, thus diverting attention from the possibility of y without x. They all rest on the doubtful assumption that y depends on x. For example, in the present case, skepticism is not always destructive; at times it is, but anything whatsoever at times may be.

To cut it down to size, in the present case the argument says, if skepticism were true I would be demoralized, and then science / religion will vanish and society will too. This is an error: doubt does not demoralize, much less destroys. The view that doubt demoralizes does that. My society lives with permanent doubt.

X

Science, then, does not console by adding much to faith. It helps eliminate faith in some items that do not deserve it. As Popper has emphasized, science is a Socratic dialogue just because it may eliminate some of our worst errors. It thus opens the road for novelties. These might be still worse errors. Though scientific technology opens the road to new and marvelous improvements, it also opens the road to new and unheard of global disasters. In particular, the hope that science will insure that the sun will rise tomorrow turns out to be the discovery that the sun may explode any minute, that pollution may alter the biosphere irretrievably, that traditional society will be irrevocably replaced – possibly by some unstable and unpalatable social order.

But science comes to satisfy a different and stronger yearning than that for certitude – it is the attempt to cater for the quest for knowledge. It is futile to ask, does science contradict religion? Suffice it that to the scientific temperament religion is uninteresting. Religious scientists are the very proof of this: they accept religion and practice it for any reason on earth, except for the reason that it tickles the appetite for knowledge.

As long as science is but the quest for knowledge, its advocates need not quarrel with the advocates of religion but ask them for interesting information. Even when science and faith clash on practical matters, the normal scientific response is to seek additional relevant information and raise new interesting questions; there is no need to fight the windmills of religious doctrine or the fortresses of religious authority. The power and the consolation of science are in the ability, hopefully, to act responsibly in efforts to satisfy and occasionally to gratify curiosity, and in the service that science might offer the soul – more than anything else. Much more, that is, than adherents of either fideism or scientism offered or even requested.

And so the consolation of science is the teaching that a quest may be rewarded in a life that might be satisfying – so that we may try to make the best of it and in our different ways. The ultimate consolation of science seems to be the view of ourselves as having tried our best to the best of our understanding.

There are almost no scientific sermons around. The most famous one in modern time is Russell's "Free Man's Worship" (26) that advocates consolation in the absence of all certitude, by going proudly to our unknown destinations doing the poor best we can under the circumstances. Another frankly secular sermon is Ernest Gellner's "Prepare to Meet thy Doom", (27) whose title is its message.

How consoling this message is depends on individual tastes. Non-religious rationalists claim that (thus far) we have none better. In particular, the division of labor between science and faith is useless. Things are no better if science encompasses all nature and leaves things above nature to religion. As Pascal has noted, this is no religious consolation, as it drains religion of value. The question remains, is there scientific consolation? Does nature offer any? The answer is, only as an object of contemplation; and thus, only to the contemplative the answer is, possibly yes.

The question arises at once, as to the consolation of action – as proposed in the celebrated conclusion to Goethe's *Faust*. To this philosophers (Plato, Spinoza) replied thus. Only the contemplative is truly active. But this is another story.

NOTES

1. See David Knowles, "Boethius," in Paul Edwards (ed.), *The Encyclopedia of Philosophy* (1967), 1, 328-30; "the *Consolation of Philosophy* was one of the two or three books of universal appeal throughout the Middle Ages .... It was favorite reading of Dante, Boccaccio, and Chaucer, and inspired numerous imitators."
2. The best history of this point may very well be John Owen, *Evenings with the Skeptics, or Free Discussion on Free Thinkers* (1881). See also my *Science In Flux* (1975), 455.
3. See my *Science in Flux* (1975), "Can Religion Go Beyond Reason?" 469-514 and "Assurance and Agnosticism," 515-24. My later views on the topic are expressed in my "The Future of Berkeley's Instrumentalism," *International Studies in Philosophy*, Vol. 7 (1975), 167-78, esp. 168-69, reprinted in my *The Gentle Art of Philosophical Polemics* (1988), 55-69. See also "On Explaining The Trial of Galileo", in my *Science arid Society* (1981), 321-51.
4. Richard H. Popkin's presentation of the fideist view is the best, especially in his *History of Skepticism from Erasmus to Descartes* (1968). See also his superb essay, "Fideism", in Paul Edwards (ed.), *The Encyclopedia of Philosophy* (1967), 3, 201-02, and the bibliography there. See also my "Irrationalism Today", reprinted here as Chapter 2.7.
5. For Bellarmino see my "On Explaining the Trial of Galileo", "The Future of *Berkeley's Instrumentalism*", *op. cit.*, and "The Two Books", Chapter 3.6 here.
6. Sir Leslie Stephen, *History of English Thought in the Eighteenth Century*, (1876) and *English Literature and Society in the Eighteenth Century* (1904).
7. The pamphlet is the third part of Friedrich Engels' *Herr Eugen Dühring's Revolution in Science* (1878; 1934). It is better known as *Anti-Dühring*. Lenin declared it obligatory reading for a Marxist, along with *The Communist Manifesto* and Engels' *Ludwig Feuerbach and the End of Classical German Philosophy*. The pamphlet is clearly the latter day substitute for Boethius' *Consolations*. This explains its being the most popular single Marxist text. Before that, incidentally, the substitute used to be Condorcet's *Esquisse d'un Tableau historique des progres de l'esprit humain, 10me époque*, the final chapter concerning the future, in which he proved that the future must be rosy. Russell, incidentally, cites Alexander Pope's *Essay on Man* as the substitute for Boethius' *Consolations* on account of its celebration of Newton's triumph. See his *History of Western Philosophy*, 2nd ed. (1961), 368.

   This is not to endorse Engels' claim. On the one hand, Marx's claim for scientific status is no longer acceptable. See also Karl R. Popper, *The Open Society and Its Enemies* (1945), and his *The Poverty of Historicism* (1957). On the other hand, contrary to Engels' claim, Marx had no priority for this claim. See John Plamenats, *Man and Society*, Vol. Two (1963), *The Early Socialists*. As noted above, already Condorcet sought to prove that the future must be rosy.

8. For Hadamard's refutation of Laplace's proof see Pierre Duhem, *The Aim and Structure of Physical Theory*, 1914, English tr. by P. P. Wiener (1954), 139-41. Rom Harré's "Laplace" in Paul Edwards (ed.), *Encyclopedia of Philosophy* (1967), 4, 391-93, though critical, presents Laplace's proof with no comment.
9. Laplace's rule of succession is so popular that its history invites an extensive survey. It is the basis of Roy F. Harrod, *Foundations of Inductive Logic* (1956). (See also the exchange he had with K. R. Popper in the *British Journal for the Philosophy of Science*, 9 (1958-59), 221-24 and 10 (1959-60), 309-12.) See also Rudolf Carnap, *Logical Foundation of Probability* (1950), for a variant of the rule. (See also K. R. Popper, "On Carnap's Version of Laplace's Rule of Succession," *Mind*, 71 (1962), 69-73.)
10. Newton claimed that the law of conservation of energy – or rather its predecessor – does not hold, so that the solar system is unstable. Leibniz claimed that it holds and renders the system stable. Kant contradicted both. He said, it holds and renders the system unstable. (See note 16 below.) The claim that it holds within Newtonianism is due to D'Alembert, 1743, at the earliest. (A better proof had to wait for the discovery of potential energy by Lagrange and by Laplace, 1777 and 1782. Lagrange presented it in 1788.) Kant's proof rests on the conservation of energy. He presented his argument in 1754. See Gabriele Rabel, *Kant* (1963), 9-10, for the translation of the original text. Stanley Jaki has discussed admirably Kant's standing in the world of physics and its history – in his annotated edition of Immanuel Kant, *Universal Natural History of the Heavens* (1981). Mario Bunge asserts that Kant was scientifically ignorant.
11. See Isaac Asimov, "Social Science Fiction," in R. Bretnor (ed.), *Modern Science Fiction, Its Meaning and Its Future* (1953), 157-76, and his "The Magic Society," *Boston University Graduate Journal*, 15 (1966-67), 9-14. For more references and discussion see my *Towards a Rational Philosophical Anthropology* (1975), 13ff.
12. Martin Buber, *Two Types of Faith* (1952). Some circles refer to the study of belief or disbelief in a proposition as to the logic of belief or to inductive logic. Buber says it is originally Greek. Inductive logic is Greek because belief or disbelief in a proposition is Greek. If belief and disbelief in a proposition is Greek and in a person Hebrew, then what are beliefs and disbeliefs in other cultures? Claude Lévi-Strauss offers the following answer (*The Savage Mind*, 1966, concluding paragraph). Simple cultures exhibit a systematic confusion of words and objects. It is, oddly, a confusion of what Buber calls the Greek and the Hebrew. (The prevalence of the view of the logic of belief as limited to inductive logic shows how narrow all current views on belief are.)
13. See opening paragraphs of Martin Buber, *I and Thou* (1970). There is hardly a comment on the linguistic aspect of Buber's presentation. It is not clear why. Perhaps commentators took it as but a metaphor, and perhaps rightly so. See Erazim V. Kohak, "I, Thou, and It: A Contribution to the Phenomenology of Being-in-the-World," *Philosophical Forum*, 1 (1968), 36-72; see, esp. the second paragraph and note 7 there.
14. See Georg Simmel, *Die Religion* (1906), translated as *The Sociology of Religion* (1959), second and third paragraphs of Section III. The question is, does Buber's link of the two types of faith to the dichotomy between the I-thou and the I-it relation hold? Michael Polanyi, for example, recognizes that the I-thou relations are deeper than the I-it relations (*Knowing and Being*, 1969, 149), but he rejected the dichotomy between them since, he claimed, one cannot doubt everything even within science. See his *Personal Knowledge* (1964), 269-98, "Critique of Doubt". He denied more forcefully that science could do with only inductive logic, since, he claimed, science rests on tacit faith in its leadership. But can one deny the dichotomy and endorse the division of faith to two types or *vice versa*? I do not know his answer.
15. Blaise Pascal, *Pensées*, 77; M. Bishop, *Pascal, The Life of Genius* (1968).
16. For the debate Newton had with Leibniz on this issue see A. Koyré, *From a Closed World to an Infinite Universe* (1956). See also my "Koyré on the History of Cosmology," *British Journal for the Philosophy of Science*, 9 (198), 234-15.
17. Bertrand Russell, *Religion and Science* (1930), 175. See also his *History of Western Philosophy* (1961), 720.

18. See Élie Halévy, *The Growth of Philosophic Radicalism* (1955), Chapter 1. See also my *Towards a Rational Philosophical anthropology* (1977).
19. For a succinct presentation of Niels Bohr's extension of the principle of complementarity see Loren R. Graham, *Between Science and Values* (1981), 55-62, and the detailed references there. See also there, 360. See, in this connection, also, Gerald Holton "The Roots of Complementarity", in his *Thematic Origins of Scientific Thought: Kepler to Einstein* (1973), 115-161.
20. For Arthur Koestler's term "controlled schizophrenia" see his *The Sleepwalkers, a History of Man's Changing Image of the Universe* (1959), 73 and 518; see also 100-3 there, and his *The Ghost in the Machine* (1968), 259-65 and references to his other works there. See the discussion of it in my *Towards a Rational Philosophical Anthropology* (1977), 337.
21. See note 4 above. Fideism is an attitude intellectually favorable to science, though to science as based on faith. Instrumentalism is not. Religious thinkers have the choice between the two. Endorsing both is confusion.
22. W. W. Bartley, III describes the situation forcefully in his *The Retreat to Commitment* (1980). Russell has observed (*Religion and Science*, 1930) the trend of seeking reconciliation between religion and science in the early twentieth century. Instrumentalism seems to effect such reconciliation but it deprives science of its informative content and so of its intellectual value. Fideism is a better reconciliation, which, however, has all the defects of scientism and more.
23. Kant's epoch-making *Religion Within the Limits of Reason Alone* clearly exhibits the traditional condescending attitude towards religion. See my "Can Religion Go Beyond Reason?" *op. cit.*
24. Charles David Axelrod, *Breakthrough: Freud, Simmel, Buber* (1979), 52, rightly presents Buber's views as intentionally simplistic, and their simplicity as their forte.
25. Peter Ustinov has put this succinctly when he said, We are divided in our faith and united in our doubt.
26. Bertrand Russell, "Free Man's Worship", *The Independent Review*, 1 (1903), 415-24, and many reprints (see the Schilpp bibliography), including his *Mysticism and Logic* (1959), 46-57.
27. Ernest Gellner, "Prepare to Meet Thy Doom", in his *Contemporary Thought and Politics*, (1974), 1-7.

Chapter 1.3

THE MORAL BASE OF SCIENCE, OR, THE ARCHITECTONIC OF OPEN-ENDED REASON

0. THESIS

Traditionally, philosophers looked for philosophical systems. By definition, systems are (sets of) theories that include answers to all important philosophical questions, or at least methods for finding them. Contrary to received opinion, it is not advisable to offer a philosophical system: it amounts to the offer to relinquish all moral autonomy. Moreover, no one ever proposed a philosophical system proper. All known theories have huge gaps in them. Traditionally philosophers are hostile to philosophies that are frankly open, thereby expressing – knowingly or not – preference for a closed philosophy as a preference for a closed society. This causes much frustration. Openness is essential to modern pluralist democratic society – and to science too.

1. INTELLECTUAL FRUSTRATION AND ITS CAUSES.

One of the most influential texts in modern philosophy is Immanuel Kant's *Critique of Pure Reason*. It is a thick, difficult, pedantic book. Few have read it through. This is a pity. The end of this book is in a different style. It is easy and pleasant. It ends with a presentation of as comprehensive a view of our world as Kant ever had. He called it "the architectonic of Pure Reason". As his image of the world was one that he thought is in our mind form the start, it was naturally extremely narrow. He followed Robert Boyle in viewing the mind as possessing two qualities, reason and moral sentiment. The task of the architectonic of pure reason was to unify them into a closed system.

The reaction to it was often hostile. Many found it a convincing argument against engagement in philosophy. Others, who cannot let go of philosophy, tried to improve on it, and their first task was to open it to new ideas or even just to the mere possibility of new ideas. This is open-endedness. Some like it, others hate it. Some think it is avoidable, some think it is not. This, of course, is a question of cost: at what cost do we want to keep our system closed or open, as we may wish? Some of those who hate openness admit that it is unavoidable and only wish to limit it. For me there is never enough of it. And the question that this raises for me is, how can we chart the image of the world, and of the human spirit within it, and yet keep it open. Many find this whole issue extremely frustrating.

My intent here is to contribute what little I can to the reduction of the unnecessarily high degree of frustration in the modern world in general and in intellectual life in particular. Briefly, the frustration that accompanies attempts to talk to people who refuse to listen is unnecessary. We can drastically reduce the degree of frustration at relatively low cost. Anyone familiar with the Book of Jeremiah is familiar with the frustration that this prophet has suffered. But one need not be a prophet to suffer such frustration. Even simple, everyday academic experience will cause it. Let me mention one personal example, my own work. To narrow matters down, let me mention my book on the famous nineteenth-century physicist Michael Faraday. Though I studied his ideas, I could not avoid discussing his person. I showed that frustration caused him much agony and that he became mentally ill. In my view for what it is worth, what made him ill was his refusal to admit that the scientific community may choose to be deaf to new and exciting scientific ideas. When I wrote the first draft of this book, I was a student and disciple of Popper. I had a model for the frustrated thinker before me: my own mentor. Like Kafka's hunger artist, Popper excelled in the art of frustration. He always looked for newer ways of penetrating the wall of silence that surrounded him. Unlike Faraday, Popper did find audience eventually – to his amazement. He found eager audience even in the scientific community that had refused to listen to Faraday and for a while also to his disciples Kelvin and Maxwell. Popper was not sated, however. He received many honors from many circles, but the philosophical establishment, whatever this is, refused to recognize him to the last. His frustration was somewhat alleviated by other forms of recognition, but it never died out.

As it happened, let me confess, I developed an allergy to this sense of frustration. I would rather not speak at all than speak in frustration. To avoid frustration I decided early on to avoid talking to the deaf. After all, I am no prophet. I have no sense of mission, so I can easily afford to avoid trying too hard. I admit, in my opinion, for what it is worth, I do have something to say, and I usually do say it. I admit, editors often frustrate me, or rather the incompetence of so many editors and referees does. But, as luck would have it, I have much more opportunity to speak than I have dared hope for in my highly frustrated youth or ever since. My aim is usually to be helpful to my audience. And so, obviously, I say what little I have to say only to those ready and willing to listen, perhaps only to those among them who seek help in whatever is in my power to do for them.

Few cases, like that of the prophet Jeremiah, are simple and clear-cut. He demanded of his audience to assent to the message that he had to deliver. He had neither need nor wish to listen. He wanted, and to an extent expected, one-way traffic. What he said was compelling, as he was a messenger of the supreme power. The enormous value of the messages of Faraday and of Popper makes their cases also relatively simple, even though they were not as simple as that of Jeremiah. The very fact that the messages they had to deliver were important and interesting caused the frustration. They could not say to themselves, had my message been more important, they would have noticed it. Their frustration was less than that of the prophet Jeremiah, since, unlike him, they did not demand assent of their audiences, and only wished to have their ideas on the agenda for public debate. They both wished for some two-way traffic. And so they listened, but no one listened to them. And so they found what they listened to increasingly boring, yet they kept listening attentively, always trying to engage others in some conversation, but to almost no avail. Naturally, they found this increasingly frustrating. The efforts to speak to the deaf frustrated them, as they did Jeremiah. Their frustration was less then his in the sense that they did not speak with divine authority. It was more then his in the sense that they vainly tried to listen in order to engage others in some dialogue. This frustration is common to people who have anything to say and no audiences to share it with. This frustration demands attention.

The catchword of our current intellectual milieu is pluralism. This is in sharp contrast to the rationalism of the Age of Reason, when philosophers condemned differences of opinion as divisive. In particular, they viewed religion and politics as divisive. They preferred to concentrate on science and ignore religion and politics, especially political ideology (to use a later term). This is true even of the religiously and politically committed. Some of them tried to prove their views rationally. Others recognized that their peculiar convictions do not speak to others, and so they were quite willing to put them aside when stepping out of their specific communities to address a wider public. Things were not that simple, of course. The greatest assault on religion aimed not at its theology but at its anthropology. The earlier learned culture, the humanist tradition, was profoundly religious and stressed that its new accent on the human qualities of human beings is in no way in conflict with any religious tradition. Not so the Enlightenment movement. Its members ignored the question of its accord or discord with the faith of their ancestors yet they tacitly undermined it by offering a new, worldly anthropology.

Medieval ideology has a single simple idea. It is that evil is natural and common to all flesh; evil rests on irresistible, unreasonable impulses. It is most irrational to sin, as the benefit of sin is small and short-lived and its cost is huge and everlasting. The Enlightenment view was the reverse. Since people are rational, they suggested, those who do not evade eternal damnation do not believe in Hell.

David Hume succinctly expressed the ideology of the Enlightenment Movement. He said, people are much more to be trusted to act on greed than to act on good will. His disciple and friend, Adam Smith, wrote his classic economic treatise based on this advice. In that book, there is no religion to speak of, and no discussion of evil, only of conduct in self-interest. It is worth mention that Smith clearly presented his system as a recommendation, not as a description. Like his contemporary Rousseau, he knew that if a recommendation is not practicable, then assent to it carries no responsibility with it, and hence securing it means little. Like Rousseau, he mentioned an example that comes close enough to his recommendation to show that he was not recommending a distant Utopia but a practical system. Rousseau said, his system was akin to the democracy as practiced in his hometown Geneva (*The Social Contract*). Smith said the same of the Hanseatic League (*Wealth of Nations*, Book V).

This idea underwent a profound transformation with the rise of positive economics, so-called. We should return to it the prominence that it deserves. With but a slight twist, it turned Smith's prescription into a description. His followers translated his ideas from the prescriptive mood to the descriptive mood. This change is very basic to the whole of the tradition of economic thinking. Marx took Smith's ideas to be descriptive as a matter of course. He probably was aware of the variance in the text, but then he assumed that industrialization rendered Smith's prescription the accepted norm. Bernard Shaw ridiculed Marx (though he called himself a Marxist) for his failure to see the reforms that the law underwent under his own nose, especially regarding child labor and compulsory education. No less important is the fact that positive economics to date is but a variation on Smith's ideas, including neo-classic economic theory and even Keynesian economics. Indeed, Keynes himself stressed in his major work that expounds his ideas (*The General Theory of Employment, Interest and Money*) that his major change was his addition of liquidity preference to the list of individual preferences that was traditionally unspecified and not under discussion. This was a sleight of hand. Traditionally, the presumption was that preferences are not under discussion because individual preferences do not matter as they concern tastes that are outside motives for economic conduct, not integral aspects of it. Not so liquidity preference. Being a preference for money over stock, it is not outside the economic system (and Keynes did not suppose it to be). So his presentation of it as but another item of taste misleads. Yet in this he did not deviate from Hume's dictum that it is best to trust people to act on greed. It only shows that greed has new, significant economic ramifications. Consumers' preference theory, by contrast, is the view of them as only matters for the economy to take care of, not as causes for its collapse. Preferences are presumably exogenous, to use the economists' jargon term, not parts of the system.

Is the theory of consumers' preference successful this way? Obviously not, yet economists assume that it is successful, if not entirely, then as a legitimate idealization, as a simplification near enough to the facts and interesting enough to take at face value, at least for a while. This is fine, except that it forces consumers, those engaged in economic activity, to the mold of an ideal type. This ideal type is Economic Man. The discussion whether Economic Man exists or not is not the discussion of the question, is greed our only motive and is it only the wish to eat and drink and be merry. This question is silly. Only silly preachers seriously assume that members of their flock are so crude. The discussion whether Economic Man exists or not concerns this question. Does the Ideal of Economic Man depict us well enough for the purposes of economic theory? Should this theory also take other factors into consideration, for example, the will to power? (This expression became famous as it has served as a title to the last book of Friedrich Nietzsche.) The will to power may induce individuals, and even whole nations, to curb their greed for goods and services for the sake of glory. It is wrong to consider glory as a good like any other, as it leads governments to interfere with their own markets in order to increase the nation's power ("guns instead of butter").

This claim is under dispute. Smith's point was that interfering with the market is unwise, and so perhaps he was against the use of power. His recommendation not to interfere with the market rested on the recommendation that only greed should operate in the market. Yet, a government may decide that the wish for glory should replace greed. Smith did not see any attraction in glory and he found warmongering abhorrent. But here he argued against the use of brute force for economic ends – as economically inefficient. He said, putting people in chains do not raise their incentive to work. He argued economically against putting people in chains, though he opposed it on moral grounds too.

Not long ago, this came to a very important political consequence. To begin with, in the nineteen-fifties the import of scrap iron threatened the American heavy industry. The then Republican leader of the Senate, Senator Everett Dirksen of Illinois, said, of course I believe in the free market, but I will not let America be the dump-heap of the world's scrap iron. This is absurd. By the free market theory, the market mechanism prevents undesired effects by itself. The senator knew that it would cause unemployment. True, believers view this damage as far lesser than the one that intervention will cause. He, however, being a practical politician, had to cater for the problems of his constituency right away. So he recommended intervention (by the way of imposing import duties) and he paid homage to the free market as lip service. Lip service is expensive, but usually politicians postpone payment for its costs.

Other problems soon loomed larger. For decades American politicians were worried about the imbalance of trade between Japan and the United States. They tried all sorts of ways to alter this. The simplest way to do so is to impose import duties on Japanese goods. This is discrimination, so it is possible to impose stiff import duties on goods that are prominent in the Japanese export to the United States. This runs contrary to Adam Smith's basic recommendation. Following it, one should not worry about any balance of payments and let the market mechanism take care of things. This is not so simple.

The motives behind classical economics were noble. Sticking to theories while ignoring the noble motives behind them, may lead to objectionable conduct. Decent economists cling to these theories for many reasons, but not for the classical noble motives that were behind it: in the mean time, these motives left the scene.

2. PLURALISM.

Pluralism developed outside philosophy in the post-World-War-II political arena, much as the result of the lesson drawn from the Shoah. The philosophical community responded to pluralism partly by ignoring it and partly by endorsing it uncritically. Both responses ignore the question, how much consensus do we need in order to maintain pluralism? The old Rationalists recommended the exclusion of all differences of opinions, leaving no room for pluralism. Nevertheless, most philosophers of science still cling to the old rationalist views.

Some of them add inconsistently pluralist politics to Enlightenment philosophy of science. Hopefully, the current agreement to allow for religious and ideological diversity within the limits of liberal democracy is here to stay. We may wish to revise the old idea that we should keep our discussion out of the areas in which serious disagreement will remain in the foreseeable future. Even the idea that we should keep politics out of the university must undergo revision in the light of the attitude prevalent among professors in many parts of the world during the Vietnam War. This received a most forceful expression in the United States, where conscription was universal and where it was possible to postpone military service by registering as a student in some institution of higher learning. Professors found it then their duty to give a passing grade to every male student of military age.

This raises the question, what are the prerequisites for putting on the public agenda an idea, and what decision procedure should decide the order of priorities of items on the agenda? Since such decision should not be in the hands of one party, it is the consensus that determines it. And thus, the limit of pluralism lies in the consensus.

Is there such a consensus in philosophy? What is the minimum unanimity among those who share the agenda of philosophy? Popper offered an answer in his classic *The Open Society and Its Enemies* that he wrote during the War, before the rise of pluralism. It is, he said, mutual respect and shared interests. This is necessary, of course. Also, he said, this should do. This is a very strong demand. It is highly controversial. We may view it as a proposal; as a description it is false, as the frustration that he suffered shows. It is the demand to limit the consensus, to have it demand no more than moral and intellectual autonomy – to the requirement of no more that respect and sharing concerns with opponents. This minimal demand already excludes too many philosophers. Consider a more advanced question. If a worthy idea is not on the agenda, what remedy does this require? If the public in question is autonomous, then one may try to interest some individuals in the new item, so that after sufficiently many become curious, the item will hopefully find its way to the agenda. Not so if the public in question is heteronomous and prefer to follow leaders on whom they happen to rely.

And so the primary question is, what does it mean to confine the discussion to the public whose members are autonomous? What measures does this situation require? This query invites both factual and moral considerations, as well as their combination. This forces us to contemplate factual and moral considerations together. This may raise the question, how do we do this? How do we combine science and ethics? Is this at all possible?

The attempt to bring science and ethics together is a part of the naturalism that all philosophers share today, religious and non-religious alike. We may wish to have both the scientific theory of our ethics and the ethics of science, namely, the moral basis of science. Consider the view of ethics as resting on science. This view is highly problematic, since norms are possibly human creations, unlike the laws of nature that are there, if they exist at all. (The philosophers who say that humans make the laws of physics mean to say that formulas that we view as laws of nature are not such laws, since they are formulas made by humans rather than found in nature.) Rather than suggest that ethics rests on science, then, it is much easier to suggest that as a human activity science rests on some moral judgment or another. This is how Stefano Gattei and Mariano Artigas read Popper (approvingly). Still some thinkers insist on both theses: ethics is scientific and science is moral. This sounds like placing the two on a par – a move acceptable on the authority of Immanuel Kant, no less. This is Kant's concept of the architectonic of pure reason.

Kant's world-view is rather disappointing, as it is nothing more than his transcendentalism, the sum-total of all that he suggested is known without recourse to experience, and regardless of how the diverse pieces of that knowledge cohere with each other. Still worse, as they are all allegedly certain, they must at least be mutually consistent. As is a very well known, he noticed that they are not. He made light of this, thus signaling the end of the rationalism that he so vigorously represented. We should not blame him. The situation was very problematic and his attitude was scarcely avoidable since in the Age of Reason nothing met public notice unless it appeared with the presumption of certitude. Kant tried to advocate some ideas despite the absence of proof for them, yet the public did not take them seriously, much to his frustration.

In the Age of Reason proposals had to rest on science – knowledge, certitude – and so all large-scale reforms were supposedly assured of success. Diverse thinkers, from Bacon to Condorcet, declared even the progress of science assured. Today we know that all large-scale proposals rest not on science but on nescience, on learned ignorance (to use the apt expression of Nicolas of Cusa). This is not merely a philosophical recommendation. It is also a point of fact regarding causing unintended harm. This fact requires a complete revision of philosophy – of both science and morality, not to mention political and educational philosophy – a revision that Popper began. He did not design his philosophy to fit large-scale measures, as he recommended that we should prefer small-scale measures as much as possible, since these are more easily reversible in case of error, hopefully with not too much harm. (He labeled this kind of reform "piecemeal".) His philosophy requires a modification, in order to fit the measures that must be large-scale to meet urgent global problems. The task of rectifying Popper's philosophy – or any philosophy for that matter – requires proper and sympathetic examination, and this requires a total reversal of the verdict that the philosophical establishment has passed on it. Recognition of its great value and revolutionary character is imperative. The philosophical establishment is not yet up to this task, as they still suffer from the scarcity of the sense of disinterestedness that is so important for a serious leadership. My concern here is less to analyze this scarcity than to discuss the policy towards it. Still, let me offer a few remarks by the way of a sketch of a description of the current situation.

A few substitutes for the idea of scientific certitude are circulating. A few substitutes for the idea of the function of scientific certitude are also circulating. In the era of classical rationalism, science appeared as the bastion of moral and intellectual independence. These days science appears as justifying moral and intellectual servitude. Briefly, science is the highest level of human intellectual ability unaided by religious traditions and institutions. Its very existence, then, is proof that we can make it on our own; proof that humanity can make do without the aid of religion. Traditionally, the idea that science is the highest level of human intellectual ability was the idea that science rests on proof. The traditional, classical concept of scientific proof was extremely shaky, as no one knew what proof is. The first rudiments of proof theory appeared only early in the twentieth century. Yet the very idea of proof came to boost human autonomy. Since we have proof, rationalists suggested, following only what we know sets us free. On this matter, then, the means were flawed, but the end was noble. This matter now stands in utter reversal.

Noble as the argument was, and commendable as its outcome was, it is invalid. Even were proof available, it would not be right to suggest that it guards freedom. What guards freedom is valuing it above all and willing to take risks in order to protect it if need be. Today rationalists are not as valiant as their predecessors. Some leading philosophers advocate the abandonment of moral autonomy, of moral freedom. And some of them do so quite openly. (Let me mention four leading philosophers who openly repudiate autonomy, three (ex-Marxist, repentant) Catholics, Leszek Kolakowski, Alasdair MacIntyre and Charles Taylor , and one Marxist, Jon Elster.) Others do this under the guise of vague pretentious terminology, advocating relativism. It is less obvious that even philosophers of science can decry autonomy. Most of them almost invariably follow the scientific consensus in a rather servile manner, and they recommend that we all do the same. It does not matter much what is our theory of proof in general, or of scientific proof in particular, not to mention any theory of proof-surrogate. No such theory can possibly help settle any dispute as to the validity of any specific concrete case of scientific proof. Any specific claim for scientific proof invites specific examination. Nevertheless, most philosophers of science use some lame idea of scientific proof-surrogate against freedom and for the advocacy of the compulsory endorsement of any idea that enjoys scientific consensus.

The reason for this blind endorsement of science has little to do with the nature of science and of scientific proof. It concerns the social structure of the scientific community and the role of philosophers and historians of science within it as the servile public-relations officers of science. The concern of philosophers with proof or with proof-surrogate today is usually a matter of revalidation of the consensus, as Gellner has argued repeatedly.

Larry Laudan has criticizes Popper's criterion of demarcation of science (in his contribution to the Grünbaum *Festschrift).* He criticizes also Popper's question, not only his answer. The right question, he says, is, why should I believe my colleague the professional scientist, who tells me whose theory has won the scientific consensus? This is hilarious. There is no theory, and thus no consensus, as to what question is right. Suppose there was such a consensus. Why should I believe my colleague (the philosopher?), who tells me whose theory of the right question has won the (philosophical?) consensus? Laudan exhibits here the highest degree of servility – to philosophers and to scientists. He takes it for granted that the consensus always deserves endorsement, and he merely seeks the justification for his servitude. He does not have to understand a scientific theory in order to endorse it; he does not even have to understand why it has won the consensus (or else he would not need opinions of others). The brave Italian socialist Ignazio Silone expressed a different sentiment (*The God that Failed*). When he was a member of the Communist International, he disagreed with Stalin. He received advice to desist, on the ground that Stalin was winning the consensus. He condemned this with the observation that at that time Italy granted the consensus to Fascism.

3. OPENNESS AND PLURALISM

Popper presents science as critical rather than as imposing. This is highly polarizing, as it may appeal at most to the autonomous – morally and intellectually. It would be naïve to expect the establishment to endorse or even to appreciate it. They offer criticism of his ideas, though. They naturally express heteronomy. Discussions and criticisms that express heteronomy may still be interesting and possibly deserve serious response, but as expressions of heteronomy they must be suspect. Autonomy always comes first.

Popper's theory considers science learned ignorance, nescience. This permits approaching it autonomously. Science is not a set of prescription for what to believe in. Current science is a set of problems, proposed solutions that are putatively true, plus past attempts at their criticism, successful or not as the case may be. Future science is open. The theories and the criticism that science comprises are not obligatory. Those not interested in criticism will naturally ignore science or deny that it is critical despite its fairly evident and familiar critical stance. It is not very efficient to listen very attentively to what such people say about science (except when they come up with some new idea, of course), much less try to convince them about the freedom of opinion that science fosters. They overlook this freedom because they preach heteronomy. And discussing autonomy always comes first.

Does nescience imply autonomy? Is autonomy the best expression of learned ignorance? I do not know. Autonomy is the mere rejection of external authority (and so it is better to refer to it as self-reliance, but let us bow to custom). If there were an authority to trust, we would not know what it is, or how to find and follow it. Thus, nescience leaves us with hardly any option other than autonomy (as we read in Plato's early writings and in Galileo late writings). Another option is the popular view that science is the best authority, because of its conspicuous success. This success became increasingly the ground for the increasing authority of science. Internal strife and recurrent revolutions shake this authority, success notwithstanding. For, if the success of science is due to its internal strife, then it should not serve as an authority except, at most, on what it has reached agreement. The freedom of dissent is thus unscathed.

Revolutions do not deplete the authority of science: people do not care much about the internal affairs of science and are content with the recognition of its external success, its success in delivering technological goods. This argument should not work. The most impressive success of science has brought about also a threat to our very survival. At least superficially, the threat seems to be due to the growth of science-based technology. Hence, technological success is no longer so obvious an argument in favor of accepting the authority of science. Nescience, then, today almost implies autonomy, but not quite. (Definition: "a almost implies b" means, "a implies b in the presence of an uncontested supposition that can be specified on demand." The supposition that makes nescience imply autonomy is that placing trust in an unexamined authority is irresponsible. Perhaps this supposition is not that uncontested, though. Its opposite is fideism, the idea that, unable to prove every view we hold, we must take some basic assumptions upon trust. It does not repel Popper's obvious criticism from morality. It is immoral, he said, to endorse any thesis uncritically without any attempt to examine its practical corollaries.)

What, then, is the rule of conduct, intellectual and moral, adequate to the autonomous in ignorance? What does learned ignorance demand of its adherents? That they should not play superior to others. It thus "almost" advocates democracy. (There is, then, a near-implication. Nescience entails democracy in the presence of the maxim that only the knowledgeable has the right to rule and stay above critical scrutiny. Critical scrutiny is characteristic of democracy alone. Different philosophers have endorsed this maxim, including Plato in his *Republic* and Russell in his *Skeptical Essays*.) Hence, nescience invites agreement about agendas and these are essential to democracy. It thus allows a consensus-without-authority, especially as to what we do not want to discuss. We do not want to repeat the few errors that we have learned to correct. Hence, it is ill advised to do away with past errors. They are a major part of our intellectual and moral heritage. Hence, we may wish to exclude from our agenda the proposals that we have on record as already rejected on the ground of some valid criticism – as long as there is no reasonable criticism of that criticism. Hence, excluding an item from the agenda, just as including it, must be tentative and remain tentative come what may.

The openness of the agenda to revision is basic to both democracy and science, and for the following simple reason. It is unhealthy to have an agenda that suits only the autonomous nescient, an agenda for autonomous individuals who share learned ignorance, and only for them. It might separate the autonomous nescient from others. It is thus dangerously close to esotericism. And, obviously, esotericism is anti-scientific and anti-democratic as it may all too easily serve as means to block the criticism that outsiders might offer. To dismiss criticism from an enemy of science or of autonomy may be risky, even though their challenge to autonomy and their rapport with science need not be too serious. We may consider our traffic of ideas and opinions with the heteronomous and/or the hostile to science or to democracy on a different level of discourse, as engagement in education rather than engagement in critical debates. In critical debates one may stand to acquire some new knowledge, some insight, from listening to the opponent, and when opponents advocate heteronomy and hostility to science and to democracy we may engage with them even when we do not expect debates with them to be enlightening. The same holds for racism, for a conspicuous example. We may wish to educate the racists, but we need not expect our discourse with them to be enlightening and scientifically challenging. This is a bit dangerous, as a racist may develop ideas and challenges that deserve taking seriously and that should be high on the agenda. Konrad Lorenz is the paradigm case here, as he was always a racist, yet to ignore his ideas and criticisms is to suffer a loss. Some of them are truly enlightening. (Even his racist criticism is more intelligent than what most racists offer, and it does invite an anti-racist response. I do not even know if Popper's blanket response to the racists will do, though I hope so. Were racism true, he said in his *The Open Society and Its Enemies*, then it should be incumbent on the more fortunate to compensate the less fortunate.)

Despite its scarcity, we can and should take anti-racism as uncontroversial, as the consensus, though tentatively so, of course. The idea of the siblinghood of humanity is not very popular. This should not deter us from taking it for granted, though, as always, only tentatively so. We have on our hand here then a new kind of *a priorism*. It is the license to take certain ideas for granted as trivial or as uncontested and to refrain from defending them or examining them until someone manages to challenge them in a new way.

In addition to heteronomy, to anti-science, to anti-democracy, to discrimination by race, gender, and other forms of discrimination, also discussions of a different kind of ideas should stay off the agenda, such as those that concern the super-natural. This exclusion is naturalism. It is traditional since the modern scientific revolution of the seventeenth century, the one that the Royal Society of London instituted. Its declared aim was the advancement of natural knowledge. Naturalism is still the most popular aspect of science. Within philosophy it now goes under the names of W. V. Quine and Mario Bunge, except that they are hostile to super-natural theories, whereas traditionally naturalism entails neither hostility nor even indifference to super-naturalist doctrines as such, as some very religious individuals share it. What naturalism requires is the suspension of judgment regarding the supernatural, yet without curbing any naturalistic studies of the fact that supernaturalism is much more common than naturalism. Debates about the supernatural continue and there is no reason to prevent the use of arguments from science in them. Nevertheless, it is better to ignore them. It may be of concern to some heteronomous people, but the autonomous will find them pointless. This is reasonable as long as contestants do not discuss the question, why are debates on the supernatural important. To make naturalists reconsider the naturalist tradition in science, defenders of supernaturalism should raise their interest in such debates.

We still do not know well enough what naturalism is. Quine took behaviorist psychology as its truest expression, contrary to the views of most of his admirers and of most other leading naturalists who strongly disagree with him on this. His reason rests on his denial of the need for a new logic of intensionality. Many invest efforts in attempts to develop new kinds of alternative logic. This suffices to show how open the issue is. Especially challenging is the proposal to develop a theory of meaning that does not assume a Platonic Heaven, of the kind that Frege, Russell, Gödel, Quine, and others have endorsed. Perhaps this invites a new kind of alternative logic. Perhaps also the need to have new theories of the mind as an aspect of the body invites a logic of emergence and a logic of existence (a free logic; it is already available). Unfortunately, formidable obstacles are here in the way. Clearly, even the old logic is scarcely in use in discussions of ontology. Quine's arguments from logic for his ontology are not conclusive. Also, he ignores the fact that this logic is extensional: it does not apply to sentences with terms whose reference is unclear, including all sentences that use words like "know" and "believe". Though referential opacity is an obstacle to reductionism, it does not impress reductionists. Even were institutions merely sets of people, the anonymity of these people and their use of institutional terms prevent valid inference from aggregates to institutions. Those who do not take arguments from logic seriously justify their conduct by their search for alternative logical theories. Yet, they should acknowledge this and discuss it critically.

These guidelines concerning the agenda of philosophy are in line with the guidelines that are traditional within science since the scientific revolution. Nevertheless, it is naïve to expect that the establishment of philosophy will endorse them, or even that the establishment of the philosophy of science will. This establishment rejects even the most basic rule of the scientific tradition since the scientific revolution, namely, the demand to suspend judgment regarding the truth or falsity of any empirical information except for repeated and repeatable information. The establishment of the philosophy of science cannot endorse this rule since it will thereby dismiss the whole literature on inductive logic, which discusses the foundation of science in singular individual experiences. (Popper and Bunge are the exception: they demand repeatability as a precondition for testability.)

The establishment of philosophy will not endorse these guidelines. It deems the rank-and-file not autonomous, and these guidelines appeal to autonomous researchers who can decide on what they wish to see on the agenda. This is true but not sufficient as an explanation for the opposition of the establishment. Democratic parliaments cater for autonomy even though they rightly deem most of their populations not exceptionally autonomous. How does one go about in such circumstances? The scientific revolution met similar conditions, as academics were not morally and intellectually autonomous. The situation altered by the establishment of new clubs of amateur researchers, the scientific societies that carried the day. They became professional in the nineteenth century, but without loss of appeal. And then the Cold War changed all this as it gave rise to what Thomas S. Kuhn has christened "normal science". So perhaps now is the time to establish a new philosophical system with its own communication system that should follow rules for the autonomous.

This proposal seems to me erroneous. It is much easier to stay within the system and communicate to all autonomous – or even merely possibly autonomous – students of philosophy that they should learn to scan the current communication system and learn to be selective. It is also possible to argue that way about the rules and about the guidelines for the agenda of philosophy within the established system and hope to alter the rules and thus liberalize and pluralize the existing system. The commonwealth of learning is not sufficiently open, but it is the very best we have.

Let me conclude by returning to the most obvious and central item for the agenda of learned ignorance. We can now commit errors that we cannot afford to make even once, yet we can hardly avoid them anyway. We now have to perform dangerous experiments, as we cannot let the situation continue as it is and we do not know how to alter it without taking even the greatest risks. The risks are already present. They are the Proliferation of nuclear weapons and other means of mass destruction, Pollution, the Population explosion, and the abject Poverty of the majority of humanity (the four P's).

This situation calls for new canons of proper experimentation, concerning global experiments that are unrepeatable. Philosophers should contribute to this discussion, as it is a major item of scientific method. In bio-medical ethics they already do that, and rightly so, and the field of ecology and of large-scale risk assessment is more challenging and more important. I should mention here the pioneering efforts of Christine Schroeder-Frechette.

The idea that philosophy should take the rescue of humanity from itself as the first item on its agenda is due to the Norwegian philosopher Arne Naess, who in old age decided to devote his energies to the rescue of the eco-system. I wish to observe that his contribution is not merely a matter of common sense, since survival is always the first item on any agenda. Also, it rests on his endorsement of the idea of learned ignorance, of nescience, as a matter of both ethics and the philosophy of science, so that it is the right point to start any architecture of open-ended reason with – on the understanding that reason is inherently open-ended, that science is nothing if not learned ignorance. This is the proper point of departure.

Chapter 1.4

THE THEORY AND PRACTICE OF CRITICAL RATIONALISM

0. ABSTRACT

1. It is better to study applications of different theories of rationality to the life of reason than to seek the response to the skeptic critique of rationalism.
2. The two standard theories concerning the rational social order are erroneous: Enlightenment quasi-anarchism and Platonist technocracy.
3. The need to answer the skeptic sounds reasonable because of the erroneous idea that justifying one's views and caprice are the only alternatives.
4. Theories of rationality concern belief (Parmenides, Plato), or the choice of a community or of its lifestyle (Buber, Polanyi), or attempts to improve (Popper).
5. This last proposal conflicts with the traditional defunct dichotomy between nature and convention that limits the choice of the social order to neo-anarchism and technocracy, to exclude the democratic way of life.
6. Most views about lifestyle ignore the social order. Yet, Weber went too far in his censure of them as indifferent to social affairs.
7. Critical rationality is essential for democracy. It is inherently social and political. Yet, it is insufficient for facing the risk of global annihilation.

1. WHY ANSWER TO THE SKEPTIC CRITIQUE

Rationalism is the advocacy of rationality, of the life of reason. Somehow, this was lost in the thicket of the historical discussion about rationality: the concern of most rationalist philosophers is with the answer to the skeptical criticism of it, as if this is all that stands between rationality and perfection, as if this is the only problem left. Perhaps this is not the best policy. It may be better to seek the import of the different theories of rationality for the life of reason than to continue to seek a response to the skeptic critique of rationalism. Possibly, one might want to spy on the Promised Land before refuting the proof of the impossibility to enter it. After all, the attempt at a disproof has not succeeded despite centuries of hard work. Why should we wait, possibly for a few generations more, for the study of rationality? And if the chosen land is not perfect, then possibly it is an error to invest all effort in the single, highly theoretical task, rather than seek some practical tasks regarding it. Many scientists propose, rather spitefully, I am afraid, not to waste time on the philosophical discussions and do research – to turn to philosophy only in old age. Is that right? This matter should perhaps be high on the philosophical agenda. For, the proposal is wrong if philosophy can help solve practical problems here-and-now, assisting research work, promoting world peace, *etc.* The proposal to postpone philosophical discussion may apply to philosophers who respond to the skeptical critique, not to those whose studies may help solve practical problems rationally.

This is a challenge. There prevalent answer to it is, possibly the answer to the skeptic critique will open the best way to solutions to practical problems. Even so, as some progress is there, it might help decide between seeking more progress and fighting the old frustrating battle. Is it not wiser to invest some effort on the question, what is the life of reason? After all, this question, too, has a tradition ever since Socrates has posed it as Plato has reported it in his *Apology*. Moreover, the discussion of this question, what is the life of reason, may be a better tool for the quest for the answer to the skeptic than a direct attack.

2. WHAT CONSTITUTES A SOCIETY LIVING THE LIFE OF REASON?

Tradition presents a simple view of the life of reason: it is the life devoted to study – perhaps to study combined with worship; in Jewish tradition, derived from Plato, no doubt, study is a cardinal item in the life of worship. Spinoza integrated this into the modern tradition. In one sense, it has won unanimous endorsement; in another, it has received unanimous rejection. As the claim that rationality favors study, it is universally accepted. As the claim that there is no rationality except in study, the consensus is that it is much too narrow. Respect for learning is a fixture of all theories of rationality; each ancient school supplemented it with some proposals regarding the good life. Max Weber has noted that except for some religionists, they all made proposals that are objectionable for the same reason: they speak of individual salvation, not of social progress: they are oblivious to matters social. This is only approximately true, as we shall notice, but let us endorse it for a while. What then is the society living the life of reason? Plato and St. Augustine asked this question and most of the Judeo-Christian tradition rejected it nonetheless, as peripheral to individual salvation.

The question is central in the modern world, in whose tradition there are two popular (sets of) recommendations / answers regarding the task / question at hand, both within the ancient rationalist tradition. The one is (Sophist) anarchism, comprehensive or tempered – the total or almost total independence of the individual, hopefully coupled with attempts to advance one's material interests through a personal submission to science as best understood. The other is (Platonic) technocracy – the total submission to the expert scientist specializing in any matter that happens to be under consideration.

All versions of anarchism are tempered: they allow for some minimal policing, usually in viewing the state as a night watchman. The starting point is simple: we are born free; to live in society we voluntarily relinquish some freedom – as little of it as is utterly necessary. Anarchists further temper this idea by advocating some compulsory education, national defense, constraints on immigration, and some government intervention in the economy. Since they seldom discuss these concessions, it is only fair to ignore their views here – until they are ready to debate the question, how much should the state interfere in the lives of citizens. When they will be ready to discuss this democratically, they will have joined the democratic camp that recognizes freedom as a universal ideal tempered by the necessary evil of state intervention.

Technocracy is usually misconstrued as an enlightened variant of democracy, since it champions scientific technology as central to the modern way of life. This is an illusion: the delegation of important decisions to technicians abrogates all personal responsibility and turns politics into expert public administration. The coordination between technicians is a super-technique, whose practitioner is the notorious philosopher-king. As Popper has argued, Plato intended the philosopher-king to possess perfect knowledge. Since Einstein voided all claims of science for perfection, democratic leadership is preferable to that of philosopher-kings.

3. JUSTIFICATION PHILOSOPHICAL AND COMMONSENSICAL

The demand to answer the skeptic first sounds reasonable because we contrast proof with sheer caprice. This is an error: the opposite of caprice is far from being proof or proof-surrogate. The traditional theory of rationality does contrast them this way. Hence, the traditional theory of rationality is an error. We should not defend it against the skeptic critique. We should seek a better theory of rationality.

Traditional rationalism, the majority party within rationalism, is excessive. Popper calls classical rationalism comprehensive, as it requires proof for every move, including proof of the validity of the proof procedure itself. This validates the skeptic critique. To insist on abiding by the choice between traditional rationalism and traditional skepticism is to insist on extremism. William Bartley and W. V. Quine call this extremism justificationism.

As Mario Bunge says, some justification is quite in place. The asset of the demand for rational justification of views is obvious: it is at least in intent an advocacy of life in the light of rationally justified ideas, the life of science – hopefully within democracy. Yet this rests on an erroneous view of science and of democracy, as Watkins has argued: it disallows disagreement – permitting it to appear only temporarily and marginally or on the forefront. The strongest argument in favor of comprehensive rationalism is simple and powerful: one is not rational who holds views arbitrarily, who behaves erratically and / or arbitrarily. Such arbitrariness or waywardness is crazy; the sane, then, can provide some explanation for their choices; whatever is the explanation, it may constitute a justification of that choice. This is the central argument of all justificationists.

This argument is valid: the sane can explain and thus justify. Bertrand Russell stated this clearly. He declared the task of answering the skeptic imperative, as they contended that no justification is ever available. Russell also viewed the idea that empirical evidence participates in rationality as the principle of induction, of founding rationality on evidence. He therefore proclaimed Popper an advocate of induction; after Popper corrected him on this, he proclaimed Popper a defeatist (in his preface to the reprint of Nicod's book on induction). As it happens, Popper's answer to Russell is extremely simple. Some situations invite decision, even urgently, and yet with no suitable familiar line of action available. It is then rational to try some new mode of action with no justification for it other than that the circumstances seem to demand some new action. As this claim is then justification enough, Bunge's demand for justification is right; Russell's demand is excessive. Justification seems to serve as the only alternative to craziness only because we pass the sanity test easily as we provide justification. This is a confusion of two modes of justification. Traditional philosophers viewed only science as able to provide justification. This view amounts to the rejection of most justifications that ordinary sane individuals offer, especially those who are ignorant of science. Two standards of rationality conflict here. One of them is too high (all justification must constitute scientific proof); the other is too low (the rational must pass a simple psychiatric sanity test). Confusing them renders vicious the innocuous, sane argument of Russell that we all justify except for the crazy, so that sanity equals possible justification of sorts: we all justify, but our justifications are question begging, and so both the demand for justification and the skeptic critique are valid.

Justificationism recommends science as the standard of rationality to live by; tests of sanity reflect the standard of rationality of a given community. Evidently, we want something in between. Popper accepted Russell's challenge and offered his in-between standard. He then declared that he had answered the skeptic. This is an error: the skeptical critique is of an ideal standard, not of any lower standard. As to the ideal standard, we should discard it anyhow, as it useless. Nevertheless, Popper agreed that, to begin with, the recommendation of any standard is open to the skeptical critique, and he therefore attempted to answer that critique. This way he would have justified his view of rationality. Alternatively, his acceptance of Russell's challenge is the admission that his criterion of rationality allows for insanity. Some justificationists, David Armstrong for one, used this explicitly and emphatically; he considers me, for one, an advocate of insanity. David Stove considers Popper an irrationalist on this very ground. Yet, the ordinary justification that ordinarily exempts people from the charge that they are insane is definitely less than the justification that justificationists ascribe to science. Commonsensical justification is not comprehensive and so it is not justificationist. Oversight of this simple fact is a source of endless confusion.

One might have hoped that the philosophers of commonsense would have cleared the confusion. The legal question of justification is different from that of scientific proof as it recognizes different standards. Suppose that justificationist philosophers will identify the principle of induction with the legal principle of responsibility. This identification should lead philosophers in search of a principle of induction to law-courts. They will find out fast enough, then, that the law is different in different times and places and is nowhere above criticism. Philosophers seek the universal rule of legal responsibility that is above criticism. This will then take them away from the law-courts and back to the search for the true, immutable principle of induction that justifies science as above criticism. These do no exist. Skeptical leaders forged the real standards of science and of historical studies, as skeptical historian Arnaldo Momigliano has observed. The search for worldly standards of criticism will be empirical and show that different societies have different standards of rationality, and that in the same society there exist different standards of proof. Standards of proof applicable to capital punishment should be most stringent (hopefully too stringent); proofs of grounds for suspicion of guilt that are acceptable to the police may follow much more lenient standards; and the law specifies still other standards – for journalism, legislation, diagnostics, and more.

This may seem relativist, yet it is not, as it allows for comparisons.

4. THE PROBLEM OF RATIONALITY SOCIALLY RECAST

The theory of rationality may be that of a choice of or the belief in a hypothesis or a view (Parmenides, Plato) and it may be the choice of a community or of its lifestyle or of its system of beliefs (Buber, Polanyi) and it may be the attempt to improve upon the best standards available (Popper). The problem of rationality has both a most abstract and a most concrete presentation. And these two are often confused. Abstractly, the problem is, what thought / action / way of life is rational? Concretely, as there are different canons of rational thought / action / way of life, the problem is, how are we to choose between them? The abstract approach is traditional ever since Plato presented comprehensive rationalism: rational thought is scientific proof, rational action rests on rational thought, and the rational way of life is the life of rational thought and action. The concrete approach to the problem developed in diverse attempts to eschew Plato's prescription, perhaps chief among them is due to Hegel. He had no new critique of Plato's rationalism. The target of his critique was the rationalism of his immediate predecessors, who had added to Plato's views some liberal practical suggestions. He criticized their liberalism. His criticism of justificationism as such does not go even to the length of the skeptic critique. (See the end of the Introduction to his *The Phenomenology of the Spirit*, passages on different kinds of science and on science as a phenomenon.) He rejected the idea that the rational way of life is scientific and so open to all. He advocated instead traditional society. He dismissed all ideas as insignificant and advocated action, especially political action, especially big, impressive political action, especially large-scale military adventures that he saw as antidotes to stagnation. He claimed that people have roots in their native societies and cannot move away from them, much less create new ones. He permitted only geniuses to alter tradition occasionally. Not ideas move them, he said, but irrational passions, the ambition to stand out. He viewed choice as impulse, as gut-feelings. Russell and Popper have called this idea action philosophy.

Ian Jarvie and I have argued against all traditional views of rationality from the obvious fact that there are different degrees of rationality. It is very hard for some philosophers to absorb this, judging by the many comments that our papers were fortunate to receive. And it is an understandable difficulty. Admittedly, the standard of rationality is a standard of propriety of sorts, and we are well aware of the fact that different cultures offer different standards of propriety. Yet endorsing the different standards simultaneously seems relativist and so irrationalist, and endorsing rationalism is admitting that rationality is a peculiar, universal standard. If this may be an ideal, it can never be a reality. The very core of western philosophy is the recognition that as standards of propriety differ we face the choice of a way of life, and that we should make this choice rational. The fire burns the same way everywhere but the law of the land differs from place to place, said old Protagoras. This observation signifies as it instituted a search for a better standard than that justified by local tradition: the search for rationality was the search for a universal standard of rationality akin to the universal laws of nature. Hence, there can be no diversity of criteria of rationality and Plato's solution is the only possible rationalist one. This argument seems impeccable. It invites scrutiny of its presuppositions and the consequent rejection of one or more of them.

5. REPLACING THE CLASSICAL DICHOTOMY WITH A TRICHOTOMY

The idea that rationality is graded seems untenable as it conflicts with the central idea of western philosophy, the dichotomy between truths by nature and truths by convention. This dichotomy rests in the basis of all traditional philosophical discussions – of science, of morality and of politics. It is grandiose but it is obsolete. It is the source and the distinctive characteristic of western philosophy but it must now go. It leads to the justificationism and to the reductionism that have made great contributions in the past, but that now cause stagnation. They must go and give way to better philosophy. The off-hand claim that truths by convention are not binding is naturalism. The claim that there are significant truths by convention was repeatedly the unfailing source of cynicism (signifying disrespect for the truth in every case when science had offered no substitute for it), perhaps even the view that leads one by inexorable logic to cynicism and to nihilism.

The first beak away from the dichotomy is in Popper's *Logik der Forschung* of 1935. There he rejected both naturalism and conventionalism. He admitted the existence of conventional elements in science but rejected the view of its theories as conventions. A number of writers declared him a conventionalist – reasonably in view of the absence of a clear rejection of the dichotomy in that book. His essay "On the Sources of Knowledge and Ignorance" (reprinted in his *Conjectures and Refutations*) rejects the theory of truth as manifest, the idea that nature is truthful and error is due to convention. Science does not capture the manifest truth; nor are its theories arbitrary: they are the best we have, and hopefully they are reasonable approximations to the truth by nature.

It is easy to envisage truth by nature representing the reality behind the veil of the appearances; it is easier to comprehend truth by convention, the truth of mere working definitions; but what are the levels of reality that correspond to the levels of approximations to the truth? This is a tremendously difficult matter, and this explains much of the hesitance to follow Popper. I suggest that it is easier to approach matters through social studies. Popper said (*The Open Society and Its Enemies*. 1945) that institutions are hypotheses. Yet they are conventional. So, I suggest that we may better understand his philosophy as a modified conventionalism, since for him the accepted truths, the instituted truths, the traditional truths, are not truths by nature, yet they are not arbitrary either. This makes rationality concern not with the rational way of life itself, but with the society that practices the rational way of life. What kind of society practicing the rational way of life should rationalist philosophers advocate and how should they link it to a theory of rationality?

6. THE SOCIAL COMPONENT OF TRADITIONAL PHILOSOPHY

To begin with, theories of rationality did not pertain to the choice of a social order but to individual salvation. Max Weber condemned them as utterly selfish. He was in error: they care about social problems, but having nothing to offer to society, they tried to help individuals. His own theory was more vulnerable to his criticism. He took for granted comprehensive rationalism, or justificationism, by which standard all societies are irrational. He applied his theory of rationality only to modern society. For other societies he offered his ideal type theory that took rationality as operative differently within different societies. Irrationalists like this. They prefer to discuss the choice of ways of life of given social orders: this choice concerns attitudes to already existing social orders: it is both more manageable and more likely to be reactionary. This is a major philosophical constraint on the practical discussion of rationality. Our task is to recognize reactionary attitudes while diffusing them.

The pioneering effort in this direction is present in outline in Popper's "Towards a Rational Theory of Tradition" (reprinted in his *Conjectures and Refutations*). How does one choose from among the given ways of life without being an irrationalist? The skeptics have nothing to offer in this matter, the radicals complain. Yet the radicals are in the same boat. They recommend an allegedly science-based social order. This is the modern world – allegedly or in truth – and it did not exist before World War II, or, at the most generous estimate, before the industrial revolution. How did radicals and skeptics differ before that? The answer seems to be, inasmuch as there was a difference, it concerned hope: radicals had and skeptics lacked hope – the hope to see the immanent rise of scientific society (Bacon). What was the skeptic offer? Leading skeptics, such as Sextus Empiricus and Pierre Bayle, answered the claim that skepticism violates commonsense and is thus absurd, by the claim that the skeptics are none the worse than the dogmatists, since no philosophy offers anything practical.

Whereas many dogmatists suffer the torment of doubt, they added, most skeptics savor it and the peace of mind that it provides. Stoicism, too, was a doctrine full of personal advice unrelated to the social order. And, indeed, what social order did any philosophers, the dogmatist, the Stoics, the skeptics or the religious devout, what social order could they offer? This is a rhetoric question, and as such most unfair. The early dogmatists were brave: dismissing all truths by convention they were opening the road to the true unity of the human race. One may speculate about the motive that made Alexander the Great make his generals marry local princesses: after all, generals and princesses have truth by convention to thank for their rank. But it is not easy to overlook the attempts of philosophers to live in the midst of their own societies in disregard for the truths by convention of their time and place, and that some of them were exceptionally brave. The accusation that philosophers were demoralizers is thus not baseless.

The word "jusitificationist", by which Bartley chose to designate the classical rationalist, is synonymous with the ancient Hebrew word "Sadducee". When I teach in Israel and use this word, my students find it hard to take; they use different vowels to differentiate the ancient Sadducee from the modern one. The reason is plain: their heritage rejects it in the religious context and favors it in the scientific context. They learn philosophy from justificationist teachers, and so they tend to view justificationism with favor. They learn about the ancient Sadducees from their rabbis, heirs to the Pharisees, and the Pharisees never offered detached presentations of views they opposed: they never admitted that the Sadducee tradition recommended only that part of Jewish tradition that they could rationally justify. Rather, they said, the Sadducees were ashamed of their heritage. The radicals who founded the new Israeli society were often ashamed of their heritage, and so the modern debate is in a sense a repeat performance. Israeli students who learn this naturally find it disturbing.

The rationalist way of life was not practicable before the industrial revolution – a point repeatedly and forcefully made by Gellner. No philosophy, he observes, could offer the social component required for the enlightened choice of a social framework as a frame for a reasonable way of life. The reasonable choice for the individual always was – as Martin Buber has stressed in all of his writings – the choice of ways for integrating in a given society, or in the best subculture it offers, and retain one's critical abilities. At times the choice was heartbreaking, as was the choice between staying Jewish and joining the promise of Enlightenment. So is the choice between staying in one's tribe and moving to the modern world of plenty. Clearly the most exciting is the most basic choice: between truth by nature and truth by convention as offered by the Pre-Socratics. (Even in the modern world, Jews were more attracted to the promise of the naturalists than to the plenty that the conventional Christian world could offer them.) I found this point stated only in one place, and very forcefully, yet with little notice from the learned world. In his historical writings Sir Lesley Stephen, a critical rationalist, I think, has noticed that comprehensive rationalism, the justificationist optimism of Spinoza and Leibniz, was a great polarizer: a doctrine very difficult to practice properly and very handy to excuse and justify helplessness and inaction and folly of all sorts.

7. RATIONALISM IS A DEMOCRATIC SOCIAL PHILOSOPHY AB INITIO.

Critical rationalism is essential for democracy. It is also inherently social and political; its democracy is an integral part of it. Today justificationism is still triumphant. Admittedly, the paradigm of rational justification of a doctrine was the justification of Newtonian mechanics, and all the justification it received did not prevent its overthrow. This calls for some fence-mending. Perhaps there is a need for a justification that will allow a revolution now and then to keep science on its toes, perhaps the scientific community can demand loyalty no different from other societies, as Michael Polanyi has observed. One way or another, it seems foolish to give up the venture just when science has come of age and has been able to offer a respectable, rational way of life. If we managed without having silenced skepticism, and if we have succeeded magnificently until now, surely we need not rock the boat.

This analysis is wrong not so much because the skeptical critique is unanswerable. Everyone knows that it is unanswerable, but few find it as impressive as the critical rationalists do. All logical arguments are forceful at the cost that all logic has to pay: the slightest alteration in a refuted doctrine may overcome the refutation and totally immunize the doctrine to further refutations. The analysis is wrong because of the *ad hoc* way in which the social component enters it. By traditional rationalism the problem, the choice, the concern, all were of the individual: it ascribed to society no role except that of a problem for the individual. Perhaps even the individualism that is so characteristic of western society, culture, philosophy, is but a by-product of the appeal of the Greek ethos to the individual – simply because it was an appeal to be critical of conventional wisdom. (We have a similar aspect in Hebrew tradition: prophets who attacked convention had nothing stronger to fall on, and tended to appeal to individual conscience. In both cases the problem and the discussion of it center on the individual; the social component of the individual's choice entered it as a moral burden and an after-thought. This will not do.)

No traditional theory of rationality supports democracy. Critical rationalism is its only reasonable foundation and it has appeared very late in the day. This is odd. Democracy and society in general should have entered the discussion at the very start, at the setting of initial problems. Historically society entered the stage of philosophy with the slogan, the fire burns the same way everywhere but societies differ. What of it? Why did this trouble the ancient Greek sophists so much? Because they discovered options they had not known exist. Herodotus explains the general blindness to the difference of diverse societies by reference to the view of one's own customs as "naturally" superior. This was a blind spot disappeared with the discovery of the diversity of customs *of equal validity*. He explains this by narrating a myth: the institution of rationality was the act of the expulsion of the magi – from Persia!

All this may have been revolutionary in Antiquity and even in some not so distant past: today it is neither very new nor very exciting. However difficult it is for an individual to break ties with tradition, it can be done, and at known costs, and while raising certain social and political and educational problems, not philosophical ones. Each of the different extant attitudes to custom is philosophically rather problematic. We have learned to circumvent the problem: out of so much frustration, perhaps, we have found it better to attack social problems by new technology rather than by forging and / or implementing new attitudes. In my book *Technology* I argue against this view. I elicited no response. The long and the short of it is that now all traditional societies, primitive, advanced, and in between, are helpless in the face of the coming cataclysm that the ecological and the peace movement are discussing at such a great and colorful detail. If the future is anything like the one that they predict, then we are doomed. If polluting the ecosystem or the enormous overuse of fossil fuel will not destroy it, then the population explosion will. Alternatively, the population explosion will lead to reckless wars, perhaps with the use of nuclear weapons; and even if the population growth in the poor parts of the world can be contained, there is no reason to assume that the poor nations will be willing to tolerate the *status quo* forever. In brief, we need a new plan – as new as our boldest imagination can offer. There is no way back, and slow progress is as dangerous as the *status quo*.

What are the philosophers doing about it? What they have been doing all the time and less. What can they do? What can anyone do? How does the proposal to do anything relate to the theory of rationality? I do not know. If there is any rhyme or reason in current global politics, it is its being so very marginal: global political activity is marginal, and almost all of it serves local interests: politicians pay heed to global politics only through the prism of national and narrower interests. Few politicians ever have the resourcefulness and vision to develop long-range plans, anyway. Progress nevertheless takes place: politicians from time to time do recognize the requirement to come up with long-range plans and engage increasingly wider outlooks. This is how the modern nation-states developed out of small local interests. Once a politician saw the advantage of a wider outlook, of planning for larger systems, the ensuing success forced competitors to follow suit. This is no longer the case. If the whole world will not mobilize in order to put global issues high on their agendas, then nothing will ensue: only global solutions may meet the global problems of the survival of humanity; local solutions will not begin to touch them. If anything, if some but not all parties are reluctant to use resources for fear of harming the globe as a whole, then those who do not share the reluctance benefit from it.

How then can the world leadership on all of its variety agree that certain global moves are essential for global survival? How can we ever hope to alter the habits of millennia and usher more rationality into global politics before it is too late? Is it rational to exhibit a reluctance to use certain resources for fear of extinction? How can we test claims for the existence of a causal relation between the overuse of fossil fuel and extinction? How much testing is required? Of which kind? Is that not the traditional problem of induction run wild? The assumption that the present problematic situation is a case of an inability to justify, to apply the principle of induction, is an admission that inductivism is now bankrupt – regardless of the status of past claims for it. The question remains, what is the situation from a critical rationalist viewpoint. There is a clear sense of crisis in the air. Whether the future of humanity is gloomy or not, it clearly is severely threatened. If survival is the topmost item on every practical agenda, then the survival of humanity should be the topmost item of every agenda, private, philosophical, national, global. Does any philosophy address this matter? Not that I know of. Back to the drawing board, and fast: the time may be too short, or we may just barely come up with a solution in the nick of time. For my part, I can only pose the problem. What I have ventured to illustrate is that the concern for a rational social order was never far from the deliberations on rationality, that these days we have to make it both explicit for intrinsic reasons and paramount for extrinsic ones. The problem of rationality has radically transformed: it read classically as, how do I know? Popper and Polanyi changed it to, how do we know? It now reads, how can we save ourselves from destruction? How can we realize global security?

Chapter 1.5

SCIENCE AS TRAINING FOR AUTONOMY

0. PREFACE

Philosophy should promote science as art for the lovers of art and art as science for the lovers of science: any reason for loving the one domain is good enough for falling in love with the other. This should improve the quality of life. For this task, philosophy needs promotion too. And the audience of the promotion thus need not be lovers of any of the three domains. They have to be independent of mind, however, and able to decide for themselves. In the received jargon, they should be autonomous. This idea clashes with the view that there is an obligation to endorse science – or art, or anything else. This move is thus revolutionary.

1. IN PRAISE OF AUTONOMY

Philosophy should promote autonomy. This is not the received opinion. As John Locke described the task of philosophy, he declared that the philosopher is an under-builder, that is to say, a building-site garbage remover. He expressed that way a deeply felt humility towards students of nature in general, and towards Robert Boyle and Isaac Newton in particular. He also expressed the hope that the task that occupied him, though it is the mere clarification of concepts, is of some aid to these researchers. As his task is the clarification of concepts, he admitted, he could not make a positive contribution to an inherently empirical venture. But he could, he hoped, contribute by removing the negative, by serving as a garbage remover. He was a great supporter of the autonomy of the individual. He saw in confusion an impediment to it and in clarity the tool for the support for science. This support he deemed essential for autonomy: clear minds are free and they freely recognize the truth. This, however, hides a problem: does science impose its ideas on us? Is science obligatory? If so, does the obligation to endorse science not limit freedom?

When Wittgenstein declared the task of philosophy as mere clearing of confusions, it was not garbage collection but window cleaning. He obviously saw himself in the tradition of Locke. Yet, he was expressing no humility. He considered himself a logician and, as such, he deemed his work a positive contribution and his theory of confusion he viewed as a part of logic. He saw as a part of logic a general theory of meaning. His theory of meaning entailed the claim that philosophical or metaphysical propositions do not possibly have meanings. He proudly based on this claim his idea that the task of philosophy is the mere clearing of confusions. Now, perhaps he had no theory of meaning; it is hard to say: we do not know what questions such a theory should answer. But let this ride.

Behind all this technical material stands the old idea that religion or theology opposes autonomy. By declaring metaphysics meaningless Wittgenstein banned theology and thus opened the road to freedom from its tyranny.

Wittgenstein declared that all propositions about the world belong to science or to common sense. This sounds trite, but it is a highly novel idea, and it is quite controversial, to say the least. These days probably most students of the theory of science as well as most students of the theory of meaning reject his central idea together with its rationale. Wittgenstein's disciples do not admit it openly, and they might even contest it. But at least Carl G. Hempel, one of his last disciples, said openly that even if his banning all philosophy was excessive, the corollary from it that bans theology is beneficial.

The scientific tradition deemed science the collection of demonstrably true propositions. The negation of a true scientific proposition is false, then, and hence, by the scientific tradition, it is non-scientific. Wittgenstein viewed as scientific – potentially or *a priori* – any statement about the world, including obviously false ones. Prior to permitting empirical information to decide the matter of the truth or falsity of a proposition, that proposition already is within the domain of science. Thus, the only possible metaphysical propositions would comprise synthetic *a priori* knowledge. Such knowledge he and his disciples declared logically impossible.

This argument is not new. Classical metaphysics is mainly ontology, namely the theory of what there [really] is, the theory of substance. Its elaboration is the theories of the different (possible) substances, God, mind and matter. These are, then, rational theology, rational psychology and rational physics, where "rational" stands for first principles. First principles of a field are metaphysics in the narrow sense, *metaphysica specialis*, in contradistinction to metaphysics in the wide sense, *metaphysica generalis*. The current view of the justification of the non-existence of philosophy is the one that Quine has issued; it is not rooted in the impossibility of any kind of propositions, since Quine exploded all meaning theories (or claims that such theories exist) current in his day – without offering any alternative. But he did insist that there is no synthetic *a priori* knowledge and he even questioned the possibility of analytic *a priori* knowledge. He did endorse, then, the idea that philosophy should not guide science, and that it should not compete with it either. He put it very oddly in the form of the thesis that science and only science prescribes his ontological commitment. I do not know what commitment is, ontological or any other, nor what kind of prescription this can be. I take it that he suggested that the best available views on any matter are those that science imparts. He claimed that since he endorsed whatever science has to offer and refused any substitute, he needed nothing else. This runs contrary to Russell's observation (*Problems of Philosophy*) that every field of study begins as metaphysical (speculative or philosophical), and that as it progresses it cuts its umbilical cord to become a science. Of course, often a science evolves by mitosis; at times it appears out of the blue, and at times it is a synthesis. The question to ask Quine, however, is this: if the task of philosophy is not garbage removal (Locke) and not window cleaning (Wittgenstein) and not running a nursery for future sciences (Russell), then what is it? Quine did not view his task as his forerunners do. He viewed his work as science proper, the science of logic and of mathematics. Some writers object to the inclusion of logic and mathematics in science, on the ground that their propositions are analytic, not empirical. Ones who assert this ignore Quine's own view on the matter; they dismiss it out of hand. He was a logician and, as such, he deemed himself a scientist, especially since he said there is no clear-cut dividing line between the empirical and the logical.

Suppose this is so. Assume for a moment that philosophy can offer the science of logic (which includes mathematics and mathematical logic). What else, according to Quine, can philosophy offer in a positive vein? This question matters – to Quine and to many others. Excluding Wittgenstein's exclusion of all the conceivable propositions of philosophy as inherently meaningless, what propositions can philosophers *qua* philosophers make and to what end?

The most obvious answers are two. Philosophers can try to accept Russell's challenge and open nurseries for future sciences. Or, they can discuss ethics. Both options do not appeal to members of the Wittgenstein school. Their interest is not in future sciences but in distancing their role from that of preachers. They allow discourses on ethics only in the analytic mood, in the traditions of Locke and of Wittgenstein, not as admonitions and not as policing. This is how analysts ignore the traditional task of philosophy as criticism of basics – not to mention the concomitant advocacy of autonomy.

This is the background to the extremely popular works of Richard Rorty. He says, philosophy is not policing but conversation. This requires more specification, since obviously not all conversation is philosophy. Should philosophers be well-versed in what they converse about? Should they be autonomous? The specification has to relate to his problem, of course, which is obvious: now that Wittgenstein has outlawed all philosophical propositions, what task has he left for professional philosophers? Philosophy in the tradition of the persuasion of Locke or of Wittgenstein has left philosophers jobless, with no task except the odd task of thought policing that Rorty rejects. What else can philosophers do? They can provide intellectual entertainment. Is this Rorty's view? The ascription to philosophy of the role of policing rests on a dislike of the inquisition, Catholic or Communist, and the wish to replace it with something reasonable. What difference will there be between the old and the new thought police? Rorty does not examine this question, perhaps since he opposes policing and defends autonomy. He denies that philosophy is policing just because some have affirmed that it is. Indeed, Wittgenstein did so. (1) Popper has admitted: Carnap has put the fear of God in his breast. (2) Quine describes in his obituary on Carnap the appearance of Carnap in a convention of the American Philosophical Association in the thirties as a kind of an entry of a sheriff into a saloon. (3)

Stories are circulating on disciples of Wittgenstein arousing a sense of terror and raising the level of precision and clarity of discourse in philosophy seminars. And then, surprisingly, in that very camp Continental philosophers and their confused speech are gaining high prestige.

Rorty swears allegiance to American pragmatism, to analytic philosophy and to Continental philosophy. His heroes are John Dewy, Wittgenstein, and Martin Heidegger. What do these celebrities share other than the limelight? They did not attempt to prove, he says. They are not foundationists, to use his terminology. What does he think they were doing, then? He does not say. He offers hardly any history of philosophy either. And the little he does say about the history of philosophy is obviously false. Dewey rejected certitude, but he did want philosophy to have foundations: he founded philosophy on practice. Wittgenstein alleged that he founded his philosophy on the rules of grammar. (He never said what these are, though.) It is hard to say what Heidegger based his philosophy on. Yet, he did express contempt for those who adhered to religion just because there is no foundation for it. Rorty considers all-important the idea that philosophy has no foundation. He ascribes it to people who have rejected it and ignores those who endorsed it – Buber, Popper and Polanyi. This situation is very serious. As Gellner has observed, it is no big deal to praise the king's new clothes when the king is sumptuously clad. Doing so when the king is naked is a most significant sacrifice: it is the relinquishing of one's autonomy for the sake of loyalty and as a proof of it. And, as it happens, Rorty is naked. Perhaps in the kingdom of the dependent souls, one must be naked to be a king.

It does not matter who said philosophy needs no foundations. What matters is the role of philosophy. Assume that it is garbage removal or conversation; assume that this activity needs no foundations. It follows that philosophy needs no foundations. Hence, Wittgenstein and Rorty have justified their dispensing with foundations. So did Popper as he said, the role of philosophy is to criticize. The others on Rorty's list said nothing like that.

2. IN PRAISE OF THE PHILOSOPHY OF AUTONOMY

To repeat, Locke saw philosophy as garbage removal and as a prelude to scientific research. Wittgenstein saw it as window cleaning and cared for the clarity that he expected to remove metaphysics, not for the advancement of science. He said that he found clarification an important policing job, (1) and he admitted that possibly he did not know what clarification is good for. (4) Rorty follows Wittgenstein's hostility to garbage while approving of the garbage that Heidegger has produced. This may disturb one who wishes to know what Rorty cares about and how this care appears in his writings.

Rorty says, philosophy is conversation. So did Socrates. We all deem Socratic conversations philosophical. This is not what Rorty has in mind. Will he consider intellectual talk-shows philosophy? Will he recognize the conversations of Goethe or of Dr. Johnson as philosophy? Or the output of Mark Twain, Will Rogers and Groucho Marx? If yes, why? After all, their output differs greatly from his. Will a conversation count as more philosophical perhaps if participants in it exhibit familiarity with Plato or Wittgenstein? Rorty has never mentioned a paradigm of a genuine philosophical conversation – not even Plato's. Nor did any other philosophy professors who take conversation as seriously; they scarcely give an example of current philosophical conversation. And for good reasons: once we have examples, someone will attempt to analyze them and prescribe recipes for their replications, the philosophical analogue of the cookbook. This may be quite all right, yet Rorty follows Michael Oakeshott in considering genuine conversation inimical to all recipes. Claiming to be a philosopher, he could at the very least engage himself in a conversation about this: what conversation is philosophical? Does he? I do not know.

Suppose Rorty is right in endorsing the view that Wittgenstein's philosophy is cavalier about assertions. Possibly, as Wittgenstein said, his task is to raise the level of verbal performance. Conversation of the style that he approves of may be confused, he implies. Nevertheless, he says, he does not want to police verbal conduct. Did he agree with Wittgenstein in considering clear lies worse than confusions? Probably neither has shown any interest in lies. Wittgenstein was satisfied with his strict demand of his associates that they should be starkly frank. His biographers declared that he had a high moral standard and strongly demanded it from his disciples and friends, both in and out of philosophy. So, frankness is not specific to philosophy. Demanding it is not yet a theory of morality, however. Jean-Paul Sartre made it so. He compared a self-aware crook, one fully cognizant of being a crook, with the waiter who earns his bread by honest toil but who pretends to be not a waiter but a person who pretends to be a waiter. And he considered that waiter morally inferior to that crook. He said, the worst thing is self-deception. (He called it bad faith.) He considered all self-deception, including the innocuous one of that waiter, as the only real crime. Yet all that the waiter was concerned with is the preservation of his dignity while performing a task that he felt was compromising it. Probably Wittgenstein did not agree with Sartre. But we do not know. Does Rorty agree? We do not know this either.

Socrates, the model philosopher, was concerned with the moral conduct of his fellow citizens and he expressed this concern everywhere he went, in street-corners, in gyms and in banquets. This supports Wittgenstein and Sartre, as well as Oakeshott and Rorty, perhaps too well for comfort. Thus, it is not that people may be engaged in philosophical conversations in cocktail lounges that matters; what is objectionable is the proposal that philosophical discourse should confine itself to insignificant exchanges, such as normal cocktail-lounge chatter or talk-show dialogues. Russell admired a person whose philosophical ideas were not particularly interesting but who was particularly concerned about philosophy to expound his ideas to Russell while the two were on a battlefield, supervising an unsuccessful truce and in danger of catching a stray bullet. What matters, then, is not where a philosophical discourse occurs but what import, if any, it may have. Oakeshott, and following him Rorty, declares that cocktail-lounge chatter has a great merit even though it is hard to say what this merit is. And Rorty couples this with what he has learned from Wittgenstein: discourse on moral issues is less valuable than cocktail-lounge chatter. I may have misread Rorty, of course. I checked it with him personally. This is no guarantee against a misreading, but it is the best I could come up with.

Can philosophy play some role of policing without becoming a thought police? The faults of thought police are political and moral. It is politically undemocratic, since democracy respects the freedom of conscience and of opinion as a basic right of the citizen. And it is illiberal, as liberalism is the ethics that takes the exercise of and the respect for this freedom as a basic duty and responsibility of every individual. The philosophers who have rebelled against traditional philosophy were not concerned with the politics of thought police, perhaps because hardly any of them was truly a democrat, perhaps because they take democracy too much for granted and so find it reasonable to ignore thought police. In any case, they are concerned with the ethics of the matter, not with attempts to indoctrinate. Wittgenstein said that he had no doctrine and that he could convince only those who already had thought his thoughts. Sartre preached that any idea is right for those who hold it honestly. Wittgenstein and Sartre deemed immoral what they preached against, be it unclear thinking or self-deception. But who were they to proclaim on matters moral? Did they have the right to impose their judgments? What grounds did they have to legislate as immoral confusion, self-deception, or anything else? Perhaps the specific to these proscriptions is that they are not ethical but meta-ethical. There is a better meta-ethical maxim: be as rational as you can! This is a good one, and perhaps it entails the maxims that Wittgenstein and Sartre advocated, against confusion and against self-deception. But to try to impose any of them is an error; certainly, no intellectual police force can do this. We have the right to cling to errors, and even act on them – as long as public responsibility permits it.

Socrates had to fight against confusion and self-deception as he fought for rationality. So, the same objection may hold against him as well. His response to this objection would be that the only police over a citizen that he recognized was that citizen's own conscience. Does this hold for Wittgenstein and for Sartre too? Should there be legitimate policing regarding the conduct of the intellect other than one's own conscience? Is this not an invitation to institute some thought-police? In democratic society, the police system proper is a democratic institution. The freedoms of conscience and of opinion are not at risk from the police that chase some sort of confusion. Examples of such policing abound where the law forbids certain confusion and certain forms of vagueness: in many kinds of activity, the law condemns these as irresponsible. Policing does take place, then, but with no philosophy and with no philosopher in sight.

This shows that whatever is the role of philosophy, and whatever policing is required for the discharge of this role, it is not what Sartre or Wittgenstein said it is. Rather, it is what Socrates said it is: legally permitted but morally forbidden, as harmful to others but more so as harmful to one's own soul.

Wittgenstein said many times that his aim was to argue against confusion, not against or for any doctrine. He fought for clarity. What is clarity good for? I suppose that, as with Socrates, it has to do with honesty. Much confusion is indeed dishonest, to oneself or to others. This is why Wittgenstein was interested in Freud's ideas. The kind of conversation that Rorty advocates as philosophy is different. What is it? He does not say.

Can the role of philosophy as Socrates has envisaged it become a social institution? Answer: yes, since it already is an institution. It is the education for autonomy, and it rests officially in the hands of parents, teachers, and preachers. Regrettably, however, preachers seldom preach autonomy, and often they preach against it, indirectly but still clearly enough against it. Parents in our society support autonomy to the extent that they know what it is, or even to the extent that they know anything at all about it. In any case, they clearly know not very much about it, at least not as much as is desirable, since we are all too ignorant about it. And at least they know about it and act in accord with it, to the extent that the system does not interfere with their efforts to educate. Perhaps more people learn about autonomy from teachers than from parents, but not enough either: alas, few teachers advocate autonomy, and fewer know how to teach it. (Nor is this easy. And they are brave and admirable, and they deserve more acknowledgment.) Are academics better than other teachers?

3. THE ROLE OF THE ACADEMY

Universities are supposedly non-doctrinaire. Whereas religious seminaries may teach given Gospels dogmatically, universities claim to do so with no dogmatism, namely, with no wish to convince and perhaps also with the wish to help students develop a critical attitude to what they study. They do not claim to be teaching the one true Gospel. This is the ideology of the modern university, not its practice. Science departments expect students to believe their preachers as a matter of course. In the humanities these days, some teaching is openly dogmatic with no reference to science or even in downright opposition to it. Philosophers and psychologists usually teach either the religion of their forefathers, or the doctrines of their supervisors. (These all too often coincide.) There are exceptions, cases of clear-cut conversions. What we need, however, is not faith or its absence, but the ability to think critically about ideas. Alas, very few philosophy professors even try to teach doing justice to competing philosophies; very few psychology professors even try to teach doing justice to competing psychological doctrines; and so on. It is not easy, and I for one cannot boast much success in this respect. But at least philosophers concede that schools of thought exist. Science professors do not concede even that much and psychology professors usually mention the opponents only derisively. (Some exceptions are terrific, of course, like the classic J. C. Flugel, *A Hundred Years of Psychology*, 1933.) They take their indoctrination of their students as both unavoidable and desirable.

There is such a thing as a loose public philosophy and a public policy that loosely rests on it. Some public figures, such as journalists and talk-show regulars, socialites, celebrities, politicians and political thinkers, preachers and sociologists and anthropologists, have tried their hands in this concoction. The ingredient missing in it, says Popper in his celebrated *The Open Society and Its Enemies*, is the Socratic care for souls, the help to citizens for their self-education in autonomy. Perhaps Wittgenstein meant this, and perhaps he thought that traditional philosophy is the very obstacle for that Socratic task. Regrettably, though understandably, he did not say so. He could not openly express his views on autonomy as long as he held a doctrine of meaning that allegedly has caused a revolution in philosophy by banning all philosophical disputes. He could not defend any specific doctrine, and so he alleged that he had none. But now that this allegation of his has lost its popularity, it is the time for his fans to speak clearly. All those who think that he taught autonomy should say so. Perhaps this is what Rorty too has in mind. If so, he should also speak more clearly.

There is one obstacle here. The theory of autonomy is not as simple and obvious as it should be. Advocates of autonomy are rationalists, and their view of rationality is that the autonomous individual should submit to science. This runs contrary to autonomy and to the scientific attitude. But let this ride. What is the submission to science? I do not know, yet I can report that this did not stop me from accepting the authority of my university science teachers even though I was autonomous then. I was in error: I took it for granted that I needed to absorb scientific information and ideas uncritically for a long time before I could weigh it in detail and judge for myself. Perhaps I had no option but to submit to the authority of my science teachers, but I should have realized that this was expediency, and that only expediency required receptiveness to science, but in principle we can learn science with no authority at all. This does not preclude careful listening to presentations of ideas: the critical attitude does require the best understanding of ideas before criticizing them. As a student I was aware that autonomy requires a critical attitude, but not that this attitude requires sympathy that requires the suspension of criticism for a while. But the presentation of scientific ideas is better and easier to comprehend if done historically, as the story of ideas and the successful and unsuccessful criticisms that they met. Most of my teachers did not know how to teach except as authorities. Philosophers of science encourage authoritarian science teaching.

There is ample justification for this, and it is in no need for the heavy apparatus of the philosophy of science. First, science courses impart competence and much information. True, this need not be the best, and usually it is not. But it seems the best available to the students who take the courses – or else they may take other courses. They do need some idea as to how to judge the use of the courses, and this they do not have. Nor is it easy to find. The first thing to seek is an answer to the question, why is the course-material important? This quickly translates in most cases into the question, what questions do the ideas in the center of the course-material answer – what problems do they solve. This is not as easy a question to answer as it sounds. But it is the first step, and it is an expression of autonomy in the study of science, in the acquisition of past knowledge. It is valuable both as scientific training and as training for autonomy. The two should come as close as possible, and this is the best way to bring them together.

NOTES

1. David Edmonds and John Eidinow report (*Wittgenstein's Poker*, London: Faber and Faber, 2001, p. 200) that Wittgenstein said, "Bad philosophers are like slum landlords. It's my job to put them out of business".
2. The opening of Karl R. Popper, "The Demarcation between Science and Metaphysics", in P. A. Schilpp, *The Philosophy of Rudolf Carnap*, LaSalle IL: Open Court; 1955; reprinted in his *Conjectures and Refutations*, , 1963.
3. W. V. Quine, Homage to Carnap, *Boston Studies in the Philosophy of Science*, 8, 1971, xxii-xxv, reprinted in his *Ways of Paradox*, revised enlarged edition, Cambridge MA: Harvard University Press.
4. Edmonds and Eidinow, *op. cit.*, p. 87. See also the end of Wittgenstein's *Lectures on the Philosophy of Mathematics*.

Chapter 1.6

SCIENCE AND THE CALL OF THE WILD

For Tirzah

Till we have a new light, we shall maintain ... that a supernatural narrative ... always implies credulity or imposture.

Ernst Renan, Preface to *Life of Jesus*

Our credulity is grosser than that of the Middle Ages, because the priest has no such direct pecuniary interest in our sins as the doctor has in our diseases.

Bernard Shaw, Preface to *Saint Joan*

0. THE ATTRACTION OF THE OCCULT

This is a review of a remarkable book by Hans Peter Duerr, *Dreamtime: Between the Wild and the Civilized*, 1978. It is over-scholarly, but not by received procedures. I will overlook this, except to say that it makes the book somewhat tedious. Its success in Germany made him a cult figure and had English and an Italian translations. His chief idea is that there is no gap between magic and science, dream and awareness, the wild and the civilized.

Duerr justly complains against the hostility that advocates of science show to the world of magic. It is uncalled-for. It is also contrary to the scientific etiquette that requires indifference to sparks (suspension of judgment) – not on account of their being sparks but on account of their being elusive. From time to time, we all experience the elusive, the wondrous. This experience is recurrent. So is the normal western response to it. At times, it captivates for a blessed moment that gives way to shaking of heads and returning to daily chores. Duerr says, this passing experience is of enormous value. Perhaps. His claim is friendly and so it is better than the hostility that he censures. And I see no arguments for or against it.

1. THE GREAT DIVIDE: MAGIC HERE AND NOW

There is a realm of experience that sparkles with excitement, for children and for the superstitious, at least, but also for poets and even for scientists – the world known diversely as the world of mystery or the occult, the world made of dreams, hallucinations – induced by drugs, by the full moon, by religious trance or by sheer love – visions, hypnosis, telepathy, magic black and white, alchemy, astrology, numerology, divination, super-natural beings, whether apparitions and the undead or our ancestral ghosts, not to mention goblins and fairies. And oh, yes, the latest: extraterrestrial visitors. It is no use declaring the world of sparks nonexistent. It is no use demanding that science alone adjudicate all phenomena, since we are here in a twilight zone where science protests ignorance. Remember mesmerism and the unjust hostility that it met for ages. Broadminded incredulous advocates of science readily concede this, yet they refuse to wander to the world of elusive sparks. It may, they admit, and at a rare moment it does, contribute to science. But they leave it at that. Some people sense a difficulty here: does the refusal to acknowledge the existence of the elusive not preclude its occasional pinning down, capture, and legitimation?

No. Even were science less exciting than the sparks, it still is tenable: it neither denies nor asserts the elusive. A researcher may assume tentatively the existence of something to pin down, capture, and then examine. Can we, then, dismiss all the elusive sparks we cannot capture? Certainly, we can: most scientifically-minded westerners protest that they do. At what cost does one live with no attention to the world of sparks? At the cost of declaring, out of hand and unhesitatingly, utter inability to believe an (American) Indian who told his ethnologist-apprentice that a deer came to him and whispered in his ear,

Hallo, friend! (p. 129)

Hallo to you too!

Hallo, friend! is the hum running through Duerr's book. He attempts to be friendly to narrow-minded incredulous ethnologists who declare our Indian friend hysterical, schizophrenic, lunatic, or plain liar. The book has a lot of notes – the text is about 160 pages, the bibliography about 60, and most of it, about 180 pages, is notes – and these are penetrating and at times sharply critical and even derisive, yet they always exhibit wild humor and often good humor and their aggression is usually naughty rather than vicious. That he loses his temper with the Russian Communists who force primitive people to become scientific-minded is another matter altogether, of course. In the world of scholarship, his attacks are iconoclastic rather than ill-tempered, and very shrewd they are – except for his hostility to skeptics (the good skeptic is a dead skeptic, Chapter 9, Note 93) and to critical rationalists (Chapter 10, Note 21 and elsewhere). It so happens I am both. Does it not strike you as odd that, while permitting our Indian friend his little innocuous chat with a deer, Duerr forbids me my doubts? Even if all consistent doubters are necessarily inconsistent, as Duerr insists they are, why should he discriminate against me and allow only mystics, mythologists, and magicians their little inconsistencies and their huge, systematic doubts? This discrimination is unfair even were Duerr able to prove my inconsistency. But he cannot, as many skeptics have shown, from Sextus in antiquity to my own self – in living memory, I hope.

Dear friend! You were both mistaken and unfriendly. Do you know that? Do you possibly regret it? Will you visit me in my dream, perhaps, to make amends? If you do, I promise I will try to reciprocate. And if we succeed to meet each other in a shared dream, we can publish a new report on a dream-encounter for all those broadminded incredulous advocates of science who deny telepathy, dreams, and all that, and we shall say to them, see? It can happen! And they will smile benignly and nod their heads wisely, and say, yes, indeed it can. And it will lead to nothing. Sparks are elusive.

– Why?

– Does it really matter that someone reported a dream encounter, or a chat with a deer? Perhaps it was a joke: the reporter says, I told the Indian, this is not a very clever conversation, and he said, what can you expect, I am just a dumb Indian. Or something like that. These stories are maddeningly elusive; you can always say, perhaps it was a joke, or a trick, or an error, or a coincidence.

Evidence about elusive experiences is ubiquitous. Some find this sufficient ground for conceding at least this elusiveness. The very same reporter, the apprentice to the Indian who converses with deer, says his Indian master was not dumb: he taught him to fly like a crow. "There seems no doubt that in some real sense" – notice the tortured language; it is our author, Duerr, citing another author, citing our apprentice reporter, to say that he, the apprentice, really became a crow but only elusively so: "though what the nature of that reality may have been is hard to say." Hard enough to remain impossible.

Scientific etiquette demands the suspension of judgment in all elusive cases. We may deem a spark report metaphorical, of course. Since here our author explicitly excludes this reading, we may take the case as a true report of a phantom experience. Imaginary experiences of flight are familiar to all children. This is not far from the metaphor of a soaring spirit, or the high of the high on drugs, akin to the heat of hot pepper and the cold of cold opium, that in the second half of the seventeenth century the great Robert Boyle reports in his book on heat and cold. Our apprentice-reporter does report that he was high on drugs. Except that he and all those tuned in to sparks, magicians, parapsychologists, mediums, visionaries – the lot – deny the report in all of these senses, the imaginary, the metaphorical, and the analogical.

Still, who cares?

Many scholars do. Ethnologists and other students of primitive cultures, recorders of myths and historians of cults and of religions – they all care about stories about miracles, strange incomprehensible events and their associated myths and rituals. The interest is large and growing. Remember Martin Buber. All his life he recorded a magical literature lovingly, but with no trace of faith in magic.

And still, why?

The interest is in part an attempt to comprehend our fellow humans and show more compassion. Duerr quotes a terrible and chilling sentence from the private diary of the leading social anthropologist Bronislaw Malinowski: "I see the life of the natives as utterly devoid of interest or importance, something as remote from me as the life of a dog" (Chapter II, Note 2). Of all the many wrongs in this nasty little sentence, we can take up the least and last. We can reproach Malinowski, that very ancestral ghost that still roams the British Isles and commands great respect and strict obedience in anthropological sessions there:

But dear Professor Malinowski, we can say to the ghost, have you never encountered a dog in the neighborhood (for, obviously, there was no dog at your home) approaching you with a greeting? Clearly, that made you, dear Professor Malinowski, a bit neglectful of facts about dogs. Oh, they surely are important. You wrote a lot of exciting stuff about the magic that all these boring natives practice; but you did overlook, then, those animals, wild and domestic, but mostly the in-between, that visited them, and gave them greetings becoming to interesting and important neighbors,

Hallo, friend! they would say.

Hallo, to you too.

Do we need intra-species human conversations in order to stay human? Duerr says, yes. But this can wait. Malinowski comes first: he was not callous, intra-species human conversation or no intra-species human conversation.

Malinowski's researches are humane and humanizing (I. C. Jarvie, *The Revolution in Anthropology*). He refused to describe native cultures from the western viewpoint, so he advocated cultural relativism. A scientist, he says, must study alien cultures from within – by going native. How far? All the way, says Malinowski, and his disciples submit: this renders quite schizophrenic the incredulous scientific social anthropologist: one half gone completely native, the other half staying completely scientific. When an American anthropologist recently joined a New Guinea headhunter chieftain's harem – utilizing the advantage of being a female, I hasten to add – there was an outcry. How far, then, should social anthropologists go native? How far did Malinowski? His diary tells us: he was very messed-up about sex, wanting to lay every female on his way but suffering from strong puritanical inhibitions. I read his expression of contempt that Duerr quotes the way Duerr reads it, as rather phony: you don't really want to sleep with those nigger-women (for this is how his diary refers to them), he probably was telling himself; they are so boring and unimportant! In brief, not dogs but native women disturbed him. He feared that a native woman might approach him and say,

Hallo friend!

Corollary: Malinowski never practiced the schizophrenia he preached, being unable to quit his puritanism or leave his sexuality with his Penelope while on a quest he joined the Argonauts of the Western Pacific. This is the critical rationalist critique of the Malinowski lore and its cultural relativism, as Jarvie and Gellner present it. Duerr unjustly ridicules them, although they are friendlier to him than the deer and the antelope.

How does he dissent from critical rationalism? He endorses its opposition to relativism and its preference for the science. He rejects Feyerabend's anarchism of the happy-go-lucky type as facile and sterile. He does disagree with critical rationalism as he offers a theory of meaning, akin to those of Collingwood and of Wittgenstein. But this is *finesse*. His chief point is his logical struggle with the experiences of the elusive world of sparks, like the experience of an articulate animal greeting a human with

Hallo, friend!

Hallo, to you too!

Duerr insists on his facts and presents his logic perfunctorily. This invites critics to concede his facts and question his logic. We do meet dogs that speak to us, we do meet Doppelgangers in the mirror, and they say, hallo, stranger! We hear television announcers talk to us personally in many a reasonably well-done commercial, and we do talk back to them. Our computers often take us by surprise when they sound so human to us: hello, there, says the computer echo. And we see the sun ascend majestically in the east, like Jason of the Argonauts of yore, and descend in the west, as he did, in a glorious sea of flames. Why does Duerr insist about all this? Who cares?

Gellner did. He spoke of the Great Divide between the world of science and all else. He saw the elusiveness of the spark, the incomprehensibility of myths, life in the twilight zone, as radically different from science. Nothing we can do can heal or obliterate the split between these, he said. Duerr ridicules Gellner, and, like Malinowski, he would like to give the elusive equal status besides the empirical. But he cannot. He does behave differently on the different sides of the Great Divide: he is sharp, iconoclastic, naughty, on the science side, but sensitive, friendly, understanding, permissive, on the other side. Also, he insists, it is no good transporting the talking deer from the elusive side to the side of the blah, of the humdrum, of the quotidian, since it will land right in Disneyland.

Nevertheless, this is interesting. Older theories are explicable by newer ones, as the new correct or modify the old. This holds for information too. Old information at times requires modification, and at times it remains stubborn, and stubbornly accords with the old theory. When we go back in time, perhaps back into antiquity, perhaps back into the archaic preliterate levels of our culture, do we find there more and more primitive theories explaining more and more vague facts? Not if there is a Great Divide. The Great Divide splits primitive preliterate thinking, perhaps, and presents both the steady world of Blah and the elusive world of Sparks. Jarvie and I have denied this (*Rationality: the Critical View*). The Great Divide does not split the primitive world, we say, it does split ours, we admit. The Great Divide shows up when rationality evolves to the level of declaring explicit criteria for empirical information, and thereby dismissing the elusive, the wondrous, the spark. The Great Divide took place as an act of exorcism, or of disenchantment, to use the term that Gellner has astutely borrowed from Max Weber. Contrary to most writers, we must agree with Gellner. Rationality has no beginning, and even an Indian master, self-styled as dumb, is very clever indeed. Nevertheless, science does have a beginning; a cultural earthquake, an awakening. Dreamtime is over and done with.

All this leaves Duerr cold. He wants the elusive spark. The elusive does not yield to repeatable observation. As repeatability is essential for science, the spark cannot be scientific. Duerr does not mind this. He refuses to split his loyalties to science and to magic; he refuses to be a cultural relativist; and so he has hardly any choice but to defy science to take cognizance, not of this or that elusive spark, but of the world of elusive sparks, of the world beyond the Great Divide, that dream-place that all of us frequent come dreamtime. And we do that with or without the aid of hallucinogenic drugs, long fasts, or rites. Duerr challenges the scientific exorcism of magic.

We all acknowledge what he demands, but he is not content. Never mind this or that elusive spark, this or that incomprehensible myth, this or that symbolic ritual that has lost its meaning eons ago. It is the world of the elusive that all of us experience. Anthropology has conceded this already long ago. Gellner said, magic formulas inherently include mumbo-jumbo. Alexandre Kojève said (in the Leo Strauss *Festschrift*), myths are incomprehensible and inconsistent by intention: their very inconsistency makes us think. Claude Lévi-Strauss has attempted to show how they make us think – by their offer of opposing poles.

2. IS INTEGRATED SOCIAL SCIENCE POSSIBLE?

Duerr's book begins with the best-known magical episode in western history: witch-hunts, witches flying on broomsticks, witches' Sabbath. We know about them next to nothing. Did you know that in order to fly, a witch had to grease her broomstick? I admit I had no idea, though it is not hard to find: the literature of the time is full of it. Bacon transcribes: the ointment that has magical effects is allegedly "made of fat of children dug out of their graves. ... But I suppose that the soporiferous medicines are likeliest to do it; which are henbane, hemlock, mandrake, moonshade, tobacco, opium, saffron, poplar leaves, *etc.*" To this Duerr adds an ethnological report about ointments that modern African flying witches use. Witches fly high in hallucinations, then, and this is common since hallucinogenic drugs are extremely accessible to all but the urban modern westerners who depend on massive imports. Witches also fly high in the fantasy of bloodthirsty witch-hunters. Researchers overlook the drugs: they want to exorcize the magic that witch-hunters believed in. It is better to add, then, the hallucinogenic experiences to the list of items transplanted from the world of sparks into the world of blah. Duerr has no objection. Nevertheless, both hypnosis and hallucination at times – only at times – turn up in a cloak of mystery, deliver strange sparks, and spread the scent of forbidden fruit. Duerr's book offers largely the wondrous elusive residue of the world of sparks that regularly sneaks into the world of blah.

This is of some philosophic and sociological interest. Pierre Bayle was an arch-skeptic philosopher who lived in the second half of the seventeenth century, when witch-hunt was at its peak before it suddenly almost vanished. He should have been skeptical about witches both on general grounds and because of the confusion of the age. To Duerr's dismay, he was not; he endorsed dogmatically the myth that exorcizes all witchcraft as deception and credulity. He should have given some witches some benefit of some doubt (p. 10).

Duerr knows: witches do not ride broomsticks as pilots ride planes, though some conservative Catholics still think that they do. He nevertheless defends valiantly the witches who say they do, against the charge that they are liars. The naïve endorse their own experiences unquestioningly, since they are pre-critical; the credulous do the same about other people's stories. There is a world of difference between *naiveté* and credulity. Observations, such as those of Sir Edward Evans-Pritchard present the superstitious as more critical than credulous. This amply refutes the theory that credulity is the same as superstition.

Credulous researchers who dislike superstition illustrate this since misreports and forgeries easily fool them. And so, Duerr suggests, let us cease calling witches credulous. No wonder that Evans-Pritchard had encouraged him to write the present book. We can now scan the literature in an effort to apply ethnographic techniques to our own backyard, to our own western culture.

This is not new. From the start of modern ethnography, R. R. Marrett had such an idea in mind, and Sir James Frazer developed it in order to apply it to the detailed study of the thesis – a century old, by then – that Christianity had absorbed much paganism. He found uncovering the pagan beneath the Christian veneer exciting. Despite his enormous influence, it is quite common in our culture to consider anthropology as the social study not of the magical side of the Great Divide but of the credulous preliterate. What matters is that anthropology is a mix of sociology, politics, and religion. For, in the societies that anthropologists study, these aspects of society intertwine, as do the sacred and the profane. A titillating example is the blend of sex and religion. This blend is ubiquitous on the magical side of the Great Divide even in literate societies. An amusing and instructive example is the attitude of Jessie L. Weston in her classic and very important *From Ritual to Romance* of 1920. In it, she follows Frazer painstakingly. She says, whatever may be the true meaning and the true significance of the grail, significant surely it was, that the lance always goes with the cup; and together they stand for the male and female genitalia (Chapter vi): "they are sex symbols of immemorial antiquity and world-wide diffusion." There is no Freud here: symbols here are sacred; in Freud's works, sex is profane. Profane sex-symbols denote, in some refined ways, sex or whatever relates to sex; sacred ones represent "reproductive energy" – universal fertility, fertility of the vegetation and of game animals and of livestock, and also, of course, of women. Sex, then, comes into it, of course, and even in a great way that would make even a Freud blush. As it is pervasive, it must bewitch our flying witches. They ride their broomsticks proudly quite naked. Sex and fertility are pervasive and central: fertility links with the four seasons, the life-cycle of the individual, the *rites de passage,* and hence agriculture, economics, sociology, religion, and psychology. Malinowski's ire was boundless when he tried to explain the silliness of the attempt to move Freud to the western Pacific.

As Duerr's book is full of sex, a puritanical witch-hunt against him was going full-blast in his native Germany and in Switzerland, where he taught at the time when I visited him in his small and quiet apartment in Heidelberg. Sex plays a few roles in his studies, none of them Freudian. His huge bibliography does not list Freud's collected works. It lists only four works of his, including *Civilization and Its Discontent* and *The Future of An Illusion,* an essay on the seventeenth-century witch-hunts that presents the persecuted as neurotics. These three references back up Duerr's view of Freud as a victim of the pseudo-scientific myth of the elusive as worthless. (Wrongly, I should add: the spark always fascinated him, and he never shook off his early faith in numerology. He wrote obsessively about the paranormal – as both very exciting and as a threat to science.) Dismissing the elusive, Freud viewed nature as an object of conquest (by science) rather than an ally. Duerr's fourth reference is to Freud's essay on

Diana of Ephesus, a symbol of nature as an ally. He quotes a Freudian to say, if we integrate [the id and] the ego and the super-ego, "we can achieve what seems like miracles" (p. 272).

Duerr's ambition is maximal – to integrate the two sides of the Great Divide, Science and Magic, Culture and Wilderness. He does avail himself of Freudian techniques: "in the archaic mentality the fence ... between wilderness and culture ... not only does not present an insurmountable boundary, but also ... this fence is at given times lowered" (p. 82) – at dreamtime, orphic orgies, *Mardi Gras*. In primitive and in modern societies alike, shamans, magicians, and mediums make it their lifestyle to live on the fence. They invite us all to join them. Like the Stranger in Georg Simmel's social theory, they are doomed to eternal alienation. The genuine among them do not complain; they accept loneliness as inherent to their lifestyles and find ample compensation in their peace with the world. They occasionally leave the wilderness and the fence to return to society, perhaps to pass on their treasure troves (penultimate page).

Duerr has predecessors to almost everything he says; his integrative vision is new. Even here he uses other people's tools whenever possible, and usually with the aid of an incredibly impressive scholarly apparatus. He professes to follow Collingwood, but he somehow overlooks his debt to him. In his valiant defense of the not-so-dumb Indian master and his likes, Duerr may sound like supporting unwittingly the kind of mythology, to quote Collingwood, "in which Philosophy or Thought takes the part of the blameless and innocent heroine led into errors by the villains Science and Religion" ("Croce's Philosophy of History", *Essays in the Philosophy of History*, 1965, p. 13). This is an error. Like Collingwood, Duerr endorses the view of Western history as one that adheres to the myths of cycles and of progress. Except that, as Kojève has explained, progress and cycles are in conflict, and attempting to reconcile them is letting the myth work on us and make us think. Does Duerr make us think or does he offer us the fruits of his labor? Both, of course; but I, for one, find it hard to draw a line. He moves constantly; he sees all sorts of connections, significant and trite; he sparkles whenever he can; he points at the sense of wonder, at the yearning, at the unity with nature that our pseudo-scientific hostility to sparks conceals; he observes that science is limited; that science, too, is an unending quest (if I may use an expression favored by the arch-villain Popper himself). He is refreshing in his sparkling bright intelligence, his wild humor, his iconoclasm; surprisingly often he scores hits. Not with Popper. What his end-product is I cannot say; only some results are clear and I can safely report them.

The idea that ethnography and ancient history are one is nineteenth century, and its sharpest expression is in the title of *Ancient Society*, the *magnum opus* of the first fieldwork anthropologist, the great nineteenth century American, Lewis Henry Morgan. This idea was repeatedly under attack. Marrett said, there is no survival, since whatever survives today must function today in order to survive rather than disappear – especially among the superstitious and among children. Malinowski used Marrett's slogan to say, let us ignore history, and Lévi-Strauss said, we can never reconstruct history since too many modes of reconstruction are available. Duerr, as a David, stands opposite the Goliath of tradition and returns to Marrett. Yet, the technique he uses is due to Frazer though in a version that Lévi-Strauss has sanctified. It is superimposing diverse myths in order to find the universal in them. The same technique has led Frazer and Weston to their way, and Lévi-Strauss to his. The road that Duerr treads is by-and-large rather orthodox Frazerian (his mockery at Frazer's anti-magic notwithstanding) – both in his accent on fertility in magic and in his historical approach to Western folk-culture. His real concern, however, is the one all modern anthropologists share – the advocacy of the view of the primitive and superstitious cultures as sophisticated and reasonable. This is the theme of the book. Though its title is *Dreamtime*, its chief end is to show that, come dreamtime, we all partake in dreamworlds and so magic is no argument to show that the primitive is less than civilized, no matter how different or even superior our civilization may be to theirs.

This is my reading of Duerr and his expressions of iconoclasm. His attitude to Frazer is lopsided and he does not say much about his methodology except what I cited above as clear, namely that old bottles – forms of speech – survive by the fact that people reused them for new wine – for describing new experiences. This, also, explains much about incomprehensive formulas. At times their meanings are lost and they are then open to reinterpretation. At times they are incomprehensible from the start, since they are historically associated with an elusive spark that is for ever lost. Duerr tells us nothing about these. Nor does he say in which sense he wishes us to comprehend magic, though he ridicules the ethnologists who think that magicians who use pebbles for magic rites are not fully aware of the pebbles being just blah pebbles. I have no wish to defend these dumb ethnologists, but to inform Duerr that most ethnologists bend over backwards in efforts to accommodate magic, that the dumb ethnologists who hate magic are the exception, and that they too are not so dumb. They too can and should greet him. They can say to him,

Hallo, friend!

Hallo, to you too!

Duerr delivers his critique of the dumb ethnologists with zest. At times he sparkles. Classical rationalism did create the Great Divide by exorcizing all the elusive around and by promising more than it could deliver in return. Critical rationalism enables staying on the rationalist side of the Great Divide and looking favorably on the other side of the Divide. I do not think Duerr has come up with a theory of rationality that is an alternative to critical rationalism, nor has he criticized critical rationalism, though he managed to poke just fun at it now and then. And so his integrated view is elusive, at best a program and an ideal. Does that make science itself, the idea of ultimate integration in truth not an elusive goal? Is not science itself a big myth, a myth to replace all myths? I suppose it is. Duerr claims that Popper's concept of truth is no good because it is idle, to use Wittgenstein's terminology: he says that Popper has no use for it (Chapter 9, Note 76). Not so. Popper uses it well as a regulative idea. This is the weakest and most disappointing point in this book: it really allows no virtue to science at all, as Collingwood said of Croce's critique.

Collingwood offered a very simple view of history: it is the study of problems that lead to solutions that lead to new problems, and this way our view of our history and of ourselves becomes increasingly integrated. This, and Gellner's critique of Collingwood, open the way to invite Collingwood, posthumously, to join the critical rationalist club. Duerr can receive an invitation in his life time: he should concede that his vision of an integrated social science is moving but elusive; it is so elusive that he cannot say what it is.

3. MAGIC NOW: RECROSSING THE GREAT DIVIDE

I know why I like Duerr's book so much. Primitive childhood may lead people who cross the Great Divide to ethnology. Yet attempts of ethnologists to cross the Great Divide to the culture that they study need not depend on childhood experiences. If these are essential, then they can pretend to have had them. They need not cheat, though. Carlos Castaneda, the apprentice-reporter mentioned above, is one of the worst offenders. Nevertheless he gave Duerr his valuable clues. Duerr is honest. He has not experienced magic in childhood. This disturbs him. He tries to fill a gap by remembering some wild childhood incidents and by viewing them as survivals of the pre-history of his culture. He succeeds, as childhood sparks are universal. But then I am biased: I spent my childhood in a society where magic came naturally.

Let us make a thought experiment; let us sit comfortably and then try to sink back into our own personal backgrounds as far as memory permits. Let me conjure a mood, if I can. I do not know if I can, since the details that are the elements of the picture that the mood associates with, these details come from your childhood memories, and I do not share the almost faded memories of your distant past. I therefore need your cooperation. Perhaps you find it difficult to cooperate, since, indeed, the mood I wish to conjure with your aid is complex and stirring – very dear yet very unsettling; very close, yet very far; a part of you and very much so, yet something you do not know, perhaps by choice. Do you remember, for example, your childhood games? Not those you played in the solitude of your crib, but those you learned when you first learned to socialize. Perhaps not that early, not when you quarreled with that other toddler about the right to play with some object. Rather, when you wanted to associate with the kids in the vicinity and let them teach you their games. Do you remember those childhood games? Do you remember the words used then? The phrases which you could not understand, the names which came with no context and quite fully disappeared when the game was dropped and quite forgotten when new games replaced them, when you grew a bit older. Do you remember the rules of your childhood games, the chants that went with them, the motions that ritualistically accompanied the motions of the performance of the moves of the game? Did you also learn to sing and dance? No, I do not mean the singing and dancing supervised by adults, kindergarten teachers or parents – although maids were a different thing, perhaps, I do not know – but song and dance transmitted by children to other children, and often ritualistically performed with no adults around – yes, up to and including sex games, such as playing doctors or playing parents, but prior to any explicit awareness of them as of sex games. Do you remember the rules of the games you played – rules that other players presented to you as rules, and rules that they transmitted to you during the performance of the game – do you remember these? Did they seem reasonable to you? No, don't go, and don't worry; of course, the word "reasonable" was not in your vocabulary then, but this need not worry you now. You can remember, can you not, the feeling of discomfort, or perhaps merely of disconcert, accompanying some practices? Do you not remember that some practices looked to you quite all right – reasonable, to use adult language – and some practices looked vaguely unreasonable? No, no, I do not mean to say, not for a split-second, that the rationality dimly felt then is a dim image of the rationality you exercise or think you exercise as an adult. On the contrary: you must remember that as a child you were familiar with the fact that your world had a very narrow horizon, that adults somehow could pierce that horizon and introduce into your world by some magic things, ideas, stories, nuances that excite or that calm? It is amazing how much children accept from the adult world with no comprehension yet with no question, and what they pick up as topics to concentrate on and worry about.

Do you not remember, in addition to all these games, chants, small rituals, also the whispers, the hints, the incredible, incomprehensible items that fill some perfectly clear and exciting fairy tales? Do you not remember those terrifying yet exciting footsteps in the dark, howls of wild animals in the open, hints at all these mysterious things that go on in the darkness of "the forest"? Adults always knew more than they could or would explain – concerning death, sex, religion, greed, broken spirits, wild fantasies that could not drop out of memory. Do you remember your first case of inability to obtain an adult's attention despite urgent need and no obvious obstacle? Here, in this half-faded, half-forgotten wondrous ring of magical clues, lies an aspect of the mood I am concentrating on – with your continued help and cooperation, I still hope – an aspect that people do not sufficiently attend to, and that most cognitive psychologists deny, particularly the followers of Freud and of Piaget, but also all other child cognitive psychologists. Children are fully and strongly aware of their cognitive deficiency, they experience what they regularly consider quite beyond their comprehension. And this experience, this second-order experience, this experience of not-quite-comprehensible experiences, is disconcerting, attractive, appealing, repulsive, appalling, fraught with a strong allure and a sense of terrible dangers, always giving each item the significance of a clue pregnant with meaning.

Cut! What we have here we cannot in any reasonable way duplicate in the adult world. And this is the point of the present study of Duerr's study of the area of no-man's-land between the civilized and the wild, the lawful and the lawless, the dream world that primeval humanity has in common with children of all times, and with simple folk, drug users, lunatics, spiritual giants, spiritual dwarfs, and you and me.

Amazingly, Freud missed this point despite his sensitivity. He knew that a childhood's wish for an embrace, frustrated, would never find compensation in adult life in the bosom of the most dedicated lover. He knew that some experiences that he half-remembered he could recount only because he had heard them from his mother. Yet he missed the half-remembered as never to be recovered, the half-remembered because no child can figure it out, and thus for ever lost. He failed to record the pain from the memory of a loss, pain at times strong enough to conjure dreamtime. He could not handle the revival of the old pain due to the old feel that there was, at the time, something fleeting, something elusive, some spark that just could be the cure to all your pains. He could only handle a revival of a clear memory, even if its clarity is sham.

The best way to approach the illusive is skepticism, the rejection of all doctrines that promise certitude. Skeptics doubt that we will ever recover all. Duerr both fervently embraces it and yet haughtily ridicules, derides, and curses it. For, it is the admission of a total defeat, of the total resignation to the incompleteness that adults inherit from children, the civilized from the prehistoric, the modern from the primordial. Duerr feels an urgent need to fight skepticism – without falling back on scientism. We may fight scientism rationally or irrationally. Rationally, it offers a theory that precludes sparks. Irrationally, it clings to facts trusting their verbal replicas. (All verbal is suspect, says Duerr; even what exactly words mean, must remain doubtful. This makes him a skeptic.) Scientism reduces language to facts and meta-language to meta-facts – feelings about facts, and feelings about feelings. Sensitivity replaces articulation. And we do not need articulation when we are sufficiently sensitive, since the whole of Nature can then come to life and talk to the sensitive in a very articulate manner. Nature may appear to the sensitive and say to them, for example,

Hallo, friend!

Hallo to you too, Mother Nature! I love you too.

Peace.

Peace on Earth.

No more war.

Love is all you need.

Love is all.

A powerful natural phenomenon is common to lower and higher animals, to all humans, from healthy babes in arms and to successful researchers: it is the search for meaning. Freudians call it projection, but it is not; it is a search for relevance. Animals under strain, human adults under strain, especially extreme loneliness or gripping terror, lunatics and researchers at their wit's end, all share this. Under enormous emotional strain one cannot glance at a newspaper without concluding from the half-read headlines almost automatically that it is good news or that it is bad news, and then, with enormous effort, in an oppressive sense of frustration, declare it irrelevant, no news at all, and move to the next headline, to the next item. If you ever saw a primitive person in desperate search, for example in an attempt to retrace a lost route to safety, then you saw a person concentrating very hard and observing hard. It is the noticing of every possible item, and noticing it as a possible clue, and then either rejecting it as nothing or following it in enormous eagerness and with great zest – until the next clue comes along. The experience is present even standing at a railway station awaiting the arrival of a friend. As the train pours passenger to the platform, it is a tremendous emotional effort to observe them all and seek clues that single out the one we await.

We dwell in an unknown terrain. The unknown constantly terrifies. At times we hide in the world of blah. At times this will not do; and then we seek clues. Most of the potential clues that we find are vain. Shuffling through them is a terrible bore, and boredom brings pain, especially when we have to concentrate hard and keep our antennae out all the way utterly defenseless. And then a moment comes, psychopathologists rudely call it a psychotic breakdown or a psychotic episode, when the pain overflows and then ceases all at once. And then we hear the whole world address us. And then we shake our heads incredulously, and then we continue to search. This experience is common. We usually do not remember it, because it is terrifying, because we see it as an ill omen, and because it is both very alluring and very appalling. So, we forget and keep running, shaking our heads in the refusal to admit and running.

Psychopathologists reinforce the fear of psychotic breakdown by advocating the empirically refuted theory that psychotic breakdown necessarily goes with psychosis. It is not nice of them to view all psychotic breakdowns as preludes to psychoses. All too often a terrible thing does happen, we must admit, and a psychotic breakdown does leave its host in psychosis proper. But we may experience one of the two items – psychosis and psychotic breakdown – without the other. A large literature describes the place of such breakdowns in the lives of artists (Anton Ehrenzweig, *The Hidden Order of Art: A Study of the Psychology of Artistic Imagination*, 1970). It happens to scientists too, including Newton, Ørsted, and Einstein.

Duerr agrees. He backs R. D. Laing (Chapter 10, Note 12), whose irrationalism I find humdrum. (See Y. Fried and J. Agassi, *Paranoia: A. Study in Diagnosis*, 1976.) The view of magicians, sorcerers, and diviners as psychotics is common. If we reject it, and if we reject the current view of psychosis and of psychotic episodes, then we may ascribe to them psychotic episodes very respectfully, as well as to artists, scientists, and most everyone else. We will single them out, then, as they choose to live on the fence, between sanity and psychosis.

Living on the fence is very hard. Aldous Huxley made it all blah by suggesting (*The Door of Perception*; *Island*) that we can strike a spark at will by swallowing a pill. That, responded Martin Buber (*The Knowledge of Man*, 1965, 99-102), is too facile. Duerr fully agrees. He reports that the use of psychedelic drugs [and more so, no doubt, their abuse] may be very blah indeed, though it may be a part of a spark if and when things sparkle. A drug culture is not that of taking a pill for kicks, nor is it a culture. The only drug that the West admits as linked to culture is methyl alcohol. The drug culture, says Duerr, is significant, as it can help novices learn to live on the fence.

Fence-dwellers are strangers in their own lands. They do not complain about it. They avoid the quotidian, the blah of every ordinary mortal, but they have their own blah, and it is most agonizing. It is no less hard and frustrating than scientific research, yet it is ever so much more painfully boring. Arthur C. Clark, the famous science-fiction author, has described it in a brief scene in his famous *Childhood's End*, where a real devil from outer space skims through all the extant literature on earth that deals with the paranormal and finds in it nothing of any value except for one or two clues drowned in so many irrelevancies. How much that poor devil must have suffered. And Carlos Castaneda too, reports similar states of deep boredom. And, like Aldous Huxley's pep pill, this too has its analogue in boring incantations, preferably in groups, of one sentence, phrase, or mere syllable, in concentration and great pain, on and on until ecstasy takes, as it also does in some *séances*, in some voodoo sessions, and in regular Church meetings where they sing traditional spirituals. This is quite general a phenomenon – shifting the fence between the sparks and the blah. Arch-Frazerian Robert Graves discussed the traditional contempt for this technique. "Kindling", he observed, is the name for it in the literature on classical Arabic poetry (*The White Goddess*, Chapter 1, fourth paragraph from the end).

Sparks that may make the experience worthwhile alleviate this terrible boredom. And as the sparks are elusive, they too have their substitutes in titillations, in references to fertility orgies and to dead children dug out of their graves. These are to sparks what the glamorous Hollywood gossip magazine is to intelligent reporting and what a Disneyland deer is to a real deer that can talk.

Living on the fence is an option open to all. It exacts a high price, though, in the form of the painful boredom that it incurs, the constant doubts that constantly shadow it, and the hardship of the task of the magician – never mind the alienation. Duerr expresses the conviction that without the people who live on the fence, humanity will be less human. Like that Titan who regained power when touching Mother Earth, so, by Duerr, we need contact with the wild world of the elusive spark, we have to cross the Great Divide personally and collectively; come dreamtime, we have to get to dreamplace, to hear a talking deer, a really talking genuine deer, really whisper in our ears,

Hallo, friend!

Hallo, to you too! Nevertheless, hallo!

Peace. Is there anything more wonderful than peace?

Just imagine: peace on Earth.

4. CONCLUSION: MAGICIANS IN THE MODERN WORLD

The world appears to a child very much as occult. Even to an ordinary person in a specific search under strain it appears so – you have to be really thirsty and desperate to really experience a proper mirage – except that for a child and for an occultist there is no clear specific purpose, only the abstract thirst, the yearning for the great, general purpose of peace, of peace of mind, of wholeness, even of sanctity, perhaps. How does the conduct of the searcher after a specific end compare to that of the searcher after a general end? Is the one superior or inferior to the other? In all his censure of all the ethnologists who belittle magic, Duerr never touches upon the following question: should he, for one, become a sorcerer?

Some of us hold that ethics prescribes autonomy and a sense of responsibility. Does magic enhance or reduce these qualities? I do not know, but I have a conjecture about people's views of the matter; our incredulous scientists, be they broad-minded or narrow-minded, will be rather impatient with magicians and sorcerers in our midst on account of their view of them as irresponsible. These people follow the will of the wisp, scientists say, because they have no sense of direction because they will not decide on directions. That children are free of responsibility is too obvious for discussion. That all spark experiences clearly go with an enormous sense of relief is all too obvious, yet the relief need not be from the pain of responsibility – it can be a relief from any pain, such as the pain of boredom, or mere physical pain. Yet, to show that the relief of the pain of responsibility is often the case, we may perform another thought experiment, one that many have performed anyway since it is rather intriguing and has some delicious aspects to it. Just relax and imagine yourself present, quite safely incognito, at your own funeral. Just ask yourself who are the people you see following your coffin, what they feel, and how they behave. There are many intriguing aspects to this experience, one of which is that it is a relief: the dead have no burden of responsibility. Different as this situation is from consulting an oracle, both offer the same sense: of relief from responsibility.

The responsibility mentioned thus far is individual. No less important is the matter of social or political responsibility – such as that of a head of state consulting an oracle in matters of war and peace. Responsible decisions are open to debate; oracles are not. The Greek oracles knew it and often returned the responsibility to the person consulting them – whoever that person was. Other oracles have no ability and no desire to do so. Consulting them is dangerous and irresponsible.

This does not exhaust the social and moral aspects of magic in modern society. Take, for example, astrology. The great medieval rationalist Moses Maimonides, blamed its practice for all of our woes. Take the astrology column in the evening press that is pretty innocuous and hence too trite to mention. My teachers, friends, and colleagues frown on it. Some of them say, it is a stone-throw from astrology to faith-healing that is a deadly obstacle to urgent medical treatment, especially of minors. Such cases are highly problematic. This is not the place for them since the present discussion is but a response to a fascinating and thought-provoking book.

Peace!

Nevertheless, Peace!

Chapter 1.7

SCIENCE AND CONTROVERSY

0. THE RESTLESS INTELLECT

Bacon said, Nature offers not only negatives but also affirmatives. For, he explained, the intellect must rest on affirmatives. This verb, "rest", is lovely, rest assured. But the definite article should go: "some intellects" is more apt than "the intellect". Some intellects must rest on affirmatives; others are restless. Science puts questions to Nature, said admirable Einstein, and She can say No and she can say Maybe, but she never says Yes. This is a clear-cut disagreement. It influences attitudes to any disagreement, as hearing Nature say Yes ends it. Bacon hated disagreement; Einstein loved it. Bacon saw it as expressing contempt and disrespect; Einstein saw it as expressing appreciation and respect. Perhaps he felt that parties in debates could love each other like parties to a chess game. Also, Bacon said, when engaged in dispute, we want to win; Einstein appealed to those who appreciated a good exchange. Einstein is the watershed – not only as a colossal physicist but also as a grand philosopher. Popper's philosophy of science, on the new side of the divide, is an elaboration of his, as Popper noted with admiration and delight.

1. THE DOCTRINE OF PREJUDICE

The hostility to controversy has a simple justification. Engagement in it is useless at best. It often creates personal animosity and so it is harmful. The urge to respond to critics is a normal way to prevent the impression of an admission of error. Benjamin Franklin took pride (in his autobiography) in his ability to hold a debate, and more so in his resistance to the urge to respond to a critic. He patiently waited for the truth to come to fore. And the truth of his view and the critic's error came to light anyway. Now, the truth does not always reveal itself and so some controversy lingers. Metaphysical controversy is notoriously different. Franklin did not discuss this: it did not relate to his own story, and he deemed it famous anyway.

That some controversy is useless is obvious: every kind of human activity includes much ballast. Is controversy always useless? This question does not command unanimity today as it did in Franklin's day, in the Age of Reason, when the prevalent view was that controversy leads to scholasticism, the custom of clinging to one's views at all cost despite all possible criticism. Galileo, Kepler, and Bacon, all fiercely opposed scholasticism. Bacon was the fiercest. He ousted the established scholastic philosophy of the time the way Cervantes ousted the established knightly romance of the time. He ridiculed it mercilessly and criticized it with much humor. This is why Descartes and Locke admired him, and why Hume and Kant saw him as the father of the philosophy of the Enlightenment movement.

Bacon explained the durability of scholasticism psychologically: the admission of error is a sign of weakness, and the refusal to admit it is natural, mostly for ambitious teachers, eager to found schools of thought. Kingdoms of the mind resemble kingdoms of the flesh: Aristotle and Alexander were ambitious, each in his own way; both were tyrants.

Bacon started with an observation that was always common knowledge: it is always possible to answer criticism, no matter how powerful it may be. Making distinctions, however frivolous, or making concessions, however minute, exempts one from admitting error, let alone recognizing the force of the criticism. Logic never invites radical revision of views, no matter how erroneous they may be. Standard legal practice is full of examples for this point. Bacon, who was a distinguished lawyer, added that insistence on a scheme of things might be a powerful means of interpreting all experience in its light. This way, any view becomes a prejudice.

Diverse thinkers repeatedly discovered Bacon's doctrine of prejudice. Sociologists know it as Gordon Allport's theory of discrimination, and psychologists know it as Leon Festinger's theory of cognitive dissonance. Can we free ourselves of our prejudices? How? I do not know Allport's answer; sociologist Maurice Ginsberg said, an all out attack on a prejudice from every possible angle might force the prejudiced to yield. Unfortunately, it does not. Festinger suggested that the situation is hopeless: we always extricate ourselves from cognitive dissonances without resolving them. He reported cases of dissonance resolution. Rather than acknowledge this as a refutation of his view, he dismissed it with a frivolous distinction.

Bacon's demand that we discard all prejudices is impossible to comply with. Still, not all is lost. It is possible to detach oneself from one's views; it is possible to learn that competing options are available. The ability to stand outside an intellectual scheme and compare it to another is a forceful means for the development and enjoyment of the critical faculty. And so, a little self-distance permits good debates on the relative merits and defects of competing schemes of things. Bernard Shaw noted this forcefully (Prefaces to *The Sanity of Art* and to *Back to Methuselah*).

Bacon recommended standing on guard against the natural disposition to cling to any scheme. He said, all that is necessary is the will to be free of prejudice. This comes with the realization that having no opinions one has no prejudice, and then one can see facts as they really are. Thus, the idea that research starts with observations is not an observation of fact, as in fact people start with hypotheses; it is a demand, and one that is extremely hard to follow, Bacon stressed. But once followed, he promised, it generates genuine, valid proofs of our findings about Nature. He assumed that there can be genuine, valid proofs. The question that this raises is, what is the right method of proof. On this he was hazy. He said, the question would find an answer in due course. But first, we should get free of existing opinions, as they are prejudices that impede the growth of knowledge. As things are, he observed, knowledge grows here and there, by accidents. The true research method, he promised, guarantees discovery "by streams and buckets".

2. THE IRRATIONALITY OF THE HATRED OF CRITICISM

The hatred of metaphysics is the hatred of the inability to resolve disputes about it. This is one kind of hatred of controversy. Metaphysicians may hate controversy too – at times even as they engage in it. They may display hatred of doubt or of disagreement, or even of critical discussion over speculations. So they tend to foster indiscriminating hatred of controversy. The two individuals who are widely considered today's leading philosophers have displayed it: Martin Heidegger and Wittgenstein. Devotees debate the question that I will assiduously ignore, as to what ideas have these famous leaders generated. Professional gossip includes anecdotes that I will also assiduously ignore, on the hatred of criticism and controversy that these two have displayed. Heidegger did not contribute to the discussion of dissent; Wittgenstein did. He asserted that all questions are in principle decidable (a principle, he said, that is not given to articulation). Undecidable questions, he concluded, are not questions. This aroused excitement and made him a leading thinker – so great the popular hatred for controversy is. I will assiduously ignore his efforts to preclude all disputes. It is better to admit that the prevalence of fruitful disputes refutes his view and to continue engagement in interesting disputes. (Incredible as it is, Wittgenstein criticized Gödel's proof that in principle we will always have some open questions; he oddly failed to see the obvious: to avoid self-defeat, his argument must be conclusive. It is not: the option that some questions will forever remain open will forever remain open.)

Anti-rationalists express hostility to criticism and contempt for it. This is self-serving and straightforward. So, it is unwise but understandable. It is harder to fathom how rationalists share this attitude.

Einstein tried once to explain the hatred of Jews. Speaking of Germany of the nineteenth and early twentieth century he said, at least in part hatred towards them is due to envy, envy of the success that some had attained in the intellectual world after they entered the modern world. This success rests on two factors, the clash of cultures, and Jewish respect for learning.

The Jewish respect for learning was transferable, since Jewish scholars were skilled dialecticians. Jewish controversies are limited to Jewish Law, and they do not apply to Jewish faith – whatever exactly this is. Jews who moved to secular studies enjoyed the free use of their critical acumen there. Whatever art these Jews possess that gives them an edge, surely, like any art, it is not restricted to its original *milieu*: all who want to utilize it may wish to acquire it; they may wish to find out what it is and what part of it they should acquire.

Participants in it were deeply involved in debates and enjoyed them and never took them personally as unfriendly assaults. The West valued the fruits of the continued engagements of Jews in the practice of dialectics and rewarded them for it. They may have this way also influenced their new environment by making disagreement palatable there, but, if so, they did not have much success. Jews like Sigmund Freud still took dissent as a sign of annoying disrespect. Einstein wrote to him expressing delight at his own having corroborated some of his (Freud's) ideas. In response, Freud wrote, I always knew that your earlier expressions of respect were mere lip service. For, how could you respect me and yet disagree with me?

Freud applied in this reply to Einstein two Baconian assumptions that are very common among intellectuals, although they are easily refutable. First, endorsing an idea is proper only after its confirmation. Second, disagreement bespeaks disrespect. Einstein was a crusader for the idea that disagreement can and should be respectful. In the eve of his life (1948), the editors of the *Bulletin of Atomic Scientists* asked him to respond to an essay by two physicists from Soviet Russia, who obviously articulated the official Soviet position. He complied, but reluctantly. He explained his reluctance: he said, before criticizing an opinion I try to make it my own, and in this case I found doing so difficult. Thus, it is better to avoid criticizing a view that is hard to defend first: criticism is an expression of appreciation and noticing this fact will improve standards in many ways. This idea is ancient; it is still not in style. Many colleagues tell me that they do not consider sincere my endorsement of it: they take for granted the assumption that the hatred of arguing is ubiquitous, and that so is the wish to win a debate. This, they observe, is a part of human nature. As the rejection of this view facilitates the admission of error, they understandably take my rejection of it as urging them to admit error. So be it.

3. THE DOCTRINE OF PREJUDICE UPDATED

Open admissions of error refute Bacon's doctrine of prejudice, perceptive as it is. Admissions of error may go with one's interest, as Alberto Coffa has noticed. And so, Bacon's doctrine requires modification. Suppose that it holds only for the cases in which it is against one's interest to admit error. Festinger held a different opinion. He said, people are ready to pay for any refusal to admit error. He referred to small costs, of course. In matters of life and death, we admit error, even if while gnashing our teeth. This will not settle matters, even though we are all better off admitting error – any error – than childishly insisting on it. So Festiger's view requires modification too.

The discussion on these lines goes in directions of social psychology, and even politics proper: when is it in the interest of a political leader to acknowledge having erred? Before going into this kind of empirical discourse, let us stay a bit longer with Bacon's initial concern: the growth of knowledge. Scientists can stick to old ideas. It makes their research very ineffectual. But this is not universal: researchers who stick to old ideas may find in them new and interesting aspects. Of course, most old-fashioned researches are ineffective, but then most researches in general are. Bacon promised that research will yield results "in streams and buckets", and this led many people to explain the barrenness and the mistakes of some leading researchers as rooted in prejudice. This placed in poor light the great Lavoisier and two of the greatest physicists of the twentieth century, Max Planck and Einstein, who were paragons of open minds. Newton was prejudiced, yet it is silly to explain Einstein's success by contrasting the prejudiced and the open-minded (as the once famous Hans Reichenbach did). True, Newton thought his theory was absolutely true, and it turned out to be a mere approximation to a better one, yet this explains neither its tremendous success nor its limitations.

The refutation of Bacon's doctrine comes from the success of self-criticism and the readiness to admit error. It also comes from contributions to science by prejudiced researchers. Many did make excuses for refuted theories. And some of the excuses were often true and their corroborations comprise important discoveries. Hence, examining excuses may be productive. Possibly not, however: Einstein dismissed excuses, and he must have had excellent reasons. This is still open to studies by open-minded historians.

The first to deviate from Bacon importantly was William Whewell. Philosophers deemed his ideas poor, and so they soon forgot him, although his influence on the scientific tradition was enormous and lasting. He agreed with Bacon: we see what we expect. He went further and said, with no expectation there is no observation. He escaped Bacon's conclusion that expectations lead to stagnation by an idea that became standard in the philosophy of science and that appears in all of its textbooks – without naming its author. New hypotheses are products of the imagination that come to explain and that lead to new predictions. Tests are comparisons between predictions and observation reports. Usually these differ, thus inviting new hypotheses. Eventually, some hypotheses lead to predictions that observations corroborate. This counts as verification: verified hypotheses naturally belong to the body of science.

Whewell's theory is remarkable. It is dialectical. It neutralizes Bacon's doctrine of prejudice and recognizes that prejudiced people may contribute to science. Some researchers have refuted their own prejudices and refused to endorse their own refutations, some of which are great discoveries. The paradigm here is Michelson's experiment that played a decisive role in the development of Einstein's alternative to Newton's mechanics. Michelson himself stuck to Newton's theories.

In sharp disagreement with the Age of Reason, we consider the utter freedom from prejudice impossible. Some philosophers take this as license to endorse any tradition, no matter how erroneous, and to overlook science altogether. The primary example here is Hans Georg Gadamer, the leading disciple of the Nazi Martin Heidegger. The Nazi experience notwithstanding, he supported the traditions of his native Germany. He also expressed preference for Aristotle over modern science.

As Einstein deposed Newton, he also deposed Whewell. He – and Popper – endorsed Whewell's view of test procedures, but not of scientific theories as final. Einstein deemed Newton's theory highly valuable though not the best.

Popper declared all refuted errors scientific. He also praised crucial tests as they may display the superiority of one theory over another. But he did not demand that tests should be crucial and that we should have faith in corroborated theories. His methodology is thus magnificent as the most objectivist ever. (Science is objective; convictions are subjective.) This raises anew the question about the role that conviction plays in research. Unfortunately, almost all concerned with it still follow Bacon.

The elimination of the need to determine the role of convictions in research before discussing its techniques leads to tremendous opportunities. To mention the most obvious and significant one, consider the traditional debate about the question, what comes first, theory or experiment? Bacon said, experiment must come first and prevent prejudice. Descartes said, we have to use mathematics, and so we must begin with theory. Now that we do not expect the utter freedom from prejudice, why not begin with both theory and experiment? This is a central question. The hostility of scientists to metaphysics rests on the (possible) competition between science and metaphysics. Advocates of metaphysics and of science say, they should be in accord. Yet, they often clash. Those who disdain controversy will use this as evidence against my proposal to begin with both theory and observation; those who love controversy will not.

4. WINNING AT ALL COST IS VERY COSTLY

Festinger said, people are willing to bear the cost of the refusal to admit error. I commented on this, saying, this is so only if the cost is not too high. I was in error: this is not always the case. History is full of cases where tribal or national leaders refused to accept criticism and this led them to utter inability to cope with disaster. These are clear cases where the desire to win has the upper hand over the desire for the truth. Where honor is a very high value and the admission of error is a violation of it, then clinging to errors may make sense. Maintaining it may be of great social importance. Responsibility for the lives of one's charges may require both the denial and the admission of error – in situations that require open, frank discussions. Such conflicts may be impossible to resolve. Fortunately, democracy did away with honor as a political value. And so, what remains is responsibility.

The problem persists in societies that discourage criticism. This is not so in modern societies, where doubt is permissible and even encouraged. It is, indeed, a mark of sophistication: only sophisticated religious institutions, for example, permit doubt, though on the condition that it is socially harmless. Contemporaries of the leading philosopher Benedict Spinoza considered him an atheist. This is incomprehensible, said Russell, as his writings are so profoundly religious. The explanation for this is simple: Divine providence is the divine account of the personal conduct, designed to mete individuals their just desert on the Day of Judgment. Spinoza denied that it exists. His contemporaries deemed this teaching dangerous. They considered the doctrine of providence very important as means for controlling simple people's conduct.

Respect for the social roles of ideas is still more common than respect for the truth. This fact seldom gains full articulation – for fear of controversy. This creates schools of thought. They are dogmatic, Popper has observed, and so, if they implement changes, they do so surreptitiously. This exudes permanent confusion. This explains much of the hostility to controversy. It does not hold for genuine intellectuals. They are those who can explain their attitudes and conduct, including any hostility that they may have towards controversy. They usually do so only marginally and incidentally, as Freud did in his above-mentioned letter to Einstein. They cannot go into sufficient detail, for fear that this will generate hostility.

Hostility to disagreement in politics may be different. A simpler explanation is very simple and obvious. It is genuinely philosophical. Consider the following philosophical suppositions. Rational people act on their convictions. All convictions should be as rational as possible. People act rationally as they act on the basis of their convictions. Many actions, some of them vital, are possible only in concert – especially in emergency. Together, these assumptions lead to the view that that agreement is essential for some vital activities – such as winning wars. Disagreement then is dangerous. This holds not only for dissent but also for mere doubt. Also, as most philosophers declare, doubt paralyzes. Only agreement can help then.

All this is highly objectionable, as it is plainly anti-democratic. Some disagreements have generated controversies, and some of these were fruitful. Otherwise, democratic parliaments are in principle useless at best.

There is an answer to this: disagreements and controversies are permissible on matters theoretical, not on practical affairs. This answer is still anti-democratic. It is poor in both respects. In practice disagreements are no less essential than in theory. Debates between commanders and their staffs on the eves of battles serve as a good example. (In some countries they are mandatory, though responsibility for decisions must rest with the command.) And hostility dominates even purely intellectual disputes. Freud, in particular, (rightly) saw no dividing line between theory and practice here. He (erroneously) concluded that he should try to convince his patients. So, he (foolishly) displayed hostility to dissent.

Criticism should be open since it is often valuable. Admittedly, it often is hostile – or it seems to be. The received rule about this hostility is that it should not be open. We often abide by this rule, as we conceal the name of the target of the criticism. Does this help? No: in time someone has to rewrite the criticism and report that its target is this or that opinion. When we must mention the target of the criticism for the sake of clarity, then we often praise the target for some other contributions, though they are irrelevant to the one under attack. And if we cannot do this honestly, then we stretch the truth a bit and consider this stretching legitimate. Even more crude methods are available, such as the inane hackneyed expression, "I fully agree with the learned speaker, but".

All this enhances confusion. To avoid it, we should list the received opinions that lead to the conclusion that disagreement is dangerous and discuss empirically the question, when is disagreement dangerous and when is it helpful? We can then examine the received premises that lead to the empirically refuted conclusion and try to replace them with better alternatives.

Here is one alternative. Agreement to act is a compromise between actors. Action rests not on opinions but on the consensus that is a compromise and so an opinion that possibly no one holds. We all suppose that it is better to act on the consensus, yet philosophers of science ignore it. They may do so on the supposition that the consensus rests on science. It evidently does not. They may do so on the supposition that the consensus should rest on science. Even were they right on this, as long as the consensus is not scientific, their philosophy is useless. We want a theory of rational action to explain the fact that we act on the consensus even when we dislike it, and the fact that a reasonable airing of a disagreement may make the consensus more intelligent. As John Watkins has wisely noted, dissent is the soul of democracy, and controversy well conducted is beneficial in public life. This is an empirical observation, and it merits stringent tests.

APPENDIX

The case of Bertrand Russell is very interesting: his expressions are often mixtures of the old and the new attitude to criticism. In his second book, on the foundation of geometry of 1897, he discussed the transition to the modern view of geometry from the traditional view of it as necessarily Euclidean. His thesis there is that all the important innovations in this respect evolved as criticisms of Kant. He himself later (rightly) criticized Kant's theory of space and of time as subjective. He criticized him very forcefully, often sounding as expressing contempt. It is therefore not surprising that he may have evoked a defensive attitude among Kant's devotees. He reports in his autobiography that once, in the discussion period following a lecture of his on this matter, a member of the audience said, we must remember that Kant loved his mother. His response, he reports there, was that he refused to believe that the human race is so vile that the love of mother is less common in it than the ability to develop a theory of space.

Russell's ability to appreciate ideas he dissented from characterises much of his writings and rests on his immense sense of proportion. Thus, he said (*Hist. Western Philos.,* beginning the discussion of the reliability of reports about the teachings of Socrates), "I would rather be reported by my bitterest enemy among philosophers than by a friend innocent of philosophy." This sense of proportion led him to ignore his theory of learning, as that theory left little function for criticism and for any sense of proportion. This is why both Jagdish Hattiangadi and John Wettersten have deemed Popper's methodology a correction of Russell's worldview. Unfortunately, this does not hold for some political pamphlets that have appeared under his name in the eve of his life. This lends weight to the idea that he is not their author, or at least not their sole author.

2. TOLERANCE

Chapter 2.1

FAITH IN THE OPEN SOCIETY: THE END OF HERMENEUTICS

0. INTRODUCTION

How can religion be respectfully tolerant despite its claim that its doctrine is perfect? The brief answer is, the admission of any uncertainty on any aspect of religion suffices to justify tolerance. A faith can be as democratic as the faithful wish; only fundamentalism, so-called, is hard to reconcile with democracy.

Criticism is most powerful when it is sympathetic. So, let us approach religion with sympathy. Let us suspend all questions of the validity of religion, and even of this or that religion, and adopt a neutral, preferably friendly attitude to them. The critical concern here will be with the claim that some religions make, that they are in possession of the perfect word of God. Is this claim compatible with democracy? As democracy did manage to live with this claim, then, obviously, the answer is in the affirmative. Yet, when we replace representative democracy with pluralist participatory democracy and ask whether it can work satisfactorily with those who claim for the possession of the word of God, then, I fear, the answer may be in the negative. For, unlike representative democracy, pluralist participatory democracy invites cooperation of citizens and of non-governmental organizations in daily political affairs, and it is hard to envisage cooperation between religious groups when each claims monopoly over the truth. Can religions give up this claim? Perhaps. Hebrew Scriptures do so, and so do some religious writers. Not so most theologians: they are shy of the very question. This is a poor influence that ancient philosophy still has over theology. It may be at its end. No one can tell.

1. PRELIMINARIES

Let us begin by minimizing the traditional conflict between science and religion. The conflict is a historical fact, a part of our heritage. Obscurantism – the hostility to reason – contributed to it first; next came scientism – the hostility to all that is not scientific. Obscurantism appears as a defense of religion, yet it is mostly a regressive political philosophy in search of alliance with religion and even refuge in it. Scientism denounced religion as dogma, prejudice, and superstition. It invited people to give up dogma, begin with utter doubt, and arrive, through science, at rational belief, at empirically justified belief. According to scientism, faith competes with science for the same place in the human intellect. Contemporary pluralist liberal democracy is different. It exhibits respect for both science and religion. It has thus transferred the competition, in-as-much as it still exists, out of the field of the politics, to which it assigns only the role of defender of the freedom of belief. Traditional political philosophies violate the traditional truce between science and religion, and so they are useless; in particular, they cannot apply to the best political systems around. These systems have no instituted political-philosophical basis. Their major characteristic, toleration, scientific and religious, has evolved as sheer expediency already in the distance past. Religious leaders learned to tolerate competition not because they gave others the benefit of the doubt, but because they got tired of religious wars and of fanaticism. Scientific leaders, likewise, had often little patience for religion and saw little use for it. But they found public silence to be the best way to undermine it, and spoke against it only privately or by implication. These changes constitute improvements, but they do not go far enough. The religiously-minded political leaders would do better to show sincere respect for competing religions and sincere interest in science. Scientifically-minded religious leaders may do the same. Some religion is intellectually obscurantist and politically regressive. Yet, to view all religion this way is to concede too much to the opponent, since at time religion supports progressive politics. It is politically wiser to encourage liberal religion than to view all religion as illiberal.

This is said not in defense of religion but in a neutral position and invitation to it to participate in public life on all its political aspects. This should include religious individuals, lobbies, and voluntary organizations – as long as it takes place within the constitutional separation of State and Church, of political and religious authority. This is the best in contemporary pluralist participatory democracy.

So much for background. And now to framework. The framework of this discussion is attitudes to tradition. Traditionally, both conservative political philosophers and radical ones, not to mention everybody in-between, endorse the same sane slogan about tradition: we must endorse the laudable in it and eschew the objectionable in it. Yet, in practice they differ. Radicals opt for science and side with tradition on the rare occasion when it agrees with science, and even then only reluctantly. It is like the ruling that decisions go with the wife when the family is in agreement and with the husband otherwise. Conservatives often delineate a large chunk of tradition as laudable and even as sacrosanct. They allow science to criticize tradition only in extreme cases, after they lose a battle against science, and even then only grudgingly. The roots of these attitudes are philosophical. Radicals declare all individuals autonomous and so in no need for tradition. Conservatives declare most individuals dependent on tradition and even on the traditional authority that enforces tradition. Both parties exaggerate. No individual is so autonomous as to be able to do without all tradition, and in modern society scarcely anyone is so devoid of autonomy as to rely on firm authority unqualified. Thus, we are all reformists, yet conservatives are grudging ones who do not admit this, and radicals are cavalier reformers as they scarcely respect any tradition. Reformists, on the other hand, scarcely distinguish between moderate and excessive reforms. (1) Reform is often problematic just because series of reforms create traditions, mostly flawed. Some orthodox religions claim possession of the unshakeable authority of the perfect word of God. Can they nonetheless participate in discussions and implementations of reforms? Or should they soften their attitudes? At what cost?

Liberal democratic tradition could and did make allowance for claims for perfection, as long as they accompanied political tolerance. This tolerance was pragmatic. Religious leaders said, I know I am perfect; I know that I go to Heaven and that you go to Hell; my faith instructs me to do my best to save your soul from hell-fire; but I will not wage war against you in order to save your soul. Scientific leaders often responded to this in kind: I know that I am right and that you wallow in your ignorance; my liberalism instructs me to do my best to save you from your ignorance, but not against your will. Both sides make claims for perfection, and these are obscurantist; but as long as they accompanied tolerance, they won public recognition. The attitude of religious leaders to science is passive and defensive. The attitude of scientific leaders to religion is that of indifference: they do not mind the widespread of religious practices. They nonetheless resent the tremendous control that religious establishments exercise over education and over the mass-media even in the technologically and politically most advanced countries. They resent it, but they allow for it all the same. The situation is thus of mutual uneasy tolerance. It is democratic, but it is a truce, not a normalized peace, not the kind of normalization and mutual respect of the expected high-level of pluralist participation in public life. This is especially so when the claim for perfection covers claims for the truth of certain anti-democratic doctrines – racism, sexism, xenophobia. Tolerance may reduce the ill effects of these tendencies, but participatory democracy invites more. It is better to replace the ill effects with positive ones. Is this possible?

2. WHY TOLERATION HAS NO FOUNDATIONS

Toleration and democracy are not identical. The one is an attitude; the other is a procedure of government. It requires that the majority tolerate the minority and that government protect minority rights. Toleration is not only towards the minority, but also towards others, be they strangers or dissenters or nonconformists. A few enlightened autocracies were reasonably tolerant; most democracies were less tolerant. Even British democracy became tolerant towards strangers and heathens only in the twentieth century. The closest link between democracy and tolerance is the following conjecture. As it is hard to keep opposing tendencies in balance, and as democracy has to show some tolerance, it becomes either increasingly tolerant or increasingly autocratic. In the latter case, it eventually ceases to be a democracy.

Historically, both tolerance and democracy evolved as expediencies, though both won the intellectual and moral support of some thinker or another, usually religionists and moralists. The support required philosophic foundations. Yet, since tolerance and democracy existed before they had any foundations, these are not essential. Perhaps, however, providing them may bring improvement.

Assume a theory that justifies democracy and toleration, that is compelling, and that advocates toleration and protection of the rights of others. Its application will not improve matters. In the early days of the University of Paris, as its members were engaged in an effort to prove the existence of God, the wise Bishop of Paris threatened them with excommunication. Possibly, their attempt allows for the thesis that faith depends on proof. And, understandably, he abhorred this thesis. They said they did not allow for that thesis: they declared that proof strengthens faith, but its absence does not undermine it. This kind of argument is unfair, as Galileo Galilei has noticed: it is the game, heads-I-win, tails-you-do-not-lose. Is the defense of toleration by offering it any foundation of any sort not the same kind of either unfairness or risk?

Popper has suggested that toleration rests on self-doubt plus the granting of the benefit of the doubt to the other. (2) This argument is defective. The benefit of doubt goes in all directions. And self-doubt does not lead to the abstention from repressive measures; it leads at times to resolute, desperate action. Popper admits that despite all self-doubt, regrettably we still must repress the intolerant who threatens our tolerant system, the anti-democrat who may destroy democracy. Hence, self-doubt itself is no preventive of all intolerant measures. The argument from self-doubt makes toleration passive, yet it should be active despite all justified self-doubts. It is incumbent on us to attempt to reduce political conflicts rooted in ideology, however, and here doubt helps for sure.

Popper discusses both the paradox of tolerance – we may have to be intolerant towards the intolerant – and the parallel paradox of democracy – we may have to outlaw a threatening anti-democratic party (2). The paradox of tolerance lies deeper than that of democracy, since some tolerance is a necessary condition for democracy but democracy is not necessary for tolerance. The paradox of democracy is valid simply because democracy is a safeguard for tolerance. Toleration, then, must be a basic principle – moral, social, and political. That paradoxically we must limit it at times precludes the specification of the exact conditions for its general validity – like the specifications of any other valid ground for self-defense. (Things must be open to *ad hoc* deliberations.)

This is why enlightened enemies of religion tolerate their opponents. They demand that we personally combat religion without taking recourse to violence and add to this the demand that the state should safeguard the right of religious practice and organization and that society should tolerate religion. Yet, all campaign against religion as such is erroneous. For, the standard justification of it is the view that religion as such is inherently evil. This cannot possibly have any reasonable defense, no matter how evil some religions surely are.

The obscurantist and regressive character of most religious establishments seems to justify hostility to religion as such. This justification is political and it concedes too much to expediency. It is better, easier, and more correct, to fight obscurantism and not religion. A more basic justification of hostility to religion as such rests on the view of religion and science as competitors; hostility to religion follows the love of science. This case, therefore, deserves a close examination. It is the idea that unless one doubts everything unproven, one disqualifies as a scientist. This idea, due to Bacon and Descartes, is scientism. It has empirical refutation in the many devout practitioners of this or that religion who were fruitful scientists. Nevertheless, science and religion may compete in many ways – Bertrand Russell was delighted to point out errors in the Bible. For example, the rabbit does not chew the cud but only seems to do so. There is no need to discuss this, since most philosophers have conceded the (Talmudic) principle that in any clash between science and religion, religion must yield. Robert Boyle endorsed this principle. Immanuel Kant based on it his toleration for religion. (3)

Yet, since religion is usually traditional, the discussion requires a clearer attitude towards tradition. Boyle and Kant referred to doctrinal matters only, not to tradition. Popper was the first to express a systematic reformist philosophy of tradition. (4) He deemed established doctrine, customs, and institutions, including science and its institutions, as received opinions. The difference between science and non-science, he said, concerns traditional attitudes to criticism. Within science a received opinion undergoes regular critical scrutiny. When it fails, a search for its substitute is underway. Traditions open to criticism, then, are quasi-scientific.

This is a new and exciting development. Except that this is not yet pluralist participatory democracy. It is not yet pluralist, though it could be. Popper's early philosophy of science is monistic: in his *Logik der Forschung* of 1935 he noticed that the critical attitude as such allows for variety, and so he requires that within science we agree to choose between competing theories and prefer the theory that is most open to criticism. Similarly, his *The Open Society and Its Enemies* of 1945 is, if not strictly monistic, at least monistic in tendency. This will not do. We should replace his proposal for the choice of the theory most open to criticism with a more pluralist one. In particular, we may allow for a disagreement regarding the proper criterion of choice of theories.

Later in his career, Popper opted for pluralism, both in the philosophy of science and in social and political philosophy. His early attitude towards religious doctrine and metaphysics was monistic. He often identifies religious doctrine (theology) and metaphysics and declares them pseudo-scientific. He proposes that we permit metaphysical views on the condition that those who hold them should do so without pretence that they possess scientific validity. The way to do so, he proposed further, is to recognize the status of these views as strictly private and personal. Even scientific realism, his own view, he declared permissible only as private. Yet, what should the attitude of religious people be towards tradition? It can be either conservative or reformist. (The third option, pure radicalism, is inconsistent.) In line with his methodological nominalism, which declares that there is no monopoly over any label, Popper should have spoken of conservative and liberal religion, conservative and liberal Christianity, conservative and liberal Judaism. Instead, he identifies Judaism with conservatism and Christianity with reformism. He dismisses off-hand conservative Christianity by labeling it Christianity-in-quotation-marks. To justify this folly, he appealed to the authority of Jesus Christ, declaring the Sermon on the Mount reformist. He did so in clear violation of the text that is emphatically hyper-conservative. This is the nadir of Popper's arguments. He violated here the minimal standards of scholarship.

This is not a blunder but a systematic error. Viewing religious doctrine as metaphysical and metaphysical beliefs as private, he had to dissociate the private attitude of religious people to doctrine from their public attitude to tradition. Attitudes to tradition, then, can be quasi-scientific or not, and so, as far as tradition is concerned, science competes with all else and society is split into two parts, the relatively open and the relatively closed. The split, said Popper, is not religious-metaphysical, and not even political, but moral. On the one side is the authoritarian who denies individual autonomy, and on the other side is the reformist. The reform that Popper has oddly missed is the reform he preaches throughout this great *The Open Society and Its Enemies*: the change of heart from following authority to individual autonomy that bespeaks the transition from the closed, rigid society to the open, scientific one. He spoke of the rise of the critical tradition not as the outgrowth of existing traditions that of necessity it was. Noticing this would have made his philosophy more of one cloth.

Popper concedes that we cannot have rationality without foundations, and that therefore the choice of the foundations of rationality is irrational. It is better to choose as minimal a foundation as possible, he observed, to minimize thereby the concession of rationalism to irrationalism. The minimum, then, is self-doubt and the hope to be able to improve through criticism. This is quite rational. Were the choice of rationalism irrational, he could hardy argue rationally in its favor, yet he did so laudably and ably.

All this is past history. W. W. Bartley, III, has altered Popper's view from that of founding science and scientifically-oriented society on skepticism, to skepticism about foundations. This should lead to the doing away with all foundations – to non-justificationism, as Bartley has labeled the new attitude. Popper endorsed Bartley's new view and even, incredibly, claimed priority for having voiced it. Similarly, he relinquished his scientific monism – his proposal that we should always prefer the theory that is most open to criticism. Regrettably, he did so surreptitiously, thus spreading confusion. The way he has followed towards implementing pluralism is that of dropping his own proposal to deem metaphysics private and personal. Metaphysics plays a public role in the growth of science: it plays the role of regulative principles, to use Kant's idiom. Kant notwithstanding, science does employ competing metaphysical ideas. This leads to scientific pluralism. For example, evolutionism and creationism. Creationism comes handy in the present discussion, due to the contingent dual fact that it was a regulative idea for science (young Charles Darwin appreciated it during his celebrated journey on the Beagle) and that these days it is fundamentalist apologetics. Does it still contribute to scientific research? Is fundamentalism consistent with pluralist participatory democracy? I do not know. Nor do I know how fundamentalists read the Bible.

3. THE LIMITS OF HERMENEUTICS

Hermeneutics is the art of reading texts – especially sacred. Traditional hermeneutics was the traditionalist, authoritarian view of the canonic sacred texts as divinely inspired. Modern hermeneutics began as a conscious attempt to discredit the same texts. Consider the contemporary hermeneutics that researchers practice, among them are religious, anti-religious, and religiously indifferent. It comprises attempts to reconstruct original meaning and to explain them historically. The standard view is that these original meanings reflect what their intended audiences have supposedly understood. This view is problematic and contested. Who was the intended audience of the given text in question? At times the answer to this question is too complex – especially when the texts in question crystallized and have then undergone a centuries-long process of canonization. Nor is this the end of the complication. Some texts defy all efforts, and remain enigmatic despite progress in philosophy, comparative religion, anthropology, archeology, and more. The Bible says of the manna that it tastes like something in honey. The intended audience evidently knew what that something was. Our ignorance of it is scarcely problematic. Not so, perhaps, our ignorance of the name "El Shadai" that later altered to "Yahwe": historians of religion would love to know what it means, but they cannot. (5) So, we do not know what sense intended audiences made of it.

The most enigmatic and most sacred biblical verse is God's answer to a simple inquiry as to His identity, a straight-forward inquiry addressed at a stranger, one that Moses made when the two first met, by the side of the burning bush. In mid-twentieth-century, the worship of ordinary language was all the rage in philosophy. A leading English philosopher, John Austin by name, regularly amused himself and his company then with gentle mockery at this verse that is the Lord's visit card – so much so that the joke became his visit card. (Being a philosopher of ordinary English, he discussed the English text, and amazingly he found it ungrammatical. He could learn from Hebrew speakers that the grammar of the Hebrew original is flawless. This did not interest him.) Another canonic expression of the same idea, due to A. J. Ayer, is no better. (6)

According to Ayer, Thou shalt not steal! has no informative meaning, and its use is nothing more than an expression of disgust at stealing. Kleptomaniacs who restrain themselves, then, suffer not moral qualms but mixed feelings. For, by Ayer, moral judgments convey no meaning, as meaning is limited to empirical science and to mathematics, whereas ethics is neither. The same holds for theology and metaphysics. Ayer's example is, "the Absolute is lazy". What has metaphysics and/or theology to do with this? Could Ayer not have chosen a more easily recognizable statement from theology, such as, "God is love"? Yet, the Absolute is the pseudonym that Hegel has tagged to God, and Aristotle has dubbed Him the Unmoved Mover. Since God is perfect, said Aristotle, He cannot move anywhere but down – and that is unthinkable. Hence, He is immobile. But He can move the world by being the object of love or of desire. "The Absolute is lazy", then, is a mock-paraphrase of "God is love". It is a paraphrase, not translation, as only the meaningful is translatable, since translation should preserve meaning.

Aristotle's theory is open to criticism. We do not know that perfection is a unique state, and if it is not unique, then God can move from one perfect state to another. It is not my purpose to criticize Aristotle. Suffice it to depict his metaphysics as open to criticism to prove that it has meaning. Ayer might say that perfection is meaningless. Not so. We have learned from modern mathematics and logic that the concept of perfection is inconsistent. But perfection may be a state and it may be an action. And ordinary English is full of perfectly smooth and perfectly well-aimed actions. The idea that perfection is a state of rest rests on the view that it is unique, and this idea belongs to Parmenides. Popper has argued that this is a metaphysical idea with a political implication – of extreme conservatism, since it depicts change as deterioration. Popper claims for Plato's doctrine the status not of a political proposal, but of a metaphysics pregnant with political implications. This is important. We can always deem a metaphysical assertion a mere metaphor. This is contrary to the metaphysician's declared intention. And here both Ayer and Popper grant metaphysicians their wish and do not ascribe to their pronouncements empirical meaning. But Ayer ascribed to them no meaning, since, he said, there is no meaning but empirical meaning and Wittgenstein is its prophet. He will then have Plato's doctrine lose all meaning, including its social and political implications. (R. M. Hare has noticed this but came to no conclusion. (7))

How, then, can Ayer allow for the study of history of theological doctrinal changes, the wars and the political reforms they led to, and so on? He may wish to deem the doctrines metaphors for certain empirical statements that do entail the intended political implications. This rests on a shaky theory of metaphors. A century earlier, some anthropologists had a better solution. Religion, they proposed, is not a doctrine but mere ritual. Ritual may include the narrating of ancient stories and the uttering of some meaningless sounds. (Some rituals comprise declarations that these meaningless sounds are doctrines that believers should endorse. Wittgenstein, and in his wake Ayer, deemed all metaphysics such meaningless sounds, viewing the expression of speculations as religious rituals.) Does this help understand religious development, religious wars, academic intrigues? Not yet. To that end, these anthropologists introduced another factor, the social function of the ritual. One of the better known authors of this idea is Emile Durkheim. He deemed religion but a political instrument devoid of meaning and of independent function. This way he made politics more important than religion (and academic intrigue more important than scholarship). This opens the floodgates to political religion or religion of politics or political ideology. It is the idea that the state cares about religion and so it imposes religion with no intention to be tolerant. (This does not explain why academics propound speculations. Wittgenstein asserted that philosophers are slum landlords (this is his expression), and that his task was to put them out of business. (8)) This is not a model of toleration.

C. E. M. Joad declared that Ayer's theory leads to fascism. (9) Ayer's friends and colleagues nastily took this as a nasty *ad hominem* attack. They said it was a scandal since Ayer was an active socialist and humanitarian. They thus dismissed the book with no discussion. In the mean time a detailed study of Durkheim appeared, by leading Oxford sociologist Steven Lukes. (10) He observes that, all good intention notwithstanding, Durkheim's doctrine supports totalitarianism. In the darkest days of Europe, Lukes adds, some of Durkheim's followers admitted this with regret. Perhaps some of Ayer's followers will learn from this.

The appeal of totalitarian ideology never depended on religion, much less on scientism or positivism. Yet, any idea that leads to tyranny is distasteful. Such is the ancient view of religion as a mere instrument of political control. Popper, for example, understands Plato to have held this view. Now usually the meaning of a text is what its intended audience has understood it, we remember. This rests on the supposition that an author succeeds to convey his views fairly well. This supposition is often good enough – but not when we come to the notoriously difficult metaphysics and/or theology. Popper claims that for fear that Plato's democratically-minded readers may find his hostility to democracy appalling, Plato did not put his message across plainly and clearly. This obscures things, allowing, Popper observes, well-meaning commentators to put a democratic gloss over Plato's texts.

Things get easily out-of-hand this way. If we insist on Plato's intentions, our study may become psychological, and when we replace Plato with Jesus or Moses, then the study may take us to inquire into the psychology of God Almighty, to theology. If the meaning of the text is public, why should we examine the understanding of it that the original intended public had? True, we should not confuse the original public with ourselves. But we do not know if Plato preferred his original public to us, as we are more advanced or better able to comprehend his message. Hari Seldon, wanting to talk to a future audience, could put his message into a time-vault for release at a specific date , and hope for the best. Isaac Asimov fashioned Seldon (in his *Foundation* series) in the distant future, in order to enable him to have this luxury. Not so Marx, after whom Asimov fashioned Seldon. And though he spoke to contemporaries, he no doubt wished to address the future communist society too, and may have used the one message as a vehicle for the other. This would be mere vanity, perhaps, but authors who live on the verge of a cataclysm may intend their audience to be a future audience and write sufficiently clearly to enter a public library and use that library as a time vault substitute. Popper feared that the Nazis would win the war and cause such a cataclysm. He wrote his *The Open Society and Its Enemies* to both the audience of the immediate future in case of a happy ending and for an audience of the distant future otherwise. Plato could have thought likewise. The Talmudic scholars probably did. Is this too fantastic even for the word of God? Surely not!

Contemporary experts on hermeneutics claim that no author has ever written the canonic Pentateuch with a distinct audience in mind. Some believers reject this suggestion vehemently as sacrilegious. Yet they cannot deny that the last book of the Pentateuch, *Deuteronomy*, came to light centuries after the death of Moses. For, a canonic sacred text says so. Did Moses write *Deuteronomy* for his contemporaries? For those who discovered it? For me?

Suppose he wrote it for me. While disregarding the antiquity of the text, and my reading it in translation – and this is the case even for Modern Hebrew readers (since ancient and contemporary Hebrew differ) – does not the mediation of translation raise problems? Even if its plain verse is unproblematic, is its theology also unproblematic? How are we to understand the claim that the Bible is the perfect word of God?

These questions justify those traditional hermeneutics of the Orthodox Jewish and Roman Catholic schools, who demand that tradition should mediate between the text and me. The Talmud and/or the Church Fathers, too, invite hermeneutic exegesis. Even Ayer does. The problem is, what exegesis is right?

There is no limit to hermeneutics in the sense that a study of a text in a limited period cannot be exhaustive. Hence, we must choose what to study, what to interpret, and how much effort to invest in any given interpretation. At times, such matters are clear: we have a word we do not comprehend and a philologist manages to find its origins and close the issue. I dare say this kind of success is rather limited. The manna tastes like something in honey, we remember. We do not know what it is and we hardly care. Martin Buber makes sense of Jehovah, declaring (5) that He is the Arabic Ya, Hua! Ya is an evocative akin to o! in English. Hua is a third person singular. Buber says, some dervishes use this expression even to this day. Assuming that Buber is right, he hardly closes a chapter in hermeneutics. The question is, were we to comprehend the Bible fully, would we come up with a much better theory? It is hard to say. Will we know then whether St. Paul meant his instruction to treat women as inferior only for his contemporaries or also for us? We cannot quarrel with people who interpret him to have never recommended treating women as inferior, though I personally do not believe them.

We see here, clearly, a pattern emerging: we may find in the writings of Plato or of St. Paul a message that we reject. We may then try a different reading or reject the author; or, in a compromise, we may find the message suitable for another audience, not for us. We may read the message as intended for an immediate public, for another public, or for both. These are nine options, each open to many variations, leading to different ways of deciphering enigmatic texts. This is too difficult and too serious. I understand Austin's rejection of all this with a quip: it was too much for him. It is too much for anyone who takes matters seriously.

How then do believers who claim possession of the word of God take all this? They must treat the Word of God seriously! Do they?

4. FAITH IN THE OPEN SOCIETY

Religious style hermeneutics is no risk to science; scientific style hermeneutics is no risk to faith. This is traditional. Nevertheless, most orthodox teachers, Jewish, Roman Catholic and fundamentalist Protestants alike (I am too ignorant about others), allege that since the word of God is perfect, an inspired text is not open to scientific critical examination. Most of the others insist, on the contrary, that a historical religion must keep changing. The orthodox recognize the change, of course, but they claim that only the circumstances change, not the law that is applicable to them. St. Paul endorsed this as he explained the abolition of Jewish Law: the crucifixion has altered the circumstances. Reformists reject this orthodox claim and ignore its traditional character. They may explain why traditionally religion was authoritarian, and why authoritarians cannot allow explicit reform, and how this forces them to pretend that reform does not happen. But before they do this, reformists must be frank and acknowledge that traditional religion was authoritarian. Reformist believers have to recognize that their traditions were authoritarian, and their reformism allows this; traditional believers cannot consider its liberal strain as legitimate, and they find no need to do so.

Oddly, although reforms are common, attitudes towards them do not reflect this. Some of the greatest religious reformers that the West ever saw preached against all reform. Martin Luther claimed that all he was doing was the restoration of ancient authority by canceling a tradition of unwarranted reforms that Rome was guilty of. More generally, discussing the Cabbala, Gershom Scholem spoke of conservative radicalism. (11) It is radicalism in its very rejection of extant tradition, but conservative in its claim to be reverting to older ones. This, incidentally, fits Copernicus. It also fits the Meiji Restoration.

Fear of hell-fire and brimstone may block religious reform. So let us consider political reforms first. The view of parliamentary democracy as inherently open to peaceful reform is now very popular. I do not know who has originated it and I have found it in the works of Ernest Barker and Popper. (12) Of course, it has a clear expression in the constitution of the United States of America. Its authors justified the practice of constitutional amendments by reference to human fallibility. Nevertheless, the political philosophy of the United States then and since saw democracy not as a system inherently open to peaceful reform but as government of the people for the people by the people. Bernard Shaw had to protest (13) and say, the people do not govern. Today this is generally agreed upon.

The justification of legal reform was usually no part of any reformist philosophy. As Benjamin Akzin has observed, usually the implementation of a reform rested on the claim that the new law was a natural law. (14) It is amusing that this way every reformer fully justified the new law, thus promising never to repeal or modify it. One would not suggest that such pretence can go on for generations, but it did. Thus, the first wave of reformers comprised mostly conservatives and the second wave comprised radicals. But this excludes reformers who were not philosophically inclined – at least not towards the dichotomy between conservatism and radicalism (Disraeli, for a conspicuous example).

Hence, the claim that democracy is inherently a system given to peaceful reform, even if true, does not make all democrats reformist. Democracy as such is not in theory reformist even if it is in practice. Popper's theory, even if true, does not characterize democratic philosophy. Popper himself rightly stressed this. In his essay on public opinion, (15) he declared the democratic theory to be the view that public opinion justifies democracy since public opinion is usually right. ("You cannot fool all the people all the time.") One might ask, why does it matter if democracy had the mistaken theoretical foundation if it was, in practice, democratic all the same? It matters, because the false theory endangers and limits democracy. Jacob Talmon discussed the danger to democracy at length. (16) The limitation that I should mention now is on pluralism.

It is not that pluralism has evolved out of a reformist ideology. The erroneous theory of democracy led to the American official theory of the melting pot that was anti-pluralist with a vengeance. The demise of the melting pot was an unintended consequence of the conduct of the Jewish lobby in the United States after World War II. That lobby developed out of the public opinion there against the cruel indifference of the Allies to the mass murder of the Jews of Europe when rescue was possible. And once American Jews gained recognition as a legitimate ethnic religious group with political significance, ethnic pluralism, religious pluralism, and pluralism as such, were not far behind. This was possible due to the practice of democratic reform. Without the criticism of the democratic theory of public opinion, the pluralism of the United States post World War II era could be but a passing phase. As it was, it consolidated everywhere in the democratic world and opened the aspiration for a united Western Europe.

Not that pluralist democracy is secure; far from it. All extant democratic ideas and institutions need a parallel reform, particularly the educational system. The very idea of democracy as open to peaceful reform requires a reassessment. For example, though Popper admits that reformism had not won acclaim even in recent democratic society, he ascribed it to Jesus Christ and to Karl Barth. And he distorted the text of the Sermon on the Mount in a much more obviously untenable way than his predecessors distorted Plato's texts. Popper's reformism is still not sufficiently pluralist, as he was still concerned with foundations when he wrote his *The Open Society and Its Enemies*. To repeat, it is easy to modify his view so that it will incorporate pluralism. The distinction between reform and reformism, between plurality and pluralism, should serve this end. Public participation in government is common, and so is the participatory democracy that encourages such participation. Yet, the distinction is vital between a theory and a practice that conforms to it. A theory that explains a practice and a critical discussion of its merits and defects is the best known way to secure it. The next step is the public transition from one theory to another. It is a complicated process and some authorities should undertake it with some awareness of some historical perspective. The same may hold for the role of religion within democracy – at least to the extent that a significant portion of practiced religion is democratic and progressive.

This, at last, has brought me to my present topic. I will conclude it with a discussion of religious foundations, of secular theory, and of their combination. The word of God properly read may be hostile to our kind of democracy. After all, the Bible supports the Kingdom of David. To reconcile the Word of God with democracy we may have to reinterpret the text or to say that the Bible has a new message for every new public. This is the meaning of the religious phrase, "the living Word of God". Assume that this is so. Then, religious leaders who take the Word of God seriously have to reconcile their view with contemporary liberal pluralist participatory democracy and demand from their congregations not merely passive toleration but also proper positive respect. They have diverse ways for doing this. But I need not go into this: it is their job, not mine.

So much for religious foundations. As to secular theory, we should not consider its foundations or justification. It has a more important role. The improvement of social or political settings is often the outcome of sheer expediency. It is then easily lost after the expediency is over – unless it wins recognition as an improvement. This recognition may rest on an old theory or on a new one. The new theory will serve as a new ground or foundation for the improvement and will entrench it. But to open the road to further improvements, we should not take the new theory as an axiom. The new theory or alternative theories or further expediencies may suggest new experiments. These may help devise newer theories. Hence, new theories are at times contributions to progress. If they are, then it is very important not to apply them indiscriminately in retrospect. Democracy is a process, a matter of degree. In particular, liberal pluralist participant democracy is an improvement over old-style representative democracy. Old-styled democracy fits theories that are less satisfactory than the current theory of liberal pluralist participatory democracy.

Until recently, the better political philosophers were engaged in the defense and promotion of democracy. To that end understandably they often offered a characterization of democracy as such and used it as a justification. This is an error – understandable but regrettable. If we do not take toleration as fundamental, then we may lose it: it is the only thing to take as fundamental – but as a regulative idea, not as an axiom. We do not know the proper wording of the principle of toleration. Democracy is a safeguard for toleration, but there are many theories and many styles of democracy. All future theories of it should take account of its varieties and of preferences concerning them. As pluralist participatory liberal democracy is the best known thus far, religious communities should participate in politics while respecting other people's faith. This must limit the way they read what they deem the living Word of God.

It is easy to see the point, so that even ardent fundamentalists need not stand on all of what they think they read in the Bible, so that they may endorse liberal rules of hermeneutics. For example, the Bible suggests that the heart is the seat of emotions. We do not think so, yet we have no qualms about setting our hearts on whatever we like. The Bible couples this idea with the one that the kidneys are the seat of reason. This sounds to us bizarre, because we never set our kidneys on what we set our minds on. Do fundamentalists stand on their principle so far as to insist that we think with our kidneys? Or will they agree that the use of ideas after they are defunct renders them metaphorical? In principle, they should agree, there is no difference between the two ideas or metaphors: if we feel with our hearts then we may just as well think with our kidneys. Thus, we have on our hands the modern view on hermeneutics (heralded by Spinoza). That metaphors can be bizarre is obvious: we can have ideas on hands and at the same time we can stand on them too.

NOTES

1. John Wettersten, "On Two Non-Justificationist Theories", in Joseph Agassi and Ian Charles Jarvie, *Rationality: The Critical View*, 1987, 339-342.
2. Karl R. Popper, *The Open Society and Its Enemies*, Chapter 24.
3. Robert Boyle, *Things Above Reason.* Immanuel Kant, *Religion Within the Limits of Reason Alone.*
4. Karl Popper, "Towards a rational Theory of Tradition", reprinted in his *Conjectures and Refutations*, 1963, 1965, 120-135.
5. Martin Buber, *Moses*, Chapter on The Burning Bush.
6. A. J. Ayer, *Language, Truth and Logic*, 1936.
7. R. M. Hare, "A Question about Plato's Theory of Ideas", in Mario Bunge, ed., *The Critical Approach to Science and Philosophy: in Honor of Karl Popper*, 1964, 61-81.
8. David Edmonds and John Eidenow, *Wittgenstein's Poker*, 2001, 200.
9. C. E. M. Joad, *A Critique of Logical Positivism*, 1950.
10. Steven Lukes, *Emile Durkheim, His Life and Work*, 1973, Introduction, Civic Ethics, The Dreyfus Affair, State and Church.
11. Gershom Scholem, *On the Kabbalah and Its Symbolism*, 1965, Chapter 1.
12. Ernest Barker, *Reflections on Government*, 1942, Chapter 3 and 275-6.
13. Bernard Shaw, *The Milliolnairess*, Preface.
14. Benjamin Akzin, "Legislation and Natural Law", in *Encyclopedia for the Social Sciences*, second edition. 9, 1968, 230-231.
15. Karl Popper, "Public Opinion and Liberal Principles", reprinted in his *Conjectures and Refutations* 1963, 1965, 347-354.
16. Jacob Talmon, *The Origin of Totalitarian Democracy*, 1955.

Chapter 2.2

THE FUNCTIONS OF INTELLECTUAL RUBBISH

FIRST INTRODUCTION: MY OWN CONCERN

Discussing anything related to intellectual rubbish, to the intellectual standards that it falls short of, involves emotional charge. It is one thing to refuse to share tastes with my neighbors, and hard enough and alienating enough at that. To declare intellectual rubbish what they appreciate is much worse. Admittedly, I may try to escape trouble: I may try to find out what are their tastes, and avoid talking about intellectual rubbish except in the company of those whose tastes I share. This will not do. First, word goes round. Second, if two people agree about one thing and then their conversation shifts to talk about another, they may then find unexpected strong discrepancies. Finding in our cultural milieu more rubbish than things of value is common. This makes it hardly possible for anyone to express freely opinions about tastes without the fear of offending.

So be it. There are a few suggestions for the avoidance of annoying others by upholding high standards. One of the commonest is to be cautious, to limit one's conversations to one's field of expertise. This does not help much, and it is very limiting. Worse, it causes a loss of the ability to learn from others: learning requires free exchanges of ideas, and diplomacy disrupts it. A better suggestion is at hand in studies of the mass media, aimed at the elimination of the ground for hostility for expressions of views. (See Judith B. Agassi, "The Worker and the Media," *European Journal of Sociology*, Spring 1970.) My concern here lies in a more limited social arena, as it centers on one rather widespread situation: young people who try to find their way in the world suffer from the defense of high intellectual standards and against intellectual rubbish. We should ameliorate their condition.

When one does not know how to speak – in any sense of the word – one cannot but stutter. Beginners cannot avoid stuttering, and the custodians of high standards prevent them from stuttering in public. They demand that novices acquire better training before they enter the public arena. It is a major concern of most training, in most fields, to train people to speak without giving them a chance to stutter: trainers devise simple and easy exercises in the hope that even utter novices can execute them without a stutter. One exercise leads to the next, slightly harder, but still performable reasonably well. The theory and practice of this idea is didactics.

Didactics is well-constructed rubbish. The five-finger exercises of the Karl Czerny type are the paradigm for it. Claude Debussy hated it avidly, and tried to show that it is unnecessary torture; he has recently won some popularity.

Didactics has a social function that makes it objectionable. It singles out the newcomers to the field who have benefited from it. Some outsiders may even have acquired proper training, but it is still insufficient: they are ready to do things for which they have no training. This means, at times, that insiders have permission to do only what their teachers already did before, since the training they can supply is limited to what they are already familiar with. But this is not necessarily so. For, the old teachers who do things the old way need not oppose new ways; they may merely insist that they train the young ones in the very new ways which the young ones themselves happen to choose. This will insure that they are in control, that the young innovators rise from the ranks, that they do the new things properly too, that continuity prevails.

Continuity be damned, say I, and so be the old teachers. The young outsider innovators deserve all the breaks that they can get, even though they may stutter. They may never learn to speak and they may only learn to speak crudely. But they deserve the breaks anyway, at least the benefit of doubt. The standard objection to them is that their stutter fills the world with rubbish, when we already have too much of it anyway. So, I wish to examine this claim.

But first I wish to eliminate from this study all emotional components, including the one I have just introduced, and offer an analysis of the situation instead.

SECOND INTRODUCTION: TO FORGIVE IS PRESUMPTUOUS

It is not true that to understand is to forgive. If anyone ever understood Descartes, surely Pascal did, yet he said, in a monumental and oft-quoted passage, I cannot forgive Descartes. Pascal was a younger contemporary of Descartes. He improved upon him in the theory of method ("The Spirit of Geometry"), in mathematics (his invention of projective geometry), and in physics (his invention of barometry). Yet it was in his theological work (*Pensées*) that he showed his best feel for Descartes. Descartes presented his philosophy as Christian. Pascal knew that by destroying Providence, by placing God outside our ordinary world, by putting Him outside the space-time manifold, Descartes made Him irrelevant to humans. And for this, he said, he could not forgive Descartes. Thus, it is where Pascal understood Descartes best that he could not forgive him.

Perhaps forgiving a shallow thinker differs from forgiving a profound one like Descartes. I do not think so. When you read Martin Heidegger, you do not comprehend him and suspect that he is very deep indeed, and so you are prepared to admire him, to believe your friends' assurances that he is admirable. But suppose you do comprehend him and realize that he offers a mixture of trivial and outrageous remarks, embellished with pseudo-scholarly comments. You understand him, and may be indignant; you need not forgive him. Moreover, his philosophy is almost all political, and it is to one degree or another a defense of the Nazi regime (in what he deemed an improved version). Two victims of that regime, a famous philosopher who was his disciple and a famous poet who barely knew him, forgave him. But not in my name: I need not forgive him, and I do not.

Now this discussion is of little import. It was nothing to Descartes whether Pascal forgave him or not. He was offering his ideas to the public and everyone was at liberty to take them seriously or not. And one the like of Heidegger is utterly unmoved by contempt for his writings. He would easily and lightly dismiss any comment of a liberal thinker. Thus, indignation is of no use. Writing as a social reformist, I deem social analysis the best tool for reform: unless we see the positive roles played by the institutions we wish to abolish, we are going to cause hardship to those who benefit from them and thus push them into active opposition. My concern here is with intellectual rubbish. I take it for granted that it is harmful, only hinting what damage it causes people and how. I shall center instead on its positive functions, and on ways to transfer these to other parts of our intellectual sphere so as to be able, not to abolish rubbish, but to reduce it to more reasonable proportions.

My thesis is, I am afraid, that the main positive role of intellectual rubbish is to protect the incompetent famous from the wrath and indignation of our intellectual police force, and they need no protection, as their positions are secure. The main victims of the anti-rubbish police are young hopefuls, the possible future pioneers. I propose that we should develop a higher level of toleration for intellectual rubbish, and let invite our educational system to help beginners develop better tastes and better selection filters. This will reduce the demand for rubbish and thus the supply of rubbish too.

Let me conclude this introduction by two examples – of the kind of reaction to intellectual rubbish that I dread most, namely, the indignation at the prevalence of much intellectual rubbish.

The first is a sincere and well-balanced paper about the research literature of the day. It contains hard evidence, serious statistics, and many references. It is "Peer Review: Quality Control of Applied Social Research," by John H. Noble, Jr., published in a prestigious magazine (*Science*, 185, Sept. 14, 1974, 916-925). It recommends peer review and quality control as cures to the ills it reports. There is no analysis of the reported situation there and no attempt to explain the strange fact (cited on page 920) that "51 percent of projects ... $45 to $50 million spent for evaluation research in ... 1970 by Federal agencies – fell below 4.24 on a seven-point scale ... (where "6" stood for the minimum standard ...)", except to say that "it provokes outrage and demands strong corrective action". The author then moves on to the making of a proposal for peer review and quality control. As it happens, peer review is common and useless (Joseph Agassi, "Peer Review: A Personal Report", *Methodology and Science*, 2, 1990, 171-180). And quality control cannot apply to research. But these are asides: I should denounce *a priori* these proposals as premature: social scientists can understand my wish to have an explanation, at least some discussion, of the present sad situation, before entertaining any proposal. But the reader need not agree. My discussion here serves merely as token empirical evidence for my claim that intellectual rubbish is ubiquitous, and that – regrettably – the default response to it is indignation.

My second example concerns mass culture. It is the suggestion of the great liberal philosopher Popper. He has suggested that legislation should put the mass media under strict censorship, since it poisons our wells. The less said about this the better. (He was my admired and beloved teacher; his pronouncements on the media embarrass me.)

1. THE PECULIARITY OF INTELLECTUAL GOODS

We have a lot of intellectual rubbish – rubbish, for short – of all sorts. We also have oftener items whose quality is mediocre. What is so peculiar to intellectual commodities, as opposed to other commodities, is their great availability, at least by comparison. Though books are scarce, and though their prices are staggering, at least the best novels are not much harder to get than cheap ones at comparable prices. One may recognize the superiority of the expensive car or restaurant, yet opt for the cheap one for mere financial reasons. This seldom holds for a book, as most people can easily afford good books but prefer poor ones. It is true that some cultural treats are rare, and so seats at the best theater or opera may be too expensive for the poor students who are their best customers. Yet, the comparative prices of excellent and poor entertainment, theatrical or operatic, are quite comparable. At times poor entertainment is more expensive – a visit to a cabaret or a leg-show may be much more expensive than an opera balcony tickets; a classic paperback is usually cheaper than a glossy magazine; and the cost of viewing the best movies is usually average.

All this, the basis of our problem, is superficial observation. No doubt, it does not hold for commodities that fail to reach the market: the demand for rubbish often makes the best commodities around scarce to the point of unavailability. For example, some excellent movies have never reached the market. Many good manuscripts and some excellent pieces of music had the same ill fate. Some excellent doctoral dissertations, submitted to some leading universities, were failed – as hopeless or as in need of some improvement. The problem at hand is that quite often the public willingly prefers rubbish to excellence. Many academics train their students to produce rubbish, or at least stuff inferior to what they could produce were they not trained: some academic training is for the reducing of standard of production – either in order to sell or due to a mistaken idea about what the demand of a novice is. In some cases, the fault is in the trainer's wrong sense of discrimination. The commonest source of this is in their efforts to adhere to the view of science as inductive. Contrary to what most philosophers of science say, the faith in the idea that scientific theories proper emerge from large stocks of information is both popular and harmful. I will not develop this idea here.

I shall discuss here, later on, the case of the good stuff that does not reach the market. This happens to be my chief concern in the sociology of science. It has brought me to this field in the first place. When I try to defend and support and help young – and not so young – unrecognized hopefuls, my first obstacle is repeatedly the intellectual police force. We have too much rubbish, they say, and accuse me of trying to bring in more. When I presented my standards for a good history of science (*Towards an Historiography of Science*, 1963, Conclusion), I observed that, as a matter of course, no historian of science could avoid making a few mistakes that they could eliminate with little additional care and effort. I added that this is quite innocuous. Almost all my reviewers – and they were sympathetic – understood me to be an intellectual police officer who, at the last moment, chickened out. Among these reviewers were some leading historians of science, including Charles Gillispie and Thomas Kuhn. Years later, they kindly invited me to speak before their graduate seminar in Princeton. I took the occasion and responded to their reviews. I said, I had with me examples from the pens of my hosts of minor errors that invade even the best items in the literature without doing much harm. I said I would show them the list at their request. They never made such a request.

I am offering now a thorough examination of their position, and I begin by conceding their major thesis: intellectual rubbish abounds. Readers who do not share this thesis may find the present examination of it not very interesting. So be it. Some readers may, on the contrary, find it self-evident that intellectual policing is unavoidable. They, too, may find the present examination uninteresting. I beseech them to pursue it and give me the opportunity to shake their opinion.

Those who demand some action to change the situation, generally oppose censorship and constraints. They may differ widely in philosophy and mentality. Michael Polanyi and Popper agree on the necessity of some form of censorship or another within the commonwealth of learning. They feel more threatened by rubbish in the learned world than by rubbish in the open market. Why? The answer lies in the maxim that the learned world could implement centuries ago but not today. It is that scholars should be familiar with everything published in their fields. I. C. Jarvie has called this a thoroughness mentality. It explains the tendency of scholars to ever reduce the sub-sub-specializations that they claim expertise in. This is inefficient: it is easier to combat thoroughness mentality than the demand for rubbish. So let us discuss that demand for a while.

2. THE HIGHBROW ATTITUDE

What exactly is rubbish? An economic definition of rubbish is straightforward: it is that which has no potential buyers, no demand for it, perhaps no utility for anyone. Ecologically, rubbish is anything causing public disutility, anything that has a public nuisance value and so should be disposed of. The two definitions are not coextensive, as much of the demand in the market is for what ecologists condemn as rubbish. Still, after all, ecology brings up to date general economic concern. Therefore, ecology and economics should not diverge. The difference between them is that ecologists pose the problem (what is rubbish?) in broader terms than economists usually do. Both refer to the same situation, though in different contexts. Economists may find it difficult to broaden their context not so much because transactions often rest on immediate concerns, but because the concern for the ecosystem may force economists to transcend their traditional equilibrium models of nonintervention and make them recommend limited intervention instead. But this is a different matter. For our present concern we may assume that in principle economics and ecology converge, as they should.

Both economists and ecologists declare rubbish to be what has negative utility value. Interventionist economists declare that there is no way to stop the manufacture and consumption of commodities – commodities that have positive utility – even if these will in the long run turn into rubbish and have negative utility. There is no disagreement between economists and ecologists because they both assume that whatever the public purchases now is, *eo ipso*, not rubbish. Our initial question was, why do people purchase rubbish. Evidently, "rubbish" in this question cannot mean what economists mean by "rubbish", as what is rubbish to economists is that which nobody will purchase, perhaps only what people will pay for its removal. Ecologists might look at this differently and declare rubbish what harms the environment.

We can tentatively define intellectual rubbish as intellectually inferior material. This, of course, has to be relative to a given standard. What standard?

Old-fashioned Whigs, chiefly economists, but also social and political thinkers with leanings toward classical economics, refuse to answer our question, why do people choose rubbish? They refuse to define a standard for it. They dismiss the question by the rejection of its very presupposition, by the denial that people freely purchase rubbish: whatever the public purchases is, *eo ipso*, not rubbish. Those who deny this assertion are highbrows who wish to impose their own values and tastes on others, who seek ways to prevent others from spending money any way they wish.

It is not clear who is a highbrow, even when there is no reference to intolerance. Possibly a highbrow is one who loves Shakespeare and Bach and Picasso; alternatively, a highbrow is one who also despises detective novels and popular music and television. Let us distinguish between these two kinds of lovers of high culture, the pure highbrows and mixed-brows. Let me use the label "highbrow" to designate the pure highbrow only. Let me also say I have no more objection to highbrows than to lowbrows: in each category, highbrow as well as lowbrow, even middlebrow, I find good products. This makes me a mixed-brow, but I do not quarrel with those who confine their tastes to high, middle, or low culture.

I wish to join the condemnation that the old-fashioned Whigs address towards the intolerant – highbrow or not – and their contempt for others. There is nothing good to say for any contempt. Intolerance – highbrow or not – is condemnable as immoral, unaesthetic, and asocial. It is misanthropic, snobbish, and élitist. This judgment is the official verdict that almost all totalitarian governments pass on highbrow art. But then, this is lowbrow intolerance. It is inverted highbrow intolerance, and so it is equally condemnable as misanthropy, as inverted snobbery, and as *Gleichschaltung*.

I do two things here quite without any debate. First, I dismiss intolerance. I know that the question of the limits of what we must tolerate and of what we may not tolerate is open and serious. Herbert Marcuse and his friends (*Critique of Pure Tolerance*, 1965) argue that as tolerance is always qualified, there is no qualitative difference between my qualified tolerance and his qualified intolerance. The question of the limits of tolerance is irrelevant here despite instances such as the right or wrong of selling pornography to children. I dismiss the argument of Marcuse without debate on the supposition that we all know that it is possible and desirable to limit intolerance within reason. The limitation should be to attitudes towards the intolerant, and even this not with vengeance but sufficiently to satisfy the need to defend democracy. (For details see Popper, *The Open Society and Its Enemies*, 1945.)

My second dismissal, given my dismissal of intolerance, is of highbrow intolerance. I take my cue from Voltaire, who said there is no need to demand of anyone to tolerate the wise, so that the demand for tolerance is the demand to tolerate the fool. And of the uncultured and of the devoid of good taste. Highbrow intolerance is just intolerance as some highbrows practice it.

All this applies to highbrow and lowbrow intolerance alike. Art, science, religion, and any other intellectual undertaking have a great variety of social, aesthetic, and intellectual dimensions and functions. Let me repeat that here my concern is not with high standards. Not only highbrows have standards. Also, universal standards are impossible (yet relativism is false, as standards are relative not only to accidental, given factors). Every category has low and high standards. The question is, to repeat more clearly, why do most people prefer – in their own chosen category – the poor to the good?

3. THE USELESSNESS OF INDIGNATION

That indignation is intolerant is obvious. Hence, the old-fashioned Whig's indignation at the highbrow's indignation, though quite understandable, is erroneous. Let me criticize here all indignation, including that of the old-fashioned Whig. Any public exercise of indignation is inefficient and so unwise. Why are people indignant? Whigs have always agreed that indignation is unwise – an attitude that protects the indignant person's narrowness and ignorance. If so, then Whig indignation at the indignant is unwise. Of course, Whigs may reply that whereas indignant highbrows (or lowbrows) are unaware of the fact that their indignation is protective, the Whigs themselves understand this, and they protect toleration, and so they are not unwise. This is very obviously a logical error that Whigs commit as regularly as highbrows: the highbrows fully understand the indignation of lowbrows, but not their own indignation; Whigs fully understand the indignation of highbrows but not their own.

Whigs are right in saying that (a) indignation is a manifestation of intolerance, (b) it is protective, and (c) it is a weapon, a means of pressure, of imposing the indignant people's views on others, at the very least on their charges and students. Whigs are also right in saying that their indignation may be a counter-pressure and so it is intolerant only of the intolerant and so as morally excusable. Yet the social function of the indignation of Whigs as a weapon is the same as that of the highbrows or the lowbrows, and its use as a weapon is lamentable: it grossly misfires. Indeed, it is a part of the whole fabric of social interaction. Lowbrow TV comedy hack-writers know this. They see in the defiance of the indignant a major part of the fun of any game they enact. It is a point central in the excellent, lowbrow, best-selling *Games People Play* of psychologist Eric Berne.

All folk wisdom tells us that the misers are fools, since they cannot take with them the wealth that they hoard. Freud says, misers are obsessive. Like everyone, they want love; like many people, they are obsessive about it, and what is specific to their obsession is that they have learned early in life that they must purchase it. They fail, but they are afraid to revise their view drastically. They therefore modify it: they still try to buy love, so they conclude that they need more money than they had thought. And so hoarding becomes their habit. The mentality of preachers is identical: they hope to convert misers by scorn and indignation; they hoard scorn and indignation as means of winning misers. Scorn and indignation become habits of the preachers against misers as money is for misers. The *status quo* thus lasts.

Why then should people use the obviously poor weapons that indignation is, when better weapons are available? One possible answer is this: they do not wish to win. The function of poor weapons is often precisely that of the maintenance of the *status quo*. A conclusion from this is that the function of indignation as a weapon is in their users' wish for it to be not very effective. This conclusion rests also on the premise that there are more effective means of fighting intolerance and of improving public taste and of propagating highbrow – or lowbrow – tastes and material. I shall presently discuss this premise.

Hence, not only the intolerant highbrows or lowbrows, but also the tolerant Whigs, are maintaining the *status quo*. They do so by condemning the intolerant instead of following the enlightened policy of examining the function of intolerance. For, the true maxim of the Enlightenment is not "to understand is to forgive"; rather, it is "it is not for us to judge; it is for us to try to understand".

4. WHY RETAIN THE *STATUS QUO*?

Whigs insist on accepting the market as it is; they demand that governments refrain from interference in the market. Even the interference of intolerant highbrows and that of the intolerant lowbrows, as parts of the free market, also invite no interference. What do indignation and scorn sell or buy? I say they buy the maintenance of the *status quo*; they buy stability.

How and to what end?

I am discussing the market as a reasonably free market within a democratic society. All the problems which young aspirants face there are much less severe than those that their counterparts in undemocratic societies suffer. Even in democratic societies there are limitations that exclude certain potential customers from the market. Such are young potential buyers of pornography; students who might purchase good books instead of the dull ones prescribed to them; even people interested in Marxism during the McCarthy era when the mere display of interest in Marxist literature required more courage than most intellectuals possessed at the time.

I limit this discussion to the case of the reasonably free market since it is relatively simple: it is an idealization that goes with the hope for improvement.

And so I postpone the discussion of good material that has not gained entry to the market as yet, in the hope that the result of the analysis of the existing market will lead to the improvement that should be precisely the increase of availability of material thus far not available. (See L. A. Boland, "An Institutional Theory of Economic Technology and Change," *Philosophy of the Social Sciences,* 1971.)

To return to my hypothesis, the scorn that all sorts of people pour into the discussion at hand is a poor instrument that does not avail itself for crucial battles but is good enough for the maintenance of the *status quo*. To argue my case, I wish to examine first the various parties in the cultural market – the consumer, the producer, and the large organizations that are involved. I will return to the creative intellectual soon afterwards.

Let us consider the consumers. Generally, they prefer the *status quo* to loss, but not to gain. Yet, I contend, in the intellectual sphere this is highly questionable if not simply pointless. For, we remember, in a democratic society consumers are reasonably free to ignore whatever intellectual material they wish to ignore. Therefore, still speaking within the framework of reasonably free markets, the complaint that there is supply of so much rubbish is really neither here nor there. Clearly, the existence of non-highbrow material does not limit highbrow consumers, nor does rubbish. This assertion tallies with classical economics in its disregard for the dog-in-the-manger stance (= the wish of misanthropists to make other people unhappy). To this misanthropic concern of some consumers of highbrow material we must add the straightforward economic concern of its producers. Given, that is, the limited budget for the consumption of culture, the producer of one kind of culture competes with the producer of another kind of culture that might possibly be its substitute (even when it is a poor substitute). The way adherents of classical economic theory see it, then, is not the way concerned highbrow consumers see it, but the way producers of highbrow material do, in their wish to retain their shares in the market.

A certain portion of sales of intellectual goods is admittedly assisted by indignation and scorn – at times customers purchase the same commodity repeatedly in a vain hope to consume it. This explains the mass sales of unreadable comprehensive books, like histories of all political processes or of all the arts or sciences from antiquity to date. This sale is the result of the pious hope to read, a hope that customers never realize. Yet this empirical observation is of little significance for the market; rather, it is regrettably a significant encouragement to the generators of highbrow indignation and scorn.

The availability or scarcity of a given kind of material in the open market depends largely on its success or failure there, namely, the success or failure of its producer at marketing it. Yet the effect is here rather marginal for both consumers and producers – in that producers are much more concerned with selling more of the same, say more copies of a book, more seats for the same show, *etc.*, rather than more masterpieces. As to creative intellectuals, they are also fighting for higher positions in the highbrow hierarchy (or middlebrow or lowbrow one), for better pay, for a grant; they have no concern with the market as a whole. To have effective influence on the market as a whole, one must create means such as, for example, a lobby in Washington or in private foundations. The wish to enable lobbying and such to take place often leads to the creation of new organizations – of artists, scientists, inventors, journalists, rabbis, *etc.* They fight for group interests. How does the group distribute its income? How does the organization contribute to this distribution?

All establishments want things to stay more or less as they stand. National organizations and other large-scale bodies defend the *status quo*. Consumers are likely to express indignation at the substitutes to the products that they consume, regardless of whether the specific consumers' organization and specific markets involved cater for highbrows or not. The indignation serves the function for a class of consumers of singling them out as consumers of whatever they consume, as co-consumers who may wish to share experiences of both delight and indignation. This highbrows and lowbrows share.

The coordination of co-consumers is important: consumers of one kind of novel may make sure that they only exchange views with consumers of the same kind of novel, or with consumers of novels: they may wish to avoid talking about books to anyone who does not share their tastes, and they may use indignation both as a means of scaring off those with whom they wish to have no exchange and as a means of enhancing the pleasure of exchanges with people who do share their tastes. They meet in conventions or in support groups of all sorts.

This function of indignation, however, is chiefly a part of a more important function of exchange among people who share tastes: it is that of guidance of a public toward new products. Indignation scares away those who might offer misleading counsel as to further purchases. Highbrows and lowbrows alike, the more assured they are of their way of going about finding new materials to purchase, the less indignant they feel toward those who do not share their tastes. This is an empirical finding. The assuredness may be due to a strong independence and it may be due to familiarity with the authority that decides matters – the old pro. One significant and central function of indignation, then, in the absence of direct communication with the old pro, is the creation of an indirect communication, of the mass of consumers who share tastes, through those who do have direct communication. Not that indignant consumers know that old pros coordinate their taste and tamper with it; rather they see themselves as consulting reliable old pros. Let us consider the old pros, then, whoever they are.

The analysis of old pros is fascinating. Jarvie has offered a glimmer of old pros in the movie world (*Movies and Society*, 1970). He describes old pros as producers of the gossip that appears in the same issue of the publication in which its denials also appear. These are at times a public-relations hack in the pay of studios, at times they are gossip columnists that the studios denounce. The gossip column, or the movie magazine, the fanzine, is not reading material for all who share tastes. It guides those who guide their peers. The same story recurs in matters of highbrow cultural commodities and the manufacture of popular tastes for them, except that here the participants, the co-consumers who exchange a sense of indignation, feel obliged to get into the act and sincerely believe their own illusions. Among them are also creative intellectuals of considerable size, famous as caustic critics.

I cannot go further here into this topic – I have a detailed study of it in a volume that is unfinished because of discouragement; it is not likely to find a publisher *(Academic Agonies and How to Avoid Them: Advice to Young People on Their Way to Academic Careers,* unpublished). Let me cut things short and identify the old pros of the highbrow and the lowbrow culture. They are public relations professionals, even if their garbs differ in highbrow and the lowbrow worlds. They often are stars. They then can and often do serve their own public relations as well.

The highbrow old pros are critics or administrators or both. They occupy high social or organizational positions, and these are constantly threatened by competition: the free market in commodities that scarcely ever undergo rapid fundamental alterations differs from the free market in fashionable commodities – and all intellectual products, high or low, artistic or scientific or religious, are at times given to very rapid alteration of the tastes of their intended consumers. This frightens both producers and consumers; they need assurance; the old pros tend to provide it.

5. THE STRAIN OF RAPID CHANGE

Whereas lowbrow fashions are refreshingly unpretentious, highbrows pretend they can identify rubbish. Yet, they often extol rubbish as material of high quality. We can understand the conservatism and poor taste of so many established highbrow commodities at least as much as we can understand the fickleness and poor taste of high fashions in clothing. This is of little interest, and it is important only because in ousting what they think is rubbish they often unintentionally oust pearls, or throw away the baby with the bath water. By any metaphor they hurt young beginners and their potential audiences.

There is one important difference, not reflected in economic theory, between fashions proper – clothing fashions, hair-style fashions, and their likes – often known as fashions, and intellectual fashions, especially in science and in high art, where change is not by mere whims and unknown trends but in accord with rational and objective and fairly rigid standards of excellence. Repeatedly, physicists find that they have to know a branch of mathematics that previously was not deemed important. This is something that physicists can easily explain by reference to a new physical theory that makes use of this branch of mathematics that was previously not necessary for fashionable physics and that has become fashionable for some good reasons.

Economically speaking, tastes are exogenous. That is to say, they are given and they are not subject to discussion within economic theoretical discourse. And so, changes of taste too never enter economics. Fashion designers have no theory to follow, only hunches – they take great financial risks and the profession repeatedly adjusts itself to it. This is different for the clergy of every religion and denomination, except for the newest ones. They can take it for granted that tastes will seldom change so rapidly as to create problems. This is less obvious than it seems, since speed is a relative matter: people do complain about the speed of change, say, in the Catholic Church. Ever since the Vatican II Council, some say it is too fast, others that it is too slow. Perhaps the regulation of speed is simply the crude balancing complaints about a change. In the arts, highbrow and lowbrow, this problem looms large. Middlebrow works, including old warhorses, are perennially quite stable, of course. (Old warhorses are, by definition, the well-known, ever popular, works of art. They usually serve middle-brow tastes.) Moreover, slow or rapid, changes of taste are institutional: this regardless of whether they are rational. Some organizations have to recognize them and their degree of rationality, whatever it is, if they relate to mass markets. It is in the interest of both producers and consumers to keep things in check to some extent. The market does change; producers and consumers may fall out; but within some limits. This calls for some regulations.

Looking back on the history of any major revolution, we see certain simple patterns. The giants who have made the revolution have made – in the abstract – its major step that required a clean-up operation and much work ahead. The market need not reflect this. A large portion of producers – of art or science, highbrow or lowbrow, but particularly middlebrow – usually produce commodities of obvious mongrel origins, halfway between the old and the new. They never meet high standards, as they fall between stools, but the demand for them is strong. Here then is a clear function for intellectual rubbish that comes in the wake of all sorts of revolutions: it serves as means for transition over the revolutionary hiatus.

Hindsight misleads. It almost entirely obliterates the multitude of these mongrel works. Historians hardly notice them. They notice the explosion of the new, and at times they make an honorable mention of the die-hard who hold the fort to the bitter end. But they systematically and cruelly ignore the halfway house, even though it is very substantial. Noticing the halfway houses is, indeed, the mark of the thorough historians. Exceptions to this are the historians who advocate continuity. They have a wealth of material before them, and they often place the transition material before they present the revolutionary material, often insinuating that the transition material has paved the road to the revolutionary one, thus making revolutions appear much less revolutionary than they were. (They also make them appear more thorough than they really are.) This is intriguing since what they offer is fake history that contains no explicit lies. It is the kind of history that rests on the obviously false notion that continuity always obtains. When continuity does obtain, establishments, not the free markets of the ideas, maintain it and even then it breaks down when new establishments compete with older ones and win. Karl Marx masks this. Real revolutions, he said, are almost inevitably bloody: they comprise civil wars. His very usage of the word "revolution" suggests this error, and it caught on, although terms like "quiet revolution" and "peaceful revolution" belie it.

A glaring example of a masked discontinuity is the Copernican revolution. Copernicus, Kepler, and Galileo dominate the history of physics and/or astronomy of the century preceding Newton. Historians of science seldom notice that at the time Copernicus played a small role; they barely mention even the mongrel theory of Tycho Brahe. When they discuss the controversies that Galileo generated, they present them as Church politics, unrelated to the scientific establishment. (The University was then largely a Church institution, but that too these historians ignore.) Similarly, historians hardly ever mention the decline of Copernicanism into the state of almost total oblivion during the half a century or so between the death of Copernicus and the telescopic discoveries of Galileo. Students and young colleagues learn that in the first third of the seventeenth century Kepler was a leading scientist. This is a false impression that survives despite the detailed study of the leading historian of science, Alexandre Koyré, of how astronomers discovered Kepler decades after his death (*The Astronomical Revolution*, 1961, 1973).

A remarkable modem example of this kind of misleading historiography is the famous historian and philosopher of science, Thomas S. Kuhn. His still popular *The Structure of Scientific Revolutions* celebrates mediocrity by declaring normal scientists obedient people who tow the line. He noticed the die-hard and the few leaders, but he deemed followers the vast crowd of scientists. Leaders make them follow, he said, by persuasion or by force. There is no room in his philosophy for the halfway. Yet, he was a thorough historian and his study of multiple discoveries, so-called, records the multitude of halfway studies around the period of scientific change. He says these all contribute to the change. I say, It ain't necessarily so.

Kuhn called a bunch of near-contemporary papers on the same topic simultaneous discoveries, and offered no criterion of sameness. He saw sameness ever so often as he ruled out the half-hearted admissions of revolutions. He had no room for rubbish in his picture of science. This is foolish, especially since the half-way rubbish is useful for those who need help to comprehend the revolution. (See my *Faraday as a Natural Philosopher*, 1971.)

Most textbooks of classical physics published these days are variants of Max Planck's texts. They are no improvement on the original. When we come to modern physics things are no better. One of the leading texts on the theory of relativity throughout the first half of the century was a halfway text between Newton and Einstein. Also, Nobel Prize laureates, the Nazi physicists Philipp Lenard and Johannes Stark, were halfway physicists, as they could not admit as significant the work of a Jew. Lorentz and Planck, and finally Einstein himself, turned conservative. No doubt, the great thinkers among these five Nobel prize laureates had intellectual reasons, perhaps even Lenard and Stark had some, despite the racist arguments for their efforts to suppress ideas of Einstein; yet most science texts, professional, popular, and in between, are more often halfway than not. Robert Cogan has forcefully argued that this is so in music education (in *Perspectives of New Music,* 1973). The fact that teaching is often from halfway and muddled texts is not a matter of purpose but of weakness: not everyone can cross the gap between the old and the new in one step and so they fumble and will not give way. Mediocre producers also have to make a living, and they have many mediocre customers.

Here is the difference between discriminating and non-discriminating markets. The latter cater to a lower common denominator of their customers and so are coarser than the former, which are more diverse, expensive, and demanding. This difference holds generally, for the highest of the highbrow and the lowest of lowbrow, for art and science and journalism and religion. Discrimination requires much more than the investment of money and effort; it also requires talent – more than we have the right to expect. And so, establishments impede growth in order to protect the slow and the not so talented among those who occupy the higher echelons.

The specific to these markets is their mode of production: its products reach the trade market from intellectual professionals – artists, scientists, academics, and clerics – through a net of administrators. Excellent professionals are not likely to be good administrators too. The administrators of professional organizations are often poor all round: they prefer to neglect their own work and so they have great incentives to move to administrative positions. The pressure on administrators is usually unjust and they usually meet it in part by the age-old bureaucratic technique of all-round, systematic procrastination. This holds particularly for all sorts of editors of professional learned journals. This is very hard on young academics whose careers hang in balance.

So much for the clear-cut cases, where the able know what is what but others are slow to catch up. The more frequent cases lack clarity in advance as to who represents the new fashion that will soon become popular. (See my "Genius in Science," *Philosophy of the Social Sciences*, 5, 1975.) I was very impressed by reports in the leading French *Journal de Physique, Pure et Appliquée* in the first decade of the twentieth century on Planck's progress then. Clearly, those in the know knew he was worth watching (most physicists did not; the *Philosophical Magazine* was frankly hostile), but they could not possibly know that he was a winner. This transpired a few years later, which was soon enough. Many important physicists – James Jeans is conspicuous here – tried first to deride his ideas, and then they tried to dilute them to boost their own shoddy output.

It is difficult to see things clearly; time, effort, talent, and luck help clarify them. Here the mediocre may help. At least we all need the ability of the mediocre to comprehend. This is the best evidence that the king's new clothes are really there and worth looking at. Hans Christian Andersen noticed that the mediocre – the vast majority – would rather declare the naked kings well clad than admit inability to see for themselves. Without denying this, I wish, somewhat paradoxically perhaps, to add to it that the best test of the power of ideas is the common observation that to become popular is to become an institution; that this holds for all sorts of ideas; and that the process simplifies ideas and lets their chief characteristics stand out.

6. THE MAINTENANCE OF POOR TASTE

Finally, there are two kinds of answers to the question, why does the public prefer poor stuff to the good stuff? One of them is simple, one complex.

The simple answer is this. Developing good taste requires education. Superior material means more to the better educated. At times the poorly educated use superior material – notably, the Bible – as if it were mediocre; usually they hardly notice it.

The fairly complex answer is this. Producers of inferior materials look for the lowest common denominators of the purchasing public and thus they have a vested interest in their poor education. They are in cahoots with the innovators' organizations, and these protect their members at a cost. The cost is the permission to have poor creators appear as good ones. The profit from the maintenance of backward, poor, and militant educational systems outweighs this cost. It is no accident that the most advanced individuals in our society, the intellectuals, the academics, the members of the free professions, are organized in semi-medieval guilds and educated in schools to which access is limited, and where entry to the guild is through its educational system. There are exceptions, to be sure, since organizations of inventors are much less organized guilds. And the computer business on all of its many vast branches is still delightfully open, though less than it used to be.

And so, poor intellectual production methods dwell in educational systems, in professional systems, and in their interlocking. Their function is to justify the guild's school by making schooling obligatory, and these schools justify the guild by pretending that its leaders produce superior material when it is often mere rehash or even plain rubbish.

All this interlocks with one further function, the satisfaction of local demand, in all sorts of senses of "local". It may be for fashionable material in national garb, or for products of mediocre innovators – artists, scientists, priests, or journalists – of a regional origin, descent, or habitation. It may be for ethnic stuff, or for products of "our" innovators. There is the need to adapt general new ideas to local practices (local in many senses), with the aid of many illustrations that are small variants on them. This matter is very important but quite obvious. It plays a particularly important role in education and in the gathering of the membership of a poor local branch of national organization. It is backward and tribal.

7. A PLEA FOR TOLERANCE

Everything said thus far suffers from one great oversimplification, and it is time to dispose of it. I divide the representatives of the public interest into ideologues and established. The established are academics these days, in addition to functionaries of national associations and guilds and foundations and government agencies. They go for stability. Ideologues pour scorn and indignation and try to raise the standard or at least keep it high. As the role of indignation is to maintain stability, the two groups overlap. Often a person is both a representative of the Establishment and a guardian of high standards. The two cooperate: ideologues invent new techniques for maintaining and raising standards; the established try to help – and by compromises, if need be.

The victims of high standards may be some makers and producers of rubbish. Yet, as there is so much intellectual rubbish around, they are not doing too badly. Perhaps, then, the producers of good stuff are the victims, but there is no evidence for that. There remain, then, the innovators – in the arts and the sciences and their likes, good, bad, and indifferent, but particularly good. We need not worry about them either. It is the ambitious young hopeful, the one who is as yet neither good nor bad, whom the net catches first.

This solves another pet problem. There are golden ages, we know. What produces them? Wealth and fame, said Hegel, attract all the ambitious young people in the vicinity, those whose passions force them to try and try and try again. What makes a place wealthy and famous? Military success, said Hegel. This is false. The Florence that started the Renaissance was not more famous than Rome. What creates golden ages is a proper attitude towards innovation: encouragement. In all likelihood, young aspirants create mere rubbish. Yet, they need much encouragement and more tolerance. Florence was encouraging and tolerant, because its traditions and institutions were encouraging and tolerant. At least there was more encouragement and toleration in Florence than in Rome.

What makes for a Golden Age is encouragement, and what enables the old to encourage the young is not a good sense of discrimination and not a peculiar sense of kindness or generosity, but toleration and the understanding that young innovators do not threaten either accepted standards or conservative producers – not half as much as is normally feared. The mediocre need no protection against innovators. Also, in Golden Ages establishments produce incentives in the form of all sorts of rewards, not only to young aspirants whose output just may turn out to be significant, but also to those who encourage them, to talent scouts of all sorts. (See my "What Makes for a Scientific Golden Age?" in my *Science and Society*, 1981.)

This, then, is my plea for tolerance. It is for skepticism. We shall do better to remember that we do not know whether an odd young aspirant is going to be good. We swore that if we meet another Mozart, another Schubert, another Van Gogh, we will treat them better. We do not. We did not know. Many of those who were *avant-garde* in the last generation, whom we consider today leaders, died almost totally unrecognized or won recognition in old age. (Physics may be an exception.) Establishments still put obstacles in the way of innovators, and they advise all sorts of well-meaning people to have too much psychological self-doubt and not enough intellectual self-doubt. (We should always be open to criticism but this should not encourage hesitation.) Institutions of professional ostracism encourage this psychological self-doubt. We shall do better to observe that if intellectual rubbish were as dangerous as material pollutants are, we would have culturally suffocated long ago. So let us live and let live, and let us recognize and praise the talent scouts who were quick to support and encourage young hopefuls. Let us create incentives for talent scouting. To this end, let us abolish all the many penalties for having supported promises that did not materialize. We should not fear boosting both toleration and active support to the limit. If we will have reached the limit and find that it is dangerous, then we may decide then what to do about it. Meanwhile, we want incentives for creating incentives for encouraging young hopefuls. The future depends on them.

Chapter 2.3

SCIENCE AND THE INTERPERSONAL

The major shift in recent years in the literatures about science is from the view of science as the knowledge that an individual possesses, to the view of science as a social institution. Different social philosophies judge science as a social phenomenon differently. In particular, the social philosophy that deems society a mere collection of individuals – reductionist individualism or psychologism – will make the least of the recent shift, whereas the social philosophy that deems people mere members of society – reductionist sociologism or organicism or holism – will make the most of it. Without elaborating on this, we may note that both of these extremes play down interpersonal interaction. The various interactionist schools should make the most of it, but one need not be a member of any of these schools to appreciate interpersonal interaction. Indeed, Popper has stressed this as early as in 1935. How does the recognition of the presence and importance of scientific interaction alter our view of science?

1. A SCIENTIFIC IDEAL

Ever since science became a familiar aspect of society, there was fascination with the question, what is it? The answer was already there. Long before science existed as a traditional social entity, as an accepted social institution, there was the traditional, institutionalized ideal of science and tradition that generally identified science with that ideal as a matter of course. Hence, the question, what is science, was not what really troubled those who pondered on it. Perhaps that question may stand for another one: does science as we know it fulfill our ideal of it? If not, does it at least approach that ideal? Alternatively, we may be asking, what ideal does institutionalized science endorse and/or practice, and, are the two, the endorsed and the practiced, the same? The question, what is the ideal of science, fits very well with a general sociological approach, as one of the founding fathers of modern sociology, Max Weber, has presented it: ideal-type theory. Weberian sociology distinguishes regularly between what one endorses and what one tries to implement. Roughly, Weberian ideal-type sociology says, it is best to characterize a society, a sub-society, a culture, or a sub-culture, by both the ideals it endorses (and the expression of that endorsement in daily activities) and the ideals it tries to implement (in some activities).

A serious difficulty is lurking here. Take the ideal type of early Protestantism. It is the most famous one, since it was the object of Weber's best-known study (and indeed the paradigm of all specific ideal-type theories). Possibly, the idea it openly endorsed and the one that it lived by were identical. And then the true believer is distinguishable from the mere conformist by the fact that the true believer provided and sustained the openly endorsed and practically implemented ideal, whereas the conformist merely emulated the true believer. The opposite may also be the case. Perhaps Protestantism had a declared ideal that was radically different from the one it attempted to implement. So, it has two ideal types, the true believer and the conforming one. In any case, it demanded strict conformity. Its success in imposing it is what mattered. The declared ideal was just an idea that justifies conformity. It may well have been cosmetic, so that only the naïve true believers and the dim-witted endorsed it literally. The choice between these two options is often the outcome of an ideological prejudice of the one who makes it. This is very embarrassing.

Emile Durkheim, another founding father of modern sociology, took away this embarrassment. He considered all religion and ideology social cement. He considered its endorsement the mere *post hoc* justification of social cohesion. He deemed any ideology and any religion, whatever it seems to assert, in reality only an expression of the value of social cohesion. When Americans say they are one nation under God, then, they do not speak of God but affirm Durkheim's thesis that national unity is very important and that declaring belief in God is a way to serve it. Hence, one who is willing to compromise ideals for the sake of social cohesion is right, and one who clings to the letter of the ideal and causes social schism is wrong.

Bronislaw Malinowski, one of the founding fathers of empirical (field-work) social anthropology and the most influential social anthropologist to date fully endorsed Durkheim's idea. Loath to ascribe to primitive people the silly superstitions that they expressed when asked to present their views on diverse matters, Malinowski applied to their views a type of Durkheimian analysis. Both of them left open the question, could they regard science too as an ideology. They had to leave this question open, and for obvious reasons. An affirmative answer makes their own views no more than a kind of religious or magical ideology, a mere *post hoc* justification of their own activity that makes sense only as a social cement within the social setting in which they lived and acted. A negative answer, however, makes science the only true ideology, and thus, by a clear implication, it makes all nonscientific ideology inferior to ours – quite the contrary to their intentions. Durkheim and Malinowski wanted it both ways: they did not wish to dismiss magical and religious views as "wrong" yet they wished their own views to be scientific – more than mere social cement. They thought that what they said was true in an objective sense in which magical views are not true. (1)

No matter how, if at all, Durkheim and Malinowski accomplished their goal and had it both ways, the very wish to have it both ways indicated their awareness of the difference between a true believer and a mere conformist. For, they regarded conformity to religion as mere means for social cohesion. While regarding their conformity to science as true faith. (2) If ever a society existed whose ideology conforms to science, then, by this theory, that society would be the only one with an absolutely true ideology, where conformity and true faith are one. Perhaps in their view science is such a society. Hence, for every other ideal type, true believer differs from mere conformist. Nor is it hard to find out who is who. This raises the question anew: is the true ideal type the true believer or the conformist? Since conformists in matters scientific abound, the question also obtains for the case of the faith in science. Is the true practitioner of science the contributor to the cohesion of the scientific community even at the expense of the true faith? Or is the true practitioner of science one who strives for its ideals even at the cost of strife and schisms? (The admirable Joseph Priestley, the discoverer of oxygen, is the paradigm for this.)

The answer to this last question depends on the answer to the original question of this discussion. It is this. What is the ideal of science? Even a very superficial empirical investigation will easily reveal that scientists profess diverse ideals and conform to still more diverse ones. Perhaps the ideal-type theory is simply not good enough for the investigation of science. (3)

My idea of science is not an ideal type theory. So, I oppose Weber. I oppose the popular sentiment that he represents and his theory. Let me illustrate this.

2. THE SCIENTIFIC ETHOS IN ACTION

Consider a group of *bona fide* scientists – let us isolate them as much as possible, say a group of astronomers working in an astronomical observatory on a remote mountain. Of course, not all people in the observatory are astronomers. The observatory is a part of a station, and the station inhabits a few members of the nonscientific crew – maintenance crew consisting of caretakers, transporters, ground workers, and providers of food and of services. The two sets are worlds apart. The maintenance crew seldom mingles with the astronomers. The astronomer are, first and foremost, very busy, since they spend only a part of the year in the observatory according to a fixed schedule, and they must make the best of their short stay, especially since they are very ambitious and hard-working by disposition. They constitute an all-male or an almost all-male preserve, which enhances the distance between the two sets. The scientists are usually sexually abstinent for the duration, whereas members of the permanent crew have their own arrangements, which vary from abstinence to living with a family (especially if the couple are both members of the crew). Even the abstinence of the abstinent crew members differs from that of the scientists. The sexual abstinence of the scientists is free of ideology, but only apparently so. It seems free of ideology since it stays only for the duration, since most of them are family men and since, though avowed sexists, most of them are sexually quite liberated. Yet the ideology that brings about abstinence in the circumstances is a high morality that precludes sexual adventurism and fosters a dedication to work – the Protestant ethics as Weber has described it and as Robert K. Merton has applied it to the community of science. This is not to deny Lewis Feuer's critique of Merton, as it rests on the claim that the official ideology of the scientific revolution and of the Enlightenment Movement was hedonistic and utilitarian. Rather, hedonists often practice ascetic morality of one sort or another; the work ethic of the modern middle class permeates the community of science, much to its detriment and to the detriment of its scientific output. (They maximize input instead of output.) In any case, Claude Lévi-Strauss has opened our eyes to the possibility that scientists follow at times Merton more than Feuer and at times the other way around. When our scientists do get together, they talk shop, mainly astronomy, but also scientific gossip. The maintenance crew is much less educated, and hardly interested in science. They talk a lot about sex and sports. The two teams are, finally, hardly on friendly terms, though relations are always cordial as a matter of course. They must be if a small community is to keep efficiently busy.

From time to time there are gatherings. There are gatherings when shifts change, when large transports appear, and when someone heralds an unusual find. One morning, for some reason, practically all the astronomers sit together in the cafeteria and the conversation is rather animated. Somehow or other, one member of the maintenance crew happens to participate in the conversation. He is an odd ball, rather elderly and usually reserved. Yet now it seems that he is fairly familiar with astronomy and even has a somewhat clear idea of what the scientists are doing. He even follows details of their specific shoptalk.

The conversations across the large cafeteria soon die out; it turns out that they all find the situation rather unusual and they want to listen to the odd ball. As reticent people often do, he cannot resist the temptation of the spotlight once it rests on him. Before anyone realizes what is going on, he is addressing the august scientific crowd that just happens to be there, still munching on a last bit of breakfast or sipping a second or a third cup of coffee after a long, tiring work night. His speech is a clear exposition of a heresy, and not a new one either. He says, "You all endorse the theory of so-and-so, but I have reasons to think otherwise." And he offers to give his reasons.

The reactions to the little speech, once it is over, are varied but quite normal. They range from the very polite through the noncommittal to the rudely dismissive. What all members of the august audience share is the utter loss of interest. Nor can one blame these people. They are experts and he is no more than an animated amateur. Clearly, were he better trained, he would have started not by declaring his convictions, much less by dismissing their convictions to the contrary. Rather, he could easily find one of them who would tell him why they preferred their opinion to his.

Nevertheless, a second unusual event follows the first. One astronomer protests. "Let us hear him out", he pleads with his peers, "let us have him explain himself. Perhaps he has something interesting to say after all."

The rest of the story is sheer embarrassment. Most people say they have reasons and they do, but their reasons are usually poor. Unorthodox views have little chance for public exposure and thus for the development of good reasons in their favor. Orthodox views find their defense repeatedly on each street corner. They are orthodox because many reasons in their favor are popular and even good. Moreover, holders of unorthodox views have additional tasks that are far from easy. They have to discover good arguments in defense of their unorthodox views and then they must learn to articulate them well and, what is hardest, learn to find a way to catch the attention of their elders and betters.

Our elderly reserved maintenance worker is no different from other amateurs. He has a chance to give his reasons, and with all good will, one can only conclude that he had neither good reasons nor the gift of articulation. Oh, quite a few people in his audience tried to help him articulate his reasons – from discomfiture and out of good will. Yet, the more they tried the more the poverty of his ideas and his reasoning became apparent. It had to come to an embarrassed end. An uncomfortable silence prevails. As usual in such cases, the embarrassment turns into hostility – a very mild one, no doubt, but quite unmistaken, and the crowd directs it not at the elderly amateur but at the odd colleague who has insisted on the elderly fellow's right for a chance to explain himself. As for the odd astronomer himself, he sits in his corner, his eyes fixed on an empty plate before him, his hand fiddling with a fork, seemingly gathering some remaining crumbs. From time to time he shrugs a shoulder as if to deflect the glances that his irate peers dart at him now and then.

Of the whole learned scientific crew gathered that morning in the large cafeteria in the isolated observatory on a distant mountain, only one is true scientist. And while he sits there, in profound discomfort, absorbing the silent but forceful public censure that he has triggered, he might lose this fine quality, his partaking in the scientific ethos. I do not quite know what meaning to attach to his shrugging of a shoulder. He may be overcoming the censure of his peers, learning to live in an even greater isolation, yet keeping the spark of human curiosity. Or he may be succumbing to pressure and saying to himself, "They are quite right. We do not have the time necessary to listen to every passing amateur or dissenter." I hope not. I wish to encourage him and tell him how much I hope he remains a scientist to the last. Scientists, after all, are so rare, that the loss of even one of them is a great loss to the scientific ethos as a whole.

3. THE SCIENTIFIC ATTITUDE

Let me report that some scientists have found the situation as I have just described it intriguing and entertaining. They sympathize with my hero yet they tend to endorse the attitude of his disapproving peers. What does this signify?

In order to see this we may wish to take a step or two backwards. Matters of this sort are not very easy to view comprehensively. One may hold a broad commonsensical view of things, or one may hold a view that is more philosophical, more explicitly and consciously anchored in some principle, in some received philosophy of science.

The commonsense view of the matter is very commonsensical. It runs as follows. Although cranks have an unreasonable demand for our attention, at a rare moment they may have some interesting ideas and observations. Paul Feyerabend is a philosopher who has won much acclaim because he has noticed this. Every crank may have something worthwhile to offer, he rightly observes, and unless we study their contributions we cannot be sure that they are worthless and that we should dismiss them. Therefore, everyone has an equal right for our attention. The absurdity of this conclusion is manifest to everyone who knows that there is enough literature on parapsychology or on astrology or on flying saucers to engage anyone for life. Of course, the commonsense view of the matter thus invites us to glance at the literature before we judge it worthless and keep an open mind about it afterwards. But common sense offers no criterion about what level of investment of effort in that direction is wise. Nor can we expect common sense to offer a comprehensive view. For this, some analysis is necessary. The analysis may lead to a criterion that may help. The philosophy of science may have such a criterion in store. Importing it to common sense would be a revolution.

Classical philosophy of science was very much bound up with a certain attitude that the fathers of modern philosophy of science have advocated. Chief among them were Bacon and Descartes, in the first half of the seventeenth century. The Royal Society of London fostered their views in the second half of that century. It was this society that prescribed the scientific *bon ton*. The attitude they prescribed, however, was entirely individual, not at all interpersonal. It was open-minded eagerness. It was doubt about current knowledge and hope about future knowledge. There was a political aspect to this, no doubt, an egalitarianism as far as the human ability to learn was concerned, and non-sectarianism in intellectual affairs and in the choice of membership in a learned society. Tolerance followed non-sectarianism, on the understanding that everyone has the right to opinions and no one has the duty to listen to them. The willingness to learn does prescribe listening, but listening must be selective (for many technical reasons). What was the selection rule?

No discussion of selection rules for listening ever took place. The question arose first late in the nineteenth century and achieved a modicum of prominence in the early twentieth century. But the discussion on it centered on the selection of speakers in the public domain. It won little attention, since public servants, editors, committees arranging public lectures, and their like, supposedly catered to a multiplicity of tastes. The rule of the Royal Society, and thus of the commonwealth of learning, was simple. Learned periodicals were always open to reports of new, repeatable observations. Editors never kept this rule to the full, as they could not tell what observation is new, yet the rule was reasonably operative. Following it, however loosely, was beneficial. Theories were always welcome, if they were scientific, authoritative, deserving of attention.

This was – or should have been – the appropriate background for discussions of science and scientific character: was this theory or that theory worthy of the honorific status of science? The scientific attitude was at the background when discussions turned on the possible scientific character of theories. The mixture of a broad-minded skepticism and a profound quest for knowledge that supposedly characterizes the scientific attitude is precisely what made so typical such questions as, which extant theory is scientific and why? For, these were the theories to select as worthy objects of attention.

From time to time, science writers have noted a tension or a polarization between the skepticism and the optimism built into the scientific attitude. Skepticism may undermine optimism. This may serve as evidence that the scientific attitude served as background. But such evidence is hardly necessary since the scientific attitude – and the resulting tense polarization – is precisely what imparts value to the discussion of the demarcation of scientific theory.

The chief quality of the scientific attitude, then, is neither optimism nor skepticism, but the two checking each other.

The picture changed when the demarcation of the scientific attitude began to intervene with the demarcation of scientific theory. This happened early in the twentieth century, when (following Pierre Duhem) Henri Poincaré expressed an unorthodox view about science. He said, scientific theory stands in regular need of corrections and adjustments. He added (and here he made an astonishing contribution) that a theory remains scientific if and only if we correct it openly and not surreptitiously. Here a new factor joined the scientific attitude, openness, honesty, or transparency. Yet, one may cogently reject this claim and argue that this factor, the demand for openness or honesty, is traditionally a part of the scientific attitude and thus no innovation in the least. This is true, yet innovation took place all the same. Innovation is often the case of a known fact playing a new role in an argument. The innovation was in the claim that scientific status depends on attitudes, on the readiness to admit openly and honestly some shortcomings and to present contributions to science as modifications of its theories. This idea of Poincaré was new and hard to endorse. Not only is it hard to stay open and honest; it is hard even to see that openness and intellectual honesty is what makes science what it is. Wittgenstein expressed in his best-selling *Tractatus Logico-Philosophicus*, views similar to those of Poincaré, yet his attitude was naturalist, referring to statements in the abstract, leaving no room for attitudes. His philosophy of science was more popular than that of Poincaré, as it looked revolutionary while it was backward looking. When Popper agreed with Poincaré about attitudes, Wittgenstein's followers read Popper's ideas naturalistically and continued to ignore attitudes.

It is hard to pin-point the place of the scientific attitude in the philosophy of the leading twentieth-century school of philosophy of science, the logical positivists or the logical empiricists or the Vienna Circle and its derivatives, or in the philosophy of any other of the schools infected with Wittgenstein. His works display the attitude of a total commitment to utter clarity and to the hope that that clarity will lead to certainty. Any expression of skepticism and of hope that his devotees had, they directed not at scientific theories at all but at clarity. A philosopher had to doubt that philosophy has attained clarity and to abandon all hope that philosophy might ever attain it. This sounds as if the Wittgenstein devotees ignored philosophy and looked at science optimistically. This attitude is the classical attitude that classical positivists advocated, from Bacon to Ernst Mach. Positivism was science-oriented and thus optimistic until Wittgenstein took over. He and his diverse devotees addressed the internal affairs of philosophy. Even those who studied induction, showed hardly any interest in science or its workings. (They studied the method of justification of theories, as divorced from the method of their discovery.) Yet, they performed the shift away from the concern with science in the name of the continuity of the traditional positivistic hostility to philosophy and their endorsement of clarity as the heart of scientific rationality and as utter impersonality. They bamboozled almost all the philosophers who cared for science, as they managed to appear as concerned with science, despite much evidence to the contrary.

4. SCIENCE AND THE INTERPERSONAL

Popper's *Logik der Forschung* of 1935, or its expanded translation, *The Logic of Scientific Discovery* of 1959, always caught readers' attention. They found it hard to comprehend. His explicit and quite systematic coupling of attitudes and theories puzzled them. He demarcated scientific theories, not attitudes, yet he added a condition relating to attitudes. (He acknowledged the priority of Poincaré.) Theories are scientific, Popper said, if and only if they are refutable by empirical means (if they are false). This, however, depends on the willingness to subject theories to empirical tests and the readiness to admit their empirical refutations. If theories are abstract entities, then their status should not involve attitudes. Alternatively, if theories are crystallizations of attitudes, of views and values, then there is no need to mention attitudes. Yet, Popper claimed that theories are abstract entities that attitudes accompany and together they fix meanings. Today, only a few decades later, attitudes are very fashionable, perhaps because sociologists have made them respectable, perhaps because science as a social phenomenon has become the vogue in many circles. Attitudes about attitudes, too, are now legitimate. We have Poincaré and Popper to thank for this.

(Parenthetically, it was easier to conceive of Poincaré's philosophy of science as attitudinal than Popper's, since Popper is a realist who has to explain how our attitudes matter to Mother Nature, whereas Poincaré was a conventionalist, and unlike truth by nature, truth by convention is a matter of attitude.)

Popper's theory of science prescribes upholding the best available views and the best available information, while trying to overthrow them. This attitude seems more intolerably polarized than the classical attitude of doubt plus hope. For, it is that of hope plus doubt – hope that current theory is true, yet also doubt about it – hope and doubt about the same theory. The classical model resolves doubt and turns it into knowledge. Popper's 1935 model dissolves obligatory endorsement of a theory into its obligatory rejection, perhaps its obligatory endorsement as true coupled with the obligatory attempt to prove it false.

Is this not a neurotic nightmare? How did the cold clarity and comforting impersonality of science become a nightmare? The answer is in the very rejection of the impersonality of science and its replacement by the (conventionalist) concept of inter-personality. Putting Popper's theory in terms of interpersonal attitudes, as series of dialogues, as he did later in his career, seems a very adequate escape from this nightmare. Some people endorse some views, others endorse other views, and differing parties may attempt to engage in a critical debate. When they do, says Popper (*Conjectures and Refutations*, 1963), science may flourish. Seen thus, the tensions are no more than in any game. The game of science can be as serious as a war game. This is so when the loss of a debate leads to a loss of a job. So science must institute rules against such cruelties. They are due to misconceptions. Correcting misconceptions will positively contribute to the rise in the quality of life in the modern world.

This is the shift from the public or the impersonal to the interpersonal. It raises forcefully the question of the selection of people to engage with, of occasions for engagements, and so on. Whom should I take for a partner for a critical debate? What should be the agenda for our debate? And this question arises both individually and publicly. Whom should we hire for a research position? There is no need to assume that the public choice is no more than a multiple individual choice. The public selects editors, and they use their own judgment as to what the public should debate. The same goes for the elections of democratic governments. (Hence, both classical social philosophies – reductionist individualist psychologism, and collectivist holism – are in error: neither pays attention to problems that stem from attention to the interpersonal.)

More important, the interpersonal attitude is much richer and more significant for science than anything that writers about science have presented thus far. We may notice this while considering the attitudes preceding or heralding the scientific attitude. This consideration turns out to have quite a rich background.

The discussion of prescience is very popular ever since the advent of science. Historians of science are quick to observe that pre-scientific technology, as well as folk wisdom, and even popular mythology and superstition, contain much that agrees with the best scientific knowledge currently available. Sir James Frazer of *The Golden Bough* fame rudely set aside all this. At the turn of the twentieth century he branded magic pseudo-science and deemed religion intellectually inferior to magic. He viewed the intellectual development of any culture as the move from magic to science. Paradoxically, he viewed religion as an intermediary stage. He was extremely influential – especially in anti-religion positivist-rationalist circles. Frazer did not intend to belittle pre-scientific knowledge. How then did it fare as science evolved? Perhaps in opposition to him, perhaps not, a few writers declared science to have emerged out of magic. Some of them are very fashionable today, especially D. P. Walker and Frances Yates. Popper, too, has contributed to this trend, under the influence of Robert Eisler, a rather obscure post-Frazer historian of ideas. Rather than contrast pre-scientific magic and scientific technology, as most writers on the matter do, or magic and science, as Eisler, Walker, and Yates do, Popper contrasts myth and scientific theory. He advocates the view that science is refined myth. The conflict of opinion on the origins of science is going strong. It is manifest in the works of Claude Lévi-Strauss, the latter-day Frazerian, who agrees with Frazer and not with Popper as he distinguishes sharply between myth and scientific theory proper. He views myth as “the science of the concrete”. Similarly, Gellner, whose views generally accord with Popper’s on scientific method, sharply dissents from Popper on science and myth, declaring that there is a “great divide” between the enchanting world of myth, magic, and superstition, and the world of plain, “disenchanted” science. (The term is Weber’s.)

5. THE SCIENTIFIC TRADITION

The philosopher Michael Polanyi, an older contemporary of Popper, presents a very unusual view of science. His fame is now posthumously and correctly on the rise, especially since Thomas S. Kuhn has popularized his views. In his view science is a body of theories that the scientific community endorses. We cannot demarcate these theories in any other way than by the claim that the scientific community endorses them. Nor can we characterize the scientific community abstractly. We should observe it as a living tradition. A major debate in traditional philosophy is whether science begins with thinking or with observations. With neither, says Polanyi. To become a scientist, one must apprentice oneself to a scientist, live in the master's workshop, and absorb the master's attitudes, techniques, and knowledge. Were there a recipe for research, then one could perform it successfully by following the recipe. Were it possible to characterize science, then it would be possible to acquire it by implementing that characterization. The only way to join science is to join its tradition.

Polanyi's view is sociological. It is the view of science as an autonomous guild and a community (in Ferdinand Tönnies's sense). Two interesting questions regarding Polanyi's theory are worth asking. What are the antecedents of science? Clearly, guilds, guild mentality, and so on, have preceded science. So did, in particular, guilds with high expertise, of the arts, of nonscientific learning, and of clerics or orders or church hierarchies. Consider, then, the transition from these pre-scientific communities to guilds, and from guilds to scientific communities. What has effected this transition? What interpersonal relations characterize a guild in general and the scientific community in particular? Polanyi's theory offers no answer to these questions, except to say that the master-apprentice relation is that of an authoritative transmission of expertise, and that the scientific community includes the leaders and the led.

Not surprisingly, Polanyi's views find no easy, comfortable place within the community of positivist-rationalist philosophers of science. Adolf Grünbaum, a leading philosopher of science, has declared him an irrationalist and thus an enemy of science and of scientific philosophy. Yet he is hard to dismiss. Why?

Polanyi seems to say very little, and intentionally so. He says that science is a living tradition that we should not force into a simple schema. Not surprisingly, then, he offers no simple schema to characterize science. Yet, somehow he seems to accommodate many important factors that other philosophers of science ignore. In particular, his stress on the interpersonal aspect of science (in one conspicuous place he uses the term "I-thou relations," thereby alluding to Martin Buber (4)) is contrary to the traditional, positivistic philosophy of science as impersonal. This endeared him to the concerned with the social and political aspects of science.

What Polanyi does say, and he seems to be quite clearly right on this point, is that one cannot have it both ways. Philosophers traditionally viewed science as impersonal and abstract and utterly clear and certain and utterly this or that. And this may be very satisfying to all those who do not care too much about the social and political aspects of science. But science does have these aspects. Neglecting them was permissible in times when science was solely the concern of the individuals devoted to it. Today, when the scientific community is very powerful, it handles science in its own ways and governs it by its own traditions. To Polanyi's mind, they do so quite adequately. Since the world of politics got wise to the social and political importance of science, politicians have attempted to take care of these social and political components – with the intent to serve interests alien to science and to the scientific community. In self-defense, then, the scientific community must undergo a transformation, become aware of its social and political traditional autonomy, and, Polanyi declares, protect this autonomy from intruders. To that end, the scientific community has to relinquish its traditional view of science as impersonal and notice its highly interpersonal tradition.

Thus, Polanyi did attempt to effect a change; indeed, he attempted a revolution within the community of science. Yet it is no more than the revolution implicit in every counterrevolution. It is fortifying the traditional awareness of tradition.

Take the public agenda of any scientific discourse. Who determines it and how? Common sense takes care of this piecemeal and with no hard-and-fast criterion. Most philosophers of science ignore the public aspect of science and with it its public agenda. Polanyi has argued that any criterion can, and at times does, clash with scientific practice and with the commonsense of the leaders who carry it out. And then, in his view, this common sense should win. Not that leaders are always right. Polanyi himself was once in the right yet his new idea failed to register for a few decades. When finally it did, he did not say, I told you so. He said instead, this is as it should be. (5)

Nevertheless, Polanyi is in error on the very aspect on which his contribution is valuable, namely, on the social and political aspects of science. The tradition of science is open and democratic, whereas he saw it as an elitist workshop system open only to young, dedicated apprentices ready to accept and uphold the tradition unconditionally. Elitism brings about power and power corrupts. (6) Social science examines the common sense of diverse social sets and subcultures. Popper's pluralist theory of science is a good shot in that direction – especially his theory of the interpersonal relations in science as friendly-hostile. The defense of the common sense of science is either common sense or science. And when common sense and science compete, common sense yields. (This is so by definition: our concept of commonsense is normative.)

Different kinds of interpersonal relations among scientists obtain, indeed a great variety of them, some more becoming to science, some less. Hence, it is impossible to decide by observation alone which relations among scientists are more scientific, which are less, since such a judgment is more normative than factual. A social scientific study may help. Polanyi should have agreed but he inadvertently did not.

6. SCIENCE AS CURIOSITY INSTITUTIONALIZED

Some interpersonal relations are more becoming to science than others. We may examine all the aspects of the attitudes that we ascribe to people in pre-scientific society – toleration, patience, non-defensiveness, and a disinterested readiness to consider ideas that at first look rather absurd. Certainly, all pre-scientific wisdom literature, religious, and secular alike, commend these attitudes.

What constitutes science cannot possibly be exclusively a matter of interpersonal relations and attitudes. Curiosity about the nature of things, for one thing, is a personal sentiment, and it is essential to science. Only intense curiosity leads to readiness to listen to "crazy" ideas (to use Niels Bohr's apt term). Is curiosity sufficient for the development of the attitude of disinterest? Since the interpersonal aspect of science is not new, nor is curiosity, one may follow Popper and Polanyi together and consider science as the crystallization of certain attitudes, personal and interpersonal, into specific social institutions and traditions. When the institutionalized scientific attitude has any measure of success, it has the fate of any other institutionalized arrangement and accrues additional characteristics and draws different kinds of people – different in attitudes and in interpersonal relations – some beneficial, some detrimental. Science as an institution, then, is a mixture of curiosity and conformity. (7)

What, then, has happened to myth, magic, and all that? Do they vanish when the scientific attitude crystallizes, as Weber and Gellner have suggested? Is there a level of sophistication and of critical attitude that utterly disenchants? This is questionable in view of the fact that science and magic do mix – even in the best scientific circles. Yet if the scientific attitude is not only the readiness to be self-critical but also the readiness and the eagerness to entertain other people's views, then this in itself causes disenchantment of a new kind. Anthropologists exhibit this attitude towards the magical superstitions of pre-scientific cultures. They find that the very arbitrariness of so much that is magic gets in the way.

The addition of the interpersonal aspects of science to the personal and the social thus helps solve old problems and raises new ones.

NOTES

1. For the philosophy of Malinowski and his egalitarian impact, see I. C. Jarvie, *The Revolution of Anthropology* (1964). A school of thought, if they merit this honorific title, called the Strong Programme, claims for science the status of both an ideology and the truth. This, of course, is a part of every ideology. So it does not count. All of their output and all the many debates around it have almost no value at all, especially since their writings are often so obscure that it is hard to fathom their meanings.
2. See the concluding remarks in Emile Durkheim, *The Elementary Forms of the Religious Life*, trans. Joseph Ward Swain (1947). See also my *Towards A Rational Philosophical Anthropology* (1977), for a discussion and a lengthy criticism of both individualist and collectivist reductionism, Ernest Gellner, *The Legitimation of Belief* (1974) and *Cause and Meaning in the Social Science*: ed. I. C. Jarvie and Joseph Agassi (1973), and Mario Bunge, *Social Science Under Debate: A Philosophical Perspective* (1998).
3. See J. W. N. Watkins, "Ideal Type and Historical Explanation," reprinted in an expanded version in *Readings in the Philosophy of Science*, ed. Herbert Feigl and May Brodbeck (1951, 723-43). He identifies two Weberian ideal-type theories, individualist, and collectivist. The two cannot merge. See my "Bye-Bye Weber", *Phil. Soc. Sci.*, 21, 1991, 102-9.
4. See Michael Polanyi, *Knowing and being* (1969), 149.

5. Michael Polanyi, *op. cit.*, Ch. 6. For more details and a critical examination see Chapter 15 of my *Science and Society* (1981), a volume dedicated to his memory.
6. Lord Acton's celebrated dictum, "Power corrupts; absolute power corrupts absolutely," refers to the power of the papacy. He was a committed liberal, and an ardent but critical Roman Catholic.
7. Bernard Shaw said, every bandwagon attracts some rabble and mixed multitude. See, for example, the end of his preface to *Androcles and the Lion* (Selected Plays, 1:848-49). More important, successful movements alter as they succeed. Science militant speaks more of curiosity and science triumphant speaks more of achievement. Science triumphant is an unavoidable social nuisance, yet it has the same appeal as other successful movements. See, far example, George Orwell's comments on the appeal that Stalinism had for intellectuals and Damon Runyon's discussion of the underworld complex. (Sheldon Richmond aptly refers to all this as "pistol envy".) Polanyi's view of science as a tradition enables him to observe the negative in science. Regrettably, he dismissed it, saying that imperfection is unavoidable. Popper also observes it, but he could not possibly ascribe it to science in the first place. On the whole, Popper's transition from the monism of his *Logik der Forschung* (1935) to the pluralism of his *Conjectures and Refutations* (1961-63) is not clear-cut; he died before he finished his task. Clearly, his view of science as a tradition could make him notice the undesirable aspects of science – especially the triumphant accent on successful prediction. I have discussed all this in my *Science in Flux* (1975), a volume dedicated to Popper, which includes a detailed, sympathetic, but critical examination of his contributions. It is a constant source of regret to me that he chose to ignore my writings altogether.

Chapter 2.4

SCIENCE AND PLURALISM

... the only consensus we ought to be worrying about is ... heterogeneity or "dissensus".

Jean-François Lyotard

0. INTRODUCTION AND A DETAILED SUMMARY

Question: What renders a difference of opinion reasonable?

One unreasonable answer and a few reasonable ones: take-it-or-leave-it is unreasonable; reasonable are classical monistic scientism, classical relativism, monistic progressivism – inductivism or critical rationalism – and, finally, pluralist critical rationalism.

1. Take-it-or-leave-it – the demand of bullies to not dare dissent and to consider those who do as having vanished. (German philosophy, so-called) Bullies do not tolerate dissent.

2. Classical rationalism or the scientific tradition or scientism – the recommendation to adhere to science and only science (Parmenides, Plato, Bacon, Descartes, Kant, Laplace). At most one answer for any question is legitimate. Its competitors are all equally worthless. Dissent is tolerable only until the discovery of the right answer.

3. Romantic reactionary or traditionalist Relativism – the view that there is no finality, that legitimacy is relative to space-time regions, each with its own criterion of choice. The choice of the criterion rests on faith (Montaigne, Bayle), on sheer arbitrariness (Vaihinger, Feyerabend), or on tradition (Hegel, Cassirer, Malinowski. Evans-Pritchard) as repeatedly modified by wise leaders (Buber, Polanyi, Kuhn) who impose it on the public (Hegel, Kuhn). Hence, dissent is permissible only between parties separated by space or time, and none within one party.

4. Monistic progressivism, whether inductivist or critical rationalist – the alternative to relativism: choice alters, but not due to arbitrary criteria of choice: experience confirms the theory that becomes the received opinion. A new theory replaces the received one only when it received higher empirical confirmation. The last modification of this is due to Popper: confirmation is corroboration, namely having stood up to severe tests.

Unanimity is thus not arbitrary but in recognition of progress. The best theory prevails as long as it is the best. The great advantage of Popper's view as he propounded it in his early works and in some of his later ones, is that he discussed refutations. He proved that the calculus of probability cannot serve as a criterion of choice of scientific theory. Instead, he offered for corroboration other *desiderata*, particularly openness to criticism and having successfully stood up to it.

Progressivism is also the view that Thomas S. Kuhn endorsed when he insisted that the scientific revolutions that he called "paradigm shifts" are rational – he never said how – but not when he declared himself a relativist. He tried to fuse relativism with progressivism. He had to fail: the two are in conflict.

5. Pluralist critical rationalism, pluralism for short, is the view that different simultaneous answers can be legitimate (rational, reasonable, putatively true), but not all: only those meeting current *desiderata*. Those that are current in science are unusually difficult to meet, and so science looks monistic. Also, different *desiderata* may exist, and then, again, more than one may be legitimate. One *desideratum* characterized pluralism: there is one final answer to every question – even though disputes are seldom totally settled and finality is seldom reached. This *desideratum* is prevalent in many concrete cases in operative ways. This invites a discussion, one that is not easy to provide.

Pluralist critical rationalism, pluralism for short, is the only view that has as a feature the refusal to impose belief in the name of rationality. It does recognize and value agreement on matters of fact and of action – the desirability of a consensus in matters practical – but not unanimous belief.

Those who advocate rational unanimity disagree about the criterion for it. This is amusing. Relativism allows different criteria as long as they do not mix, but it cannot guarantee their separation. (It may have trouble with their interface.) Unlike monism, pluralism does not address the choice of objects of belief. Monism is a disapproval of disagreements: it requires their quick settlements. Pluralism favors situations with more than one rational answer to a given question, as grounds for rational debates, without demanding their rapid settlements. Pluralism differs from relativism and agrees with monist critical rationalism in proposing that some rejected answers are not rational (even after new and better criteria of rationality emerge). Pluralism agrees with relativism and differs from monist critical rationalism in allowing for different criteria of choice. Pluralism differs from relativism: the one grades criteria of rationality, and the other refuses to do that. (The incommensurability thesis of Duhem and Evans-Prichard declares it impossible.) Recommending the grading of criteria, pluralism differs from all other versions of progressivism. Pluralism takes rationality as a characteristic of actions (including deliberations and debates), never a matter of choice of belief (Jarvie).

This raises a few intriguing questions for pluralists. As views are not candidates for belief, in what sense can they be rational? What makes entertaining a view rational? How does entertaining a view differ from belief? Is there a rational choice of a belief? Why is science apparently monist? How and to what extent can and should pluralism be the approved practice? Finally, is the choice between monism and pluralism not a matter of rational choice of a view to believe in? If yes, are we not back in square one?

These questions acquire a new color once we allow rational disagreements. These lead to debates that accord with recognized rules that accord with the best available *desiderata*. (The commonsense rules of debate of normal democratic societies should suffice.) Supposing all this is initially reasonable, how does it work in practice? We do not always have alternative answers to given important questions, particularly not where the standards of requirements from alternatives are very high. This is the case particularly in physics. After we take care of these safeguards, pluralism becomes for a while compelling or unproblematic. I will now discuss this compulsion.

1. PLURALISM VERSUS THE TAKE-IT-OR-LEAVE-IT ATTITUDE

An idea may be compelling or unproblematic: it seems convincing. Failure of efforts to criticize it may enhance this sense of conviction. Whatever this sense is, it vanishes in the light of new critical efforts. This remains even after the criticism meets with a reasonable answer: a message may remain convincing even after it becomes controversial, but, as long as reasonable controversies regarding it go on, it is not compelling or unproblematic. Racists, sexists, and their likes, may dispute compelling or unproblematic messages nonetheless. Newtonians, Marxists, and others have expressed eminently reasonable dissent from the consensus. They may express some reasonable dissent in some future day. Not today; not as long as they repeatedly approach criticism defensively.

Some public figures manage to convey the idea that their views are compelling or unproblematic: any dissent from them, they insinuate, is offensive: it is ignorant, wicked, stupid or irrational. Those who take this seriously should notice that it invites an obvious sort of questioning: why must dissent from them be ignorant, wicked, stupid or irrational? Nevertheless, at times they manage this way to suppress questioning and criticism. And as long as no one questions an idea, the claim that it is compelling or unproblematic may help possibly restrain questions and criticism. Definitely not otherwise, though. Hence, a claim that an idea is compelling or unproblematic is never compelling or unproblematic. At most, it is a fact. (Let us ignore here mathematics and logic.) The claim that an idea is compelling or unproblematic may serve as a challenge to render it weak and problematic. These public figures may relent or ignore the criticism. The oversight of critics and the critical attitude by suggesting that one's view is compelling or unproblematic, Popper has labeled "the take-it-or-leave-it attitude". This label is apt: authors cannot force readers to agree, and readers are therefore at liberty to disagree. Authors need not explain, much less justify; they need not attempt to convince; they need not argue at all. Readers may still want explanations. To refuse them, says Popper, is irrational. Rationality, then, requires of authors to explain why they say what they say, in order to enable their audiences to judge for themselves. The error of the take-it-or-leave-it attitude, then, is the insistence on the view that some ideas are compelling or unproblematic: it is an empirical matter that some statements are compelling or unproblematic and that some presentations of them are compelling or unproblematic. It is an empirical matter, not a matter of one's say-so. Readers who can judge for themselves are intellectually autonomous. Refusing their request for explanations is ignoring them. It leaves authors with the heteronomous readers only. This suits the authors who impose their judgment of what is compelling and unproblematic. This is their liberty.

That some assertions are compelling or unproblematic is no news: G. E. Moore used it repeatedly as the cardinal argument against skepticism. He constantly harped on compelling or unproblematic assertions of one sort, those which everyday experience is full of. (The external world exists: there are at least two external things, my right hand and my left hand.) Avrum Stroll has recently suggested that as scientific assertions are precise, they are less compelling or unproblematic than everyday ones. Copernicanism is a clear-cut counter-example to Stroll's suggestion, as are all scientific assertions that run contrary to everyday ones. In the whole history of science, there is only one compelling and unproblematic basic claim about scientific character: reports of well-attested repeatable observations that are not overturned by other reports of well-attested repeatable observations are compelling; their observed repetitions and imputed repeatability make them scientific (Popper, *Logik der Forschung*).

The ancient skeptics of Pyrrho's school discussed the compelling or unproblematic character of some theories. They recommended that, against any compelling or unproblematic theory, one should seek a contrary theory and try to render it equally compelling or unproblematic. They did not disagree with the ancient dogmatists about the facts of the matter; but whereas the dogmatists preferred situations in which doubts are settled and opinions are set, the Pyrrhonists preferred the contrary option. Unfortunately, the root of this is not a matter of preference. Both parties were after the same result, namely peace of mind. The Pyrrhonists asserted dogmatically that the suspension of judgment brings peace of mind, and their dogmatist opponents asserted the opposite equally dogmatically. It is a compelling or unproblematic fact that, of those who practice dogmatism and of those who practice Pyrrhonism, some but not all have attained peace of mind.

It is not peace of mind but rationality that is at issue. Some dogmatists are irrationalists. The rationalist dogmatists (including Catholics) agree that we should settle doubts: we should not suppress them but discuss them openly until they vanish. Some problems stay.

2. PLURALISM VERSUS MONISM

Popper's opposition to the take-it-or-leave-it attitude, then, is for the adoption of the rationalist attitude. The take-it-or-leave-it attitude is irrationalist because authors may have to explain themselves. The idea that an explanation renders a message compelling is monism. Initially it was scientism, but some monistic schools retained it while foregoing scientism. Popper advocated monism in his first book on scientific method, his celebrated *magnum opus*, his *Logik der Forschung*. In his *The Open Society and Its Enemies*, he demanded explanations, and even good ones – so good, that they should impose consent. In some of his later works he said, I need not explain myself, I only need open myself to your criticism: if you disapprove of what I say, please try to show me my error. Obviously there is a middle way. We should explain ourselves as monism requires, and open ourselves to criticism as critical rationalism does.

I do not know if I need discuss Popper's view here or move on from his trailblazing critical rationalism to the pluralist critical rationalism he also advocated, but let me dwell on his view for a little while for the sake of contrast. It is easy to reconcile the two attitudes I am ascribing to Popper, the one that requires explaining oneself and the other that requires inviting criticism. I shall do so by offering the following *desideratum*: the best answers to given questions are those that are open to criticism. Popper's refusal to answer those who asked for justifications of his view in order to help them reach agreement, and his invitation to criticize him, thus becomes an adequate friendly justification rather than a harsh demand to offer proof or an aggressive take-it-or-leave-it attitude. And there is much to say for this reconciliation, since it replaces the harsh attitude and the aggressive, anti-critical one, with a friendly invitation.

Nevertheless, Popper's answer invites modification. To begin with, it is most obviously unacceptable in logic and in mathematics, where proof is positive justification, not a justification of a view merely as open to criticism. Mathematics is the field where scientism seems to work, and in his earlier writings, Popper endorsed this view. Admittedly, scientism in mathematics, usually called foundationism, was never free of trouble. Nevertheless, until Imre Lakatos appeared on the scene, no one seriously questioned foundationism even while admitting its problems. (W. V. Quine has called himself a non-justificationist and offered what he deemed the severest critique of justification in mathematics, namely the absence of a clear-cut division between the analytic and the synthetic. Even he did not attack the requirement for positive justification or proof in logic and in mathematics, though in his review of Lakatos' classic *Proofs and Refutations* he expressed agreement.) We should exclude logic and mathematics from the present discussion. It is not easy, as Lakatos has shown. Mathematical proof, he says there in effect, is the positive way of provoking criticism. Merely expressing an invitation for criticism, then, is not enough: one has to offer incentives for it. A theory presented as good is more of a challenge to a critic than a theory presented as merely open to criticism. A theory is good if it complies with certain *desiderata*. There are examples of claims that are open to criticism, but are not worth criticizing (are beneath criticism). Scholastic literatures of diverse times and cultures are critical in some ways and yet are of little value. They do not merit critical attention. The Catholic Church, the Communist Church, and advocates of positivism and scientism use precisely this difference: they provoke critical debate about their defunct views: it keeps them in the news and gives the impression that criticizing them is a worthwhile activity. The same holds for advocates of induction who keep churning out old rubbish in the leading philosophical periodicals. It is like playing chess after the checkmate. Even Popper fell into the trap: he kept inventing new criticisms of scientism in general and of inductivism in particular, without asking whether this activity deserved his attention. Initially, he determined his monism by his chief *desideratum* that orders priorities: choose among the available theories the one that is most open to criticism, he advised. This includes a part of the proper demand to reject *ad hoc* amendments. But the demand holds even when there are no alternatives to them: Einstein ignored the extant correction to Newton's theory of gravity as too *ad hoc*, even though it is also open to criticism.

Popper saw himself as a pluralist. I do not know what this means in the light of his not having withdrawn his requirement to choose from among competing theories the one that is most highly testable. As long as his requirement of the choice was of the choice to test, it is fine. But at times he spoke of other choices. This requires a contrast of the pluralism he later advocated with the monism that he had advocated in his early masterpieces, his *Logik der Forschung* of 1935, and in his *The Open Society and Its Enemies* of 1945. He never spoke of a change of mind on this matter. He assumed that we can or should settle all doubts by the use of some ingenious criteria. This is an excessive demand: as he stressed, criticism is inspired and will not turn up to order. In any case, fighting doubt is always unwise.

Scientism and the take-it-or-leave-it attitude are the old extreme alternatives. Scientism demands the acceptance of one and only one justifiable answer to any given question. It is unwise. Justification, any means for making the choice of one theory obligatory, makes rationality compelling and unproblematic. The skeptic critique, then, defeats it. (This critique does not aim at any criterion of justification that is not compelling or unproblematic. We will return to this.)

Scientism has a simple message: skepticism is answerable. It presents this message as compelling or unproblematic. Millennia of failure to answer it, then, do not count. Suppose that this is right. As most views are not compelling or unproblematic at all, what is the rational attitude to them? Can an answer to a given question become compelling or unproblematic? If yes, then the question is, how? This is the problem of induction. It is a compelling or unproblematic question, yet Popper dismissed it, and even for reasons that he deemed compelling or unproblematic. There is some justice to Popper's dismissal. Suppose that the empirical evidence of empirical science is compelling or unproblematic. Does this hold for theories? Lavoisier says, do try to criticize my views as your failure to criticize them will impose them on you as compelling or unproblematic. Popper seems to have agreed. But is it compelling or unproblematic, asked Heinrich Heine, to admit every compelling or unproblematic theory? Or is the endorsement of scientific ideas compatible with attempting to criticize them, as Popper has implied? This is open to doubt, and hence Popper's dismissal of the problem of induction is not compelling or unproblematic.

We are badly stuck.

Admittedly, in ordinary circumstances we seldom worry about answers that are not compelling or unproblematic. Here is a compelling or unproblematic fact: most ordinary thinking involves taking for granted much that is not compelling or unproblematic in the least, merely because it is common practice. This has roots in common law, as it condones error that is widespread in the community. Most ordinary thinking, thus, is not good enough. It is a plain, commonsense, empirical fact, compelling or unproblematic, that some fairly accessible expert advice can help improve upon many important items of ordinary thinking. This leads some thinkers to the suggestion to listen to experts more, to learn more, to help disseminate scientific knowledge more. This is lovely. Yet, it raises a question: does science justify its claims better than commonsense? By what criterion? Is that criterion justifiable? Is it compelling and unproblematic?

There is no need to go into such matters. Suffice it to draw the terribly important corollary that is at hand already as to the matter of choice of answers to given questions: some of the answers that science offers and that serve as grounds for decisions are not compelling and quite problematic. We may hope that as science grows, it will answer unambiguously an increasing number of questions. This leads straight to the question, will science have all the answers one day? The classical positivists said, yes. Modern, logical positivists went further and said, whatever question science cannot answer, cannot have proper wording and so shall remain for ever unuttered. This is wild fantasy (thinly masked as logic), and it is besides our present concern that is pluralism.

The pluralist answer is commonsense. The take-it-or-leave-it attitude is objectionable. The alternative to it is to explain one's position reasonably well (Bunge). Reasonably good explanations are not that good as to exclude dissent. The question, then, is, how good should an answer be? This question, too, needs reasonably good answers (Jarvie). This does not lead to an infinite regress as long as we do not demand a perfect answer (Settle).

3. RELATIVISM IS NOT PLURALIST ENOUGH

Pluralism has entered in the middle of the twentieth-century stealthily. It permeated first politics and then political thought and then philosophy. It is extremely popular, yet it has no claimant: it has no declared originators. There are a few reasons for this. Let me discuss only those that have to do with the philosophical aspect of the issue. Let me begin with relativism as a forerunner of pluralism.

The first appearance of some sort of pluralism in philosophy was in the writings of the pragmatists, especially William James. He wanted to permit religious freedom. He noticed, all too painfully, that the criterion of rationality that science employs – as he understood it (in the strictest empiricist fashion, by the way) – precludes any diversity of opinions and so by implication it is religiously intolerant, permitting in principle one religion at most. He offered relativism to permit each culture its own choice of religion. This left him with the obvious question, how does scientific monism square with his relativism? He was aware of this question, of course, as it is too obvious to overlook and as he answered it: Darwinism applied to philosophy permits only those systems of thought that are not too destructive for their holders. This is a compromise between monism and pluralism – exactly what James wanted. Whether this compromise is satisfactory, I will not say. Rather, I will attack it politically in the manner that made it clear that pluralism must be more than the diversity of societies with different civilizations / cultures / religions: the idea of relativism permits different systems side-by-side, but forbids their intermingling and it forbids their improvement through criticism and comparison: it does not allow individuals to dissent from their own civilizations / cultures / religions, and so, it is not tolerant, all good intentions notwithstanding.

The case of the melting-pot versus ethnic pluralism, taken philosophically, is very interesting. Scientism requires a universal melting-pot. Relativism allows and even requires local melting-pots. Pluralism allows and even recommends diversity. One may object that ethnic pluralism has to do with mere lifestyles, not with views and values. This objection runs contrary to the received opinion that all too often the clash of lifestyles is a clash of views and values. This is painfully obvious from the case of the struggle against the plight of women who suffer severe discrimination and bodily mutilation on the excuse that the injustices towards women are parts of the complex lifestyles to which they belong, against which we claim that women are equal.

To show the specific to pluralism, let us notice one more example, more hilarious than tragic. Followers of William James prefer general relativism to his monism regarding science. They also see democracy as relativists, as good only for democratic societies. They take the toleration of democracy for granted, but they do not preach it, least of all for societies with non-democratic traditions. Nor do they view controversy as essential to science. On the contrary, they assume that science commands unanimity. It is easy to argue against this from historical evidence. Nineteenth century physics included vital studies of electricity along two lines, the Ampère-Gauss-Weber ruling idea of action-at-a-distance, and the Faraday-Kelvin-Maxwell minority view of action in the vicinity, of field theory. This is a bold example for pluralism in science.

The pursuit of the relativist line of thought may help clarify the contrast between it and absolutism. Relativism may allow for disagreements as characteristic of transition periods. It would be cheap to respond to this with the (true) observation that all periods are transition periods. Instead, the response might be this. There is no relativist theory of transition: by relativism each system comes with its own criteria and rules by virtue of these criteria. Transition periods abide by neither the old criterion nor the new. Possibly, all changes are subject to a meta-criterion; but this idea is absolutist. Possibly, old criteria give way to new ones in recognition of their rational superiority. This idea is absolutist too. Possibly, then, all changes are irrational. This is the conservative view. Its application to science is unpleasant: each move within science is supposedly rational, especially if the same move also happens outside science. To say that changes are generally not rational should lead to a new theory of change. At the very least, it is eminently rational to view as rational one transition outside science, the transition from magic to science. This is extremely compelling and unproblematic. Gellner has argued against relativism from the rationality of this transition. Relativism renders the rationality of this transition impossible, since the magic system is hardly changeable and being pliable it is capable of absorbing any change with minimal effort, so that there is never a need to relinquish it. By the relativist theory of rationality, it is irrational to abandon a worldview which always recommends itself and is intellectually very satisfying come what may.

The only way out is the adoption of a meta-criterion that is a-rational, like plain survival. An unacceptable criterion perishes in the struggle for survival; but more than one criterion may survive, and more than one system; thus, relativism allows for the survival of competitors – meta-criteria, criteria, and systems.

How does a scientific system vanquish its predecessor? Suppose it is superior in some sense. Then in that sense we can compare systems and in what sense is this relativist rather than absolutist? The answer is obvious: it is relativist on the assumption that the truth is relative. Is this truth (that the truth is relative) absolute or relative? This question relativists deem faulty, since the relative truth is not a criterion but a meta-criterion. Fine. Are meta-criteria absolutely or relatively meta-true? No answer. It is a case of take-it-or-leave-it and one that looks rational as it is initially rational but when the discussion about it progresses far enough, it becomes a case of take-it-or-leave-it.

This explains the enormous rational appeal of many irrational practices, such as the challenge to criticize an increasing number of arguments in favor of a defunct view. According to the pluralist critical rationalism that is publicized here, one may advocate rationally erroneous views, of course; one may advocate rationally even views that were subject to effective criticism – as long as there is no clear idea as yet of how strong the criticism is and how radical. The fact that relativism may be sophisticated and require lengthy discussions before its weakness becomes clear, makes advocating it eminently reasonable, especially to those who do not have a sufficiently long breath to run the whole course of the debate and witness its eventual collapse.

This explains why it is rational to spend efforts on rendering a critical debate as short and comprehensible as possible. And briefly, given that classical or uncritical rationalism (the doctrine of rationality as proof) is false, relativism is the simplest alternative to it. Clearly, an alternative was desirable ever since Einstein's theory of gravity replaced Newton's theory of gravity. It was clear then that Newton's theory is not the last word in physics yet it is far from being worthless. The competition between the two theories was short lived. This suggests an argument in favor of relativism and against pluralism: monism explains why once Newton's theory reigned and then gave way to another theory. Relativist monism suggests that Newton's theory never had a competitor. According to it, as soon as Einstein's theory appeared, it won without contest – simply because it did not oust Newton's theory. This is simple, but false: a crucial observation tipped the scale. Einstein did oust Newton's theory, but not its value – intellectual or practical.

4. PLURALISM CONSIDERED

Let us start with examining pluralist critical rationalism by checking its ability to solve the pragmatic problem of induction and/or rationality. Some followers of scientism, notably John Worrall, have claimed that it fails here. He rejects critical rationalism, claming that it does not support the preference for walking down stairs over jumping off windows. This preference is compelling and unquestionably rational. Hence, either Worrall is seriously in error, or he identifies a serious defect in the theory. For, a valid argument that appeals to a compelling, unproblematic fact is compelling. For Worrall's argument to be valid, his demand should be valid. He demands of philosophy to justify compelling facts. Is this demand compelling? If not, perhaps he should justify it. How can he justify this demand? Why does philosophy have to justify compelling and unproblematic facts? What counts as such a justification? Considering the compelling and unproblematic nature of some facts, should philosophy explain or reinforce it?

This is a crucial junction in philosophy: if philosophy should reinforce the compelling and the unproblematic, then philosophy performs a gratuitous and possibly harmful task while ousting pluralism. The wish to explain a compelling or unproblematic fact, however, raises the question, what are the *desiderata* for such an explanation? In particular, different reasonable explanations for it may compete. Such explanations cannot all be compelling or unproblematic. Explaining a compelling or unproblematic fact may even render it not compelling or problematic at all. This is a point of Lakatos: a way to criticize a theory, he said, is to try to explain it first. For, it is a compelling and unproblematic fact that the compelling and unproblematic is not always true, and explaining why some instances of a case are true may point at instances in which it is not. Thus, being compelling or unproblematic may very well reflect some level of knowledge. Increasing knowledge mounts problems.

That some assertions are compelling and unproblematic is compelling and unproblematic. Scientism takes this as the compelling and unproblematic proof of the solubility of the problem of induction. This amounts to the (false) identification of the compelling and unproblematic with the true and the true with the provable. Since compulsion is relative, relativism seems thus compelling despite everything compelling and unproblematic to the contrary. Above all, the ubiquity of the diversity of opinions is compelling and unproblematic. That some statements are compelling is fortunate: otherwise, diversity would make us islands. Relativists suggest that societies are islands, and the Pyrrhonists, who recognize no compelling statements, suggest that individuals are. Relativists rely on tradition with no reservation; Pyrrhonists reject it the same way. Both parties ignore the desirability of diversity as a means for rational debate. Scientism rests on the wish to preclude dispute by appeal to universal unanimity, and relativism and Pyrrhonism rest on the wish to preclude dispute by appeal to insularity. Pluralism is the readiness to engage in disputes by taking the unproblematic side-by-side with the problematic. Some disputes about some compelling and unproblematic statements render them problematic.

Pluralism is thus a combination of the best of both scientism and relativism. It is tolerant, in accord with promises of relativism, and it avoids insularity, in accord with promises of scientism. And all agree that some statements are compelling and unproblematic, but only pluralism allows for this compulsion to be provisional with no wish to justify all compelling and unproblematic statements by a single argument: every one of them invites a different treatment.

Consider Worrall's attack seriously. Why do we find compelling the proposal to use stairs rather than to jump out of windows? Worrall knows the answer, of course: we wish to reach the ground uninjured. He suggests that this refutes critical rationalism because critical rationalism takes theories of gravity as open to criticism, not as credible. He wants to see the credibility of some theories justified, those that are compelling and unproblematic. He suggests that the compulsion is rational and that a theory of rationality has therefore to explain the rationality of compulsion. That is to say, we all agree that some theory of gravity is compelling; Worrall insists that this compulsion is rational.

The compulsion is a fact. What, if anything, makes it rational? Worrall has a compelling and unproblematic answer to this: the argument from sanity: to behave in disregard for the compelling or the unproblematic is irrational; it is crazy. According to Worrall, critical rationalism is plain crazy – not in intent, but in fact: unintentionally it violates the argument from sanity. One may be sane yet espouse views that unbeknown to one are contrary to some compelling or unproblematic information. But as one sees that, one desists. Does critical rationalism violate this argument from sanity? On the argument from sanity, possibly monist critical rationalism as young Popper advocated it agrees with Worrall. Feyerabend cites Popper to advocate both the argument from sanity and its contrary. Where are we then?

The argument from sanity, to repeat, is very compelling: it is crazy to jump off a high window in rejection of induction. Is it acceptable then? If it is recognized as conclusive, then not; otherwise, yes. Its status as conclusive is far from being compelling and unproblematic. It would be conclusive, were all compelling and unproblematic assertions true. But, to repeat, some are not. (It is terrible and all too comprehensible fact on record that quite a few people jumped out of high windows in the New York World Trade Center on the September 11 tragedy.) How then is the argument from sanity still valid?

William Bartley, who quoted *Don Juan in Hell* of Bernard Shaw to that effect, tackled this point beautifully. Shaw says, reasonable people adjust to circumstances and unreasonable people attempt to adjust the circumstances to themselves. Hence, he concluded, we owe all progress to the unreasonable. We may put this directly and without the air of paradox. Overturning of a compelling assertion, rendering it problematic, showing that it is only a qualified truth, amounts to a revolution. It is central to the present discussion, then, that critical rationalism and only critical rationalism endorses the compelling only as provisional. Worrall's dissatisfaction with the argument that we refuse to jump from rooftops because we think that this will kill us, his desire for proof that this is so, is the demand that the argument should be clinching.

Feyerabend has rightly observed this. The view that jumping off a rooftop is deadly is not quite compelling and unproblematic: had we grown wings, things would be different (or if we had parachutes or hand gliders). Does Worrall consider the safety of the parachute as compelling and unproblematic as that of stairs? It is a compelling, unproblematic fact that attested unsafe staircases look compellingly safer than attested safe parachutes. Some users of parachutes never learn to see their safety as compelling as it deserves to be; their intuitive preference for staircases remains unshaken, and it backs Worrall.

Is the compelling and unproblematic merely personal? Is it merely psychological? These are important questions, since it is repetition that makes observation reports compelling; it is the inability to refute them that makes them unproblematic.

Perhaps not: suffice it to view the compelling and unproblematic as engaging the efforts of researchers to test it or to explain it.

This is obvious, and it renders dispute friendly and dispels the need to endorse the compelling and the unproblematic as true. Some compelling or unproblematic assertions become then problematic for those who develop ideas that conflict with them. This is the discovery of Galileo. He developed ideas that he found compelling and unproblematic, and these were in conflict with other compelling or unproblematic ideas, so that he found it a challenge to refute them. This is how new ideas help render older ones problematic. Some philosophers, including Kuhn, Feyerabend, and Lakatos, advocate the opposite view. They have declared it necessary to replace compelling and unproblematic ideas with alternatives that are no less compelling and unproblematic. Otherwise, the old ones simply will not go away, they said. There is some truth in this anti-critical view, as new ideas render it easier to criticize old ones the way Galileo illustrated. But the opposite idea is more important and is true. When sets of ideas become highly integrated, the line between theory and observation blurs, and so the extant theory looks increasingly fact-like in its very compelling character, so that a new theory is called for. Popper has observed this early in his career. It is awe-inspiring.

5. THE NEW PLURALIST ERA

Pluralism has sprung on us unawares, some time after World War II, around 1950. *The Origins of Totalitarian Democracy* of Jacob L. Talmon appeared in 1955, and caused a stir, as it presented the scientific monism that the fathers of the French Revolution attempted to apply to politics as the root of chauvinism and worse, as *Gleichschaltung*. Strangely, already in the middle of the nineteenth century Max Stirner, Alexis de Tocqueville, and John Stuart Mill voiced objections to political monism. Talmon caused a stir nonetheless, as he declared scientific monism the root of the political monism that is the root of the evil. Monism may all too often be a form of dogmatism rather than scientism. Talmon's message was not pluralist; it was the rejection of the assumption that politics is scientific. This seemingly went well with Popper's unmasking of the doctrine of historical inevitability as pseudo-scientific a short while earlier, in his *The Open Society and Its Enemies* of 1945. Yet, in that book Popper advocated scientific politics, contrary to Talmon. The enemy of both of them was the scientific pretense of earlier utopians. Popper attacked their politics as pseudo-scientific and Talmon attacked it as monistic. Both considered scientific politics monistic, but Popper defended it whereas Talmon attacked it. Pluralism would have resolved their differences, but neither of them was a pluralist then, and neither presented science as pluralist then. They did so later.

Talmon's case study required analysis. He drew a moral against all demand for uniformity. Contrary to this, Gellner has claimed, in his *Nations and Nationalism* of 1983 (and in earlier writings), that nationalism requires uniformity as a basis for a common national culture that serves the nation and its economy. Chauvinism is thus a mere exaggeration. The rejection of nationalism, then, is also an exaggerated rejection of chauvinism – especially if it rests on scientific monism. Scientific monism is erroneous even in science. Gellner claimed that the need science has to prevent what Popper calls reinforced dogmatism, which is a version of the take-it-or-leave-it attitude, suffices to impose some sort of uniformity, at least in intent. The question then is, what is the opposite of the take-it-or-leave-it attitude. Scientism suggests that monism is. This is true, but monism is inaccessible. This led to relativism and to its islands of monism. Pluralism is the right option: not all alternatives are equal, and we all have to take for granted the unproblematic, although only until someone turns it problematic.

Diversity of customs does not suffice to generate pluralism. The idea that it does is a dangerous error. Diversity needs political safeguards to insure its durability. Scientific monism suggests that the uniformity of science should help decide, once and for all, which set of customs is the best, and goodbye to pluralism. Hence, it is important to criticize scientific monism, set scientific pluralism as its substitute, and see to it that it rests on clear legal foundations. This is the new democracy of the post-World-War-II era.

The compelling or the unproblematic are in constant tension with the possibly objectionable. This initiates the pluralist search for new alternatives – on all levels, factual, theoretical, methodological, and metaphysical. This is seldom easy, and this means that it is never easy to challenge the compelling and the unproblematic, that it is even harder to offer an alternative to a good idea. As science is full of good ideas, it is hard to offer alternatives to them. So, science looks monistic. Establishments make demands on dissenters, in order to eliminate dissent. The institution of safeguards against this rests on agreement. The agreement should be to protect pluralism and especially the ability to institute reforms, since pluralism requires a reform.

The ability to institute reforms rests on the distinction between uniformity (of the endorsement of what is compelling and unproblematic) and the consensus. Here is a clear-cut instance of the difference between them. Consensus seldom has agreement of opinion behind it: even two partners seldom share views relevant to their ventures: they usually compromise or else they break their partnership. National consensus is less simple, as it is never easy to break up a nation. (Otherwise, relativists will be unable to view nations as islands.) Since the consensus is a sort of agreed-upon middle, no one has to endorse it but many must endorse it as a *modus vivendi*. Monists say that when a belief is rational, the consensus on it is likewise rational. This is an obvious error: the two obviously differ, since rationality is widespread, at least in science, and agreement is rare, even in science. People who say that all physicists agree about, say, quanta, either do not know or prefer to ignore many controversial details about quanta. Agreement belongs only to repeated, uncontested testimonies; on theories, it is at most only on broad outlines. Even in broad outlines, on what quantum theory says, controversy often rages.

To conclude, the chief argument for monism is from sanity. It is the argument that obviously only the insane (and the philosophically inclined, Wittgenstein added) would dispute some compelling and unproblematic statements. This argument is a boomerang. Critical pluralism thus is – that is to say, should be – compelling and unproblematic. But only in broad outlines.

Chapter 2.5

WILD GOOSE CHASE

Geoffrey Dean, Assisted by Arthur Mather, *Recent Advances in Natal Astrology: A Critical Review, 1900-1976*, London: World Astrological Association, 1977.

0. NO HOSTILITY, PLEASE

The editor of *Zetetic Scholar* has challenged me to apply Popper's criterion of demarcation of science, or any alternative to it, to the claims of this book that current astrology merits scientific status. As it has the stamp of approval of the World Astrological Association, I agreed to review it. I could not read all of its nearly 600 compact pages, full of snippets of history, astronomy, astrology, statistics, and other miscellaneous items, with little links and no focus. If the editor rejects this review on this ground, I will abide by his decision.

Paul Feyerabend, Arch Advocate of Academic Unreason, has thrown a gauntlet. He says, rationalists condemn superstition without knowing much about it. He condemns science and rationalism, but not out of ignorance. I agree with him about condemnation: I regret it. Let there be no condemnation of folly, be it rationalist or not. The freedom of thought, as Voltaire said, is the freedom to think foolishly, as the freedom to think wisely is obvious. As to knowledge, I disagree with Feyerabend. I need not be much schooled in what I consider superstition, since I have no wish to condemn it and may ignorantly regret what I consider people's waste of time on it, perhaps also suggest that it may raises false hopes and hence painful disappointment. By the same token, I have no objection to Feyerabend's view of science and of rationalism as a waste and as a false hope: I only object to his aggressiveness and promise to try to be gentler myself – with folly of all sorts. True, some folly has harmful consequences. We should fight them; the only weapon for fighting folly itself is education.

Much of early modern science was the disposal of ballast. Bacon demanded it, asserting that ballast is a serious impediment; Galileo, Descartes, and Locke worked at it. Bacon identified as ballast a few separate items: scholasticism, metaphysics, pseudo-science, dogma, prejudice, superstition, error, and insanity. This identification has echoes even in the works of Popper, the first liberator from Bacon's doctrines. The situation is vague. When Frazer called magic pseudo-science, he knew that most magicians cannot possibly pretend to be scientists, having no idea about science; he meant that they pretend to know.

This is false: superstitious people are often more skeptical than scientists (and for good reasons, need one say). Nor is this all. Once we admit that there is such a thing as a scientific error, and it becomes relatively easy to refute Bacon's identifications of the diverse kinds of non-science. Here let me discuss only one case, his identification of superstition with pseudo-science.

1. THE CURRENT THEORY OF SUPERSTITION

Superstitions are ubiquitous. The great Niels Bohr used horseshoes for good-luck charms, and when asked whether he believed in it, he smiled wryly and said, he had it on good authority that it helps anyway.

Bacon offered an explanation of superstition, of its prevalence, persistence, and attraction (*Novum Organum,* 1620). Superstition is still popular. The mind is lazy and mental inertia reinforces what one has learned, he said. Society deems the readiness to change one's mind as an admission of failure, and so it offers an incentive for stubbornness – especially to teachers, more especially to teachers wishing create schools of thought. When one sees the world through the spectacles of a conjecture, one sees the facts as the conjecture distorts them and forces them to appear in agreement with it. Thus, evidence always confirms conjectures. For example, the lucky survivors of shipwrecks who came to make good their vows for sacrifices to Poseidon, write evidence for the power of vows on the columns of the temples in honor of Poseidon. Those who comprise contrary evidence cannot show up. As we confirm conjectures, they become dogmas. Bacon identified dogma, superstition, and prejudice.

Bacon proposed this rule: begin with observing facts, do not select evidence, and make no conjectures: let the facts speak for themselves. Two centuries later, Dr. William Whewell suggested an alternative. Begin with a conjecture, he said, and select one that is explanatory. Design observations with which to test the conjecture and to possibly refute it. If tests do not refute the hypothesis, they validate it – once and for all. This once and for all clause was superseded when Einstein superseded Newton, despite all past validations of Newton's theory.

The disagreement between Bacon and Whewell is about the way science tests its ideas empirically (and superstition does not).

What then is the allure of superstition? Why do people go for arbitrary ideas? Bacon's theory still is the only contender: people see the world through the distorting spectacles of their opinions and then they see only confirming facts, not refuting ones; they may see even refutations as confirmations. This led to two responses, Popper's and Maurice Ginsberg's. Popper said, marshaling validating evidence, as both astrology and psychoanalysis do, is pseudo-scientific. The scientific way is to subject hypotheses to severe tests or to admit honestly that they are metaphysical. This, however, applies only to views that we consciously assent to, he added. Most prejudices and superstitions are not conscious. Ginsberg said, it is no good to offer a refutation of a prejudice, since the prejudiced will explain it away. He recommended simultaneous attacks on it from many angles.

Ginsberg's proposal often fails. When a prejudice is under attack from many directions, its adherents diffuse the assault by taking it item by item, separately, and painstakingly slowly. The rejection of criticism rests on excuses, and these may be true. They may also contradict each other. Taking criticism slowly, item by item, may obscure such contradictions; confusion may cloud them. Superstitious people are often confused, and confusion helps to cling to a superstition despite criticism.

Not so the volume at hand. It is unusual. It is relatively free of both confusions and excuses. Hence, it invites a new attitude from Bacon's followers. They may be ignorant of the present book and choose to remain so; they may dismiss it as superstitious; they may judge the hopes it expresses for a future scientific astrology not superstitious at all (Bacon's doctrine opposes unwise opinions, not unwise hopes). Two of these options are rational and revolutionary. Admitting that this book is rational and scientific because of its admission of the negative verdict of the facts is revolutionary in a minor way. Admitting that this book is rational and scientific because of its effort to rehabilitate astrology is revolutionary in a major way. A shift in the discussion of rationality has taken place, as some philosophers discuss the rationality not of theories but of research programs. Imre Lakatos has noticed that this shift is a revolution, and, were he not high-handed, or had he lived longer, he might have contributed something to the debate on research programs. Anyway, we may acknowledge the shift from the view of rationality as the mere assessment of views to the view of it as the assessment of research programs too. This shift demands an assessment – regardless of our view of this book or of any superstition. This book made me see this clearly and it has alerted me to my want of a comprehensive view of the matter. I will try to make amends towards the end of this discussion. Here I will conclude with a critique of the central error of the traditional view concerning superstition.

Erroneous theories of reason are less harmful than erroneous theories of unreason. An example is the traditional theory of rationality as proof. It makes all disagreement unreasonable. This is inapplicable to reasonable disagreement. No one says, every dispute between scientists is a proof that at least one party is unreasonable. Yet, this very theory – that all dissent is unreasonable – is the basis of Kraeplin's theory of paranoia that is still official. (See Y. Fried and J. Agassi, *Paranoia: A Study in Diagnosis*, 1976.) And it is easier to declare people prejudiced or superstitious than to certify them mentally ill. Historians of science have traditionally judged as prejudiced all medieval scientists, all eighteenth century chemists except for Lavoisier and his followers, and all those who rejected Mendel and/or Darwin. Many historians of science today still hold this view. (See my *Towards an Historiography of Science*, 1963.)

The identification of superstition, dogma, prejudice, metaphysics and unreason, polarizes ideas and decisions to wisdom and folly and excludes middle options. Rationalist philosophers recognized exceptions: tastes, love, and friendship. They repeatedly state that these are not wisdom and not folly, even though harmful taste is foolish, love is blind, and as friendship includes trust, it may be credulity. Attempts to apply the classical theory of reason and folly systematically cause extremism in many ways. It is a part of rational procedure to be critical, even at times optimally critical, but it is unreasonable to be very critical in matters of love and friendship and toleration.

2. IMPRESSIONS ABOUT THE VOLUME AT HAND

Consider the signs of the Zodiac. Everybody can guess that two waves stand for Aquarius; but few academics are able to identify Virgo. Aquarius is probably the sign of the flooding of the Nile in spring. Virgo is the Egyptian goddess of grains, and her sign is in the fall, the time of harvest. The signs of the Zodiac are twelve mainly because of the twelve lunar periods making a solar one. Well, not quite. But then the signs do not divide the sky into equal parts either. It is no good to have twelve equal parts in the sky and allow each of them 365:12 days, as the solar year is not a round number of days, and the months are not of equal length. This, and the precession of the equinoxes and geography hinder astrological judgment. Besides, when does a Zodiac month start? At sun-set, at midnight, or at sun-rise? Does it start at the moment of its appearance of a sign in the east, at mid-point, or in the west? Is the division along the northern hemisphere's tropical zodiac or the sidereal, since both differ from the constellational zodiac? No answer. The authors frankly admit this, offer no guideline, and express hope nonetheless. This hope merits examination.

Condemnation of decisions of astrologers on these matters as arbitrary is common. This is high-handed: both science and superstition mix precision with some juggling of data in order to fit facts and conjecture well. Much depends on the theory and on the available instruments of measurement. Little discussion is available on this. Popper is the first to have offered a correlation between limits of precision of instruments with limits of required precision, enabling a decision procedure as to when the precision of an instrument is sufficient. As always, decision depends on the purpose at hand: calendar calculations must always be precise to a day, and the degree of precision of astronomical data for calendar calculations depend on the number of years ahead that they project. The decision to reform the calendar every thousand years determines a limit to the required precision. In applied science, theory provides the limits of precision: theory is usually continuous, evidence is always discrete. Consider a horoscope. If it says, those born under the sign of Aquarius are likely to be sailors, then in order to apply it to anyone we need to know one's birthday. If it says, those born under Mars are likely to be soldiers, it depends what "under Mars" means. If it means the night Mars appears here or there, then suffice it to know the night of birth; and for those born during the day suffice it to know whether the crucial night comes before or after that day. If it means when Mars rises, sets, or is in conjunction with a constellation or with another planet, then its application requires much more precise information. As for tests of theories, the more precise a theory, the easier it is to refute it, but also the greater is the required precision of test measurement. The claim that Aquarians tend to be sailors clashes with facts. Refutations exist of all of its versions. But the need is not for refutation but for explanation: why should careers depend on the flooding of the

Nile and not of the Amazon? (This question must have troubled the ancient Greeks already: Hesiod identified Virgo as Dike, the goddess of justice.) The authors note the two traditions on the Zodiac: opening of discussion of the Zodiac, on page 47, they say, "It is not impossible that both Zodiacs are valid," that is to say, "both tropical and sidereal". They mean, some theory may be true. How? This book, thick and full of detail, says no more on this.

The authors take it for granted that luck, prescribed by the moments of births, has some link with the very constellations that have reached us from the depth of the ancient Eastern Mediterranean basin. We know now for sure that there are no constellations: stars that look like neighbors are light-years apart. The authors know this. Yet they never take cognizance of it, much less of the possibility of abandoning astrology. The apparent constellations depend on time and place, and tradition does not say how to adjust natal data to variations. Did you know that astrologers in the Far East use the sidereal system (of the fixed stars) and in the Near East they use the tropical (solar) system? Did you know that the number of signs was not universally twelve? Did you know that harmony – between a planet and a constellation – reflects Ptolemy's intuitions about how the (alleged) four elements (earth, water, air, and fire) relate to the symbolisms of the heavens? Can you correlate planets and metals? The golden sun and the silvery moon are obvious. So perhaps is quick-silver (Mercury). But should we not replace the iron that is Mars, the God of War, with Plutonium? Or should Plutonium go to Pluto? I shudder at the proposal: the authors take seriously every such option, including Arthur Koestler's humorous critique of astrology (p. 22): he said, we may want to correlate newspaper headlines with the signs. No matter how harsh is their critique of the gullible on point of empirical evidence, the authors are gullible whenever scant facts and fuzzy theory allow. At times, they also go for maximally available precision, though at other times they are sensitive to the limits of useful precision for given ends.

Can / should we make a fuzzy theory precise? How? The authors notice how hopeless the situation is, yet they do not consider aborting the project. The degree of precision of a system can increase in different directions. We need guidelines to decide between competing options. We can move at random: we can choose professions, say, seamanship; or personal traits, say, courage; or dispositions, say, homosexuality; or positions, say, appearing in the social register. If more seamen or brave people are born on a given month, day, time of day, conjunction, aspect, house, eclipse, harmony, if more gay or famous ones are, if any correlation transpires, it will delight the authors. Some correlations will appear, since now astrologers use computers. With enough trials, by Bernoulli, the appearance of correlation is virtually certain.

The authors record problems to solve. Many correlations undergo tests, and these require rechecking, cataloguing, and surveys. Ever since the Copernican Revolution, each scientific development raises new difficulties for astrology. Some developments may be astrologically indifferent, especially if they relate to the observed fixed stars alone. Others are not. The geocentric and the heliocentric system determine very differently the astrologically important modes (points at which planets pass the ecliptic; which ecliptic?). Eclipses, conjugations, comets, and even planets multiply, not to mention metals, chemical elements, personal traits, and events at large, to correlate to the larger heavenly events. (Despite a promise, the astrology here is not restricted to natal astrology.) Mathematics plays havoc. Analytic geometry has destroyed houses; spherical trigonometry and statistics destroy aspects. (Already Kepler viewed them as too coarse and he modified them to keep up with the times; p. 278.) Incidentally, though Kepler shared to the last much of the outlook of our authors, at least he decided that all traditional astrology was utterly worthless: he advised to ignore it. He wanted a fresh start.

3. THE THEORY OF THEORY ASSESSMENT

The traditional theory of scientific method concerned itself with ways of generating scientific theories and of checking their scientific status. Things are different now. Most students in the field concern themselves not with assessing methods of generating theories but with assessing methods of checking their scientific status: given a theory (a set of statements), how can we decide its status *vis-à-vis* the body of existing factual information, and decide that it is or is not scientific? What is the criterion of theory assessment, of demarcating the scientific from the unscientific theory? We may narrow this discussion, if we want, and seek a demarcation not of all non-science but only of pseudo-science. Regrettably, such a demarcation involves denunciation.

Tradition claims that science validates its theories empirically. This is a superstition, still immensely popular, perhaps more than astrology. Both defenders and opponents of astrology appeal to validation: they want the facts to decide the fate of theories. But validation is tricky: we decide what to do with information, before it can decide for us. We weigh evidence, dismiss it, endorse it, cross-examine it, defer to it, abuse it, *etc*. By received standard, validation is inductive. It says, thou shalt not advance a view unless you validate it empirically first! It rejects all non-science as dogma and superstition. This is scientism.

Popper, and more so his leading disciple W. W. Bartley, III, made a different proposal: never mind the ground for my having advanced a doctrine, they suggested, as long as we make it open to criticism and are willing to accept criticism. Thus, Popper expressly declared Carnap a rationalist thinker because he presented his superstitious view (logical positivism, but this is irrelevant here) as open to criticism and accepted criticism and proposed ever newer variants of it. These variants are only slightly less boring than astrology.

Perhaps I am unfair to Popper. I discern at least four conflicting variants of his view. First, science is conjectures and refutations. It vindicates the astrology in the present book as well as other superstitions, such as logical positivism, as rational and scientific. Second, science is satisfactory explanations, where one of the requirements that explanations must satisfy, if not the only one, is that of independent testability. Third, science is conjectures, their corroborations, and the refutations of some of them. Fourth, science is explanatory conjectures, their corroborations, and the refutations of some of them. By Popper's second and fourth views, as neither astrology nor positivism explains, neither is rational, much less scientific. The third view makes it an urgent problem for all lovers of astrology (or of positivism) to ask, does it have any corroboration? The answer still is, no. But one can look for them. This book, that I consider obviously unwise, conforms to Popper's third criterion. Its authors agree that astrology has not come of age, has as yet undergone no test, and so is not yet scientific. As long as one lives one can try to render an unscientific theory scientific. Is there a criterion for the rationality of such attempts? Did Popper discuss this question? Perhaps, yes. For, I have not exhausted here the distinct variants of Popper's views; to answer my question I should add the fifth: Popper's latest view of science is that science is solving problems in a manner open to critical empirical assessment. This way not only the rationality of theories but also that of research programs may be open to assessment. Consider the present book, then. Does it present a problem? Yes: why does astrology persist? Is there any objective ground for this persistence? The authors take it for granted that any possible validation to astrology would explain its persistence and validate it; their sifting of all the vast body of evidence in search for a validation makes sense, then. But it is painfully weak. I cannot compare astrology east and west, Ptolemy's astrology with Kepler's, Kepler's with that of this book's authors, nor either with the vague notions of the many gullible believers. How can we view such a variety as one phenomenon?

The persistence of a view may, at first approximation, rest on either gullibility or facts. Contrary to the current theory of theory assessment, both these categories, gullibility and factual support, are complex, and they overlap. Nevertheless, we can begin by sorting views into these two categories and take it from there. From time to time, we may wish to test even what we consider the most absurd hypothesis. Extra sensory perceptions and spiritualism are sillier than astrology, yet there is no need to dismiss all tests to them, not the tests that Michael Faraday and W. Grey Walter have conducted, for example. But, we should be careful not to waste our lives on these matters. It is much cheaper to dismiss all superstitions on the ground of the theory of program assessment. People cling to superstitions, among them scientists. The life-work of the great social anthropologist Sir Edward Evans-Pritchard is a breathtaking, monumental defense of the rationality of magic. Until we take such matters seriously, we will superficially dismiss the prevalence and perseverance of superstition as mere gullibility. This will not do. Even if we combat superstition – and I would rather not – we may first want to properly assess its strength.

4. IRRATIONALISM

Some people deem religion superstitious yet they tolerate it more than astrology, psychoanalysis, or Marxism. Other people hold the reverse attitudes. Tradition on the problem of the limits of toleration rightly focuses on the toleration of the intolerant. We refuse to allow the law to assess the damage that education causes, and for obvious reasons. (Much upheaval follows even the rare cases in which the law must interfere with the process of education – say, in cases in which teachers are clearly abusive.) Yet, we put enormous efforts into education that subtly expresses our fears and hopes about what we want to transmit to the young and how. Miseducation is harmful. This is indisputable, yet we want to tolerate the miseducated even as educators. Honesty requires that we should confess ignorance on such matters. At the sight of the younger generation becoming superstitious and losing their ability to think critically, toleration raises a conflict: regret requires action; without invitation to act, tolerance requires inaction. This imposes taking superstition somewhat seriously. To take it seriously we should study not this or that superstition, but superstition as such, the approval of it, irrationalism.

Evans-Pritchard spoke seriously of the rationality of the magical system of the Sudanese tribe known as the Zande or Azande, on whom he wrote his classical *Magic and Witchcraft among the Azande* of 1937. He said, science is in principle unable to explain coincidence, yet magic does. All superstitions, perennial, ephemeral, or private, concern coincidences. Arthur Koestler expressed this view in his book on extra-sensory perceptions, *The Roots of Coincidence*. Evans-Pritchard added that the Zande are critically minded: they test their oracles. They never criticize magic as such, because they cannot; their whole thinking apparatus is occupied by magic; it is the whole of their transcendental apparatus: without it they cannot think.

One of the reasons for the spread of irrationalism is the unwise attitude that many rationalists exhibit towards it. Thus, the august audience of a lecture of Evans-Pritchard that I once witnessed, welcomed and cheered it as a mere hilarity. He left the lecture hall exasperated, with tears in his eyes. "They will never take seriously what I say," he explained his exasperation to me; "they will never believe that I mean it seriously." Many did, and these were all on his side. Can there be a dialogue between rationalists and irrationalists? Both parties seem to respond in the negative, and they are both at fault: such a dialogue happens, and at times it is fruitful. Irrationalists are skillfully finding defects in rationalism and quickly applying their findings in their teaching. Thus, not only astrologers, but also other superstitious groups are recruiting new believers by training them to see the world through the most general abstract principles of their sects while discouraging criticism. This last ingredient, the skepticism about rationalism, is absent from the present book: it appears as rationalist. Yet, its authors cling to an unspecified astrological research program, perhaps under the guise that any decent empirical research is laudable. It sounds rational in its recognition of the poverty of the evidence in favor of astrology, but as this reinforces its clinging to research in astrology, it is irrational.

Evans-Pritchard discussed the difference between innocent and dogmatic superstition, as well as between superstitious and innocent dogma. He noticed that science pays little attention to coincidence whereas superstition concentrates on it. He stressed the importance of the fact that magic – as any other superstition – injects meaning into events rather than explains them. And herein lies the peculiarity of superstition. Opponents call it wishful thinking; sympathizers call it yearning. I count myself among the sympathizers. The atmosphere of sympathy with superstition is with the great yearning of the soul, with the quest for meaning. Interested readers may enjoy the works of the greatest student of superstition, Arthur Edward Waite, who wrote in the first quarter of the twentieth century. In his book on the Holy Grail, in its conclusion, he confesses he too is in quest for a new mass. (He was a Roman Catholic.) Regrettably, too many scientists are stuffed-shirted professors who insist that they are content with the state of their knowledge. Some of them are addicted to the inverted-superstition that denies the very possibility of yearning, the very existence of the puzzle, of the riddle of the universe, of the riddle of life, or of the meaning of life. This inverted superstition is logical positivism and like all inversion it is parasitic, and so we need not take it more seriously than the superstition that is its host.

Science grew out of medieval superstition because, more than any other superstition, it fed on the yearning for the revival of old glory, the yearning that ended up as the Renaissance of culture and of science. Not accidentally, superstition was more common, not less, in circles that made the Renaissance of science. Numerology and astrology reigned there supreme. The name for numerology was "kabala" and changed into "Pythagoreanism" (since Pythagoras learned its secrets from Moses in person, as Renaissance cabbalist Regiomontanus narrates). The long line of Pythagoreans begins with the followers of the Christian cabbalist Giovanni Pico della Mirandola whose "Oration on the Dignity of Man" is the opening of the Renaissance of science – because it promoted dignity and because it distinguishes between the evil magic that Moses forbade and the good magic that we now view as the roots of the scientific tradition. Who was the last Pythagorean? Perhaps it was Newton; perhaps the twentieth-century Brouwer and Bohr shared it. Yet whatever it meant for Galileo, he opened his first great book with an admission of guilt and the promise to clear the Pythagorean house of all mumbo-jumbo.

Logical positivism too rests on yearning. It survives all criticisms of it by remaining a hope, a program for research. But if the hope for it is acceptable, why is it not also acceptable for astrology or any other irrational idea? Philosophers of science spill so much ink on the theory of rational belief, whatever it may be, and then they allow their own irrational ideas not as beliefs but as hopes: not able to support a doctrine rationally, they may hope to do so one day. And then, farewell to the rationalism of the positivists and of all other traders in rational belief. Does this vindicate astrology? I do not think so.

Perhaps, one might object, what is wrong with irrationalism is not hope but mumbo-jumbo. This is not sufficient. The irrationalism of Evans-Pritchard and of Polanyi is no mumbo-jumbo. Of all contemporary academic writers on the subject, Feyerabend is perhaps alone in boldly inviting mumbo-jumbo – all of it, old and new. He legitimizes it all *a priori*. Irrationalists pretend that any system is *a priori* as good as any other, since thinking begins and ends within a system. Bacon said, we can live without a system; Kant said, we all share one system; most of irrationalists today are relativists who deem all systems equally valid. Some discrimination is badly required. And we begin with the ousting of superstition not because it is a mere hope and not because it is mumbo jumbo, but strictly because it is too elusive to be any good (though historically it may have been important). The preference of the illusive over the refutable is a matter of decision, said Popper.

5. SUPERSTITION, THE ILLUSIVE

The traditional discussions of the demarcation of science from pseudo-science or from non-science, is between sets of hypotheses. Popper's criterion of demarcation is within this traditional framework, but it is radical all the same: he proposed to view as scientific refutable hypotheses, ones that we may conceivably refute by some conceivable observation reports. For example, when we say, as we do, all living things include proteins, and then we can easily imagine a living thing with no protein in it, with other compounds replacing all amino-acids. We know of no observation report of such a thing. But we can imagine it, and so the hypothesis is scientific. Yet, Popper has noticed, something is missing here: we can change the meaning of our hypothesis to exclude any refutation of it in its present sense as soon as we meet such a refutation. (The meaning of the word "gene" altered radically a few times and the impression persists that there is a single unscathed theory of genes that dominates genetics.) He therefore requires, in addition, that we avoid such moves, that such moves – rescue operations, *ad hoc* amendments, doctoring, tinkering, juggling – invalidate all claims of hypotheses for scientific status. This means that the discussion concerns hypotheses of fixed meanings: once the meaning of a hypothesis changes, it simply becomes a different hypothesis that should be testable and tested too. Then the move is not juggling. Hence juggling is the evasion of refutations.

Meanings often shift gradually and surreptitiously. At times we like this. It may sound clever, but it is unwise; it is a subtle form of (self-) deceit. When people are willing to project meanings into the world – as Evans-Pritchard says magically-minded people do – then it is easy to confuse them. They are gullible, and this is but one of the many ways to deceive them. All attacks on gullibility are useless. The hypotheses that our authors refute are fully articulated and clear-cut; they advocate in their stead suggestions, hints, not clear-cut hypotheses. Hence, it is pseudo-astrology that they refute, pseudo-superstition, and they advocate genuine superstition. Popper's claim that astrology is pseudo-science is false, then, though he follows a tradition that rationalists and astrologers share, of viewing astrology as illusive rather than as clear-cut. Clear-cut astrological hypotheses are refutable, and the authors present them as refuted. The usual astrology tantalizes; it is no food for thought, but it is no stones instead of bread either; it is the smell of bread, perhaps the promise of bread, perhaps the claim that bread is in the oven. Yet just as the superstitious put meaning into magic, so they put meaning into astrology. Frazer's fusing of pseudo-science with superstition was an error, even though an idea may be both: astrology is, and our authors have managed to separate its two aspects. Having discarded the pseudo-science they cling to the superstition.

Magic as pre-science cannot be pseudo-scientific, since it preceded the advent of science, before pretense for scientific character is possible. Also, a prescience may be of honest use – even today – as a heuristic device. All in all, confusing the innocent with the regressive and the parasitic is expensive. Innocent superstition is easy to discuss, though criticizing it destroys its innocence. Superstition that is parasitic on science is pseudo-scientific. Being regressive, it is a shifty part of a complex process; like many a superstitious pseudo-science, it is a promise in the guise of its fulfillment.

The contrast of the regressive with the innocent may be informative. The regressive is infantile, fixational. In infants, as in lower animals, those feelings are visible that are elementary according to Watson's behaviorism. These are, first, clinging or fixation; second, aggression or fight or hostility; and third, fear or anxiety, or flight. Clinging is common even in adult animals in times of stress and great needs. It is regressive. The great stress or need that leads to it may come from objective conditions and from states of conflict (internal or external); anxiety and regression go well together. As Konrad Lorenz has noted, anxiety causes confusion in all animals. Even under stress, humans can make efforts to stick to a point, focus their attention on it, and be clear within it. When we see nothing but what is at the center of our attention, we are prone to some sense of magic, like children ignorant of their environment. The situation is conducive, however, not only to superstition, but also to (ambulatory) paranoia, fanaticism, or dogmatism. There are variants, but we do not know how to grade them on a scale since diverse factors present themselves here. Dogmatists are clear when their dogma stays in the center of their attention, when they try hard to ignore its periphery. They cling to the center of their attention aggressively. Fanatics are confused about the center of their attention; likewise, they ignore its periphery; and they are more aggressive than clinging. Paranoiacs are clear about the center of their attention, increasingly confused about their periphery, cling, and very anxiously so. The superstitious are confused, and if they are wise about it, they allow themselves the luxury of superstition only in the periphery of their thinking and they act more out of anxiety than out of clinging. These characterizations make it hard to grade paranoia, superstition, fanaticism, and dogma. Many paranoiacs are superstitious – so much so, that the literature endorses Kraeplin's definition of paranoia as an erroneous *idée fixe* logically sustained. But there is a limit to this. The dogmatism of Newton was not regressive. Nor were the naïve superstitions of Pico della Mirandola and of Kepler; or the paranoia of so many prophets, lay and religious. The fanaticism of ever so many leaders, military, political, or religious need not be regressive either. Notice also that prejudice is altogether missing here. It is too coarse a category to take seriously. The interested reader may want to consult Michael Banton's various enlightening works on it.

Evans-Pritchard identified the Zande magic system with Communist dogma, and even with his own Anglo-Catholicism – all as self-imposed intellectual systems. Systems are frameworks, he said. Frameworks, he added, are not open to criticism and replacement. (Popper called this idea the myth of the framework.) They can, and at times they are. They may generate research programs, good and bad ones, or, as Lakatos called them, progressive and regressive (or degenerative). What makes a program progressive? Popper did not discuss this question and Lakatos said we can discuss it only in retrospect, not in prospect. The present book illustrates a declining program in action. Or is it? Has the sun of astrology set four centuries ago, or not yet arisen? How do we decide? Shall we cast a horoscope for astrology, perhaps?

No: we do not know the moment of its birth.

Astrology has a glorious past that is dead; should it have a future? There are more interesting programs and perhaps more promising ones. The value of this book is in its sorting out astrology as a pseudo-science, which its authors reject, and astrology as superstition, which they share. The critical attitude of science – the readiness to declare a theory erroneous rather than to take the blame for ill success of predictions – is absent from magic, superstition, prejudice and dogma. Yet, the world of the pre-rational permeates with magic and superstition, whereas prejudice and dogma, even paranoia, are accretions of modernity. The very admission of the possibility of scientific error, perhaps even of rational error, breaks the mold of traditional theories and invites a new look at these categories. Some writers have started this, notably Michael Banton on race prejudice. But the challenge is still there, and it deserves a response.

Chapter 2.6

SCIENCE AND ITS PUBLIC RELATIONS

0. ABSTRACT

Science is honest. Its Public Relations system is not. Indulgence towards it is an error: it is hard to prevent the deceit common in the public relations of science from spilling over to science itself. The qualified will eventually learn to correct the resultant inaccuracies, but the cost is high.

The public relations of science are a complex affair. In charge of it are professional popularizers of science and ill-fated academics – philosophers and historians and sociologists of science and science education experts and failed scientists. They make exaggerated claims, supposedly to improve the public image of science and thus to serve the public. They undermine the traditional view that science produces autonomous thinking, not persuasion.

1. SCIENCE INSECURE

Rubbing shoulders with the famous is not frequent, but when one hangs about academic circles as long as I did, one is bound to meet with some famous colleagues. In my vanity, I try to keep tags on them and find information about their affairs. Some of them are famous also for their ability to make enemies. From my scant experience, I can report that even they are fascinating, but the nice and unassuming ones are more so. One famous physicist who was particularly nice and reasonable raised my curiosity. From early life he was dedicated to science and showed talent. People helped him find a place in the academy, but he felt undeserving. They showed him how to further his academic career, but he found this distasteful. His interest was in the advancement of science, not in his career. Finally, he made a bold move and switched from the faculty to the dean's office. He did well. He found out that he could further in good conscience the careers of peers who, as scientists, were not much more able than he was. Not interested in power, he achieved it nonetheless. He soon found himself in the position of a dean of science. As a dean in a reputed institution, he had no need to boost his colleagues. Some junior members of his office did that. He found to his surprise that he could assist them in this and even tutor them in the gentle art of the public relations of science that he had never mastered. It came to him naturally. Gently, he explained to his juniors, easy does it. Do not exaggerate. Just allow your audience to do this for you. Just report any find, and stress differences between it and its precursors, no matter how insignificant these may be. And report all the information that may support their ideas and their conceivable future applications. And put aside the problematic. You can mention past hostility, but only on two strict conditions. Name no names and no arguments, and always report a happy end. He was always afraid of exposure, but he knew that this fear was unreasonable, a projection of his frustrated wish to advance knowledge.

One day his old boss entered his office in a huff. I am on my way out, the grand old man said, home or to the better place. And you are my only replacement. This was an utter surprise. He tried to explain to the old man why he had somewhat neglected his research. He was surprised to learn that it was no news to his old mentor. The old buzzard knew the details of his activities. He asked him, why was he reluctant to blow his own horn but not to help his peers do the same. He pressed him on this, and thus forced him back to science full-time. He said, he had no more need to blow his horn. He was in error.

The day he won a Nobel Prize, he was in a shock. He knew that he had to go to Stockholm to justify the decision of the Swedish Academy to honor him. He read speeches of his predecessors and decided that he could emulate the justly humble among them. Being a nice fellow and knowing what he had to do, he did the right thing. He stayed cordial despite the prize.

Let me mention some factual background information. The traditional, modernist view of the rationality of science is that of perfect proof. The task most common today among philosophers of science is to dilute scientific perfection minimally, to render it near-perfect in a sense that would further justify its reliability and thus the demand to yield to its authority. Behind this venture is the most common idealization of science as (nearly) infallible and always the right choice, so that it is always better to endorse its authority than to make autonomous decisions. Philosophers of science fail miserably in their persistent efforts to prove this. This is interesting, especially since there is such a simple alternative to it, namely, Popper's view of science as the most critical venture, and so as the ally of individual autonomy. It is the first attempt at a break away from the traditional image of science as perfection, pure or diluted. Paul K. Feyerabend and I have claimed that Popper's image of science suffers from some idealization too. Yet, even if all of our strictures of it are right, it still remains the least idealized and idealized in a new and interesting manner.

(The contrast between the authority of science and individual autonomy is not always obvious. In Isaac Asimov's science fiction, *I Robot*, robots neutralize autonomous public decisions and impose the most rational ones.)

The criticism of Popper's image of science as idealized is very marginal as compared with his criticism of the traditional image of science as perfect and so as authoritative. The survival of the traditional image is an interesting and important social phenomenon that invites careful study. Those who wish to demote science need a new theory to explain the observation that the products of science are often more reliable than products of alternative ventures. Feyerabend denied this observation in his later phase. He was simply dishonest about this.

My own efforts in this direction led me to develop a critical theory of technology. (See my *Technology*, 1985.) Very briefly, contrary to what philosophers traditionally advertise, science does not impose its authority beyond what the democratic principle of the freedom of opinion allows. Rather, it rests on the idea that industry uses a different concept of reliability, and that responsibility demands the severe testing of an innovation before allowing marketing it. In some cases the law requires the responsible application of some science-based technique. It cannot require belief. This invites a contrast between the traditional view of rationality as the reliance on science and the view of rationality as autonomy. The latter view invites a new view of science too. Any such view will be a variant on Popper's view of science.

This view is very important and very exciting. Despite its staggering popularity in diverse circles, there are surprisingly scant discussions of it and critical assessments of it. Most of the literature on it is incidental and marginal. Otherwise, it is expository. I will here disregard the expository altogether, especially since most of it is faulty on a vital point. It presents Popper's output as the result of yet another effort to defend the authority of science by entrenching its reliability. Most of the criticism of Popper is surprisingly cursory and mostly worthless. Were it detailed and thorough, it would easily boomerang. Its advocates cling to the defunct idea of rationality as reliability and so they cling to the image of science as authoritative, and they assume that Popper clings to it too. Their assessment of Popper's alleged effort to establish the authority of science as a failure is valid. It applies more forcefully to their own efforts to defend the authority of science. Their criticism does not apply to Popper's work, since he opposed the idea that science is authoritative. My aim here is neither expository nor critical. I will not combat the allegation that Popper defended the authority of science; I will combat the image of science as authoritative – not by preaching autonomy but by discussing two of the interesting inaccuracies or misconceptions of Popper's view of science on which the alleged critique of his views rest. These relate to the asymmetry between refutation and verification, and to the prosaic nature of science.

Bacon posed most forcefully the asymmetry between evidence that supports and that undermines a given hypothesis, between positive and negative evidence. Popper has endorsed Bacon's argument, as did his countless followers – gladly or reluctantly. After all, the thesis is a part of logic: the refutation of a universal hypothesis by a piece of (negative) evidence is final, yet the support of such a hypothesis by a piece of (positive) evidence that it entails is not, as the possibility of a future refutation threatens this support. This is the whole of the asymmetry thesis. Einstein put it colorfully: scientific experiment, he said, is a question to Nature. She may say, NO, and She may say, MAYBE, but She never says YES. Bacon said, at times Nature says YES. He asserted that verification is possible; he did not say how. Whewell, the great nineteenth-century polymath, was the first to do so: he presented canons for verification. Newtonian mechanics was his paradigm. Consequently, the refutation of Newtonian mechanics refuted Whewell's canons too. Faith in verification is still alive and well. Its use is only for popular consumption. It is allegedly a form of harmless exaggeration permissible for public-relations purposes. It is harmful.

One may attack the asymmetry thesis by reference to the possibility to refute negative evidence. (This is folly: the possible falsehood of counter-evidence does not remove its conflict with the theory. But the folly stands for a valid argument: despite the asymmetry thesis, logic permits clinging to the theory by rejecting the counter-evidence.) This has led Carl G. Hempel to a verbal innovation. In parallel to the replacement of verification by confirmation, he suggested the replacement of refutation by disconfirmation. (The word "infirmation" is more apt, as it keeps the etymology symmetric: both "con" and "in" may legitimately serve as prefixes to "firm".) He never developed this suggestion: his proposal of the idea of disconfirmation is not serious. It is good enough to use to lambaste Popper with, but not more than that.

The question is not whether the asymmetry thesis is true, since it is a part of logic. The question is not even, can the asymmetry be broken? Rather, it is, what certitude does the thesis allow? And this remains a valid question, and the critique of Popper's view of the asymmetry is that the answer that his rivals ascribe to him is unsatisfactory. It is the suggestion to replace verification with refutation, allegedly because final empirical refutation is possible. This criticism is valid: as Popper has emphasized from the start, there is no assurance that any hypothesis is false. Some people ascribe to Popper another wild, absurd view that all hypotheses are false.

Some writers consider the utter uncertainty of hypotheses intolerable. They find intolerable, as they expect of a theory of rationality that it should decide unequivocally and with finality that certain hypotheses are irredeemably objectionable. (Think of racism as an example.) These writers cannot be happy with Popper's views. Nor need they be. It is their liberty to remain dissatisfied. It would have been nicer of them were they more candid and explicit about what they like about Popper's view and what not. There is little option but to view these writers as lacking autonomy. They are sure that certain hypotheses are most objectionable, and rather than object to them, they wish science to impose this objection. This is pathetic. Of course, the rejection of obviously objectionable hypotheses is eminently rational. These writers want this rejection to be certain. If this is not due to a lack of autonomy, than it is adherence to the defunct, classical theory of rationality as certitude.

Thomas S. Kuhn offered a new argument for the need for scientific authority: without scientific authority, scientific revolutions are impossible as then there is nothing to rebel against. For researchers, he said, science is bread-and-butter and so it is as ordinary as any bread-and-butter. Revolutions are delicious, he agrees, but "science is not all pie."

2. SCIENCE IDEAL AND MUNDANE

Søren Kierkegaard declared the bread-and-butter kind of religion a kind of blasphemy. Nothing disturbed him more than the act of putting on Sunday's best every Sunday and going to church as a matter of routine. He saw no possible honest approach to the Lord, except in fear and trembling. It is no surprise, then, that what notice he met from the Danish clergy of his day was hostile. At least for normal clerics, religion is as much bread-and-butter as science is for the normal researchers.

Popper found Kuhn's criticism an "eye-opener" – not regarding the phenomenon of normal science, but its having become commonplace. Earlier, he was not aware that so many researchers endorse these days this attitude towards science that he found objectionable. At least Kierkegaard was aware of the popularity of what he found objectionable. Was he in error? Was Popper? I do not know. Is it better to ignore religion or to take it for granted? I discussed the current situation with some Danish clerics. Their churches are empty these days, and this grieves them. But they have no wish to return to the past. They ruefully agree with Kierkegaard: they do want people to come to church, but not routinely. Those who feel the need to face their maker will find that the priests are there for them. The Danish clergy agree with Kierkegaard. Kuhn and his fans disagree with Popper.

The pair of science and religion invites a third party, art. What is the case with art? The traditional distinction between the fine and the applied arts comes to mind. Are the fine arts more exciting than the applied arts? No. Moreover, this is irrelevant. Sir Ernst Gombrich has asserted emphatically that fine arts are extensions of applied arts. All art is bread-and-butter for artists. Artists who always live intensely and relentlessly, experiencing constant great excitement (Van Gogh, Kafka), suffer greatly. It is inhuman to demand of every artist to suffer the burden of the demand to live daily in great excitement. The alternative idea is that great art is rare and the rest is the mere bread-and-butter that we may ignore. It is no less objectionable. It is just as extremist as the idea that all art should be exciting. Perhaps it is the same idea in a different garb. The last option that remains, then, is the idea that art is mostly mediocre and that so it should be. This idea is very unpleasant as well. Art is or should be challenging. Yet, we cannot face severe challenges every day of our lives. (We cannot even go to foreign restaurants every day, says Gellner.) It does not follow that what is not exciting in the extreme is dull in the extreme. There are degrees of challenge. Autonomous people decide for themselves when to accept a challenge and when to let it go. This holds for art, science, and religion.

It may be useful, then, to notice that not only art and science, but also religion, has practical aspects. Since Denmark is a thoroughly secular country, there is no legal obligation there to use religious services ever, so that anyone who engages a religious functionary – to perform rites of passage, for example – does so out of a simple feeling of the need for these services. Ardent followers of Kierkegaard may suggest that then the rite should proceed only with fear and trembling. Possibly so. Their counsel may prevail, but definitely not as a matter of course, since free people have the right to choose to heed it or not.

Why then does the criticism of Popper's view of science as too radical sound more convincing than the criticism of Kierkegaard's view of religion as too radical? Perhaps because it looks more like a homily than a serious contention. Popper has rejected the authority of science; did Kierkegaard reject the authority of religion? I do not know. As to the parallel case in the arts, perhaps the radical view of art looks like the typical chitchat that people expect from art critics and commentators. In any case, art has no authority. Moreover, the question at hand is logical: does the practice of normal science conflict with Popper's view of science as conjectures and refutations? If it does, then never mind the question of excitement. If it does not, then, likewise, never mind the question of excitement. How does excitement enter the picture anyhow? The answer is that Kuhn has introduced it, and it was my mistake to follow him rather than to declare his claim irrelevant from the start. He admitted that scientific revolutions do occur but he said that they are rare and brief and in other times normal science reigns. He admitted that scientific revolutions, as all revolutions, may be exciting, but he insisted that normal science is and must be boring. It is his insistence that research is boring that is at stake, then. David Budworth says in his autobiography, reading Popper makes him regret that he has left research; reading Kuhn has the opposite effect on him.

Only the prevalence of Romanticism allows so much confusion to stay – Romanticism about science, art, and religion. Not all revolutions are exciting. Normal science is objectionable not since it is dull, though we should always fight dullness. Normal science is objectionable since the demand for it rests on a dogma. Kuhn defended it; Popper criticized it. And rightly so: before imposing a boring task, we may try to replace it by a more challenging one. And even if a task in normal science rests on many questionable assumptions, there is no need to believe in them. And taking much for granted occurs even in a revolution. When Erwin Schrödinger developed his famous equation for the electron, he took Newtonian mechanics for granted; to suggest that he ever endorsed it is insulting; yet this is a corollary to Kuhn's philosophy. Schrödinger's equation is the prevalent paradigm today in many branches of applied physics, yet no normal physicist will believe the Newtonian system by which it abides.

There is more to normal science that is distasteful. Unlike proper science, it comprises research that employers impose with total disregard for curiosity. Kuhn explicitly approved of this. He views this as a refutation of Popper's view of science as exciting. Popper's lovely aversion to the imposition of routine uninteresting work is regrettably no support to his view of science. Regrettably, this too is unimportant. At times, science does demand hard, uninteresting work. But the need for it is no excuse for its rigid imposition.

Nevertheless, Kuhn has a point – not on excitement, but on revolutions. He rejected the view of science as revolutions only: we have no energy to rebel constantly. Also, a revolution is against some established order. Otherwise, there is nothing to rebel against. But Popper never said, science is all revolution. He said, science is conjectures and refutations. Kuhn understood Popper's terms "conjectures" and "refutations" as synonyms with his terms "paradigms" and "scientific revolutions". This is inaccurate. Popper suggested taking seriously and examining critically any highly testable conjecture, whereas Kuhn suggested endorsing the paradigm. Popper required of all those concerned with science to admit refutations. Kuhn rejected this requirement. He took the admission of a refutation to be necessary only when the establishment demands it. The establishment introduces a new paradigm, he said, as they find fit. He calls such events "paradigm shifts". What a paradigm shift exactly is, he added, matters little, except that it is centrally coordinated. He uses the word "revolutions" as a metaphor for paradigm shifts, regardless of whether the old establishment or some Young Turks are in command. The leadership, then, is simply those who lead, and paradigms and paradigm shifts are for the scientific establishment to determine and for the scientific rank-and-file to accept. By contrast, Popper never discussed scientific establishments, and he demanded no obedience except to the law of the land. Kuhn said, researchers are and must be dogmatic, and he approves of it. They dogmatically endorse the paradigm that the establishment imposes on them, he said. Paradigms are ideas that the establishment imposes, he insisted. This is not true. Employers, not scientific establishments, are responsible for imposition. Kuhn spoke as if the main employers in research are universities and the decisions on academic employments are largely in the hands of the scientific leadership. Yet, if you reminded him that Einstein had no hand in any academic appointment and that academic administrators constantly partake in decisions on appointments, he might have dismissed this as irrelevant.

Are there scientific revolutions? Popper offered no view on the matter. Mario Bunge has. We keep changing our views, he says. As radical changes are hard to come by, it is wise to try small changes before proposing and examining large ones. There are here empirical questions, then. Are researchers dogmatic or independent? If they are dogmatic, does this not refute Popper and Bunge? Kuhn's view is a non-starter. The scientific leadership approves of his idea as they take it to be good public relations. They do not believe it and they are not interested in its detail.

3. SCIENCE IDEAL AND IN CONTEXT

The literature whose function is to extol science has led to one whose function is the opposite. It comprises empirical studies of laboratory life that, being empirical, are not to take lightly, and that expose the clay feet of research, thus allegedly refuting all established views on science, including Popper's. Though the empirical information these studies convey is true, they are pseudo-scientific. Let me explain.

These empirical reports go further than Kuhn's report that the work of normal science is quasi-routine. They are detailed reports on daily life in the laboratory. They expose it as very far from ideal – the way life everywhere else is.

Researchers usually are less prone to misconduct than other segments of the population. We should not jump to the conclusion that science makes its practitioners decent. The conduct of Nazi normal scientists refuted this conclusion. Let us overlook this and agree that normally researchers are rather honest, though still somewhat prone to misconduct anyway. It need not constitute research – even when normal scientists perform them on the premises of their research institutions. This view is reasonable. It is comparable to the view that an act of blasphemy that some clergy commit is not religious, even if they do this on religious premises. Yet, things are not so simple. Where does one draw the line? In religion, religious doctrine decides this. (Some religions condone some blasphemy: remember the *Book of Job*.) An act of blasphemy is thus always within the religious dimension. To say that blasphemy is not a part of my religion is easier than to say that it is not a part of anyone's religion. To see this, take something less problematic, such as the stock exchange. It is hard to say that though inside trading regularly takes place there, it is not a part of it. Perhaps it is not, since it is decidedly illegal. It is harder to exclude from the stock exchange legal acts that are immoral, such as slight deception of customers that stockbrokers scarcely avoid. Dealing is as common in the stock exchange as it is in scientific research. Particularly in the act of publication, negotiations regularly take place between author and editor with the aid of referees and mediation of all sorts of power brokers. Is this science? It is, and it often stinks all the same. Science is wonderful but it can and should improve.

The literature on laboratory life has stimulated much discussion as its image of science is not idealized. Is it wrong to idealize? Consider idealizations in art. Take negotiation there. Is it evidence that art is not what one thinks it is? Is negotiation always a compromise? Does negotiation compromise art? Gombrich says, no. Art can benefit from negotiations between some artists and their customers, such as Michelangelo and the people who commissioned his work, including the Pope of his day. Moreover, the idea that research includes negotiation is scarcely new except in detail. For, the same message as this literature conveys is present in the earlier literature, in the writings of Michael Polanyi, for example, who has discussed the process of training, of hiring and of firing research workers, and the like. He saw this as central to the maintenance of uniformity of opinion in the profession of scientific research, as he deemed uniformity of opinion essential there, especially in training, and he declared this uniformity possible to achieve and maintain only with the aid of the imposition of discipline. Admittedly, some use of force is unavoidable in any society and sub-society. But the imposition of opinions is contrary to the basic principles of individual autonomy and of democracy, and is pointless. At most uniformity of conduct is; never uniformity of views.

Merton said science is egalitarian. Should we listen to Merton or to Polanyi? Do the facts that agree with Polanyi support his views? Do not believe what scientists say, mused Einstein in his celebrated Herbert Spencer Lecture, look at what they do. He said this in jest. He said, scientists defer to the authority of scientific experience, but in their research they exhibit opportunism and recklessness. Why is what Polanyi and the reporters of laboratory life say so different from what Einstein said? Because he was preaching autonomy and they are preaching heteronomy. Polanyi said, the leadership of science should block cheating. The laboratory-life reporters say, they do this only up to a point: they report that some researchers bend the truth if and when they hope to get away with it. So, they reject Polanyi's proposal. Commentators took this foolishly to be license to bend the truth.

That some researchers misbehave appears in the relevant literature as an empirical observation, in order to recruit the authority of science, though commonsense will do equally well if not better. Ever since science became prestigious, people who crave prestige as researchers with no ability to earn it may pretend to deserve it. Popper discussed one kind of this pretense, the pretense that a hypothesis is scientific. He dismissed fake information as insignificant, since it is short-lived. The scientific tradition is very strict on the demand to have information repeatedly checked before granting it scientific status. Faking a diploma is still not agenda, because until recently tradition ignored the diplomas. Their very use today suffices as evidence that not all results of research are scientific. Reports on laboratory life to that effect are redundant. The use of diplomas rests on the relative ease of checking. The use of mathematics is only a little harder to check for propriety. Its misuse rests on the popularity of the idea that mathematical symbols are a sure sign of scientific character. Fake mathematical formulas are commoner than fake diplomas. This proves that gatekeepers are inefficient. Does this refute any view of science as decent? If yes, then the very use of gatekeepers already does.

The same question put to art connoisseurs is significant, though: how do they differentiate the genuine from the fake, the great from the good? A general criterion of the genuine is easy to formulate. Connoisseurs find it a challenge to differentiate specific cases and evade it in general. This seems to support Kuhn. How does it say anything about his view of normal scientists as bored and burdened? Only because he says their products are valuable. He idealizes. Their products are often valueless. Who pays for them, then? If it is some private funds, then there is no need to discuss it. If it is some public funds, does the public care for this kind of waste? The discussion thus fizzles out again. The only heartwarming thing to report in this connection is that some researchers have claimed that the boredom of research that Kuhn reports is no different from the boredom that many workers suffer, and that we should try to alleviate all such boredom to the benefit of all concerned. Since Kuhn's fans bow to experts, and since experts are now increasingly endorsing the view of the movement for the improvement of the quality of working life, they should yield.

Somehow, it seems discomforting to broaden the discussion of the boredom of scientific research to the boredom of employment in industrial society in general. This happened in all cases in which we discuss science in concrete terms. Some important questions about science do demand airing but they are outside the normal domain of the philosophy of science and of its public image. For example, will scientific training improve if it moves from the university to industry (as Budworth suggests)? Apprenticeship used to be the rule, and now it is the exception. So much so, that schools of law and of medicine mask apprenticeship by the demand that apprentices should be armed with university degrees and function under different names, not as apprentices. The same holds for the question, is it better for science to separate research from training to some extent by the institution of research institutes? Likewise, is a scientific *lingua franca* a good thing or not? And if yes, should it be English or Esperanto?

These questions merit study. Regrettably, they suffer neglect. Philosophers of science ignore them due to excessive idealization. Kuhn's effort to avoid idealizing science as breathtaking led him to idealize every employee in the modern science-based industry as a researcher proper. What idealization is legitimate, then, and what is excessive? This question almost never came up, and there is only one known answer to it: opponents in a debate can approve of controversial moves or veto them. More specifically, when a move is a concession to the opponent, it is permissible; otherwise, the opponent may veto it. This holds for all controversial moves, and so it is applicable to idealizations.

The students of laboratory life claim that they report facts with no bias, but they apply the idea that controversies are nasty and detrimental to science. They report that often resolutions of disputes are not rational but manipulative, and that this includes rhetoric. They ignore the great value of some terrific controversies. Their hypothesis is, reality is, a social construct – whatever this exactly means. They say that their observations are application of this hypothesis. Hence, they, too, idealize. Their idealizations receive their popularity, such as it is, from their dislike of autonomy.

4. THE FACTS ABOUT SCIENCE IDEALIZED

The ideal of science as total knowledge, and even with assurance, is naïve beyond words. It hardly invites criticism beyond the mere observation that it is unattainable. Neither the whole truth nor certitude is attainable. The two ideals differ. Whereas the ideal of the possession of the whole truth is valuable in itself, certitude is not. The wish for certitude is simply to insure that what we have is the truth. Perhaps certitude can satisfy other needs as well, and thus qualify as an ideal proper. For example, certitude can provide an inner glow of satisfaction. This is hardly desirable. Both the whole truth and certitude are obviously unattainable, yet, as David Miller tirelessly stresses, only one, not the other, is a worthy ideal. This sounds paradoxical because of the erroneous identification of the normally desirable with the ideally desirable. Under normal conditions, we need assurances and national defense; ideally, they are unnecessary. Imagine cases where we want, even ideally, some sort of assurance. The case of the knowledge of the provability of a mathematical theorem, for example, is such. It is interesting because its deducibility or dependence is interesting, not its certitude. The same holds for the question of refutability. It is the question of the incompatibility of some hypothesis with some possible observation. And the certitude of incompatibility has nothing to do with the quest for certainty that plagues philosophy for millennia.

Miller is right: the truth is better than certainty. The expression of this fact in the public relations of science is simple. P. r. people claim that what science tells us is true. It follows that it is reliable. P. r. people need not argue that scientific hypotheses are reliable. This claim follows from the claim that they are true. Should we endorse this claim of p. r. people? People hardly ever question scientific statements and they hardly ever question their scientific status. This includes the general public, the students in science courses, and even research scientists when they learn of scientific hypotheses that are not of their fields of study. At most, systematic doubt and criticism are the practice within one's own research discipline, and hardly even there. There is nothing wrong with this. Doubt as such is easy and of little or no interest. Criticism is always valuable but usually hard to come by. The received view is that doubt is an essential or (at least useful) prelude to criticism. Not so. The history of the criticism of the leading scientific hypotheses is full of cases of true believers who refuted the hypotheses they believed in and thus made great contributions to science. Critics who are true believers may try hard to reconcile their own hypotheses with their own criticism. This is true of most of the best critics of Newton's system. Einstein was a true trailblazer as he was more than willing to let this grand system go and even try to replace it. Many philosophers struggle with the presence of scientific revolutions. The most popular book on them, by Kuhn, declares that superseded systems, like that of Newton, are not false. They are admirable, of course, but still false, no matter what p. r. people say.

The interest in the question at hand is purely a matter of public relations. The presentation of science as ideal is not new. Yet the view that we are in possession of this ideal has a complex history. The 1810 edition of *Encyclopedia Britannica* has an article on science and it is very brief, asserting the ideal of science as certainly true statements. It neither asserts nor denies that we are in possession of science. The next article is about scientific amusements, and it reports on a famous book on science as entertainment, which was a very popular item in the eighteenth century.

P. r. people present science as the truth, not as the whole truth. This is manifest in the usage of such expressions as "science is not yet in possession of a full explanation of ..." or "We still do not quite understand ...". These expressions are rare: p. r. people do not speak of doubt, of difficulties, of unanswered criticisms. Usually, they mention these only after a victory. "The innovation is important, as it overcame a well-known difficulty."

We might expect that at least historians of science will present as rather defective the hypotheses that were plagued with such difficulties. Not at all. The solution of the difficulty tells us that science can overcome difficulties. This requires of historians of science to write somewhat anachronistically. If this sort of deceit fails, they can consider the difficulty-plagued old idea a system, a paradigm. Systems or paradigms are never open to questioning, especially not with the aid of information that later developments have revealed. In short, they defend the defunct idea either anachronistically or in the name of historical accuracy. They thus immunize it against criticism in the name of historical accuracy. They will not mention the historical fact that criticism led to new discoveries: we should place science above criticism.

The official p. r. system admits the rationality of science its hallmark; laboratory reporters view it as vulnerable. Nevertheless, they officially support science; indeed, they are p. r. spokespeople for science. This is amazing. Do the p. r. spokespeople for science silently consider them the enemy? This is an empirical question, and I suggest as an empirical hypothesis that these people stand on the side of science, that the damage they do is unintended. I am following here the grand theory of Claude Lévi-Strauss, who says that pairs of idealizations complement each other. The laboratory reports in question are not empirical but contrary idealizations. They present their idealizations as real thus adding the confusion about idealizations.

So we need guidelines about idealizations. I propose the following. Every party to any dispute has the right to veto any idealization, and the one that we cannot veto is either one that is a concession in a debate, or one that we have introduced: we cannot do without idealizations. We may try to improve our idealizations as we try to improve any of our hypotheses and techniques. It is this critical strive for improvement that is the hallmark of critical rationalism. For this, not only science needs veracity, but also its p. r. system. Its members should cease to assume that they should bend the truth, and appeal instead to the intelligence of their public. This is both useful and honest.

Chapter 2.7

IRRATIONALISM TODAY

SUMMARY

According to classical rationalism, prejudiced people cannot conduct proper research, and all error is prejudice. Known cases of prejudiced and even Nazi scientists refute the first thesis. The crucial experiment between Newton and Einstein refuted the second: Newton's theory is not true, yet it is admirable all the same. Contemporary popular irrationalists declare crucial experiments impossible. They sound plausible because of the failure – due to cowardice – of the leadership of science to admit openly the need to replace the traditional theory of rationality.

0. BACKGROUND

Our western culture exhibits different attitudes towards rationality and irrationality. There is no agreement about the question of the rationality of many an item of our culture. But at least we agree that the scientific culture is the bastion of rationalism. We also agree that irrationalism is a response to rationalism: cultures not familiar with rationalism have no irrationalism. ("Pre-rational" is the customary label for them.) Now some parties try to raise doubt about all this. Natural scientists, and even those who only wish to count as natural scientists, often abhor all that is not scientific, especially superstitions and occultism. And they deem blasting at these from time to time a pious duty – for the sake of saving the young. Many of them follow the obviously erroneous tradition of science that identifies the irrational with the unscientific – particularly the occult, of course – despite the obvious fact that most of the occult is pre-rational. Others, though still hostile to the unscientific, are more tolerant to the occult that fills our distant past. They join philosophers and psychologists and cultural historians and historians of ideas and social and political commentators in tolerating the pre-rational, though not the anti-rational, and particularly not the post-rational. Pre-rational people are not aware of the better rationality of science. Were they able to choose, they might choose science. By contrast, the anti-rational and most of the post-rational reject reason and science, openly preferring the superstitious and occult to the scientific. Some post-rational writers differ. They have no patience for the superstitious and occult, but they see science as one tradition amongst many – a special tradition, but still not the only acceptable tradition.

Social and cultural scientists and historians do not stigmatize the pre-rational as irrational. They study some superstitious and occult belief-systems without condemning them. In the first half of the twentieth century, two important and influential thinkers expanded received views on scientific rationality. Sir James Frazer offered a general theory of magic as pseudo-science, inferior to science yet superior to religion. Max Weber developed his theory of ideal type that permitted the study of diverse religions and magic systems as general social frameworks, without comparing different frameworks, while treating them all on a par. This is his celebrated view of value-free social science: judge ye not. The views of Frazer and of Weber are elaborations and expansions of the rationalist tradition. They incorporate ethnology as merely informative and so as scientific, even when its subject matter is magic and superstition.

At about the same period a revolution took place, which introduced a new and sophisticated version of relativism. The revolutionaries were Pierre Duhem, Emile Durkheim, Bronislaw Malinowski, Sir Edward Evans-Pritchard and Michael Polanyi. They were all important and influential thinkers. They developed their ideas in contrast to the view of the scientific tradition about rationality, yet they clearly belonged to that tradition. To destroy the tyranny of science over metaphysics, Duhem used conventionalist-instrumentalist methodology. This methodology denies science the ability to impart theoretical information about the world. Metaphysics then remains the uncontested authority. Durkheim considered social cohesion a supreme good. He deemed magic and religion mere instruments for the enhancement of cohesion – ways for society to celebrate itself. Malinowski went further. We should not judge, since judgment comes from our own parochial framework. Evans-Pritchard went still further. He offered a profound philosophic reason for putting all frameworks on a par. One can judge only within a framework. Hence, no framework from which to compare frameworks is impartial. Polanyi added a new and exciting argument. Able masters cannot articulate their knowledge, he said; they transmit scientific or artistic skills in workshops. This exempts them from placing their ideas on public platforms and subjecting them to critical scrutiny. The public must accept the expertise of masters.

The two most popular philosophers of science of the second half of the twentieth century, Thomas S. Kuhn and Paul K. Feyerabend, vulgarized the new irrationalism. In accord with Duhem, they declared frameworks incomparable. Hence, there can be no comparison between Newton's system and Einstein's, since, they said, there is no crucial experiment in science. They did admit that today's physics follows Einstein. But this is not to condemn Newton, much less to consider his ideas false. The two were not in full agreement. Kuhn approved of the traditional scientific attack on magic and nevertheless he endorsed Evans-Pritchard's defense of it – as acceptable only where it is common practice, in magically oriented cultures. As a researcher, however, he shared the traditional hostility of science against magic. Feyerabend rejected it as imperialist.

The view that magic and science are equals is the mark of contemporary irrationalism. So is the view that the physics of Einstein and that of Newton are incomparable. It is simply incontestable that, intellectually, science is superior to magic, and Einstein's physics is superior to Newton's. Nevertheless, the habitual attack that practitioners of science repeatedly launch against the superstitious and the occultists is a traditional error, both tactical and philosophical. It is defensive and hostile, as it rests on excessive claims for the power of reason that boomerang.

The traditional hostility of scientists towards the occult rests on two arguments of Bacon that became traditional and characteristic of the Enlightenment Movement. First argument: unreason impedes the growth of knowledge. Second argument: the promise of the growth of knowledge is so huge, that sacrificing traditional ideas for it is worth while: soon reason militant will become reason triumphant. Both traditional arguments in favor of combating unreason are obsolete: some great researchers were superstitious and prejudiced; and there is no way to obliterate all tradition.

These refutations render the attack on unreason pointless. Hence, it is time to rethink the task of the education of the modern rational autonomous individual, and the task of rationalizing politics. Still, as far as the merely intellectual aspect of the matter is concerned, suffice it to argue against what is merely seemingly reasonable in the irrationalist doctrine – and to incorporate into a new version of rationalism whatever is reasonable in it. We should invest available energy in efforts to develop a theory of reason on such lines, a theory that should take account of the ability of irrationalists to contribute to the growth of knowledge and add to the understanding of rationality.

Here is my plan, then. (1) I shall present here the classical Baconian theory of rationality – his doctrine of prejudice, or his radical epistemology. (2) I shall then describe the way in which the doctrine still influences and confuses serious thinkers (3) with ill political effects. Next comes (4) the growth of modern irrationalism. I will conclude (5) with open practical questions on attitudes regarding irrationalism.

1. REASON REAWAKENING

In the first quarter of the seventeenth century, Bacon asked, what is the cause of the Fall of Man. There is no way to secure progress unless we know what has destroyed ancient culture, what led from light to darkness. For, no matter how much we progress, we may find ourselves on the same road as the Ancients and like them plunge into a Dark Age. Bacon's answer to his own question is very simple. Ancient skepticism caused a great discouragement. Ancient dogmatism raised new hopes to counter the discouragement and took root. And dogma is superstition and superstition is prejudice and prejudice acts like blinds and blinkers. Darkness is never in the world; it is the quality of the blind.

All this calls for comments, explanatory and critical. Bacon accused Aristotle for having caused the Middle Ages by imposing dogma on the world. And he explained this by ascribing to Aristotle a cheap motive: the lust for fame plus the inability to achieve it by honest means. Together, these two features move to imposture. The ambitious would stake claims for fame – create school dogmas in the guise of scientific truths. Hence, the very destruction of the Aristotelian school opens the road to progress, but it also opens the possibility of imposture to another unworthy person. Hence, the risk of new Middle Ages is all too real. Bacon said, Copernicus was an impostor lusting for fame. This accusation does not rest on any specific information. Even if the Copernican doctrine may be better than that of Aristotle, replacing one school's dogma with another will not do, said Bacon. Nor was he the only person who issued this warning. To take two conspicuous cases, Huygens said of Descartes that he wished to replace Aristotle. Dr. Thomas Thomson said, Lavoisier wished to replace the phlogistonist Stahl in lust for fame. Thomson offered proof for his allegation: Lavoisier's doctrine is untrue. Hence, it is a school doctrine.

Bacon had two chief problems. One was the demarcation between science and pseudo-science. The other was the explanation of unreason. And the two are complementary. Knowledge is power, said Bacon, power over Nature. Unreason replaces real power with mere fame, with power over people's minds. In terms of Bacon's disciple Hobbes, power is the ability to control. Bacon thus demarcated science from pseudo-science by viewing one as the real power that commands consent and the other as the mere reputation that is power limited to a school of thought. As knowledge is the means for controlling Nature, Bacon's advice was to stoop to conquer. This advice comes from the books of courtly love. Speaking of knowledge in sexual terms was very common, incidentally, and rooted in the Cabbalistic tradition, popular in Bacon's day. So was the idea that a dispassionate approach to love is the only way to insure success. Fear and hope are both excesses that disturb and prevent the assured and self-assured approach, so necessary for conquest. It is no accident that Bacon, and following him Hobbes and Spinoza, speak against both fear and hope and in favor of tranquility; it is no accident that they identify these attitudes as skepticism, dogmatism, and rationalism, respectively. They also all agree – with Epicurus, as it happens – that fear inspires doubts that invite dogmas with which to overcome fear by hope.

Today only experts remember Epicurus. As to the Cabbala, fewer experts are familiar with it. Yet, some of their influence survives. It is the requirement that philosophy should steer science between the Scylla of skepticism and Charybdis of dogmatism. Almost all twentieth-century philosophers share it, including Russell and Collingwood, Popper and Carnap and Polanyi. Though it is obviously out of date, it is hard to say who among the better-known philosophers rejects it –. except for Mario Bunge. Oddly, the recipe for the avoidance of the two extremes is usually the proposal to blend them in proper proportion – to be at times a bit dogmatic and at times a bit skeptic. This makes the theory of scientific rationality a myth in accord with the general theory of myth of Claude Lévi-Strauss. Bacon's view is a myth, and one that is still very popular.

Skepticism	Science / Reason	Dogmatism / Faith
Fear	Tranquility	Hope

Polarity and mediation in general – *à la* Lévi-Strauss

Bacon's declared that research is easy. So he had to explain the scarcity of science: he had to ascribe it to some non-intellectual obstacles – moral or emotional. The two intellectual violations of science, skepticism and dogmatism, reflect inadequate emotions: fear and hope. As knowledge is power, those who lust for fame appeal to inadequate emotions. The true investigator of nature is pure of heart, humble, and industrious. Truth reveals Herself to the worthy. Since we must be humble and open-minded to receive Her, She has to come by Herself, by surprise, by accident. Hence, the celebrated saying off Joseph Lagrange, accidents happen to the prepared mind. The alternative option is to impose preconceived opinions on Nature, to prejudge Her, to have a prejudice. Thus, to voice a hypothesis is to be presumptuous, to rape Nature in order to conquer Her, to put Her in chains and force Her to reveal Her charms.

Bacon's doctrine of prejudice, the view that we see what we want to see, is amazingly astute, profound, and intelligent. It also has many confirmations. Philosophers and psychologists rediscover it regularly. Students of perception ascribe it to the *Gestalt* psychologists who have lovely illustrations for it. Cognitive psychologists ascribe it to Leon Festinger (his theory of cognitive dissonance). Sociologists and politologists ascribe it to Gordon Allport. Philosophers expound it repeatedly, as does W. V. Quine in a successful popular volume on the philosophy of science that he has coauthored. The doctrine is so amply confirmed that it often goes without saying: it has become an integral part of common sense. Any obviously false theory considered obviously true because of its many confirmations illustrates Bacon's doctrine. This is also true of Bacon's doctrine itself. In accord with the doctrine, the doctrine itself has become a prejudice. Its adherents see all facts as confirming it and when they see its refutations, they declare them confirming or else irrelevant. Any resistance to any new theory confirms the doctrine. Its adherents claim that a new theory overcomes resistance not by rational means but either because the prejudiced die out or they make some irrational move. S. P. Langley has offered this idea and Florian Cajori has repeated it, and then Max Planck repeated it in his scientific autobiography, and Kuhn quoted him. Collingwood and Polanyi said so too. It is now customary to ascribe it to Kuhn.

Bacon's theory of prejudice has many confirmations, but also some refutations. Whewell noted this first. Suppose that researchers articulate hypotheses clearly, deduce from them testable corollaries and then test and refute them. If they consequently admit error, then they exhibit rational conduct. Bacon said, this does not happen. Scholastic philosophers criticize each other for centuries, he noted, and with no school admitting defeat. How then did Bacon counsel his readers to get rid of prejudices? He dissuaded them from debating with the intellectually committed: intellectual commitment makes committed people deaf to criticism. How, then, could one ever give up one's convictions and start afresh? How is the need for a radical change win popular conviction? Is a radical change *à la* Bacon easier than small changes *à la* Whewell?

Bacon assumed that prejudice is strong. Hence, nothing will counter it short of starting afresh. He called his readers to concede ignorance honestly. He suggested that all it takes is the realization that renouncing the little that we have we should regain much more. We will gladly start afresh if we realize that relinquishing a little is essential for finding much more. It would be a very profitable trade. This is radical rationalism or the doctrine of the Age of Reason.

2. UNREASON MIXED WITH REASON

Radicalism recommends an all-or-nothing policy, since a pet prejudice, however small, may act as a cancer. We have no criterion to measure the prejudice and its weight except from the viewpoint of the truth. We believe that we possess the truth, but this very belief is a prejudice that bars us from it. Radicalism is inapplicable to any one field of study. The endorsement of a prejudice in one field prevents a truly new start in another. This is so for an obvious reason. There is no guarantee that a prejudice will stay in one field. The division of learning into fields can be true or prejudicial – scientific or prejudicial. Hence, science must precede the division of learning into fields. To say of people that they are prejudiced in one field but scientific in another is a prejudice – unless the division in question is scientific.

Nevertheless, Bacon declared natural and social studies distinct and social studies subject to religion. For, in social affairs any dogma, he said, is better than none – even the Talmud or the Koran. This Bacon said only in his *Essays*, and these had the merit of popularity but thereby the defect that serious readers took them lightly. In his more serious works, Bacon only demanded the complete separation between science and religion – for fear of contamination. His disciples differed on this matter. Hobbes viewed religion as a superstition (though he supported state religion). Spinoza demanded toleration and democracy as the better stabilizers of society than state dogma (though he too supported state religion), and moral individual autonomy as the best means for coping with the vagaries of life. Robert Boyle decreed that disputes should be taboo in learned society and Newton complied for lack of desire to air his religious heresies in public. Officially, science tolerated religion even though by its official ideology religion is superstition. The influence of Bacon, Hobbes, Spinoza, Kant, and Tom Paine, who all spoke of religion proper as superstitious, made the hostility of science towards religion an open secret. Laplace's dismissal of God was excessive, since the deists, especially the fathers of the American Revolution, saw themselves as the true heirs of the Enlightenment Movement and they maintained a condescending tolerant attitude towards established religion. The change came with Darwin. His disciples felt that they were on the verge of victory – against established religion in Western society.

This was a serious error. The bastion of unreason was not established religion. The real enemy was not Christianity but reactionary chauvinist political philosophy and all that. The worst reactionary chauvinism, whose culmination was Fascism and Nazism, as well as the enlightened sophisticated variants of it, was a silly irrationalism. Early in the twentieth century, Emile Durkheim considered social cohesion a supreme good, and religion merely instrument for that. If not Hegel, then at least enlightened Durkheim, recommended taking religion as a mere political instrument. Religion taken literally, he considered too silly; but taken symbolically, as he recommended, he saw it as the means by which society worships itself. Thus, he commended politics, since politics uses unreason for a good cause, and thus he commended unreason well used. people devoted to science use ideas of this sort to permit themselves to restrict their rationality to science and to ignore the irrationalism of the politics of their own nations, even if they partake in it. They advocate a religion as a tool for activating the masses, and thus they pave the way to the use of mass irrationality to promote the worst objectives and inflict the worst atrocities ever.

The explanation of a failure is obvious. Solomon Maimon, whose autobiography of about 1800 describes his medieval-style childhood, describes how he had performed a cabbalistic magic ritual, failed, checked his performance, and became a heretic. This raises a simple question: why did other Cabbalists not come to the same conclusion? The answer is obvious. The terrible story of the Rabbi Joseph Della Reyna of the seventeenth century illustrates it. He succeeded to put the Devil in chains, but then he had a swollen head, became careless, allowed the Devil to outwit him, became a heretic, and finally committed suicide. The moral from the story is this. Only the humble can contribute. The nearer one is to success, the harder it is to stay humble. Bacon explained the significance of humility. Without it, one becomes ambitious, develops a dogma for a school, becomes prejudiced in its favor, and thus disqualifies as a researcher. The Age of Reason deemed the refutation of a theory sufficient evidence for the claim that its originator was conceited, impatient, and eager to establish a school.

All this fits the theory of myth of Claude Lévi-Strauss. He was a pseudoscientist, at times a bit of an obscurantist, and with more than a flair for that kind of a reputation that Bacon had described. Yet, he made an admirable and important contribution to scholarship. His theory of myth is also a theory of rationality. Myths think for us, is his justly celebrated slogan. Consider a quality that is neither consistently present in nor consistently absent from any item, we tend at times to say of that quality that it belongs to that item and at times that it does not (instead of saying consistently that it partly belongs there). Worse, we tend to polarize and confuse the absence of a quality with the polar opposite of that quality. For example, we identify the not-good with the bad, even though most of us are neither. As we polarize we ignore the fact that things belong to neither pole but are somewhere in between. Thus, we say of people that they are good and we say that they are bad although they are neither. We say more often of people who are more good than bad that they are good, and we say more often of people who are more bad than good that they are bad than that they are good. We are mostly neither black nor white but gray. Myths describe us as black and as white, and we mix the descriptions to get the gray.

Lévi-Strauss is extremely enlightening on the question, is science myth-ridden. Consistent with his view, he often says, science is devoid of myth, and sometimes he says, science is all myth. This is an obvious error. As Bernard Shaw noted (Preface to *Saint Joan*), science comes with myths, but they do not disturb it overmuch; they disturb its social relations. For long I was troubled, let me confess, by noticing that scientists are at times dogmatic, at times skeptic, and that engineers at times boast that they can repeat an experiment – including artistic ones – with any specified degree of accuracy, and at times they resign to inability to do so sufficiently well to preclude possible disasters. Lévi-Strauss's ideas present such conduct as the norm. What makes science glorious is that occasionally it escapes this norm and offers great, exceptional alternatives. Lévi-Strauss's theory suffers excess as it condones myth-making and invites us to indulge mythmakers. Rationality is the worthy attempt to transcend myth and rationalize our views, our conduct, our lives.

3. NO JOKES HERE

The scientific revolution declared all falsehoods unreasonable. This is false. When it became impossible to hold on to it, when fallibility proved humanly unavoidable, obscurantism claimed victory. Incidentally, Edmund Husserl partly made this point soon after the rise of Nazism. He rightly and bravely attributed it in part to the failure of rationalism. He bravely proposed that we try again. It did not occur to him that this very proposal to look for a new answer already provides the new answer: it comprises a new theory of rationality. We usually and rightly ascribe this to Popper, who was the first to have offered a fallibilist theory of rationality. The traditional and false theory of rationality had disastrous consequences for society, and so latter-day advocates of obscurantism centered on the human sciences and propounded the idea that the methods of the natural and of the social sciences differ. They claim that the idea is new. It has been commonplace since Hegel.

The response to the irrationalist philosophy of the social sciences has been immediate and it still goes on. For some part, it is a compromise. For some other part, it is more of a patchwork than a comprehensive response. Efforts never ceased to apply to the social sciences the methods of the natural sciences, namely, induction and precision (including mathematics). The trouble is, no one knows what is induction, or what precision is beneficial and how to achieve it.

As most scientific endeavors are not very successful, it is no surprise that efforts to apply rigorous methods in social science are not very successful. Yet, two developments in recent times deserve attention, because their quality is very low yet their success in the commonwealth of learning is very high. The one is the effort to improve rational philosophy – by tightening both the standards of precision of the philosophical investigations into science and the standards of scientific investigation. The other is the attempt to apply scientific method to the sociology of science. The prejudice of the one effort is that both science and philosophy are piecemeal and given to piecemeal improvements that should be cumulative. The other is that the conventions of the commonwealth of learning are just and stand behind the success of science and of the scientific culture. This amounts to the view that each success of each move within science is due to merit alone. This renders the scientific community utopian. The very nature of the investigation at hand is Baconian. It is therefore bound to end up with the result that Bacon was essentially right. Yet, there is a central difference. Bacon's teaching received its dynamics, its drama, from the tension between the search for true knowledge and thus for true power, and the search for reputation, or followers, or popularity or political power. The philosophers and the sociologists of science currently defend the power of the leaders of science by claiming that the scientific *élite* are both the best researchers and the most powerful: they make the slogan "knowledge is power" true by convention. This is absurd. Not only does it make the commonwealth of learning utopian. It makes all its dealings with the rest of the world free of all the known intrigues and injustices that beset it. Members of the scientific community, especially its *élite*, love this idea, and even the idea that were they to take over politics, it would be all to the good. The astronomer Sir Fred Hoyle expressed this idea in his terrific science-fiction novel *The Black Cloud*. Many members of the scientific elite have supported this ideology (only in conversations, though, as they do not publish on anything except in the areas of their expertise). From this the false observation follows that good scientists could seldom collaborate with the Nazis. All this is in conflict, of course, with Einstein's amply corroborated condemnation (see below) of most German physicists who worked in Nazi Germany.

This leads us to the attitude of German natural scientists towards Nazism. And, Nazism and science is a chapter in the history of science, perhaps in the social history of science, which historians still shun – because it conflicts with Bacon's cabbalist doctrine according to which only the worthy can make serious contributions to science, only the good and wise can be a benefactor of humanity. The Cabbalistic doctrine can explain away failures of good people to make a contribution; it does not explain their failure to be good.

Robert Jungk, a Jew, a pacifist, and a victim of Nazi persecution, wrote his *Brighter Than a Thousand Suns* under the influence of this mystic doctrine. The book is breathtaking in many respects, yet it rests on the radical claim that good research and Nazi ideology could not share the same person. Jungk showed the manuscript of the book to Werner Heisenberg. The final version contains a letter from him, in which he defends himself in a most self-defeating manner against the charge that he had been a collaborator, and that in his famous meeting with Niels Bohr in Copenhagen he was trying to recruit Bohr to the service of the Nazi regime. After this meeting, Bohr hurried and escaped from occupied Copenhagen to join the Manhattan project and thus help develop the brighter-than-a-thousand-suns bomb. Since Jungk blamed scientists for the production of the bomb, and since he did not blame any German scientist for efforts to harness science to the Nazi war machine, he had little choice but to blame the participants in the Manhattan project. In principle, this does not work: either scientists are above blame or they are ordinary mortals. Except that collaboration with the Nazis is no ordinary sin. Understandably, he found it easier to blame some scientists for some neglect, then to blame the Nazi researchers for their collaboration with the devil. This led to a justly enraged response, yet no respondent explained Jungk's odd attitude.

The answer to the question of Heisenberg's collaboration with the Nazis awaits the opening to the public of some British papers that will put the matter beyond reasonable doubt. Max Planck's collaboration with the Nazis is more of a challenge, especially in view of the unquestionable testimony of his best friend, also a Jew, a pacifist, and a victim of Nazi persecution, namely Albert Einstein. For, Einstein said, he found debates with Planck enjoyable, despite Planck's chauvinism, because he was so very rational. Einstein's testimony goes very far. He was unhesitatingly swift and devastating in his negative judgment of the conduct of all of his German colleagues, not during the Nazi era, when pressure to misbehave was high, but afterwards – for their not having then done the right thing, which is an admission of guilt and a proposal of some amends ("Wiedergutmachung"). He stayed friendly with very few, among them Max Planck. Also, Planck too was a victim of the Nazi regime, as it legally murdered his son. Nevertheless, there remains a fact that the literature ignores. It is on open record. Planck was the director of the research institute that dismissed Einstein. He consented to the dismissal. Of all the members of the institute, then named after Kaiser Wilhelm and now ironically after Max Planck, only Erwin Schrödinger resigned in protest. One day, when official attitudes to rationality will be friendlier, and when the German nation will be in position to come to peace with its dreadful past, they might rename the institute in Berlin after Erwin Schrödinger. Incidentally, Einstein made it easy on Planck: he resigned and went to the United States. But why did Planck capitulate?

Because he, Max Planck, held two doctrines of rationality, a Baconian doctrine for the natural sciences, and a Hegelian one for the social sciences; a theory that was partly radical and partly reactionary. To be precise, no one can keep the two fields separate: they meet, and not only in the *Keiser Wilhelm Institut* or in Hiroshima. So, we can say, Planck's view was, in effect, radicalism tempered by reaction here and reaction tempered by radicalism there. Were he a bit more radical, he would have acted differently.

This is not merely past history. Bertrand Russell has issued a warning against the trend of science joining the military-industrial complex to create a global nightmare. But the warning was reasonably only against the nightmare – he took care not to make any forecast. Nor did his reasonable warning increase rationality. On the contrary, it ended up as a support for the new special type of irrationalism, virile and aggressive, against science and against technology. This new irrationalism draws on and enhances the recent compromise between social science and the classical irrationalism of the Reaction, to which we are coming. (Arne Naess has noted this compromise in his paper on Paul Feyerabend as the prophet of the Ecological political movement: Naess remained a rationalist and decided to devote all his efforts to ecology.)

4. THE NEW IRRATIONALISM

Bacon declared that neither dogmatism nor skepticism is justified, since truth is to be had for the asking – provided we are ready to get rid of our prejudices, become humble and industrious, and so on. There was a disagreement between his followers, the intellectualists and the empiricists, whether our views are science rather than prejudices because they rest on our intellectual intuitions or on our senses. This disagreement, says Kant in the closing of his *Critique of Pure Reason*, is the center of the history of all philosophy. One might find it puzzling that the truth is manifest yet whether *a priori* or *a posteriori* is controversial and discussed for millennia. Perhaps Kant was in error. Perhaps at the center of the history of all philosophy lies the controversy about the question, is the truth manifest? Kant overestimated the controversy within his radical rationalist party, despite its casting doubt on the platform of his party, because he underestimated all competing parties. Moreover, as he claimed that he had overcome the difficulty that he was presenting, he was blowing his own horn. (His proposal was to synthesize the conflicting views, thus bringing the desired party unity. His proposal was to recognize theory as synthetic truths that are *a priori* valid. It failed miserably.)

Kant declared the difference between the radical rationalist schools fundamental and important. It was epistemic, methodological, and transcendental. Facts indicate otherwise. Empiricists were not satisfied with empirical information and stressed that science must be theoretical. Intellectualists, on their part, insisted on the importance of details. Thus, intellectualist Galileo said that a grain of sand reflects all the laws of nature, and Spinoza echoed him as he said, the more we understand individual things the more we understand God. Intellectualist Descartes had a principle of induction. Empiricist Locke accepted as *a priori* valid some important truths (including the principle of induction). They all agreed that science is both inductive and deductive. Imre Lakatos has presented this as the double-secure method. In Newton's *Principia* a mathematical system enables an empirical generalization in stages from observed fact to his law of gravity. Kant took it for a model for his theory of scientific experience as observed facts couched in a mathematical language. (He even fashioned the structure of his *Critique of Pure Reason* in as close a parallel to Newton's *Principia* as he could.) Bacon's argument for the possibility of empirical proof was *a priori*: Nature cannot hide forever, he declared. Kant's transcendental proof, so-called, the central argument in favor of his system, was empirical: science exists in fact, and only his system explains this fact, he imperially declared. He was a rationalist, of course, yet he did contribute to popularity of the arch-irrationalist writings and style of Hegel, who convinced many readers that he was Kant's legitimate heir and successor.

Modern irrationalism began with the recognition of the impossibility of utter objectivity. The most important aspect of this impossibility is in its alleged permission to government to control the minds of citizens. This is not a corollary to Kant's philosophy. It is an expression of the desire to attack the Enlightenment Movement's radical politics, as Hegel stressed repeatedly. Kant's irrationalist critics did not ask, if we all agree, could we not be in error? Nor did they ask, do we all agree? They said, the authorities force citizens to agree. (That is to say, this is what the authorities do / should do.)

We do not all agree. We may dismiss this fact with the aid of a frivolous distinction: those who disagree are prejudiced. According to the rationalists, the basis for agreement is our science. According to the irrationalists, it is our superior culture. Agreement changes, truth alters, but agreement within a culture must prevail: *Ordnung muss sein*. What happens when science and culture clash? The physicist Philipp Lenard answered this question when he denounced Einstein's theory as Jewish rather than as science. Had we been more honest with ourselves than we are, or had the Nazi atrocities not made it so hard for us to admit the existence of Nobel Prize winning *bona fide* Nazi scientists such as Lenard, then we would have debated matters. As things stand, we disregard Lenard's insults without debate. This enables the undebated theory to survive, especially after great irrationalists have made it respectable by putting a human face over the barbarian element in it.

This happened at the same time as Lenard wrote. Malinowski and his school did that. Durkheim's relativism presented religion as society's self-worship. It was *étatist*, authoritarian, and illiberal. Malinowski has turned it humane, as Gellner and Jarvie have argued. Their relativism prevents anthropologists from holding a condescending view on cultures other than their own. This looks unconvincing. Putting on a par educated anthropologists and the preliterate and technologically backward societies that they observe is far-fetched. Evans-Pritchard met this difficulty by turning Kant's philosophy as a weapon against rationalism as such. Intellectual frameworks are necessary for thinking; we think within them and so we cannot argue against them, be they primitive or sophisticated, magical, Newtonian, or Einsteinian. Evans-Pritchard and Collingwood developed this view more-or-less simultaneously. Their philosophies are more-or-less the same, at least in the reading of John Wettersten. For Collingwood, statements have meanings as answers to questions, and questions follow prior answers. The absolute presuppositions, the suppositions of an intellectual framework, then, cannot be answers to questions and thus they are not open to question. Collingwood stressed that rationality is very problematic, as it is both certitude and non-dogmatism, since non-dogmatism amounts to tolerance. He offered no way out of this problem. He never asked in what framework his discussion of frameworks takes place. Nor did Evans-Pritchard. Is what they said immune to criticism? This depends on whether what they said is a framework or within one. As they were answering a question, their answer should compete with other answers.

All defenders of modern irrationalism view intellectual frameworks as language systems. This idea is today popular among disciples of Wittgenstein, as they ascribe it to him. They usually ignore Duhem who seems to have said so first. He defended the idea that intellectual frameworks are not given to comparison, suggesting that they are languages, and that no translation is perfect. This, however, does not isolate intellectual frameworks, and so they still are comparable. Some people are bilingual or polyglots. Evans-Pritchard was able to talk like an anthropologist and like a member of a tribe that he was studying.

A physicist can speak like a follower of Newton or of Einstein. Doing this exhibits the relative merit of each language. The assertion that it is impossible to speak outside a system is outside the system, and so it is self-contradictory. The assertion that when speaking inside a system one cannot effectively criticize it is false. Joseph Roth's moving novel *Job* describes a Jew who admits from the Jewish viewpoint the advantage of universalist liberalism over its Doctrine of the Chosen People.

5. CONCLUDING REMARKS

The current fashion in the philosophy of science dictates saying the following. There is no crucial experiment. There is no way to compare Newton and Einstein. The switch from Newton to Einstein was a switch to a new way of thinking of new generation. Or, the switch was purely pragmatic, not for rational considerations regarding the truth. One who wishes to criticize all this has to see that criticism that expresses a different way of thinking does not count. In the last resort, the rejection of any criticism rests on some general presupposition (about the relativity of all general presuppositions). Blocking criticism lowers the level of philosophic debate. It is at an all time low. But then irrationalism is not bound by standards of rationality.

Irrationalism is anti-intellectualism, mysology, the distrust of reason, the despair in reason, or the fear of reason. It is attractive. This attraction, said Popper, rests on decidophobia, to use the term of Walter Kaufmann. Its adherents may use valuable arguments. In addition to the arguments against a theory of rationality that may be rational, irrationalists usually defend extreme skepticism or extreme dogmatism. It seems that these lead their advocates to irrational conduct. Interesting counter examples for that abound: advocates of extreme skepticism or extreme dogmatism can act rationally.

Explaining irrationalism psychologically, sociologically, or politically is dangerous. Such explanations may offer ways to explain away some unpleasant yet valid criticism of our brand of rationalism. Admittedly, irrationalism may be propaganda, and it may come to incite mobs to riot. We may have to counterattack it then in sheer self-defense, but only when we judge this necessary. The Darwinians broke a taboo and attacked the clergy under the pretext that the clergy were attacking science. There was no incitement to riot then. Some biologists felt that Darwin had supplied them with excellent ammunition, and so they incited their own mob to attack. The outcome was backlash, clericalism, the worst kind of scientific irrationalism, and political disasters.

Rational philosophy has survived all that not because it is strong but because democracy is. Now democracy is at risk. To lower the risk, we may fight the increased authority of science, one whose advocates act as an *élite*, and the growth of seemingly pro-science irrationalism that they tolerate. We should contrast all the way democracy and the rule of any *élite* whatsoever. Second, we must see the greatest risk of government by *élites*. We should recognize the most dangerous mistake that clever members of the *élite* are prone to make: the use of sacred lies. This is as popular today as ever. Former U.S. Foreign Secretary, Professor Henry Kissinger, freely confessed on a national television talk show (of Dick Cavett) that when in office he had the duty to lie to his underlings. (He delicately limited his confession; he did not say generally, his duty was to lie for the good of the state.) His reasons are not important; their consequence is: we cannot criticize lies since they are not agenda. And not having the benefits of criticism is political suicide – as Spinoza said in his *Tractatus*; as Shaw said in the preface to his *The Apple Cart*. To preserve the critical attitude, we should criticize lying, not this or that lie. (Commenting on the TV program *West Wing,* Kissinger spoke differently, though.)

We hopefully have still enough strength to be able to confess ignorance and speak honestly. The devil gladly lends his weapons to his opponents, but it is foolish to accept the offer. Whatever may happen, the foundation of rationality is honest speech, including frank admission of weaknesses. The reason to fight irrationalism is only political and educational. We hardly need fight interesting irrationalists, like Evans-Pritchard and Polanyi. After the battle on reason is over, those who stick to defunct versions of irrationalism become uninteresting. The same holds for those who stick to defunct versions of rationalism. There is no intellectual motive for fighting against them – only political and educational ones. The politics of untruth is the politics of unreason, even when scientific leaders take recourse to it. The only way to bring to politics back its past dignity and to find new dignity for it, and for diplomacy and for international relations too, is to return to truthfulness, to open political action and to open rational reassessment. The success of the new irrationalism, then, is the result of the new weakness of rationalism: since Einstein replaced Newton, rationalism was in need of revision, yet it still fails to admit this fact officially and blatantly.

I therefore propose following observations as major presuppositions for any theory of rationality that might claim to be satisfactory.

First, no one is always rational.

Second, some dogmatists and some skeptics are at times rational to some extent.

Third, there is no *a priori* need, therefore, to steer away from the Scylla and Carbides of dogmatism and skepticism.

Fourth, rationality is a matter of degree.

Fifth, it is irrational to pretend to be in possession of a satisfactory theory of rationality while advocating a problematic one.

Sixth, irrationalism is harmful, but its harms invite examination and the development of efficient techniques to combat it, for a better view of rationalism.

Seventh, hence, a major practical problem for a satisfactory theory of rationality to solve is, when and how should we attack irrationalism and when is it better to ignore it?

Eighth, rationality must demand honesty, particularly intellectual honesty.

3. REASON

Chapter 3.1

MINIMAL CRITERIA FOR INTELLECTUAL PROGRESS

0. ABSTRACT

For logical reasons, progress in the dispute about the choice of a criterion of progress seems impossible, as it renders any move question begging. Our criterion of progress may limit critics by permitting them only criticisms that accord with it. Openness to criticism thus becomes a criterion that on a few counts is necessary but insufficient, especially for science, where progress is most pronounced. The need for criteria for progress is the need not to fall for mock-progress. This renders criteria for progress proper unnecessary; it is enough to have criteria for possible progress. Instituting such criteria might enable the now so scarce encouragement of the young. Since much pseudo-progress is due to the confusion of a program with its execution, this change will also prevent much pretense. But the chief aim here is to reduce the harm that current practice of acknowledgment causes.

1. THE CHOICE OF CRITERIA FOR PROGRESS

Expressions of disagreement about progress may easily cause a feeling of loss of orientation. The difference between criteria of superiority of new ideas promotes a search for a meta-criterion or super-criterion, yet each criterion may serve as its own meta-criterion. Thus, for any criterion x, those who endorse it may also endorse the meta-criterion that will recommend the choice of x. Thus, every religion recommends itself, every philosophical school recommends itself, and science recommends itself. Thomas S. Kuhn has called the outcome of such moves incommensurability. Holders of different criteria of progress follow the different criteria in different directions. This doctrine is very popular these days – especially his application of it to competing scientific theories. (1)

The outcome is obvious. If Party One advocates Theory Alpha and Party Two advocates Theory Beta, then each develops the ideas, problems, tasks, set by the choice of their chosen theory. Soon members of Party One loose interest in ideas of Party Two, and in no time they cease even to know what that party is concerned with. From time to time, a member of Party One, a bright young aspirant, makes a foray into the domain of Party Two. These expeditions will, naturally, end by a return with empty hands or a total loss due to conversion to the criteria of progress of the other party. Two exceptions may occur. One of these concerns common grounds that enable the foray to end up with a triumphant return with some booty. Conscientious party members will then acknowledge that the booty belongs to the other party. Patriotic party members will act differently. The second exception concerns the mastery of the ideas of both parties that permits comparative studies. These are useless from the viewpoint of either party. They belong to a third, possibly embryonic party, however. So may be the converts. For, every case of conversion may raise the question, how did it come about? If Criterion One yields to Criterion Two or leads to Theory Beta, then Party Two wins by a knockout, and Party One is on the way out. Otherwise, there may be an (embryo of an) alternative to both, a new criterion or perhaps a new meta-criterion. Then, possibly, both parties are on the way out. Alternatively, conversions are irrational. Some say conversions are pragmatic. (2) This is possible at all only if pragmatism is a meta-criterion.

These are all the extant options, and it seems that logic allows for no other. Yet, the facts remain different from them all.

Pragmatic considerations often apply pragmatically. This is proper for applications that are only local, *ad hoc*, and temporary: any consideration may apply temporarily without the aid of a criterion. (Objections to *ad hoc* moves, however justifiable, cannot be so overriding as to prevent all of them.) The following is proof that *ad hoc* measures are better than the use of pragmatic considerations across the board. Were conversion irrational or consistently pragmatic, we may choose on occasion to convert to Party One in the Fall Semester, to Party Two in the Spring Semester, and hold no opinion over the summer vacation. This is too arbitrary and thus rather uninteresting. More interesting is the case of constraint, when a criterion shows a way, and experience refuses to go that way. Since this is a matter of brute, blatant experience, it is easy to confuse it with practical matters. But counterexamples block the identification of the pragmatic and the empirical. Consider the classical, best-known criterion for scientific progress: endorsing a theory is forbidden unless it is obligatory; it is obligatory if and only if it is scientific; and it becomes scientific as it receives empirical verification. No theory has boasted more empirical verification than Newtonian mechanics. When empirical evidence went against Newton, it thereby also went against verification. All the empirical evidence against Newton was important, but not all of it was useful. Import is not usefulness.

Thus, more options are available than earlier considered: criticism of old criteria may open new avenues. How then are criteria open to criticism?

2. THE OPENNESS OF CRITERIA TO CRITICISM

It is easy to protect every pet idea from criticism. Early in the seventeenth century, Bacon rendered this fact central to his theory of science. He said, the wish to defend a pet idea against refutations is irresistible, and it is always possible, but it is very harmful for research. (3) So stood matters until the crisis in physics of the end of the nineteenth century. Pierre Duhem and Henri Poincaré were willing to rescue a good theory. They were the chief spokespeople of the new (or rather renewed) philosophy that avoids all claims for truth in scientific theory and deems all science mere instruments for prediction. And there is no need to abandon an instrument for prediction just because of its limited usefulness. Thus, on this philosophy refutations are the result of no more than excess – of the application of the theory to areas beyond the limits of its reasonable applicability. This sounds very impressive, but what can one do when applying a theory to some cases which obviously fall within its purvey yet obtaining unsatisfactory prediction? Then one can always make an excuse, blame the measuring instrument or the theory on which it is based, *etc*. When a prediction goes wrong, some correction must occur; but, as Bacon knew, we may make do with the smallest correction and leave intact the main ideas involved. (4).

The practice of accounting for an error by small modifications became topical with the refutation of Newton's optics early in the 19th century. William Whewell showed the limitation of this practice. After a theory underwent modification, he said, it invites new tests. As test predictions repeatedly fail, the theory should repeatedly undergo modifications and new tests. Failing to notice this ends up in pseudo-science, he concluded, that might lead to confusing the difference between a theory and its modification, or, worse, to forgiving oneself for identifying them on the excuse that the difference between them is negligible.

Duhem and Poincaré limited their discourse to valuable, successful theories. They wanted them to endure. So, they suggested that the discovery of some defects in a valuable, successful theory should not lead to its rejection. It should have some inbuilt ambiguities to permit us to save it from such mishaps. But they demanded strict honesty and openness about the process.

Logic permits this. Some significant decisions within accepted research procedures are anyway immune to the pressure of empirical information. Much choice is available here: of mathematical tools, of what aspect of a refuted theory to modify, or even of the variant of its wording to start the modification with, and of certain blanket suppositions. (The blanket suppositions are frameworks – intellectual or metaphysical – such as causality and atomism.) Research has much leeway for struggle before any results turn up in the shape of empirically testable theories. These should undergo tests and with no leeway. There is nothing surprising or unusual in this. It opposes Bacon's criterion. Followers of Bacon know this, yet they will not entertain the possibility of abandoning Bacon's criterion. Instead, they offer a modified version of his ideas, which they do not examine, or, still worse, which they confuse with the unmodified original. They often waver between the original and some (often unspecified) modification of it. This is their liberty: they have made their choice and paid the price for it.

Lenin advocated the idea that criticism should be constructive: one should not give up a theory before experiencing the force of some alternative to it: we need not listen to critics, unless they offer alternatives. (5) And then, Lenin added, the best defense is the offense: we should attack the alternatives. He was too cocky to see that he was recommending a strategy that assures success to its users no matter how poor is the opinion that they defend. It is wiser to seek better alternatives to extant ideas than to defend them. What makes new alternatives better? The increased accord of the new with the empirical data does. The equal accord of different ideas with extant information invites crucial tests. Advocates of constructive criticism reject crucial tests. Do they have an alternative to them? Is it open to criticism?

3. OPENNESS TO CRITICISM AS A CRITERION

Criticism is as old as Greek philosophy and possibly older. Explicit, systematic criticism is the famous Socratic or dialectic method. Socrates did not invent it. In Plato's early dialogues, Socrates repeatedly reminds his interlocutors of the established rules of critical debate, and they repeatedly accuse him of subtly breaking them himself. In these dialogues, he considers these accusations seriously as important and deserving scrutiny. They are, as we say in contemporary democratic parlance, points of order, and so they have priority. And yet, surprisingly, only in recent times has anyone tried to reconstruct the rules explicitly. The renowned Oxford philosopher, Gilbert Ryle, did so in essays that colleagues have hardly noticed. (6) The leading sociologist Robert K. Merton has fared better in his sociological study of science. He claimed that researchers regularly obey the rules of science. This claim is easy to refute and its refutation led to the view of his effort as useless. This is sad, since, though the rules do not obtain, they are right. His not having said so has encouraged his critics to sanction every piece of violation of the rules that they can observe. They do not care about scholarly etiquette. What are its rules? Why is there hardly any explicit study of them? Presumably, these rules should serve as criteria for intellectual progress, perhaps only minimal criteria and only for possible progress. The absence of canonic wording for them has led to the present discussion.

One difficulty on the way of wording the rules of critical debates was the absence of formal tools. Traditional formal logic does this very poorly. This is where Bacon came in. He said, dialogues are endless and useless, since logic enables one to get away with frivolous distinctions. This is also where Duhem and Poincaré came in. They responded thus. It is silly to make a frivolous distinction merely to avoid admission of error, they conceded, but it is wise to rescue by such means the marvelous structures of past successful scientific theories. Einstein decided a short while later to sacrifice these marvelous ideas (that he always admired) and forge new ones. Since all this happened early in the twentieth century, and since logic evolved only then and even somewhat later, it is not surprising that only in 1935 a new option appeared. It was not utterly adequate but it was quite an ingenious breakthrough. It has appeared in Popper's classic *Logik der Forschung* as well as in his *The Open Society and Its Enemies* of 1945.

Popper performed a marvel of simplicity. He took openness to criticism, the ability to find a contradiction, as the only criterion of the ability to contribute to (scientific) progress (or to intellectual progress in general). He declared finding such contradictions progress. Even the discovery of a new theory that some future evidence may contradict is progress. To seek a theory that is consistent with known empirical information and open to refutations by future empirical information, and to seek such information, Popper saw as the sole kind of effort at a contribution to (scientific) progress. He recognized the dependence, at times, of progress in the empirical sciences on other kinds of progress. Such is progress in mathematics, in some unscientific – metaphysical – theories, and in the invention of new laboratory equipment. (Duhem had earlier recognized this last dependence. He denied the possibility of any dependence of scientific progress on metaphysics.) Popper hardly paid attention to progress in mathematics and he dismissed metaphysics as by-and-large merely pseudo-scientific. Let me skip all this now. Suffice it to observe here that everyone endorses as a necessary meta-criterion the criterion of openness to criticism. But Aristotle, Bacon, Duhem, Quine, and many others, emphatically declared this criterion to be far from sufficient. Popper declared it necessary and sufficient.

Though in need of fine-tuning, Popper's proposal is stupendous. Inability to find its revolutionary character deeply impressing amounts to inability to contribute significantly to the present discussion. In particular, all those who still stay with Duhem have simply missed the train. Unless and until they respond adequately – appreciatively – to Popper's proposal, they are not likely to contribute to progress of the theory of progress.

4. OPENNESS TO CRITICISM AND ITS IMPORT

Popper's initial idea of criticism was strict. He did expand on it, but he never quite gave up his initial strict idea. He then spoke of openness to criticism in general, and of strict or *scientific* openness to criticism, of refutability or falsifiability. It refers to what Michael Faraday and Claude Bernard called "empirical criticism". Empirical criticism is a very special, strict sort of criticism. A theory has to pass many tests (most theories fail them) before it deserves empirical tests. A search is then on its way for a new experiment with which to test the theory. This test is empirical criticism. Popper viewed empirical criticism as the hallmark of science. A theory is scientific if and only if it is empirically open to criticism. Popper introduced first (1935) the idea that the only criticism possible is either logical or empirical and both constitute the proof that a certain (set of) statement(s) is inconsistent. Later (1945), he agreed that there is also historical criticism and moral criticism, which do not live up to the high standards of empirical criticism. (Moral criticism relies on improved intuitions, says Popper, and this seems question-begging. Historical criticism is inconclusive, as it relies heavily on interpretations and so it is open to dissent, unlike empirical criticism that relies on empirical evidence that is not so easily open to dissent. And so, whereas the admission of empirical criticism amounts to the admission of a factual statement that contradicts a theory, there is leeway as to the admission of moral and of historical criticism.)

There is more to it: by Popper's canons, there is little point in a test of an already refuted theory. Yet, tests occur regularly of theories that are known to be false, such as Niels Bohr's theory of the atom or Schrödinger's equation for the electron. Their tests were significant, since their confirmations would make it necessary to consider each of them an approximation to the next stage. Their refutations would be more important as data that demand explanations on that stage. We can put these matters as *desiderata* and discuss criticisms in general in accord with given *desiderata* that, of course, are also subject to criticism – as any item within our intellectual horizon. But things are still not as neat as one may like.

Popper ascribed to openness to criticism two roles: it is both a meta-criterion by which to judge criteria, and the criterion of scientific progress. For, it does not suffice as a criterion for progress: we may want more from science than the presentation of any theory that is open to criticism. As we do need more, then, possibly we may drop the openness to criticism as a criterion and make do with it as a mere meta-criterion. Will that take us to the pre-Popper ethos? Not if we refuse to lose the benefits accrued due to his insight.

Let us consider the problem of demarcation of scientific theories again. The same theory may be scientific or not, says Popper, depending on the readiness to have it criticized. So, at the very least, theories may be scientific when read critically. Hence, the question is not just of the scientific character of theories but of theories plus our attitudes towards them, so that the philosophy of science all of a sudden turns sociological: the critical reading of theories is public, and thus social, institutional. This point is significant. Popper made it repeatedly and admirably throughout his *Logik der Forschung*, thus inaugurating a new field: the sociology of science. He dismissed all traditional theories of science as psychological in principle, namely, as psychologistic. They are thus question begging, since psychology is a science, founded in accord with the theory of science founded on psychology. His trailblazing work did not go far enough, however: the proper alternative is not to demarcate or delineate scientific theories and in addition refer to institutional means of fostering criticism, but to demarcate or delineate science as the social institution that produces sets of conjectures and refutations that cause intellectual progress.

Popper is in error. The two criteria, of openness to criticism and of progress, diverge. Not all sets of conjectures and refutations are progressive. Bacon was right in presenting scholasticism as a counter-example to this, and Duhem has not shown how we may re-institute scholastic techniques without a return to the barrenness of the medieval dialectic. Duhem denied that the Middle Ages were dark because, as an apologetic Roman Catholic, he approved of scholasticism. Poincaré aligned with him, but only after having blamed the scholastics for confusion: once we are clear and open about methods, once we openly admit error and administer face-lifts to our theories frankly and for specific reasons, then we are free of scholasticism. Duhem did not object, of course, though he preferred to be silent on this. Popper, by contrast, added a rule to his theory; though lovely, it regrettably does not work. (7) It is, always seek a highly refutable theory and always test it, and never rescue it by small *ad hoc* modifications. Later he altered his view *ad hoc*: he altered never to read almost never.

Counter-examples abound to the traditional Baconian rule against modifications. Without this rule scholasticism may indeed reemerge. Hence both Duhem's theory and Popper's theory cannot stand as they are. The counter-examples to both Duhem and Popper (not to mention Bacon) are of two sorts. First, technology. It creates today many highly testable theories and tests for them, even though they boast little scientific interest. At times they present challenges to applied mathematicians and even give rise to new branches of mathematics. The second counter-example is from science. Einstein knew how to amend Newton's theory of gravity in order to account for minor deviations from it, but, he said, it never occurred to him to examine that option. The theoreticians' reluctance to handle such theories, even though they are highly testable, sharply contrasts with the practitioner's eagerness to do precisely that. Space technology today is quite unthinkable without the regular application of ideas of the kind that Einstein found so repellent. This illustrates the distinct difference between science and technology, refuting all of Popper's predecessors who have identified the two, and refuting Popper who has tried to distinguish between the two but failed. He has described technology at times as sets of rules of thumb, at times as applications of scientific theories, and never as the development and tests of theories whose value is not scientific but practical. (8)

The confusion between science and technology is traditional. Popper tried to avoid it. He proved their distinctness by reference to the use of false theories in technology. But science uses false theories too. For example, when Schrödinger failed to create a relativistic equation for the electron he made do with a classical one. Popper discussed repeatedly the error of his competitors who confuse probability with empirical support, probability with probability increase (due to evidence). He made a similar error: he confused testability with testability increase. He stressed testability increase when observing, with Newton, that experiments not relevant to Kepler's or to Galileo's theory may serve as tests for Newton's theory. Hence, it is more than their conjunction. Yet he demanded testability, not testability increase. (The conjunction of two testable theories is testable, but this does not satisfy those who want new tests. Thus, when Newton showed that his theory unifies Kepler's and Galileo's theories, he stressed that it also explains the tides.) Science invites testability increase; technology seeks successful applications, and thus confirmations, namely, the standing up to severe tests. At times a technologically significant theory fails a test, and then its advocates may reasonably attempt to find other technologically significant tests in the hope that it will pass them.

Popper's system is admirable, but those who admire it and learn from it should admit that it is not perfect.

5. WHAT IS INTELLECTUAL PROGRESS?

Initially (1935), Popper showed contempt for all intellectual activity outside mathematics and empirical science – with some notable exceptions. Later (1945), he allowed that ethics and history are also open to criticism, though not as sharply as science. Finally (1958), he admitted that some metaphysics is open to criticism. All this is much ado about very little. Popper always took openness to criticism to be the mark of intellectual value, even though much worthless superstition and scholasticism, both traditional and contemporary, is open to criticism. Were Popper not initially so hostile to everything except mathematics and empirical science, he would not need so much effort to rid himself so slowly of this hostility by discovering the obvious. Also, his route took him only as far as he wished to go. He ignored the openness to criticism of many superstitions and of much scholasticism. When he met these, he dismissed them by ascribing dogmatic attitudes to their adherents. He thus shifted his grounds from theories to attitudes. He thus also clung to Bacon's tremendously ignorant, not to say dogmatic, view of the superstitious, in utter disregard for the obvious fact that they are all too often at least as skeptical as any scientist around. Indeed, more so, since scientists often dogmatically endorse science when the superstitious find solace in doubt. Also Popper denied progress in the arts. Lamentably, he expressed conservative hostility to almost all twentieth-century *avant-garde* art. (9)

We may take things again from the top, and seek simple and practically incontestable criteria (incontestable for the time being), perhaps merely partial ones. Let us begin with a simple thought-experiment, one often proposed but seldom followed up. Suppose we could wake up the dead, and choose to wake up some intellectual giants of the past, scientists, mathematicians, metaphysicians, moralists, prophets, artists, or inventors. How would they respond to today's situation? Would they deem us intellectually more advanced than their intellectual milieu? This thought experiment is well worth examining.

Take the position of Duhem seriously and proceed with the thought experiment in question. The geniuses of the past had their own criteria, explicit or implicit, and by Duhem's view they would stick to them and use them to judge progress. They would thus see less progress than we do. For example, as Newton appreciated facts, he would acknowledge all the many facts we know which he did not, but as he had his own view of science, he would barely notice Einstein. For a similar example, certain new techniques would have excited Beethoven – whether new sounds or new forms or even the means of extending one symphony to the length of one hour and more. But much of the *avant-garde* music performed today in the concert hall or in recitals he would deem not music. Now, this possible outcome of the thought experiment is Duhemian.

Though we still cannot wake up the dead, we have their dogmatic heirs to represent them, at times closely enough. In my *Faraday as a Natural Philosopher* I have illustrated this. The Astronomer Royal of Faraday's time appreciated his experimental discoveries, not his ideas. John Tyndall even frankly dismissed his ideas as silly in his *Faraday as a Discoverer*. He admitted and even emphasized that in the hands of giants even silly ideas can be fruitful. But it did not occur to him that possibly these ideas were valuable. The more Faraday's reluctant colleagues ignored his ideas, the more they compensated him by expressing vociferously admiration for his factual discoveries and for his genius. It meant nothing to him: he craved recognition for his ideas – not assent to them, only recognition. He never lost hope, but his peers' dogmatism made research increasingly distasteful to him. I documented all this in detail in my book. Not surprisingly, even the professional specialized literature on him still overlooks all this. (10)

To continue the thought experiment, then, are we sure that past giants would behave dogmatically? What would be the responses of a non-dogmatic giant like Kepler? Could we imagine Kepler come back to life, glance around, and stick to his criteria? Not very likely. He would have laughed his head off and alter his criteria as best he would be able to. How? By what criterion? No criteria. It takes a genius like Kepler to conjure an answer to this question.

This brings us back to our initial question. We may now begin answering it by a return to our preliminary answer. We may try different criteria for size; we may see whether the meta-criterion of openness to criticism does not help us do better, at least by comparison with other meta-criteria. But this response is only preliminary. A true Kepler, were there one living and breathing among us, might come up with a great surprise. In the mean time we may savor the preliminary response and examine the possible surprise. Can we break down surprise or is it of necessity transcendent of any description and any criterion? Can it, perhaps, itself serve as a criterion?

6. MINIMAL CRITERIA FOR INTELLECTUAL PROGRESS

The view that generally we cannot compare criteria is erroneous, since, clearly, the less a criterion requires the better. A heavily loaded criterion may be too difficult to grasp and to apply. Yet, a weak criterion may be useless. We want a criterion as weak as possible yet strong enough to perform its task. What is this task? The task seems obvious, yet often writers overlook it and confuse it with related but still different tasks.

We have a few questions. What is science? What is intellectual progress? What is the weakest criterion by which to characterize either? The answers to them seem obvious. They are not. Consider science. At times we wish to demarcate it as an attitude, at times as a body of theories, or as a tradition or a social institution or even a social class. The following questions come in a batch, yet they differ from each other. Who is a scientist? Who is a research scientist? What does scientific research produce? What characterizes the theories that the scientific tradition takes as scientific? What does this status convey? How much does science depend on its organization?

Why does all this matter so much? Because of our commitment to science. What are we committed to? What are we supposed to do in order to stand by our commitment? When do we betray it and should we ever?

Philosophers of science tacitly take commitment to scientific opinion as the most central. Many of them discuss nothing else but the way we express this commitment by endorsing scientific opinion. We should always believe the best scientific opinion extant. We should believe the best scientific opinion available and we should teach the same. This answer permeates the works of Popper's leading disciples, John Watkins and William Bartley.

This makes science dogma – the best dogma, perhaps, but still a dogma – the least defensively advocated, but still advocated. This is not healthy. Surely, we want to teach not the right theory but the readiness to rethink, to reconsider. This sounds like Popper's view of the meta-criterion as a criterion. And it is insufficient, since to reconsider may be to propose a frivolous distinction or to attempt to revolutionize our thinking. Which of these two options is better? Or should we characterize and select some middle option? Or should we leave this question open?

As long as we understand commitment to science as eagerness to endorse the best scientific opinion extant, we are in trouble. Not only do we thereby encourage dogmatism. We thereby cling to false, empirically refuted views. First, we thereby cling to the view that endorsing an opinion is a matter of decision. It seldom is. Second, we thereby demand the endorsement of scientific opinion, yet the endorsement of any opinions is a private affair, unlike scientific opinion, which is public. Progress is private or public and public progress is independent of private opinions. Contrary to Bacon's supposition that researchers who advocate error disqualify, many researchers have performed experiments that undermined their own views, quite to their own chagrin. (Michelson and his famous experiment are the rubberstamp example.) Also, Bacon assumed that promoting one's prejudices is useless. Yet, discussing prejudices at times was extremely enlightening. (Ernst Mach's discussion of absolute space is the rubberstamp example.) At times even the promotion of superstitions led to significant results. The best-known opposition to science is the early 19th century movement of *Naturphilosophie*. Most scientists in its heyday rightly spoke against it. Yet it did stimulate some significant developments, in biology and even in physics, including the discovery of electromagnetism and more. (11) (Another example is phrenology, which was utter rubbish and yet led to Paul Broca's discovery of different functions of different areas in the brain.)

What, then, is commitment to science? We need not answer this question. Enough if we have a criterion of scientific progress, or of intellectual progress of any sort. We want criteria of progress because we love progress and we want it, and lots of it. We want improvement and we regret having let improvement at times sink into oblivion and having had to invest a new effort in retrieving it or in recreating it. This seems so obvious, it hardly needs saying. What is far from obvious is that this may suffice – at least for a start. We may begin with simple unproblematic tasks that we perform regularly. Or we may prefer simple and somewhat problematic tasks that for one reason or another we hope to be able to perform. Suppose one is stuck in such a process. Suppose a friend helps. Surely that is an improvement. We usually are polite enough to privately thank the friend, in private or in public, but we usually do so as individuals, not as a public. As a public, we first examine a move and only if, by some criterion of progress or by no criterion, as a public we judge it progress, then we acknowledge if publicly to the one who has proposed it. This is quite wrong and downright rude. Individual postponement of expression of gratitude meets with disapproval; public postponement should, too. It definitely does not. This is bad.

This is wrong and a stable part of the scientific tradition. From the start, its standards had a public character. This renders the distinction between the private and the public important. Philosophers of science refuse to recognize it as they view science as resting on the individual act of commitment and of perceptions. The sociology of science began only in the twentieth century as the earlier hostility to esotericism found its expression in the psychology of the researchers. This was a part of the view that all the human sciences are psychology at bottom. It was at the time the received view. We need not discuss it now. Suffice it to agree to allow sociological considerations to enter our discourse freely, and leave the question open, whether sociology is psychology at bottom. This suffices for the free discussion of the question, how can we render the public standards of science friendlier. (12)

At what stage should we acknowledge, privately and publicly, that a certain move is a contribution? The received opinion is, a public acknowledgment should follow immediately the recognition of success. We do not know what success is. To make practice friendly and reasonable, we should facilitate rapid acknowledgment, and to that end we can view all acknowledgments as tentative. We can then acknowledge as a contribution every serious attempt to contribute and discuss its merits afterwards. I declare it a progress to replace the search for a criterion of progress with the search for a criterion of possible progress: I will explain the advisability of this change.

The exercise of caution before acknowledging progress is general. The most conspicuous example is the conduct of the committee that decides on Nobel Prize awards. One can disqualify this instance because the committee was never the paragon of proper scientific conduct anyway. At least the knowledgeable do not think of it this way. Yet, the present discussion is of science as an existing social institution, not as an ideal. And the Nobel Committee is the rubberstamp example of public approval and thus the source of the *bon ton* even more than the revered Royal Society of London. Also, in matters of caution, it is by no means unusual. In some cases, it is even *avant-garde*. Though it took thirty years of struggle before it recognized Barbara McClintock's work, her Nobel Prize did contribute to her recognition. This sharply contrasts with the Committee's reluctance to grant the Nobel Prize to Einstein or to Planck despite their tremendous reputation. Caution recommends postponement of recognition of problematic contributions to progress. Such recognition is itself no contribution (unlike Planck's recognition of Einstein that was a great contribution). Caution leads to granting recognition to small contributions in preference to big ones. Big contributions shoot through many levels, up to and including the most general criteria held, whereas small contributions are easily recognizable as they leave much intact. Duhem's proposal to take all refutations as calling for mere face-lifts employs this. In science, he claimed, all changes are small. There are no scientific revolutions, he taught: caution is the hallmark of science. (13)

The classical Baconian view of science as anti-scholastic gave way to the Duhemian view of science as continuous with scholasticism. This transition amounts to a revolution in the sociology of science. There is no literature on this fact. Michael Polanyi is the philosopher most clearly associated with the change this requires, and his approach to the philosophy of science is clearly social, as he considered the say-so of the leadership of science a touchstone for scientific value. Yet there is almost no literature on this either. (14) There is sufficiently clear-cut empirical refutation of Polanyi's view. Though he never said that the scientific establishment is perfect, he did take it to be *bona fide*. And the *bona fide* of the scientific establishment has systematic defects that urgently invite reform. It is easy to show this empirically. Let me conclude this discussion by presenting one of these systematic defects, the establishment's stone-walling the non-established *avant-garde,* its pretence that their contribution does not exist, until they join the establishment (sometimes posthumously, if at all).

7. THE STONEWALL

The worst is that official caution leads to and justifies official stall responses: no comment. This is frustrating. The pain this caused Faraday is manifest in his scientific correspondence. Historians of science still refuse to notice it. They likewise ignore my book on him and his pain. As this pain is avoidable, we should discuss it in efforts to prevent its recurrence.

Even if we cannot prevent the stonewall attitude, we can prevent the pain that it causes. No justification is acceptable for hurting individuals who attempt to contribute to progress and who badly need encouragement. Had we made public the reasons for the lack of response, then this would be a warning for the parties in need for it. If they were then still frustrated, then their frustration would be the outcome of their refusal to heed the warning. The frustration of Faraday rested on his profound faith in the received canons of scientific conduct that the scientific establishment of his day violated. They would not comment on new scientific theory unless and until it is successful. This is regrettable and wrong, but at least it is well-known and so not as frustrating as other cases, such as the Nobel Committee's refusal to acknowledge the amply confirmed ideas of Planck and of Einstein. Offering a new idea worthy of a test is progress. Thus, the refutation of the Bohr-Kramers-Slater theory of radiation of 1924 at the very first test did not deprive it of the recognition as important that it had – and still has. The importance of a theory, even if not successful, is publicly better known when it is officially recognized. Science fiction has expressed this idea ("Flowers for Algernon"). It made doyen of science-fiction writers, Isaac Asimov, move away from his early Baconian view of science to a more critical one.

All the caution in the world does not justify my postponement of an admission that you have contributed for sure at least when you are ahead of me in any of the following ways.

(1) You help me get unstuck.
(2) You have an idea that is relevant yet it did not occur to me.
(3) You have worked out an idea I was unable to develop though I tried.
(4) You have already performed an experiment that I regret I have not.
(5) You have found a new reasonable answer to an important objection.
(6) You have shown me an important question that is new to me.
(7) You have shown me that an easy question can replace a hard one.
(8) You have shown me a new and interesting perspective on my project.
(9) You can provide me with a (valid) criticism of an idea of mine

All this provides us with a minimal criterion – or set of criteria – of progress – on the supposition that it is advantageous to consider progress any case of possible progress. If we are able to acknowledge such progress every time, then we have seen land. Viewing science as an attitude, we may take these criteria as minimal too. They represent interest, openness, and good will. It is the attitude of the friends of the open society.

On the personal level, this is at times quite intriguing: by-standers may help with good advice without being able to use it themselves. They may help by refusing to accept a confused statement as clear even when they themselves cannot see the import of their refusal. And so on. But what matters, of course, is the public recognition of serious attempts to bring about progress as progress – the contribution of an individual, a team, an individual helped by a by-stander, and so on. It is not easy to translate the private to the public, and it requires a public discussion. To that end, however, we have to admit publicly the need for such a move.

Why do we not admit all this publicly? Because we are cautious. Obviously, caution is no excuse for silence, as we may speak cautiously. I can say, maybe your idea is no good, but I have to think about it. I can say, your idea looks interesting now, even if after careful examination I may think differently. I can say, I do not know if your theory deserves a test – I must consult colleagues about it. This way I admit that you have contributed. When you make me think I should admit at once that you make me think – regardless of whether the new thinking leads to valuable results or not.

Can this become a publicly recognized practice? Yes; easily. For example, experiments may cost money and so invite official approval. Usually only approved projects become public knowledge. If we acknowledge any project on the agenda a success, if we acknowledge the short-listed items as more successful than the others, if we admit that we may have erred in picking the winner, then stonewalls can crumble. Will they? After all, in the performing arts we do count nomination as success no matter who wins. Why not do the same in science and scholarship?

8. THE ESTABLISHMENT

The stonewall stands on solid foundations, and we need some strong stuff to shake them. We badly need some intellectual dynamite. The foundation is the fact that institutions outlive individuals and that we punish sins of commission but seldom sins of omission. Thus, we take pride in honoring those who have contributed to progress, and the public expressions of this pride are sweet when the contributor is still with us and bitter-sweet otherwise, but never sour much less acrid. We are glad to correct old errors but we never analyze them and institute means for avoiding their repetition. Those who have stonewalled the contribution to progress are immune. The same university that has actively discouraged its distinguished members until they became distinguished, now takes pride in having sheltered them.

What we need is not penalty for cowardice and for concealment (of our having learned from contributors) until acknowledgment was safe. What we need is, first, not to penalize those who support the wrong party. Now this is an already achieved situation. Though the Nobel Committee has often enough preferred minute contributions to giants ones, they are immune to penalty, censure, or even criticism. Yet they all fear criticism and censure and ridicule. This can change by reforming our educational system. We may, as individuals and as a movement, refuse to join belated public declarations of acknowledgment, unless they go together with some self-criticisms and some attempts at reform. It is no good honoring the forgotten heroes of the last century while cowardly ignoring the heroes of the last generation. It is even no good honoring the heroes of the last generation while stonewalling everyone around who is not established. Acknowledgment need not be too courageous, but it must be free of cowardice.

What is the place of the friends of the open society in the modern world? Of course, not all contributions to progress and not all the recognition of such contributions come from among these friends. But they acknowledge having erred with ease; they love to see and acknowledge contributions wherever they are. And they constitute, by this very virtue, a kind of a club or a workshop. It is an *élite* club or workshop whose members are autonomous and free of cowardice (though they need not be particularly brave) and frank in their recognition of their own errors and in their acknowledgment of debt – to the point of seeming to non-members breezy and over-generous, particularly towards opponents. It is the most exclusive club or workshop on earth in that it has no membership cards and no entry rules, and yet members recognize each other more easily than English aristocrats do. Anyone is able to join, simply by exhibiting intellectual frankness, curiosity, love of progress and readiness to acknowledge promptly and gratefully corrections of errors and of oversight and any idea not previously known. And the friends of the open society are not defensive.

How does this club or workshop integrate in the scientific culture and in society at large? It has no club house of its own, and no workshop proper. It is scarcely ever established. It is parasitic on the society of science and on society at large. In the modern world, some institutions safeguard the freedom of criticism and of thought. And, of course, we need not be followers of Popper or of Asimov to see that even though the scientific establishment is the very best we have, still it considers some contributions to progress as merely rocking the boat. Hence, the need for the institutions to safeguard freedom is real and these institutions are very important – even though they are far from perfect. Potential contributors often need help or at least recognition. Denying them recognition curtails their freedom. What hurts them is not the reluctance to help them but the disingenuous conduct of stonewalling instead of admitting being impressed with new ideas.

Should friends of the open society assault this disingenuous conduct or should they create new institutions for mutual support? They should try everything possible, save exclusive institutions. Even ones confined to the autonomous. Establishments are unavoidable pests. Education for autonomy is the weapon against the harm they may cause.

The minimal criterion for intellectual progress here proposed leaves old attitudes far behind and opens new avenues of research. When this addition to the debate concerning criteria of progress will be publicly acknowledged – as new and as possibly worthy of examination – then the debate will have radically altered, regardless of its acceptability or unacceptability. Relevance and raising new problems are criteria for progress that everyone accepts, even if not as sufficient. Yet, they often suffer from neglect. A debate that will take them more seriously will be different, and hopefully better. Public airings of the general aspect of the grievance of unrecognized contributors and possible contributors will reduce their frustration. The wish for more individuals to try to contribute invites more encouragement. The wish for this leads to efforts for institutional improvements; it invites public debate.

NOTES

1. See Thomas S. Kuhn, *The Structure of Scientific Revolutions*, 1962. For extended discussions of his views, see my *Science and Society*, 1981 and my *The Gentle Art of Philosophical Polemics*, 1988.
2. Kuhn's pragmatism is a meta-criterion. He ignores this. He both asserts and denies that he is a relativist. The description here of the rift between the two schools, accords with Pierre Duhem, *The Aim and Structure of Physical Theory*, 1956. Concerning conversion in science and Kuhn's approval of it, see I. Bernard Cohen, *Revolutions in Science*, 1987, and my review of it in my *The Gentle Art of Philosophical Polemics*, 1988, 123-130, esp. 128.
3. For Bacon's doctrine of prejudice, see my "The Riddle of Bacon", *Studies in Early Modern Philosophy*, 2, 1988, 103-36. See also my *The Gentle Art of Philosophical Polemics, op. cit.*
4. See my "Duhem versus Galileo", republished in my *The Gentle Art of Philosophical Polemics, op. cit.*
5. Among those who demand to limit criticism to constructive criticism are C. G. Hempel, Thomas Kuhn, Paul Feyerabend, and Imre Lakatos. See my "Kuhn's Way", *Philosophy of the Social Sciences*, 32, 2002, 394-430, 411.
6. See Gilbert Ryle, "Dialectic in the Academy", in Renford Bamborough, editor, *New Essays on Plato and Aristotle*, 39-68. See also his contribution to *Proceedings of the 3rd Symposium on Aristotle*, 1968, 69-79.
7. See K. R. Popper, *The Logic of Scientific Discovery*, 1959, Chapter 5, and my "Sensationalism", reprinted in my *Science in Flux*, 1975.
8. See my "The Confusion between Science and Technology in Standard Philosophy of Science" reprinted in my *Science in Flux* and my *Technology: Philosophical and Social Aspects*, 1985.
9. See Chapter 1.6 and Chapter 2.5 for more details about superstition; see Chapters 4.2 and 4.6 for more details about progress in art.
10. See my *Faraday as a Natural Philosopher*, 1971. See also the essay on him in the virtual *Encyclopedia Britannica*, which derides this book.
11. See Owsei Temkin, *The Double Face of Janus*, 1977. See also my *Faraday as a Natural Philosopher*.
12. See my *Towards a Rational Philosophical Anthropology*, 1977. See also my *A Philosopher's Apprentice: In Karl Popper's Workshop*, 1993.
13. For Duhem's continuity theory see my *Towards an Historiography of Science, History and Theory, Beiheft* 2, 1963, 1967, Chapter 9.
14. See my "Sociologism in the Philosophy of Science", reprinted in my *Science and Society*, 1981.

Chapter 3.2

BLOODLETTING

We still face the old question, what is it that we do when we are engaged in scientific research? We no longer contend, as once we did, that we are looking for the final truth or the last word on any matter that happens to engage our curiosity. What, then, do we seek?

There are many theories of science. The most acceptable of the modern theories about science is the theory of it as explanations and their corrections in case they turn out to be errors. It is Popper's idea. To judge its value we can examine it directly and we can use it to scrutinize some case histories, some scientific doctrine, preferably fairly well known ones, but ones with some unresolved questions associated with them. It turns out that this new theory, be it true or false, is better than its predecessors in the sense that it helps tackle these questions.

Bloodletting may serve as an instance. There are many publications about it and seemingly there is not much left to say about it. Yet, for all that, some very knowledgeable people among us find its immense popularity – despite its being utterly harmful – so puzzling as to be virtually inexplicable. The theory of explanations and corrections of errors explains it nicely. However great an error is, Popper suggests, we cannot eliminate it through rational criticism before we make it open to criticism. And empirical refutation, he says, is the highest form of criticism. In brief, we got rid of bloodletting only because we have managed to render refutable, and refute, the hypothesis that it is beneficial. Hence, the refuted hypothesis is of great value – as refuted.

Bloodletting in itself always was clear enough an affair. Its use was quite general, because until the middle of the nineteenth century all writers declared all illnesses as essentially or inherently one and the same, namely, imbalance. To put it in jargon, nosology was non-specific until then. The claim that bloodletting is beneficial could not undergo critical examination, because the medical practices with which it was combined were so complicated that any medical claim was then immune to criticism. Nosology could not otherwise have remained non-specific for so long. Eighteenth century nosology was still non-specific. Unusually, a few variants of it were rigorously developed. All of them were open to criticism. Criticism soon followed and led to their overthrow. The last of those who saw bloodletting as a cure-all was Benjamin Rush.

The practice of Rush was unusual not with regard to bleeding, and not with regard to its having rested on one simple cure-all, but in its combination of these two characteristics. The theory was non-specific but clear-cut. It therefore could not last long. The test and refutation of the claim that bleeding is beneficial soon followed. It signaled the end of non-specific nosology in general and of bloodletting in particular.

For ages, medical theory was poor, vague, and almost entirely stagnant. In comparison, the eighteenth and nineteenth centuries look radically different. Successions of increasingly definite nosological theories appeared. Stahl, Boerhaave, Hoffmann, Cullen, Brown, and Rush – they all presented their own improvements on their predecessors in rapid succession. (Both Brown and Rush had been students of Cullen.)

The case of Rush ends the western history of the non-specific school, the school that viewed all illness as essentially one. Next came the Paris specific school, whose doctrine was rather quickly overthrown. All this was genuine and rapid progress, yet historians of medicine all too often view it with dismay, as they hold a primitive theory of science as always right. They want neither stagnation nor rapid succession. In their view, each contribution should be stacked neatly in our hall of fame as the last word on its subject. Even Pasteur's sweeping generalization, his view of germs as the cause of all illness, is not the last word. Koch and his disciples refuted it. Yet, it was probably the greatest single contribution to medicine, just as bloodletting may have been the single greatest error in the history of medicine. In particular, it was Pasteur and his critics who together led medicine out of the controversy about specifism.

The idea that disease is imbalance or the lack of harmony of one kind or another is common in ancient Greece and in ancient China. Perhaps it is universal. It is so vague that even now it is hard to decide whether it is acceptable. When it becomes clear, simple, and definite, even to a small extent, as it did in the eighteenth century, then it is clearly objectionable. A long dispute took place that led to this advancement. The traditional schools of medicine, the humoralists and the solidists, simplified the doctrine of balance a little, but not enough. The humoralists viewed the harmony in question as the balance of bodily fluids. They sought to cure all ills by restoring that balance. The solidists viewed the harmony in question as the balance between the tenseness and the laxness of the semisolid parts of the body. Humoralism was dominant until the eighteenth century.

The vagueness of the two doctrines expressed itself in practice: the practices of different humoralists differed widely and the same is true of the solidists. On occasion, a solidist might prescribe more like a humoralist than like some other solidists, and *vice versa*. Both parties prescribed strange and complex treatments. They would use emetics, purgatives, and bleedings for one reason or another, depending on traditions and on the details of the various views that the various practitioners held and on their hunches. The situation began to alter with the introduction of the mechanical philosophy. Various historians have pointed out that at first mechanistic theories of physiology could not influence nosology or treatment. For centuries, no one could draw significant medical conclusions from the view of the heart as a pump, of the stomach as a mill, and of the kidney as sieves. The direct, practical significance of each mechanism appeared later, in the nineteenth and twentieth centuries. Nevertheless, important new factors arose in the eighteenth century, and they may connect with the then new mechanical philosophy and the ideas of simplicity and clarity that stood behind that philosophy. The mechanical orientation won over the biological orientation in natural philosophy, and then solidism won over humoralism in medicine. According to mechanism, solidity is essential to all matter – it declared even particles of water solid. Eighteenth-century nosology and treatment were highly simplified and this permitted rendering them much more clear-cut and austere than their predecessors. Fancy medications were (are) still popular, even with the most progressive physicians of the time, beginning with Sydenham, the seventeenth-century reformer of medicine. Fancy medications never disappeared, but leading thinkers were always against them, and in the eighteenth-century they began to have significant influence. Thus, Stahl, who opposed almost all treatment, believed in an inner homeostasis of sorts (if the anachronistic use of this word is permissible). The nosological theories of his followers were equally simple: Boerhaave assumed that the cause of all illness is blood impurity (acrimony); Hoffmann suggested nervous spasms instead; Cullen said spasms cause debility and debility causes illness; Brown said debility was the primary cause of disease. Following Stahl, both Boerhaave and Cullen advocated decreased intervention with the system, especially reducing the application of purging and bleeding.

For the first time in history, something like clinical observations on the effects of medication could then take place. This explains the rapid succession of theories. When Brown failed to see the difference between Cullen's two stages, and when he postulated that the stages of an illness are merely different kinds of debility, he recommended stimulants – any stimulant. How simple. Rush followed this recommendation in the epidemic of yellow fever in Philadelphia with no success. He then proposed the theory that the cause of illness is vascular tension, to be relieved by reducing fluid-pressure, and for this he recommended the drawing body fluids, perhaps by purging, or, more drastically, by bleeding – even up to four-fifths of the normal quantity of blood.

Bleeding, stimulating, and other eighteenth-century medications, were not new. The novelty was in the relative clarity and parsimony of the theories that accompanied them, and the subsequent austerity of the prescriptions that rested on them. Complicated ideas, complicated concoctions, and rules-of-thumb galore, were common till then, and then they were officially banished (not in practice, though). What constitutes a stimulant was not clear then (nor is it sufficiently clear now). But it was clear enough to Rush to enable him to find empirical evidence against Brown's theory. This was immense progress.

Opinions about Rush run the gamut. One extreme view says, he was a good physician. One volume that expresses this view mentions bloodletting only thrice, and briefly. Imagine a biography of Luther with one chapter on religion and theology that mentions his quarrel with Rome but only cursorily, and mainly as unimportant, and you may have a glimpse at the intellectual level of this discussion of the worth of Rush's medicine. Some writers praise Rush as a psychiatrist and staunchly refuse to discuss his medical theories and his having practiced extravagant bleeding during the yellow-fever epidemic and later. "The science of bacteriology," says one author in defense of Rush, "was not yet born". As if anyone accused Rush for not having known the true cause of yellow fever (a virus, not a bacterium, by the way). "His methods were doubtless crude and they seem to us like methods of despair." Admittedly, though treating all fever by reducing patients' temperatures is very crude, in certain circumstances it may well be the best available treatment. But when hundreds of victims of yellow fever, even pregnant women, died of the anemia caused by Rush's extravagant bleeding, the word "crude" cannot apply even crudely; "wide of the mark" is more like it.

Another author reports that in the publications of Rush there is "not one page of scientific value. There are few medical writers, and certainly none of celebrity, whose works are less worthy of perusal today than those of Rush." What this means is that if one were to ask the author and Rush intelligent questions on matters medical, their answers would always diverge. True, this attitude has kept that author rather ignorant of medical history. For, roughly speaking, he is in equal disagreement with Hippocrates, Galen, Sydenham, and Rush. He is not, but to differentiate between them, as good historians may want to, one needs more tolerance of error than he shows. Political and social historians take this for granted, but historians of science are sadly lagging behind.

Another writer who uses agreement as a criterion, tried to find some points of agreement. His paper is as sympathetic to Rush as possible, praising him as it does for his psychiatry, for his universal knowledge, for his participation in public affairs, especially in social medicine. As to Rush's nosology, it notes two points of agreement. Rush associated intestinal lesions with capillary distension, though by implication only, and in reversed causal order. His non-specifism, his view of all illnesses as due to one factor, resembles present-day theories of low resistance and of hypertension. The author finds all this rather feeble and he criticizes Rush for not having attempted to verify his theories – yet he also reminds us that in Rush's day there was no knowledge of statistical testing, sampling, the use of control groups, and similar techniques.

We can explain why one historian sees nothing scientific in Rush and another sees some science. Both consider science as infallible: they agree on principle and differ on details. Science is proof, proof precludes disagreement, and we know what has gained proof. Hence there can be no disagreement about any detail. But there is, and so the previous sentence is erroneous: science is not proof: it allows some disagreement.

L. H. Butterfield's edition of Rush's letters offers the clearest view of science as infallible, through its application to the case of Rush. Butterfield says, Rush's "fateful pronouncement of the necessity of bleeding in yellow fever … is of more biographical than scientific significance." Butterfield says that Rush was no fool, and he explains the logic of Rush's reasoning, but he considers this too as mere biography, not as medical history. He says, Rush had failed in using stimulants, succeeded in using depletion, used more depletion, and had "abundant testimony" of success. He reports that Rush "believed that experience as well as theory sustained him", yet this cannot be medical history. No matter how much medical practitioners and researchers see that theory and practice justify whatever they do, and no matter how commonplace their actions may be, if the historians of medicine disagree with them, they will denounce their actions. Let us assume that medicine will progress further and abandon some positions that win general consent today. Historians of medicine who will live two centuries from now will then reject the practices of the medical practitioners of today. If they will adhere to the principles current in the history of medicine today, then they will declare that these practitioners act in their capacity not as scientists but as private citizens. Let us hope that the history of medicine will advance too, and that the field will abandon the principle according to which today's medical research needs the endorsement of tomorrow's historians

As it happens, there are personal, private aspects of the story of Rush. Butterfield ignores them. He says, Rush was seemingly successful with one method of depletion – purging – yet he used another one – bleeding. What has happened? Doubtlessly, purges were far from successful. Yet Rush did report them as successful – even in letters to his wife. Why did he think so, and why did he not stick to his presumed success?

For one thing, Rush's reports to his wife are highly unreliable. They are full of self-deceptions, some of which he could not sustain (such as the merit of purging) and some of which he could (such as the merit of bleeding). In his letters to his wife, he speaks of his opponents' total failure and of his own total success. When he reports to his wife that one of his own assistants had died, he does not raise the question, why had he not spared that assistant by the use of the lancet. Rather, he says, he, Rush, had trusted that assistant to bleed him when he had been a victim of the yellow fever.

What caused this self-deception, this lack of self-criticism? Partly, at least, it was the pressure of the situation. On the day he issued his public declaration about the benefit of bleeding, he writes to his wife, he had a good night's rest. He writes at five o'clock in the morning, he adds, and he has already had seven calls. This report of a good night's rest may give us some idea of the pressure of work and fatigue. We still do not know just how much overwork clouds judgment. Lord Chandos raises this question in his autobiography when reporting that, during World War II, cabinet ministers were frightfully overworked and had to read daily, for instance, material of at least a novel's length. He was probably alluding to the damage that he thought overwork had caused Sir Winston Churchill.

We still overwork our interns, perhaps in order to increase their stamina, but we have scant evidence on the psychology and physiology of overwork. We still do not know what damage due to fatigue is reversible, what is not. We may be bleeding our interns the way Rush bled his patients. We know that he was prone to self-deception, as we all are. Perhaps we may connect his capacity for self-deception with overwork, but this is highly unsatisfactory. His colleagues were even more uncritical. He could not deceive himself, even in the midst of the yellow-fever epidemic, concerning Brown's method of stimulation, with which he had started, and concerning his own method of purging, which he tried and gave up as well. He stuck to bleeding not only during the epidemic, but also after a long period of reexamination and public debate. So, why was he more successful in deceiving himself about bleeding?

The idea that what is scientific is what our present textbooks express is very unsatisfactory – as anyone who has read any century-old textbook of science should know. When we speak of science as systems of theories on which we may base predictions, we usually mean correct predictions, or at least reasonably correct ones, but we seldom say so. The demand to predict only correctly is terrifying. By Popper's revolutionary view, the ability to use a theory in order to predict – correctly or incorrectly – is already an achievement. We can then check the prediction, and when it does not come true, we may try again. At first glance, this may sound fairly commonsense and plain sailing. Thus, when historians censure Rush for his not having tried to verify his theory, they obviously mean, had Rush tried to verify it he would have refuted it. And so, it seems, there is no difference between verifying and testing. But this is not true, as already Bacon pointed out in the early seventeenth century. Bacon's chief contribution to scientific method was his observation that one's efforts to try to verify one's hypothesis will almost invariably succeed, independently of how false it is. To illustrate Bacon's discovery, we may take the case of Rush as presented by H. O. Perry Pepper. Pepper discusses, in particular, Rush's claim that bleeding brings speedy relief. Rush's claim rests on the observation that where bleeding for the first and second time causes fainting, the third and fourth do not. Pepper has conducted experiments on volunteers to check this observation – probably the first experiment performed in the field of the history of science. Pepper's result is quite instructive. Bleeding may cause anemic fainting, but not when the patient lies down. Probably, he suggests, Rush's patients were so damaged by the first and second bleeding that they could not sit up, so they escaped further fainting – with the fatal consequence that Rush thought they had improved.

This illustrates Bacon's theory that it is easy to verify a hypothesis for spurious reasons. Bacon concluded that one should never entertain a hypothesis, but let facts act as guides to the true theory. Bacon considered criticism useless because it is so easy to verify one's own hypothesis to one's own satisfaction. People could then dismiss criticism as expressions of the personal defects of critics. Since people believe that in science error is avoidable, they cannot but see all criticism as the denunciation of the object of criticism as unscientific. Hence, all criticism amounts to disrespectful debunking. Hence, every debate is a war involving personal attacks, and every wrong side is obviously vicious. These corollaries of Bacon's views became the norm, and they led to two typical mode of reaction. It became the rule not to name the criticized party; to allow the criticized party to alter their views surreptitiously. Robert Boyle, the first leader of the Royal Society of London, advocated this practice, and unwittingly legitimized it. When implicit criticism was not clear enough and the inner logic of the debate required a detailed analysis of the opposite view, the response was defensive and led to personal quarrels. Remnants of this are still extant. In public, commentators do not disagree with the papers they comment on; they "fully agree with the honorable speaker, but ... ." The theory of the speaker is not false but in need of some "minor addition", "clarifications", "restatements". Any euphemism for criticism and replacement will do. These euphemisms illustrate our regrettable insistence on taking criticism as inherently hostile.

The task for a historian of a controversy, then, is often to reconstruct implicit criticism from hints, and to expurgate an explicit controversy from its personal venom. Without remembering that tradition demanded that only a theory that is above error should claim scientific status, we can hardly explain the defensive behavior of those who did claim scientific status for their ideas. Even today, such behavior is rather typical, the result of the imposition of standards of science that are much too high, and that force researchers, physicians and others, to pretend that they are above error. The widespread identification of science with demonstrability entails the identification of error with sin.

Rush did not always agree with Bacon's doctrine. For example, he wrote to John Adams (April 5, 1808) that "prudence is a rascally virtue". Yet, he was a Baconian of sorts, like most of his contemporaries. In the fashion at the time, he was cocksure and savage to opponents.

He supposed that people had quarreled more in the past because they were more quarrelsome in the past. He also related the disposition to quarrel to impiety. Certain current traditions, customs, and errors are conducive to quarrels, and the hypothesis that in the past these might have made people quarrelsome is reasonable. Obviating quarrels by suppressing differences, of concealing or minimizing them and their value, is traditional. It boomerangs, as it promotes quarrels. The ability of differing parties to take differences well renders superfluous and even silly all methods of suppressing them.

The alternative way is Bacon's. When there are no opinions, there are no differences of opinion; and with no opinions, one can observe facts independently of theory. If this were possible, then it would be possible to avoid error. And then, to be open to criticism is unwise imprudence. And then, those who are subject to criticism have the right to feeling slighted.

Popper has put things very differently. Factual discoveries owe their existence not only to their discoverers, but also to those whom the discoverers have criticized: but for their theories, critic would not have made the observations that were the discoveries. They would not have known what to observe. They would not have known which instruments and techniques to invent. So much about theory. Attitudes to practice differ too. In the field of the philosophy of science the advocacy of the unity of theory and practice is almost unanimous. This attitude leads to the view of everything else as hypocrisy, as preaching water while drinking wine. It suffers from the aftermath of the quarrel between science and religion. Priests do not practice what they preach, but scientists do, they suggest. This is silly. Scientists do not preach. They study some questions. At times, they offer to the interested some answers, usually confessing their own dissatisfaction with what they offer. This is not universally true. Some scientists, more teachers of science, and most philosophers and historians of science, are still militant. And many practitioners are much too authoritative. But this tendency is happily on the decline. Science needs no defense – least of all, militant.

We may then reconsider the judgments of Benjamin Rush. The literature combines his thoughts and activities in a most unacceptable manner. It is a secure privilege to believe in the worst superstition, but not to implement it, as it may harm others. It is a legal point that when a community accepts a superstition, action based on it is inculpable, even if it is harmful according to all scientific experts. Not so if it is idiosyncratic. This holds for superstition and *a fortiori* for scientific theory. Was Rush then in the wrong when he applied his own ideas? If we judge him by current scientific standards, then, we must answer the question in the affirmative. If we judge him by current legal standards, then we must answer the question in the negative. *Verdict for the Doctor*, by Winthorp and Frances Neilson (1958) is a study of the case of libel that Rush won. It shows clearly that he acted well within accepted legal limits. This need not arouse contempt. Rather, it may raise our hope. If improvement of the legal standards of those times was possible, then hopefully improvement of the legal standards of our times is possible too. Theories were then not as refutable as today. Irrefutability, said Popper in a memorable maxim, irrefutability is not a virtue but a vice. The distinction between the early and the advanced stages of an illness, which played an immense role for Rush and his predecessors, was so wooly, that even had modern statistical methods existed then, they would not be easily applicable to them. Yet, some refutation was possible. The result was a barrage of simple, though rather traditional theories that quickly opened the way to the clearer, more revolutionary theories of the nineteenth century. The development may be more obvious in the light of the improvement of views about criticism. Popper's view of progress through the incessant invention of testable theories and the use of empirical methods of examining them should entrench this openness to criticism.

Still, there is the question of responsibility – especially of physicians. Some errors are clearly inadmissible. To try to avoid all error is to evade the question, which error is irresponsible? The answer is, avoidable error is irresponsible. So, when a new idea appears on the horizon, the requirement is to test it before applying it. Was the bloodletting of Rush an innovation? It is hard to say. Were today's laws applied, possibly they would oblige him to test his method. The availability of test methods is not enough: their application depends on the readiness to overcome defensiveness and to recognize criticism for what it is: appreciation of its object and help to those who wish to improve upon it.

Chapter 3.3

THE CRITIQUE OF LINEARITY

0. PREFACE

"Linearity" is the (unfortunate) jargon term for proper progression. Building a house normally starts from the foundation and progresses upwards. Linearity gains its respectability from science – by mistake. Progress in science is linear only locally: theories of gravity appear in succession, each built on its immediate predecessor. But different branches of science are not linear. A scientific theory unifies some theories and some observations, and gives rise to others. This too is not linear – it cannot possibly be.

Heinrich Heine (*Religion and Philosophy in Germany*) contrasted medieval and modern thinking in a striking manner. Bertrand Russell did the same (*Marriage and Morals*). Medieval thought centered on religion and on human evil as its chief concern. By contrast, ancient and modern thought is science oriented and deems humans good. Irrationalists dislike reason in general and science in particular. They hanker after the Middle Ages and deny reason all spiritual value. They wish that philosophy should move from the concern with science to the concern with the arts and with alienation: science aims at the conquest of Nature, whereas alienation is from Nature, as well as from society, and even from one's own self, so that science cannot cure it.

The positive part of this proposal is valuable. To uphold it while respecting science is a challenge to the current *avant-garde* philosophy that advocates systemism and aims at the integration of science in pluralist democracy in a bootstrap fashion: an improvement of any part of our thinking invites reexamination of all of it..

1. THE TRADITIONAL, NARROW MODERN VIEW

By traditional philosophy science is demonstrable. The way of developing ideas and proving them was under debate. This is the whole of traditional philosophy of science. Current debates in it focus on the nature of scientific knowledge. Suffice it to realize that knowledge is not open to disagreement: we rightly ignore contests about it, unless some very strong arguments with some immediate appeal accompany them. The case of racism and even sexism illustrate this: we are rightly impatient with those who advocate them. Thus, for now we do not need the sufficient conditions for a theory to possess the status of knowledge: two observations should suffice here. 1. Knowledge is not contestable. 2. Some scientific theories are incontestably false.

Most philosophers of science are unhappy with this: they want to have knowledge at least in some weak sense. Already Aristotle introduced this idea. There is no objection to it, as long as we exercise caution not to confuse the strong and the weak senses. Knowledge in the strong sense implies truth, not knowledge in the weak sense. Overlooking this may lead to relativism. By this view, Newton's theory was true a century ago but is no longer true. This means that for a while it pleased Mother Nature to follow Newton, until She thought the better of it and decided to follow Einstein. This is not serious, yet sadly some serious people champion it.

This response angers them: they say that Mother Nature is a social construct to begin with. This is hilarious. Social constructs arise within society and society arises within Mother Nature. Identifying Her as a construct is like identifying a person as a portrait. It sounds profound but it is a simple error.

Those who say that science rests on experience suggest that it comprises a vast database plus a depiction of its structures. The idea that science is knowledge in the weak sense is the old idea that faith in scientific theory is obligatory, minus the idea that this faith is irrevocable. This is central to the philosophy of Michael Polanyi; it is now popular because Thomas S. Kuhn endorsed it and vulgarized it. They say, the scientific consensus comprises shared tentative beliefs that are essential for the organization of research.

A new tradition of the study of science in the weak sense began with the crisis and the revolution in physics of around the year 1900. Two such ideas appeared then, one inductivist, one instrumentalist. Inductivists say, a theory that evidence backs becomes (not certain but) probable. Instrumentalist say, science is still certain, but only because it is utterly uninformative. As instrumentalism denies scientific theory its informative character, it finds an intellectual vacuum that invites unscientific theory to fill it – metaphysical or religious. Those who find this impossible turn to inductivism in a pinch. As induction is impossible, they return to instrumentalism in a pinch. The two systems are thus complementary myths *à la* Claude Lévi-Strauss: they are popular because they see scientific progress as the growth of its database. How then do databases grow? By the growth of technology: better instruments bring more information.

This is hardly credible: it raises a simple question yet one repeatedly dodged: how do we improve our instruments? The answer is obvious and uncontested: new ideas improve instruments that improve data that invite new ideas. In brief, technology is applied science. This answer is a version of the linear model.

This model was at the heart of modern philosophy. Today it comes in two versions, the inductivist version that views technology as applied science and instrumentalism that views science as the mathematics of technology. Tradition was inductivist and anti-instrumentalist: first give up all preconceived opinions, second, observe, third, allow theories to emerge out of the information by itself. There are no other details on this procedure. The discussion now centers on the question, what is science – in the strong or in the weak sense of the word. Recent critics of the inductive model, particularly Popper, have repeatedly refuted it. Let us ignore this fact for a while and discuss the model itself, since the refutations leave totally unaffected most philosophers of science, such as the leading philosophers Carl G. Hempel and Adolf Grünbaum, who continued to discuss the nature of science while taking the linear model for granted and dismissing as an irrationalist anyone who rejects it.

Their approach makes sense: the linear model is the traditional theory of rationality, and so deviating from it naturally looks an abandonment of rationality. (Galileo reports – with understanding – the very same reason for the reluctance to abandon Aristotle.) The linear model is a myth, and it is hard to refute myths. Noticing any significant social factor in the growth of knowledge already deviates from the model, and even such simple things as encyclopedias that contribute to the growth of knowledge, comprise deviations from it: such contributions are evidently not linear.

2. THE SOCIAL CONTEXT OF SCIENCE

Some claim that long ago Marx, who noted the social context of science, has superseded the linear model. The view of Marx as perfect spoils the discussion of the interpretation of his works. One of his current leading exponents, Jon Elster, has ascribed to him the current most plausible philosophy of the social sciences simply because it is the most plausible. As far as linearity is concerned, Marx was emphatic: he viewed social progress as the result of scientific progress, never as its cause. This is his version of materialism: scientific progress is the basis of all social change. This renders medieval economy superior to ancient economies, and thus medieval technology superior to ancient technology, and thus medieval science superior to ancient science too. The Marxist scientist J. D. Bernal said that medieval science is superior to ancient Greek science. Sad.

Science and technology do not always advance in concert. Only to the extent that science transcends all national and linguistic barriers, can it approach global unification. Even then, one country may be more advanced in one science and less in another, since one science may be more advanced there than another. This is so for many reasons, public policies being one of them. It is hardly possible to have all nations technologically equal, nor can we always say that the developed world is equally advanced in all technologies, though, largely, societies proficient in literacy are more advanced than illiterate ones. Even this is a prejudice. Were technology as advanced in one part of the modern world as in another, investors would not go to one country to benefit from one technology and to another country for another. Advocates of the linear model say that this is inessential: in principle we are all equal. This way they dismiss all the complaints of the poor, relying on the principle we all can be rich. They say, poverty is insignificant as it will soon find its remedy through science. This is the optimism of the linear model that characterizes it from its beginning. It was endearing and even noble; it is no longer humane.

Traditional scientific optimism rests on the premise that science progresses and causes technological progress. This rests on the idea that technology is nothing but applied science. This is the linear model.

Perhaps even the introduction of the social dimension of science will not impress the advocates of the linear model. We may ask them, then, what evidence will make them change their views? If they stick to the policy of declaring marginal all information adverse to their model, then arguing against them is useless: fortunately, no law forces people to alter their views. How then do scientists come to agreement? Here the outlook of Polanyi and of Kuhn enters the picture forcefully. Science, they say, is an organized system that imposes its views on its membership on the threat of expulsion. Amateurs do not count, then, and professionals must endorse official doctrines or lose their jobs, Kuhn said approvingly.

Deviance in science refutes this. It is customary to consider deviant scientists marginal. Yet, they often initiate progress. Polanyi and Kuhn ascribe changes of opinions to decision of the leaders who know when changes are needed, and when and how to bring them about. Leaders, they add, are unable to explain their decisions. At times Kuhn adds this: refutations of the ruling doctrine lead to their amendments *ad hoc*; this renders their application clumsy; eventually, this invites radical change. The leadership decides on the change. Alternatively, those who make the change thereby become leaders – by definition. Yet, progress in technology differs from progress in science. Technology uses scientific theories, new and old alike, orthodox and unorthodox alike, and as *ad hoc* as needed – as long as they deliver the goods. One of the most commonly applied theories, in theoretical physics as well as in technology, is Schrödinger's equation that abides by the old rules of Galilean transformations (rather than the new Lorentzian ones).

How do leaders change views? Is there any chance that the philosophical leaders will give up the linear view of science? This question requires analysis of the conduct of leaders. It is tempting to try to convert them on the understanding that this will have a snowball effect. This is an illusion: there is no shortcut to education and to increasing the taste for improvement. Nevertheless, it may be useful to present the rationale of the leadership.

Science has much prestige. Those who wish to bask in prestige tend to praise it. Technologists who wish to bask in the prestige of science praise the linear model hoping that it makes them scientists. Kuhn's view appeals to them because (following Derek Price) he generously grants them the title of normal scientists: his ideas allow the personnel of technological industry to bask in the prestige of science.

Here lies the right way to advance a leadership with vision: it can grow with the development of research. This at least will be the right, honest, and straightforward conduct. Hence, the right question is, how is the rejection of the linear model going to improve significantly our understanding and application of science? If this question does not find a simple, very plausible answer, then perhaps we are on the wrong trail. To find out the worth of the alternative research we should, in line with Popper's philosophy, take cognizance of the refutations of the original model.

The linear model ignores the social character of science. This character raises the question, what kind of society allows science to flourish? What are the necessary social conditions for viable research? One reason the linear model is popular among researchers is that it suggest that science can flourish in any place with enough wealth and leisure – to allow for time for research and investment in the construction of instruments. This view leaves no room for the social problems of science, and for the need to arrange priorities in research, not to mention education in science and technology, professional, vocational and amateur, plus the education for autonomy that is the basis of all that is good in the modern world.

Rudiments of scientific research appeared in many societies without leading to scientific revolutions proper. This is so partly because even the best conditions need not deliver the goods. Yet it is reasonable to assume that for the success of the scientific revolution the essential conditions were society's endorsement of research as a value, society's ability to invest some energy in research, and the existence of some social organizations that could promote it and pool its fruits. An essential condition for this, said Popper, the first sociologist of science, is the openness of a society to criticism.

Gellner has suggested that the fathers of the scientific revolution could ignore the social conditions under which they were operating just because these were sufficiently congenial to the growth of science. Since then science was threatened a few times, in the twentieth century more than ever before, because it was then that politicians and generals learned of its importance. (The military command does not have the foggiest idea as to how much of the military budget they should optimally invest in research.) The same goes for technology much more forcefully, yet the most important item in this context is social and political technology. This still awaits attention. Social scientists feel inferior to physicists because they think no social engineering can be as devastating as nuclear weapons. This is most unimaginative: our very survival depends on the success of the next advance of social and political technology. To make it more scientific is to allow for experimentation, and the only system open to social experimentation is democracy. This is the chief lesson from our recent self-inflicted disasters.

3. THE END OF LINEARITY

Science and technology interact strongly in the new theoretical field of practical computability that fuses mathematics, science and technology from the start. Adherence to linearity leads to the oversight of the fact that all science and all technology involve some social and some physical studies. Every study as such is interdisciplinary. Systems theory explains the fact that usually natural scientific research ignores social issues: they are so marginal there as to be practically irrelevant there. Not so technology. Physical technology has a vital social part. This is clear since a technology may be applicable in one society and inapplicable in another despite the absence of differences in their physical circumstances. A society may endorse or reject the technology of another society, and then it either blooms or collapses. This is true of all societies that modern technology has invaded, from native Australian to Japanese. Alienation in these societies is much more common and painful than in Western Europe. As the post-modernists ignore this fact, their pretense to sensitivity to the human condition is sham.

There is no need to show that science and technology differ: obviously, their aims differ. Understandably, then, at times researchers pursue projects that they do not deem of much scientific interest, yet they pursue them all the same because of their presumed significance for technology. This occurs sufficiently often to have a name: basic research. Basic research is evidence of both the diversity and the cooperation between science and technology. At times the fruits of technological research and of basic research happen to be of great scientific importance; especially medical research has at times great import for the life sciences. This means that even from the pragmatic point of view we should not confuse science and technology. Since the results of research are new, it is the pragmatically wisest to invest both in science and in technology. But it is inviting trouble if the sole justification for the spending of public moneys on scientific research is that it is useful for technology.

Science strongly interacts with technology, and unless the two are distinguished, their cooperation and liaison will lead to the oversight of the conflict between their interests and then the stronger interest will win. The interest of technology is in principle stronger than that of science, and this is why it is wise to install some safeguards, like the famous clause in contracts in technological research permitting an established researcher to work on independent projects for a day or two a week.

The most unpalatable trend in science is the tough-and-no-nonsense attitude. It is not greatly objectionable, as it is simply the indifference to anything that is not a great moneymaker, and as the right for indifference is a valuable right unless otherwise specified by law. Advocates of the tough-and-no-nonsense attitude are often also advocates of the linear model. This is not easy to comprehend though it has a simple explanation. The linear model encourages the oversight of the existence of different frameworks for scientific and for technological research and training. It thus lets the interest of science yield to the interest of technology without prior debate. As a result, when the military and then industry invaded research institutions, no one could stop them. This is dangerous. It leads, moreover, to the backlash of the demand for the segregation of scientific from cultural research. This segregation will elevate culturally valueless research to the status of science. Those who benefit from this move naturally favor it. But science must be self-critical, and for this it must be democratic and pluralist and present to the general public both ever-newer popular science and its criteria of success and the difference between them and other criteria, such as those of technology. Some new science policy writers are now claiming that the only way to promote criticism within science is to have it compete in the open market. This is better than having the military usurp it but not as good as placing it in the public domain. It belongs there, and only a confusion of science with technology prevents seeing this. The difference between science and its fruits is easy to see, or else patent testers have to show it.

Without this difference, the following standard problem of management of research makes no sense. Should we relegate research to an expert scientist or to an expert organizer or to those few individuals who are expert in both? There is little literature on this, and most of it sidesteps the difficulties involved as it ignores conflicts of interests. Obviously, when a situation is unproblematic it is easy to find an answer to the question at hand; also, in such cases it matters little what answer is applied, especially since it is easily correctable then – or else the situation is not as unproblematic as it is made to appear. But usually the situation is problematic.

This matter is of appointments to specific projects. For the project as a whole, things are different. Candidates for any position of leadership or management are always volunteers. One reason Niels Bohr was a leader and Einstein was not is that the latter refused to participate in the running of the scientific community, though from time to time he intervened, to encourage an idea that looked less promising to others than to himself, and to address President Roosevelt and help create the Manhattan Project. Bohr was not ambitious, but most people in managerial positions are, and instead of displaying leadership, they rub shoulders with the powerful. As to scholars, administrative posts attract them only if they are burnt out and if they have left research for too long to return to it. They exert unhealthy influence on the young under their tutelage.

Even if the problem of filtering out people who have a great interest in leadership so as to be able to seek the right ones, even then the problem remains, how do we decide such matters? The tendency is to optimize the choice by reference to qualifications without the benefit of any institutional analysis. Good leaders are sensitive to the aims of an organization and to the wishes of its membership, while canvassing new ideas and experiments. We do not know how to draw the line between responsible experimentation and carelessness, which matters especially in academic institutions, where success fetches handsome rewards but failure hardly ever calls for punishment. This is the cause of the polarization of the academy into the lethargic and the reckless, with a few exceptions that are the true *avant-garde*. Not that punishment is recommendable, but caution is, especially where punishment is hardly an option. Even in civilized politics the greatest punishment is usually hardly more than the failure to be reelected, though political irresponsibility is much more dangerous than that of research management. This, however, may be shortsighted, as we should wish research management to improve and be able to help us build a new responsible political tradition and raise a new, responsible political leadership.

I do not have a solution to the problem, yet I have some suggestions. For example, we may discourage managers from publishing fake or recycled results and from appending their names to other people's publications. We may encourage them to report on their management experiences, and thus accumulate a significant data-base. This may stimulate some research into (the socio-political aspects of) the efficiency of laboratory management. Thus far, there is almost none. We do not even know where do the more successful managers of interdisciplinary projects originate, and what is the secret of their success. There is not even the beginning of the examination of decisions of grant-giving committees, and quite possibly this is rooted in vested interests. Only generous grants for research into the question of the efficiency of grants will possibly refute this hypothesis. Until then let me express my view of the prevalence of the influence of vested interests.

How does experience back a hypothesis? This question is the same as the question, what exactly is science in the weak sense and how is it possible? Current studies of the question concerning science in the weak sense run parallel to the old, traditional studies of the question concerning science in the strong sense. This is confusing. For, whereas there is proof that science in the strong sense is impossible, and whereas the same proof holds for some versions of science in the weak sense, there are versions to which it does not apply and these are not so problematic as to make us doubt its possibility. Making the question hold for all versions of science is then confusion. What version of science in the weak sense does not fall prey to the classical critique of strong science? The theory that science is not proof but probability is vulnerable to the classical critique of strong science. Science in the weak sense is verisimilitude. This idea is not vulnerable in the same way, since all claims for verisimilitude are hypothetical.

The theory of empirical backing of theories is thus without a question to answer. The question may be, assuming we know the difference between fact and fiction, how does evidence differentiate between theory and fiction? This matters, because we have scientific technology: any technology that claims scientific status has to have stood up to test. This demand is legal, however, not philosophical. And like all law, it is subject to specifications: the law prescribes tests and determines what results should count as success. The law is not good and we always try to rectify it. Contrary to what philosophers of science say, corroboration is license to act, not obligation to believe. Philosophical justifications of the law interfere with efforts to reform it. It is particularly sad that due to the prevailing prejudice in favor of the linear model laws about corroboration ignore their use in social experimentation in many fields from aviation policies through education to zero tolerance.

The linear model suggests that only one right sequence for economic growth is right. This supports the ideology of no strings attached, of offering poor countries food and money the best way to helping them. The food went to the ports of the poor countries and rotted there for want of transportation and distribution. The moneys went to Swiss banks. (Only the Swiss Red Cross first broke this policy only in Ethiopia during the famine there, not any government.) The poor nations are now worse off, not better off. The West blames them rather than criticize common western errors, while poverty is becoming increasingly explosive and a threat to all, since it is a major factor in the rise of ethnic wars and international terrorism. It creates tremendous migration pressures, especially in the form of migrant labor forces. These pressures incite anti-democratic movements in rich countries despite their strong democratic traditions.

All technology has a social and a physical side. At times, its physical side is so prominent that it is advisable to ignore its social side; at other times, the social side is so prominent that it is advisable to ignore its physical side. These days social technology is becoming increasingly important, with problems becoming increasingly pressing. These are internal – education, drug abuse, unemployment – and external – the Proliferation of weapons of mass destruction, the Pollution of the environment, the Population explosion and Poverty (the four P's). Certain central urgent problems are easy to solve as far as the physical side of the technology is concerned and almost impossible to solve as far as the social side of the technology is concerned. This is the challenge for all responsible leaders and for all researchers who wish to show their concern for the future of humanity.

Chapter 3.4

SCIENCE, POLITICS AND OBJECTIVITY

The equation of the idea of objectivity with that of scientific objectivity is popular. Thus, what we deem not fully objective we demote to the status of non-science. Objectivity thus prescribes unanimity. All should agree about it except for those whom we disregard: the ignorant, the vicious, or those who are otherwise at fault. Hence, if scientific politics is possible, it should prescribe unanimity among all who count. Otherwise, we should deem subjective all politics, especially party-politics. And then, the decision regarding party affiliation becomes a matter of mere taste. All this seems to be a clear-cut conclusion from the equation of science with objectivity. This equation rules out both rationality and pluralism, and, finally, democracy too. It is shaky, then, to say the least.

In one use of the words "subjective" and "objective", they are complementary. Everything is either subjective or objective but not both. In another use, a subjective decision may become an objective fact. The term "objective" denotes the given, whatever is "out there" independently of anyone's wishes, or, in philosophical jargon, the outside world. These terms are not always synonymous. The inner world is also a given. If the subjective is the complement to the objective, then the inner world is subjective. It is what we experience "immediately", or what we can freely manipulate. Liberal writings express respect for the subjective. The case that liberals take as most important here is that of tastes – perhaps because it is the only one that they deem truly impervious to an offensive by science. We may alter tastes at will, it seems, as they are not scientifically given and not open to scientific treatment. But perhaps not; and if we cannot alter them, then many liberals would respect them all the more. This, after all, is the repeated liberal defense of uncommon sexual preferences. Of course, it is better not to defend them: they are in no need of defense. Whatever our tastes are, and however we have acquired them, science accepts them as given facts and respects them as facts. Nevertheless, this is not the whole story. Moral decisions comprise a more important example. And they are problematic. Philosophers deemed some of them right, other wrong – unlike tastes that are neither. Hence, to the dichotomy their status is undecided.

German philosophers, especially Hegel and Heidegger, have managed to confuse all this – intentionally, since they craved a special status due to their being German philosophers. It is not clear whether they put to one consistent use the words "subjective" and "objective" – or any other word. (Hegel endorsed some sort of contradictions. And perhaps he viewed true subjectivity as more objective than objectivity. See Piotr Hoffman, *The Anatomy of Idealism*, 1982, 40-3.) He viewed duty as given, irrevocable, and independent on any decision, and he viewed all independent decisions as caprice, regardless of contents. To convey this view of his, he called duty "objective": it is not a matter of choice. It is a duty, he suggested, to endorse the tastes and decisions of the powerful. ("The subjective will … is … objective … in so far … as it has itself for its determination" whereas the merely given will, "the objective will … is the will of a child, the ethical will, also the will of the slave, the superstitious man, *etc.*", *The Philosophy of Right*, §26.) Thus, received morality is childish and slavish. The wills that powerful individuals impose on the world he found superior and obligatory to recognize. Thus, imposition turns the wills of the powerful from subjective to objective ("the identity in subject and object", §110; "… carrying out my aims I retain my subjectivity … during the process of objectifying them …" §112). (This is the doctrine that might is right.) He considered childish, slavish, and superstitious the liberals who respect other people's tastes enough to leave them alone. It is not surprising, then, that his texts are confusing. He called the truth "the objectivity of concept". In his writings, objectivity in this sense signaled what we must endorse, what is obligatory to recognize. Subjectivity is to objectivity what superstition is to science. He declared his own opinions scientific and objective. Other opinions, especially liberalism, he declared subjective and superstitious. The powerful who have succeeded to impose their will on the world, then, "externalize" their own subjectivity by actions and render it objective (§113). Thus, science and power become one, political and scientific authorities become one, and the endorsement of both becomes obligatory. This is not the common meaning of the word "objective", but then it is not the common meaning of the word "science" either. In science, only facts and demonstrated theories are objective and obligatory to recognize; in Hegel's vocabulary it is something else.

This is a colossal confusion. There is something to it, though. It is an echo of Plato's distinction between knowledge and opinion. For, an opinion may be true and in that sense objective, whereas its being held with no good reason is possibly subjective. And so, in a sense the objective and the subjective overlap, and in a sense they do not. Also, subjective items (like tastes) as facts are objective. The freedom to change tastes makes them subjective and the obligation to recognize and accept them is objective. So, it is perhaps better to avoid using these terms.

The situation is uncomfortable quite apart form terminological confusions. It suffers from an intolerable polarization: there is very little that is obligatory with no appeal. It is better to escape the polarization of all that is human to the objective and the subjective. We have to realize that life is not black and white but mostly gray. This is true, but it does not help. We can examine the gray and separate the white – the grain – from the black – the chaff. We do so all the time – if not in physical act, then at least in a mental act, in thought-experiments of sorts and in acts of valuation. We do so in politics and we do so in science. We evaluate research work – as colleagues, historians of ideas, or curious bystanders. And we evaluate the part or aspect of some research work that we deem scientific, and we ignore, or explain away, the rest of the research work as due to politics, to ideological factors, to academic intrigues, or to something else that we judge of little import.

It is difficult to endorse the polarization of all that is human to two extremes, to what we value, and to what we disregard. Perhaps it is impossible to avoid the polarization. Of course, the polarization to the objective and the subjective is parallel if not equal to the polarization to good and evil. The readiness to polarize to good and evil raises questions about the quality of evil. But we need not deal with all matters at once.

Let us ignore evil for now and consider the rest – including the good. To deem matters as mixtures is still to no avail, since we can sift in mental acts the good from the rest. We ignore indifferent maters of our heroes, and do not care about their daily concerns. We ignore their weaknesses, their politics, and their mixing lofty science with base politicking. We remember their lofty contributions and their important deeds.

Our forgetting the politicking of our heroes does not matter, we hope. Yet, we may wish to remember details in one context that we gladly forget in another. For example, Sir Isaac Newton, perhaps the greatest scientist of all times, was not an agreeable character. He fought his enemies doggedly. Though it is distasteful to deny this, physics textbooks need not mention it. A person interested in the personality of Newton may study his politicking and his passionate, obsessive, and open desire to destroy Leibniz, who was a great thinker too, only because Leibniz was Newton's chief opponent – and one whose politicking was nastier, truth to tell. But we need not care about the petty politicking of these two giants. We need not pretend that there was any objective ground for their mutual hostility, and we can dismiss it as purely personal, as purely subjective. The subjective does not count. Only the objective does, including their disagreements. Yet, this presents two Newtons, one objective and one subjective, whom we view as only accidentally identical. Our history of science downplays the role of the subjective so much that had we ascribed to one of them the researches that the other has made, it would make no difference to our stories. Some people like this, as their ideal is that of utterly impersonal scientific activity. Others find this disturbing.

Mixing the objective and the subjective will not help them, as long as science remains purely objective and politics remains purely subjective. The mixing itself remains then purely accidental. Those who find this disturbing do not wish to see science so divorced from its cultural and political background. Yet, the polarization to the objective and the subjective forces us to take this as a fact, and even to the extent that no matter how well we mix them, they remain distinct.

Mixing the good with the indifferent, then, will not do. Nor will the admission of degrees of goodness. We often overlook small scientific discoveries and center on big ones. The more detailed our studies become, the more attention we pay to smaller and lesser details. But we deem them all good and objective –

ever so small but fully objective nonetheless. Objectivity is no criterion of degrees of significance or importance. Rather, it is a precondition. The ideas of metaphysicians or ideologists or political visionaries might be of the highest importance, and, on occasion, they would be – on the condition that they are objective, they say, and ideas that are not objective are at best useless and often evil, they add, as most of them are. For example, were lighter skin color statistically better correlated to greater brainpower than darker skin color, it would be valuable information. And so, a paper (by A. Jensen, in 1969) that appeared in *Harvard Educational Review* caused much excitement, as it reported seemingly statistical information to this effect: allegedly, this periodical is utterly scientific. *The Bell Curve* by R. J. Herrnstein and C. Murray reported the same seemingly statistical rubbish and had a similar effect.

Who is then to decide what is proper statistics? Everyone interested enough in the question. Sufficiently many people accept the statistics to warrant a public debate about matters ("the Bell Curve Wars"). They are not serious and it is wise to ignore them, since it is easier to fabricate statistics than to check it. Example: a group that took monopoly over the care for the environment and called themselves "the Club of Rome" issued a book (*The Limits of Growth*, 1972) with terrifying statistics regarding the calamitous destruction of the environment in the near future. Debate ensued as to the accuracy of their statistics. The debate stopped when finally experts realized the truth of the matter: the statistics was no exaggeration; it was sheer fabrication. The thesis of that book is sadly true. These people fabricated because it is easier to produce the semblance of science than science proper. The presentation was needlessly sloppy, however. This shows that they were trying to impress people whom they considered fools. Hence, being impressed with it, no matter how, raises the suspicion of being a fool. Alas, many wise commentators fell into the trap and took the report seriously.

It is not easy to convince people with fake statistics. Totalitarian regimes prefer to force people to vote and to vote the right way in order to issue true statistics about their having the vote of an overwhelming portion of the population. It is much easier to force some statisticians to lie than to force a whole population to vote; yet, these regimes prefer to do it the hard way. In Lebanon, some decades ago, the government proscribed population statistics rather than issue false claims. The result was no less than a civil war. Of course, the war was more the result of discrimination and the presence of private armies. Nevertheless, the discriminated suffered patiently until they realized that they were a minority no longer.

It is easy to complicate matters. The recognition of discrimination as an objective fact is a powerful political factor, as our example forcefully illustrates. This truth may seem a support for the thesis that science is objective and politics is subjective. This is erroneous and harmful to the extreme. The victory of the subjective position of one Lebanese war-lord over objective scientific knowledge did help initiate a bloody civil war. This, however, is not to say that all politics is subjective. One may turn the tables and say, no one seriously affirms this thesis. No matter how often they may reaffirm it, they always do so thoughtlessly. Otherwise, they would have to discuss an obvious question. Is politics everywhere equally willful? After all, Lebanon's proscription of demographic studies is unusual. In some countries, politicians operate more scientifically; in other countries they do so more willfully. Speeches of national leaders reflect this: tyrants make silly speeches even though they can easily hire good speech-writers.

In sum, objectivity and subjectivity are facts and fancy. The objective is then factual, dependent on the real world, whereas the subjective is arbitrary, dependent on the willful individual. The subjective may be falsehoods, possibly even lies. The objective is true. Of course, there is a great overlap here, such as the subjective mixing with the objective, for example, the frankly semi-documentary semi-fictional narrative that is unobjectionable, and the objectionable willful distortion. And there may be the subjective attitude to the objective, to facts. And there always is the objective fact about subjectivity. Tyrants are more willful and arbitrary; and scientists are more deferential to facts and thus more objective – with notable exceptions, such as the initial Atomic Energy Commission of the United States of America. This should permit divorcing Newton the thinker from Newton the cantankerous. This also will divorce the objective statistics about skin color, which is independent of anyone's taste, from the subjective statistics that represents the tastes of pretentious mock-statisticians.

This theory is exceedingly popular. It also puts art into the category of subjectivity and science into the category of objectivity. And it makes politics a matter of taste and so of art. ("Politics is the art of the possible.") It makes one wonder why the United States of America is not as war-torn as Lebanon unfortunately is. It permits to explain this fact by the claim that subjectively – as a matter of taste – the love of peace is more popular in America than in Lebanon. Is it an objective fact? It is hard to judge, as it is hard to find why anyone likes this rather than that, or the other way around. For, if one tries to explain the prevalence of the love of peace – or the respect for facts – say, as the result of lessons from horrors of civil wars, then one ceases to view tastes as subjective and takes them as objective.

This may be a bit too fast, advocates of the current view will say: interaction between the objective and the subjective does not nullify the distinction between them. The taste of a tyrant may be arbitrary and willful, yet it does create objective facts, like the spreading of destruction and war and sudden death. This in itself does not make the tyrant's willfulness less subjective. Moreover, the tyrant may be full of good will and this good will, no less than any ill will or ambition, is subjective. Tastes are subjective, advocates of the current view insist, even if they result in facts that are objective by definition.

On the current view, the unscientific nature of politics rests on the fact that politics rests on tastes. When the taste of the American people prescribes a chicken in every pot, a car in every garage, and so on, then American politics will necessarily differ from Nazi politics that aims towards national glory and prescribing cannons instead of butter, as the German tyrants kept repeating to their public. And the fact that American tastes differ from Nazi tastes is objective and open to study. Yet, as long as politics rests on tastes, it is subjective, not objective, and not scientific.

The very idea of social science, and more so the idea of technology – physical, medical, social, or any other – runs contrary to this claim concerning tastes. For, it is the study of how best to satisfy tastes, regardless of what they are. They may concern art, social organization, or the shape of a bridge. The theory of ends and means covers all technology and most of the social sciences. Those who say that satisfying tastes is forever besides the domain of science must deny that politics can be scientific, that political or economic theory can be science, and even that architecture or any technology can possibly have scientific aspects. As researchers cannot avoid having tastes, and as there is no possible guarantee that tastes do not interfere with research, ultimately objectivity vanishes. Of course, wherever researchers of different tastes cannot find the same results, the presence of tastes is evident; but if not, then possibly this only shows the prevalence of some taste: we can never ascertain that we manage to keep all tastes out of research.

Advocates of the current view will not be at a loss for words. They use the scientific jargon and declare it an objective scientific fact: economists, who study the optimal way to satisfy the tastes of consumers, deem tastes exogenous and thus as unscientific. Consumers choose freely. Their choice is arbitrary and so subjective and so unscientific. Its outcome is an objective fact for science to admit. The basis of all economic or political consideration, then, is arbitrary, though the study of political systems or of markets should be scientific. Social science is possible, on the condition that tastes stay exogenous.

This idea is exceedingly popular, as it is in line with the democratic principles that respect every whim of every individual. It is not the only idea that tallies with liberal democracy, and we may seek the best one that does so, one that allows the scientific study of these whims without thereby putting them down. The demand that scientific theory accepts tastes as given without judgment is the idea that tastes must be and stay exogenous.

The exceedingly popular idea that tastes are exogenous is at the base of the polarization of everything human into the given and the arbitrary, the fixed object and the free subject, the objective and the subjective. A factual report may be objective because facts impose it. The same factual report may be subjective because it expresses a desire to make it. But tastes are not as willful as current views suggest. One can hardly believe that the enormous uniformity of tastes is due to sheer accident (or propaganda). Tastes depend to a large degree on prior tastes (as parts of traditions). Prior tastes are factual. Hence, tastes are never fully willful. Nor are facts entirely given. They appear somewhat differently to people with different tastes and opinions. Hence, it is better to speak of the dualism of (statements of) fact and of decisions, though even that is questionable. Further, it is better to realize that false statements of fact may be no less objective than true one, in the sense that – unlike decisions – they (falsely) refer to the given. This is crucial. Many people say, I cannot decide as yet, since I am not in full possession of all the relevant facts. They mean that they prefer the facts decide for them. And they express this preference systematically in the mock-sciences of inductive logic and games theory (in contrast to the contributions of Abraham Wald and of G. B. Dantzig that help calculate consequences of decisions).

The dualism of statements of facts and of decisions is better than the dualism of the objective and the subjective, as by this dualism only truths are objective. This is of the essence – for, it is only the truth that is truly objective, truly independent of what anyone of us thinks or feels about it. Hence, only if we can know the truth can we bridge over differences for sure. Since knowledge is problematic, we often replace it with the consensus. Objectivity thus became a matter of consensus – of all those who count, of all except the knaves and the fools, and so politics, essentially relating to differences of opinions, is subjective. And, as the truth is allegedly given, the cause of all deviations from it is always some ulterior motive, so-called, some bad taste, arbitrariness, and unsavory motives – and then those who disagree with us are knaves or fools and farewell to democracy.

The truth is not a given – not often, that is. The love of truth is a strong motive, a matter of taste, indeed, and this motive brings researchers to invest inordinate effort in their quest, and they jubilate when they find something that they hope is true or even is capable of improvement. (See Stefano Gattei, "The Ethical Nature of Karl Popper's Solution to the Problem of Rationality", *Philosophy of the Social Sciences*, 32, 2002, pp. 240-266.)

Taking truth or objectivity as a norm rather than as a given, in agreement that it is not a precondition but a *desideratum*, a desired end, invites speaking of degrees of readiness to subject views and hopes and even tastes to varying degrees of tests of objectivity. This permits parties not only in politics but also in science, not only political parties but also scientific schools of thought. These exist as long as science exists – or at least since Newton and Leibniz had their fierce and regrettable quarrels. Calling a scientific school of thought subjective – or objective – is useless, as it is no more than taking sides in disputes: calling anything objective is possibly objective, but it is also possibly subjective. (Dreams are real and they are about reality, but they are hardly ever true.) Unanimity is scarce. With hindsight, at times unanimity on one matter or another may evolve; it may also reverse itself. The difference in objectivity between astronomy and politics thus becomes a mere matter of degree. Democracy calls for the elimination of flagrant violations of the truth, and it will call for the reduction of the concealment of the truth even when the concealment should serve the national interest, and it will demand some tests of some assertions.

There is a conflict here between two standard aims of democracy. The national interest requires the concealment of some information regarding security. And it always requires the publication of information on matters of public interest. These two interests always require a balanced decision between them. The interest of the publication of information is that of the improvement of democratic public opinion. But even with the best available information, democracy will not be able to adjudicate in each controversy which party is right. There will remain legitimate and reasonable disagreements about tastes as well as about facts. Differences of opinion must in some cases be valued or else we have given up democracy. The question before philosophy today is, which controversies are rational. This opens the road to many sorts of research projects whose results may enhance democracy. What these projects should share is the recognition of the confusion that the polarization of all ideas into subjective and objective can cause. There are degrees of objectivity, and there is no utter objectivity. And if we view degrees of subjectivity as complements to the degrees of objectivity, we will find it easy to say, there is no pure subjectivity either: politics should aim at raising the degree of its own objectivity.

On one matter, then, the reactionaries are right: stability and progress may clash. True. Léon Walras has devised a system of a stable, evolving economy (an economy in dynamic equilibrium, to use the jargon term), but it is an idealization with no practical application. As any democratic system is forever in the process of further democratization, the choice between progress and regress is inescapable. And regress reaches stability only when we are in cages, says Karl Popper in his magnificent celebration of democracy, his *The Open Society and Its Enemies*.

Chapter 3.5

SCIENCE AND THE DETECTIVE NOVEL

0. THE REALIST AND THE ROMANTIC

Detective, spy, and other suspense novels come in two distinct variants: the realist and the romantic. In the romantic variant, each event, each new development, is exciting, leading in quickening steps towards the expected climax. In the realistic variant, much of the work of the hero is pedestrian. Searching endlessly one address after another in utter frustration, fruitlessly tracking suspicious characters evening after evening, this is the material that the author weaves the narrative from, until the pace quickens towards the climax. Science fiction is a parasite on the detective, spy, and suspense novel genres. This genre, then, reflects more openly the fact that the conflicting views, the pedestrian and the romantic, concern science. The common view by far is that scientific work is as pedestrian as the realist detective's work reflects it, though in science the climactic excitement is greater, as is the reward of success. The truth about science sits between the realist and the romantic. The right criterion for the scientific character of search procedures is not so tight as to dictate a search procedure, but it is tight enough to make it somewhat rational.

1. RESEARCH AS PEDESTRIAN

The pedestrian aspect of learning is in the accumulation of data and the systematic elimination of false possibilities, especially when a list of these is given. The accumulation of information about one's whereabouts in a new town may serve as an example for the collection of boring but useful details. It is valuable, but only as means to an end, and when one leaves the town for good, one readily forgets street names, addresses, and shortcuts. Our knowledge of characters, or even of the character of one given person, is similarly a matter of cumulative information. New information often leads to unexpected changes of one's attitudes towards one's associates and friends, as well as of one's opinions of them. The changes comprise replacements of errors due to having jumped to wrong conclusions. The new information corrects them. Without jumping to conclusions, views do not alter but merely mature. The accumulation of information enlarges and fortifies views. Apart from the accumulation of valuable information, there is the systematic elimination of options. When an electric light fails, we may check the bulb, the switch, and then the fuse. If all this does not help, we call an expert electrician and request further diagnosis. Diagnosticians go systematically over possible sources of any given trouble, beginning with either the most likely one or the one easiest to eliminate. They eliminate systematically false possibilities until they find the true culprit that they are after. The situation may be very complicated, as is the case of an allergy. Here the number of possibilities is so large that the job of systematic elimination is more tedious – unless one is lucky. Where science helps in the search for the cause of allergy is in grouping types of causes together, and other timesaving operations. In medicine, however, diagnosis is the tool for treatment. Crime-detection usually ends with the solution of a riddle.

The difference between mundane activities (learning street names and finding causes of people's allergies) and genuine scientific learning (discovery) lies in the fact that the result of scientific research is often spectacular. Yet most people agree that most of the work of scientific research is no more exciting than the examples mentioned here. Indeed, some would say, while learning street names may never contribute to scientific progress, the acquisition of information about allergies, together with much more medical information of the same kind, will in the end contribute towards spectacular scientific success. It often does, as the history of science shows. Perhaps the most spectacular result of tedious work was the discovery of the genetic code mechanism through the detailed study of structures of giant molecules.

Not so. The enormous databases in hospitals, weather stations, and such, can serve as the paradigm. Scientifically, they are utterly useless. Collecting them is partly for practical purposes, partly for the sake of science. In many research stations, clever and competent workers perform dull experimental tasks. I used to view these activities as always useless, but my study of the philosophy of Robert Boyle has made me change my mind. He viewed dull experimental work as the proper means to introduce novices into science. A dull experiment is far better than no experiment at all, since it may be a prelude to interesting research. This is true; he engineered the scientific revolution this way, no less. Yet, I am far from willing to recommend his policy for this day and age, since many – perhaps most – researchers today could and would do interesting work but for their submission to deadly habits of dull work on the view that scientific experimentation must be mostly dull. In any case, Boyle's concern was educational. This aspect of research aside, dull research still is futile, unless it is very specific. Practical or applied research may involve thousands of specific experiments, most of which are too specific to report, as they interest almost no one. Still, some of them may later prove to be of wider interest. Similar things happen in a detective, spy, or suspense novel. It often turns out that a valuable and conspicuous clue was at hand from the very start of the search, but its value as a clue was overlooked – to a great, almost total, loss. This is why a detective collects as much information as possible, and then surveys it repeatedly, since there is no telling what might be the telling clue. This is, historically, the very reason for the call to anyone interested in the advancement of learning to participate in the collection and publication of all data. Bacon launched that call early in the seventeenth century, Robert Boyle, the leader of the Royal Society, repeated it, and then that Society did so since its foundation early in the second half of that century.

Solid scientific progress must be largely pedestrian, said Bacon, and one must suspect exciting developments as impediments to real progress. He was not the first to present science as pedestrian, but he was the first to insist on the need to be pedestrian and to contrast uncompromisingly pedestrian with exciting research. This led him to the argument about the unsuspected significance of clues.

Bacon knew that preconceived ideas might guide experimental research. The guiding idea may be exciting both in itself and in suggesting shortcuts, he observed, but it is almost certain to be misleading, not only in the sense that it might suggest a wrong turn, but also in that this would obscure valuable clues. Worse, error clings to the searchers who employ it, and blinds them to salient facts. This, too, is standard fare in detective novels, especially where the prejudiced is the detective's boss. If no preconceived idea guides the search, then, to begin with, all facts are possible clues and all possible clues have equal value, because without a guiding idea there is no telling which of them will lead to the truth and serve as a cornerstone of a future theory. It is this reasoning that led Bacon to the proposal that research should consist in the patient gathering of massive databases and their patient, repeated examination until, perhaps with the aid of a systematic elimination of false possibilities, facts fall into a pattern by themselves.

Of the few writers who have discussed Bacon's idea of the importance of eliminating false possibilities, two stand out: Robert Leslie Ellis and John Maynard Keynes. Even they did not discuss the difference between false options and preconceived notions. They both considered Bacon's method rational, and concluded that in his view the number of possible worlds is limited, so that a finite number of eliminations must result in one affirmation. The elimination of all options but one demonstrates the correctness of the remaining one. This is the message of both Bacon and Sir Arthur Conan Doyle. This is not the whole story, of course. Models conforming to this pattern help. I have already mentioned the case of a failure in the electrical system of a common modern dwelling in which a number of faults may occur. In this model, having checked all options but one, and finding them all faultless, leave no room for hesitation: the last available option is right.

Bacon's view about science depends on one idea about the universe, as he deemed research limited to a finite number of possible worlds. (This is the idea of the simplicity of nature or of its limited variety.) It makes science unexciting. By the received view, it is the basis for all empirical science. The standard situation in detective novels reflects this. A crime occurs in a perfectly isolated place and the task of the detective is to eliminate names from a given list of suspects in order to make the task of pinning down the criminal less impossible. After the successful elimination of all impossible options, says Sherlock Holmes, the remaining option, no matter how improbable, is true.

The list of options is not always limited, and even when it is, surprise may lurk. The list is inaccurate, a name is missing, or misspelled, or added. The same holds for the simplicity of nature or the limited variety. It does not work without further information. It is no use knowing that the list of permissible laws is finite. The correct full list of possible laws is essential for the elimination of most of them by empirical means. This is possible in detection and in technology, not in science.

Bacon's image of science is thus impossible. Thomas S. Kuhn has offered a variant of it. Every period in the history of a science, so goes Kuhn's variant, confines itself to limited variety. In every given period, researchers investigate a limited set of options, and the community of researchers endorses this limitation. Research under these limitations is fairly dull and humdrum. To use Kuhn's terminology, science within a given period is "normal", its tasks are solving relatively simple riddles, and most researchers are "normal".

2. THE EXCITEMENT OF RESEARCH

This raises exciting questions about the nature of the transition from one period to the next. Kuhn says precious little about it. He says it is a period of excitement and of sleepless nights. Yet, somehow he manages to give the task of creating the transition, the scientific revolution, to the leadership. This is obviously silly, since initiators of revolutions become leaders only after they succeed. Whether this occurs during their own lifetime or posthumously, depending on the speed of the revolution, they seldom are the acknowledged leaders earlier. This correction of Kuhn's description makes his theory fit the facts pretty well. Some researchers do exciting, trailblazing work. Most of them do drab and boring work. Kuhn deems valuable the abundance of drab, dull scientific output. It rarely is.

Although Kuhn says repeatedly that the leaders make the revolution, he sometimes admits that Young Turks make it. Many youngsters aspire to positions of leadership. How do some of them succeed? Kuhn gives no answer. He says that the leaders toss in their beds and then declare a revolution – not on general principles, but for reasons specific to the situation at hand. This leaves things elusive. It may indeed be hard to explain why science takes first one narrow set of possibilities to examine and then another. It seems easier to recognize only one limited set. But then the question is, which one? Philosophers can hide behind generalities, but historians will be naturally drawn more to Kuhn's view, since we can see, historically, how in any given science, in any given period, most researchers use one set of presuppositions as the basis for their answers to given questions. Philosophers, however, may see in shifts of presuppositions Kuhn-style the vagaries of fashion and prejudice. This is an impasse.

Bacon's writings lead to this impasse. He was insensitive to it. His theory was not as clear-cut as his interpreters have suggested. True, he found only a finite range of possibilities to explore; he took the whole task of science marvelously lightly, as a matter that we can conclude satisfactorily in a fairly short time – a few years or a few generations, he was not sure, but not much more.

Taken so naively, with not much articulation and with less questioning, Bacon's theory appears in a better light. He was decidedly against formulating clear hypotheses. Though they offer guidance, he said, this guidance is likely to lead up a blind alley that the guide will declare the true destination of the voyage and refuse to withdraw. Bacon recommended instead the collection of vast databases. He was aware of the fact that these may stimulate the imagination. He recommended going slowly on this and continuing to collect data while curbing the imagination. Soon, ideas will begin to form in the researcher's mind. If the research work is sufficiently slow, most ideas will emerge while the researcher has in hand the facts that refute them. It is not that Bacon recommended forming hypotheses and then eliminating them. After all, once a researcher forms a hypothesis clearly, then in no time it becomes a dangerous prejudice. Rather, the hypothesis meets its refutation while it takes shape.

The difference is crucial. Some people try to hunt only what they see clearly; they want to make sure they aim at a rabbit. Often, by the time they have assessed the situation the rabbit has disappeared. The better hunter shoots the rabbit just when it pops up. Of course, this is rather dangerous as what pops up may turn out to be another hunter; but this is where my analogy ends. In Bacon's philosophy of science, it is imperative to shoot down the hypothesis before it is fully fledged, otherwise it becomes a prejudice; the only theory he allows to develop fully is the true one.

This is not quite accurate. He did not quite articulate his view this way, but when he did finally develop it in detail and with an example, he explicitly allowed for an informed guess. As Ellis says, this amounts to a declaration of bankruptcy. But perhaps it was just a poor presentation of a highly sophisticated and subtle view.

What raises doubt is Bacon's insistence that the collection of many clues must help. He spoke of the collection of clues as of an act of worship of Nature and of respect for Her, as opposed to making theories and forcing them on facts, which is tantamount to raping Nature, or putting Her in chains. And, Bacon writes, you win Her and make Her show Her charms by wooing and flattery, not by rape. All this sounds very vague, and it seems to make better sense as long as it remains vague. Upon the first scrutiny, it appears as useless, the number of clues being too large in almost every challenging search.

Bacon's view appears in the various types of detective novels, from the genteel tradition of Agatha Christie to the harsh realistic one of Hammett and Chandler. The list of suspects, the clues, the drab work, are all in the realistic tradition, but they are all rituals that detectives perform in order to prove they mean business. Business, somehow, starts dull and ends exciting! All this is present in Bacon's amazingly astute texts. He said, observations lead to theory because they comprise evidence of taking Mother Nature seriously, as wooing Her, as alluring Her to reveal Her charms. (The sexual metaphor is a vital ingredient in the Cabbalistic literature that influenced Bacon greatly.)

Silly as all this may sound, it sounds more reasonable in its application to amateur and dilettante researchers and sleuths. The amateur detective, spy, *etc.*, is less informed than anyone on the scene, has no access to tools of the trade, must outwit both the police and the criminal, and solve the problem first. How does the pathetic amateur become the hero? The simplest answer, given by Chandler, is that this is impossible; the amateur sleuth novel is sheer naïve make-believe, a lighthearted daydream. Even when credible, for instance when only the amateur succeeds in gaining the support of a street gang, whether as in *Emil and the Detectives* or as in *The Lavender Hill Mob*, even then the story is more entertainment than serious stuff. Chandler's detective is credible when police officers beat him up. He is less credible when he nonetheless succeeds. (Indeed, faithful to realism, Chandler often fails him.) The more realistic novels have the whole police force or intelligence-gathering agencies do the work – like Kuhn's "normal" researchers. And currently, television crime-series and crime-detection agencies spring up like mushrooms, and their chief assets are databases and computers with connections to more databases. The search for possibly relevant data is thus less of a legwork today than it used to be. And Bacon aimed at mobilizing an army of amateurs in search of a huge database. And he succeeded.

The little credibility Chandler's hero may have (as Somerset Maugham has noted) rests on the thoroughness that proves seriousness – in tune with Bacon's vision. Chandler's hero goes through the motions through thick and thin, mostly thick. He follows clues he knows to be wrong, be they some accidental red herrings or some clues that those who wish to trap him have planted. He follows clues leading nowhere and suffers the boredom gladly, and he follows clues leading to traps and is beaten up – all in order to show he means business. This is how he draws out the criminal. The serious hero magnetically attracts the truth, as Bacon has promised.

Oh, yes, the occasional lucky amateur may hit the jackpot, by an accident that in retrospect seems easily understandable – a characteristic of accidents, whether in science or in crime detection and espionage. But even in the matter of accidents, one should be careful not to exaggerate. Since the search is for the unknown, all success is in a sense bumping into the truth; each find has an accidental character. Not accidental are the effort, and the ability to appreciate the significance of an accident. This is why, generally, the accident occurs to the old pro; accidents happen to the prepared mind, said Lagrange on Newton and Hansteen on Ørsted and Pasteur on Pasteur.

Here detective novel writers have a better explanation than philosophers of science. The following up of all clues unnerves the criminal, but it can hardly unnerve Mother Nature. Why then does She reveal her charms to the sincere hardworking researcher? How does the mind prepare for discovery?

A given complete set of alternatives poses no challenge. All one has to do then is to follow each and every item in a routine fashion. Michael Crichton exemplifies this in his *The Andromeda Strain,* an exciting science fiction novel that he wrote when he was a medical student. In that story, the researchers are stupid but dedicated and brave. Their stupidity shows in their not following up enough possibilities or not doing the job thoroughly enough. (There must be an excuse here: were the researchers simply negligent, then they would prove unworthy of the solution to the riddle. The excuse that Crichton offers is this:

the failure to notice the crucial clue is due to an epileptic fit that a researcher suffered at the time of the appearance of that clue.) The setting gets its dramatic character less from the danger to humanity that the Andromeda strain poses, than from the full computerization of the research center devoted to its study. It turns out that the planners of the center forgot a few facts, including a few human factors. But, more crucial, the researchers who work there are tired and overworked so that they forget, for instance, to make some necessary autopsies. The drama of the story derives from the fact that humans beat the computer and escape the penalty that their poor planning and general failings could incur. To make the story more exciting, the planners rely on this extra human resourcefulness from the start. This is the gamble on which the real drama rests.

3. THE FASCINATION OF RESEARCH

The deepest fascination of all sorts of thrillers is that they give expression to one major emotional conflict about reason and its emotional significance. Both the view that reason is mechanical, pedestrian, and the view that reason is mysterious seem obvious. The mechanical view of thinking is the view of it as mechanical, an algorithm, a set of operations to perform mechanically, the transparent cranking of a handle of a calculating machine, perhaps too complex to envisage all at once in the mind's eye, but in principle just cogwheels. The psychology of it is obscure. The view of the logician is that reason rests on the deductive validity of inferences, regardless of its origins. Why does it seem comforting to view reason as a transparent machine and why does the view of the mechanical or automatic part of reason seem clearer than the view of the logician? Of those who love logic, some find this view of the logician disturbing, and others find it comforting. The mechanical view is emotionally inferior to the view that there is something mysterious and magical in the operations of reason. Of course, both views are mutually exclusive, and, of course, other views are possible, including the just mentioned view of the logician.

According to this view, calculations are not sufficient and mysterious magic is not necessary for bringing about the valid inference that is the vehicle of reason. Sheer luck may supplement the mechanical and the mysterious. All views of reason share the idea that valid inference is the vehicle of reason – be its source mechanical, mysterious, or lucky. Somehow, emotional attachment goes to each single option – as long as the other two are out of sight. This is ambivalence.

It is possible to resolve this ambivalence by declaring the mechanical a magic ritual preparatory to the mystic – in a manner similar to the way expressions of potent formulas and ritual baths serve the Cabbalist. Both Bacon and the author of the detective novel saw the collection of facts or the following up every possible clue as a ritual. This permits hovering between the rational and the mystical, and so it is great fun. It bridges the gap between the realistic and the romantic by prescribing natural steps and leaving room for magical ones, between the cold calculating and the warmhearted intuitive. In the famous science fiction television series Star Trek (which was held in great esteem, evidently, at

Cape Kennedy and similar centers of worldly wisdom), the skipper of the starship stands for intuition, while his science officer is an unemotional being, a hybrid born of a computer and a race resembling the human race. Yet, on occasion, the skipper thinks, and on occasion his science officer feels. Great fun for science-minded adolescents, but in essence not much different from the adult entertainment of the detective and spy thriller of the allegedly realistic but really romantic stuff.

Raymond Chandler, one of the giants of this game, wrote an illuminating essay, "The Gentle Art of Murder", which concedes that his work is romantic because its hero is romantic, since he is an incorruptible and determined pursuer of justice – a utopian. Oddly, this is untrue, since Chandler makes much use of conflicts of just causes, such as the sleuth's unquestioned loyalty to his client within the limits of the law, and his legal obligation to turn his client in to the law for one crime or another. Oddly, also, the Chandler sleuth is rather immoral in small ways, such as sleeping with his beautiful clients and beating up a small-time operator in order to receive significant information. The bosses of the small-time operator may get him killed in revenge and make the sleuth feel a moral obligation to catch them and hand them to the authorities, or, barring this, take the law into his own hands. Would you call such a sleuth a utopian?

The romantic aspect in Hammett, in Chandler, in John le Carré, and even in the rather square Agatha Christie, appears, if ever it does, with some interruption of the business of the sleuth of following up each clue. This is the arch-formula of Asimov's detective novels too, and most of his science fiction novels, even of some of his short stories, belong to the detective or spy game. He sometimes observes the more traditional Christie formula of following all clues, examining all aspects – the exhaustive routine jobs. Sometimes he observes the Chandler formula of interrupted research. In at least one, *Caves of Steel*, he interrupts the Christie formula of exhaustive search. Time runs out. At the last moment, someone secures a period of grace to allow the sleuth to complete the routine job and reach justice. The time shortage element plus the grace period was a totally new element that Jules Verne introduced as the only suspense element in his *Around the World in Eighty Days*. It is now standard equipment. It belongs to an essential formula of the famous, highly successful television series (and movies), *Mission Impossible*. In each episode there, timing with a very narrow margin is the essence of the plan. Something somewhere does not go exactly according to the formula, forcing the team to improvise and race against the clock. They invariably make it with a second or two to spare.

Suspense has its own rules. In the adventure genre, it is simple. There the hero has to save the world by preventing an explosion, or war, or the capture of a good person. In the adventure genre the suspense element has the sole function of creating tension and releasing it. In the drama of scientific and technological puzzle, the solution to an intellectual problem is a precondition to the resolution of the tension. This in itself need not contribute to the drama; it may have a seriously dampening effect. In a drama of pure suspense, a solution to an intellectual problem may be a part of the build-up of tension, not a part of it (*Torn Curtain*). It may be a part of the tense scenes, if it serves as a comic relief. In this case, the solution is often that of a nutty professor whose sheer eccentricity and/or foreign accent may suffice for getting a laugh. In the suspense drama plain and simple, then, intellect must remain marginal. In the more problem-solving type of suspense, in the detective or spy novel, it is important that the intellectual dimension of the story creates further tension. Invariably, the tension is between the mechanical or drab view of reason and the romantic or mysterious one. And the closer the two get, the greater the tension.

This receives a clear expression in stories of the aberrant conflicts inherent in reason, for example, in fiction that describes mad scientists. They believe more honestly than others in certain ideas about science. (They exhibit this way the paradox of paranoia that Fried and I have discussed in our *Paranoia: A Study in Diagnosis*, 1976.) They also have some ideas that are quite arbitrary, not to say bizarre, that they deem rational or logical. They know that their conduct is dangerous for them, but they plan lines of action in the hope that some most unlikely events will turn up just at the right time to save them. Alternatively, they will see a benign or a sinister conspiracy behind random events, and have no trouble following their initial ideas logically, since it is easy to escape any intellectual difficulty by adding odd ideas to them, ideas that are unlikely but conceivable. When a mad scientist's conduct leads to a clash with the law, it is possible to hope that just at that moment all the law enforcement agencies will be furiously busy elsewhere. Fiction shares such qualities with the right diagnosis of madness, yet it is easier to convey this description in fiction than in the learned literature.

This links up with quite a number of already mentioned points. The madness of the mad scientist is characteristic of both schizophrenia and the bad science that allows the epicycle or the arbitrary moves that Bacon has so violently criticized. This is an interesting phenomenon. The truly mad scientist is no ordinary borderline case. It is commonplace that madness and genius verge on each other. But borderline cases are not exciting, though they may be essential to the dynamics of plots that keep readers wonder which way the cat is going to jump. Rather, the common grounds, the characteristics that a kind of science and a kind of madness happen to share, is responsible for the intensity of the detective story with a mad scientist as a culprit. The culprit must look as engaged in serious research before the truth appears, or else the story is no thriller.

This intensity pertains to Bacon on two counts. First, the description of speculative theoretical musing distracts from intensity in literature, but such theoretical activity can be very intense in real life. When Einstein had his first brainstorms, they were emotionally too much for him. (He had to leave everything and take a vacation in the country.) Literature cannot convey this kind of intensity – even historians of science have not yet learned how to convey it. Detective and science fiction authors alike treat their theoretical framework as given. Second, the same may hold even for research work that is empirical in character. Thus, even when working through the finite list of possible solutions, as permitted within some fixed theoretical frameworks, thinking is not mechanical. Though the conditions permit it in principle, in fact there are too many possibilities to exclude. And then, some think intuitively, like chess players.

(The number of possibilities in chess, too, is in principle finite, but only in principle.) Thrilling fragments from the history of science, such as the story of Semmelweis, or of Paul Ehrlich, have just this kind of thrill, as many a reader of science fiction and biography should know. And so, to the very end, our ambivalence about the nature of reason persists – is it an automatic algorithm (pedestrian) or is it obscure intuition (ingenious)? Our analysis presents this ambivalence as the source and the dynamics of the tension in problem-solving novels, be they detective thrillers, or spy novels, or science fiction.

So much for the logic of works of art about thinking. Even when thinking is obviously intuitive, viewing it as at times mechanical and at times mysterious creates sufficient tension. And considering matters seriously, as when authors analyze stories, or when biographers analyze the scientific method that their heroes have allegedly employed, the conclusion still is that intuition is a mixture of the mechanical and the mysterious, or that it is in-between. In the jargon of Claude Lévi-Strauss, algorithm and mystery are the two poles that think for us, and intuition is at times their mixture, at times a mediating pole.

What do philosophers of science say to all this? According to them, thinking is at times mechanical computing, and at times intuitive, but in order for it to be rational one must at least go through the process of eliminating false possibilities. Popper said first, this will do. Others say, positive evidence is also required, in order to support the option deemed true. Later Popper mentioned crime detection as the search for positive evidence. But not in the sense of the standard philosophy of science. Rather it is a matter for the philosophy of technology. There discussion largely turns on standards for positive evidence. These are far from perfect. This fact is the standard source of tension in dramas that belong to the cross between genres – the courtroom drama and the thriller. But this is a topic for another discussion.

4. EPILOGUE: THE ORIGINS OF THE DETECTIVE NOVEL

Who was the first author of the modern detective novel? This question is the subject of a long, perennial discussion. Most writers agree that it was Edgar Allan Poe. But, obviously, both crime dramas and detectives are older even than that; it is the mystery in "The Mystery of Rue Morgue" that makes it a first. Let me explain.

The story of a wise person who knows best is common to all ancient cultures, Hebrew and Greek included. The story of a clever, thoughtful, bright-eyed eccentric who solves a puzzle, is less common but not unknown either, and, again, it occurs in all ancient cultures. (Oedipus, Samson, and King Salomon are perhaps the most famous of these.) Yet, these stories were either of mysterious wisdom, or of luck and wit – never of a systematic methodical search. With the rise of the ideology of modern science that Bacon has first offered, methodology makes everyone a researcher: just as a compass and ruler enable everyone to draw a straight line and a perfect circle, he said, so his rules enable everyone to be a productive researcher. He was the first to declare that the right method of research is systematic, thorough, and all-encompassing. Voltaire's story "Zadig" uses this. Zadig is uncommon in his ability to see through common clues to significant facts. Sherlock Holmes inherited this quality from him. Baden-Powell recommends his method to prospective boy scouts. The thorough education of a spy as Kipling describes it in his *Kim*, especially the exercises in memorizing pointless details, likewise impressed Baden-Powell and he recommended them to boy-scouts. All this presumably illustrates scientific method. T. H. Huxley has an essay on Zadig where he says so, and Conan Doyle makes Sherlock Holmes say so repeatedly.

Is the detective's knowledge commonplace, then, or does it have a special, privileged origin? Conan Doyle did not know, and made this unresolved conflict – the central theme running through his detective stories. The Sherlock Holmes movies simplify matters; elementary, my dear Watson! says Holmes to his companion there, and when Watson agrees Holmes regrets having made it so. Some elements of commonsense are commonsense only after their discovery. This idea is already in Poe, though only implicitly so. The clues are there. In principle, every reader can solve the mystery. But they do not. Our hero does. Are the heroes of thrillers, then, exceptional or not? We can see how exceptional they are in that even the fertile mind of Alexandre Dumas emulates Voltaire here (*After Twenty Years*). So did Umberto Eco (*The Name of the Rose)* quite recently. Are the heroes of thrillers, then, exceptional or not? Once a clear answer to this question is at hand, and the thrill is gone.

Chapter 3.6

THE TWO BOOKS

The Divine Author wrote two books: the Bible and the Book of Nature. This last title, "the Book of Nature", is obsolete. Galileo Galilei used it in a very significant way. Usually historians use it in connection with Galileo. Obviously, they find it exciting. No one explains this excitement. The title itself is puzzling: the Book of Nature is not a book proper. Some read the expression as a synonym for "Nature". They are in error. Galileo asserted that the language of the Book of Nature is mathematical. The Book of Nature then is not a set of natural phenomena. It is a set of natural laws. It is, strictly, a book of laws. Alternatively, "The Book of Nature" is a loose metaphor. It stands then for reason and its products. I propose that the right reading includes both the strict and the loose sense of the title. "The Two Books" is a metaphor on the uneasy relations between faith and reason.

Attitudes to the relations between faith and reason have changed radically. The most important aspect of the situation is this. Before Spinoza, all authors who addressed the issue denied that there is, or even that there can be, any such conflict. Before Spinoza, the target of criticism of the Bible was, as a matter of course, the readings of the text, not the text itself. Spinoza was the first to criticize the text. Yet, even centuries later, most of the authors who addressed the issue denied that a conflict between faith and reason is at all possible. The conflict entered the public domain only in the nineteenth century, with the rise of explicit atheism and with the assault on faith that Darwin's disciples launched.

The standard explanation for the impossibility of a conflict between faith and reason is simple. To view any article of faith as contrary to reason is sacrilege. There is some evidence to the contrary, but it is invalid. It is the famous assertion of Church Father Tertullian, "*credo quia absurdum*": it is a matter of faith since it is not a matter of rational discourse.. Some read this to mean, the faith in the trinity implies that three equals one, and this is evidently absurd, and I believe it anyway. This reading is faulty. The belief in question is in the resurrection in the flesh. This resurrection involved a philosophical difficulty. Ancient philosophy identifies body with matter and soul with form, and it deems the identity of the individual to rest in the soul alone. Resurrected individuals may have bodies, then, but not the same bodies as they had upon their demise. (Alternatively, the difficulty rests in the atomic doctrine that allows one atom to belong to a succession of different bodies.) Refusal to relinquish a belief based on such a difficulty is reasonable: many such difficulties were successfully resolved. This holds in matters theological and scientific alike.

In Antiquity and in the Middle Ages, the literature on faith and reason concerned harmonizing faith and reason almost exclusively. Its standard technique was the reading of some articles of faith anew in line with reason – never the other way around. Such studies are still recurrent. Nowadays, however, serious scholars rightly ignore such studies as well as their detailed refutations, as they are unenlightening. This was not always so. Most of the important medieval studies were in defense of faith against the claims that it is opposed to reason. (*The Guide for the Perplexed* of Maimonides goes the other way.) Around the year 1600, St. Roberto, Cardinal Bellarmino made news as he criticized the Copernican hypothesis on religious grounds (1616). This demands explanation.

Partly, the explanation rests on the novelty of Copernican astronomy. Until its advent, both faith and reason were familiar through received standard texts. As Copernican astronomy was an innovation, it did not carry the obvious authority of reason. This holds for any innovation. The Talmudic endorsement of the Julian calendar was as innovative as its replacement with the Copernican one. (Not so with the Gregorian calendar, as it was a mere adjustment.) Every innovation invites critical examination. And religious critics may use Scriptures as a tool for forging criticism of new ideas. The objection to this rests on the view that The Book of Nature is the more readable. This claim is due to Bacon. (1620) Galileo said, it is the more readable for the familiar with mathematical symbols. (1627)

Galileo and Kepler, the two leading Copernicans of the time, referred to the Two Books in their efforts to repel the criticism from Scriptures. But they did more than that. They freed science from the duty to harmonize with faith. The criticism from the Bible rested on the view of it as revelation, as Holy Writ, as the Book of Books. If anyone wishes to answer this kind of criticism from a religious viewpoint, then they have one ready. Galileo has outlined it early in the seventeenth century. Pope John Paul II has officially endorsed it late in the twentieth century. He ascribed it to Galileo and to Bellarmino, who was responsible for having burned as a heretic the Copernican Giordano Bruno in the year 1600, and who threatened Galileo with a similar fate. (1) Galileo's answer to this criticism of Copernican astronomy from the Bible is theological. We will soon come to it. Kepler and Galileo also had a less important answer. God wrote two books, the Bible as well as the Book of Nature. Since both are divine, both are true. Since both are true, they cannot be in conflict. Hence, all attempts to criticize one by the aid of the other are absurd. To use the one to check the reading of the other is rational, though.

This, the lesser answer, is evasive. Its evasive character should be obvious. The criticism was not of the contents of either of the Two Books, nor of the claim that there is perfect coherence between them. No one was contesting the claim that both Kepler and Galileo were making. The critics were challenging the truth of Copernican astronomy, of the new reading of the Book of Nature. The question was not, does Copernican astronomy contradict the Bible? Copernican astronomy was a new reading of the Book of Nature. Was it true? It is just because the Book of Nature cannot possibly conflict with the Bible, that some critics rejected Copernican astronomy on its strength. What Copernican astronomy says cannot be true because it conflicts with the Bible, they said.

Much of this dispute is lost on modern readers. The question was, as a new reading of the Book of Nature, is Copernican astronomy true? It sounds synonymous with the question, is Copernican astronomy true? Bellarmino understood these two questions as very different. As a hypothesis, he said, he endorsed Copernican astronomy with no hesitation. This made it a useful tool for calculations, but not necessarily a truth of nature. (Ptolemy's system is the paradigm of such a theory: he was a follower of Aristotle, and was therefore convinced that his own system is false. But, he insisted, it is useful.) Early in the seventeenth century, Bellarmino sent to China Jesuit monks to spread Christianity, and, with it, European culture. And so, he told them to take with them to China both their old religion and the new Copernican astronomy. His dispute with Galileo (during which he threatened the life of the great thinker although he admired him), concerned the status of Copernican astronomy as a true reading of the Book of Nature. He instructed Galileo not to teach Copernican astronomy as the true reading of the Book of Nature. In his great *Dialogue on the Two World Systems*, written long after his august opponent was dead, he said clearly and repeatedly that he observed the prohibition against declaring Copernican astronomy true. The Church was going to repeal that prohibition, he added, and then he would be the first to assert that Copernican astronomy is true – a true reading of the Book of Nature. Whatever is the difference between reading an idea as a true hypothesis and reading it as a part of the Book of Nature, both contestants saw it as vital.

There is no need to comprehend the distinction in order to understand the dispute. The controversial view was that Copernican astronomy is a true reading of the Book of Nature. (The pragmatic assertion that the Copernican hypothesis was useful was not contested.) Since it is contrary to biblical teachings, if it is true, then the Bible is erroneous, and that is all that there is to it. Of course, no one was ready to contemplate this option. And so, the critics seemed to be right when they said, as a reading of the Book of Nature, Copernican astronomy is false. The right response to the critics, then, is this. The conflict between Copernican astronomy and the Bible is only apparent. It rests on a misreading of the Bible. (It is taking a metaphor literally, Maimonides notwithstanding.)

How can we check this answer? We can try to show that Copernican astronomy is true. This will refute the reading of the Bible that conflicts with it. Even if it is the current approved reading, it is false and so it must go. Bellarmino said to Galileo, show me that you are right and I will alter the official reading of the Bible. This sounds traditional but it is an innovation: Bellarmino relied here on the Catholic counter-reformation, on the decisions of the Council of Trent; he tried to keep the scientific hegemony of the Church. (1)

The question of the truth of Copernican astronomy is thus independent of the Bible – each of the Books is open to different readings. (Theologians also insist that some readings are complementary and others are conflicting. The present discussion concerns only conflicting readings.) If the Copernican hypothesis is false, then we must abandon it, Bible or no Bible. Otherwise, we must (by the force of logic) reject either the ascription of the Bible to divine inspiration or our current reading of it. Hence, researchers may ignore biblical criticism of science and leave the matter to theologians and Bible scholars. Should Bible scholars, in their turn, revise their reading of the Bible in the light of science, or should they likewise ignore science? In case of a clash, who should yield, science or theology?

This question dodges the concern of Kepler and of Galileo. They defended the autonomy of science. They wanted to have a criterion of truth that is internal to science and independent of the Bible. Galileo offered a first move. The language of the Book of Nature is mathematical. This is not a sufficient criterion: it is but a starting point. Why is Galileo's oft-quoted statement so very relevant? I do not know. Those who quote it never explain why they do so. Perhaps they think it is an exhortation to study mathematics and the assurance that such a study is very important for science. But exhortations, even the best of them, are irrelevant in the present context. Why is Galileo's thesis so very significant, then?

Galileo's claim has everything to do with the problem at hand. To repeat, he claimed that the Nature is an open book to those who know the language in which it is written. This is false, yet critics did not question it, since they found reading the Bible even more troublesome. That biblical texts are immensely problematic is no news. The idea is definitely protestant that one can read the Bible for oneself and solve for oneself the problems that the text poses. Galileo's Catholic critics could not possibly endorse it. To repeat, Bellarmino placed the controversy in quite a different context – in the context of the Catholic counter-reformation. He said, we must stick to the official (Catholic) reading of the Bible, the one that agrees with what the Church Fathers have understood, unless otherwise explicitly and officially stated (by the Church). It is inconceivable that the Church should offer a new idea that may conflict with the Bible as officially understood and then rescind it. The Church needs assurances against waywardness. He forced the advocates of the new idea to recant unless and until the Church formally endorses their position; the Church will do this as soon as the researchers will supply a proof. Bellarmino demanded that Galileo should yield, as long as there was no proof.

As to Bellarmino's own reading of the Bible, he based its authority on the authority of tradition. He admitted that the authority of tradition is limited, though. He agreed that he could change the reading though it rests on tradition, but that he ought to do so upon decisive arguments. Nevertheless, he stressed that the traditional reading should stand until the Church officially alters it. Hence, he took the authority of the Church as basic. Galileo claimed that claiming authority for the Church in matters scientific was an innovation, and a self-serving one. We have to endorse half of this claim: it was an innovation. If we must ascribe motivation, I will ascribe to Bellarmino the concern for social stability, the fear that allowing thinkers the freedom to prefer the authority of reason to the authority of tradition creates a loophole that might easily widen, and then there is no end in sight. Later experiences bore out Bellarmino's fears: the Church has now much less authority than in his day, and quite generally so.

Innovation was then inevitable. Renaissance and Medieval thinkers shared unbounded admiration for Antiquity, when knowledge of God and of Nature prevailed. The idea that dominated the mood then was not that knowledge was lost, but that the key to it was. Knowledge is a closed book because the keys to the comprehension of ancient books are lost. Therefore, what we need learn is not the truth, but the proper code to the true books. This permitted the expression of criticism of both the Bible and Greek philosophy. Medieval thinkers saw this only as criticism of traditional readings of these books. And they took it for granted that these readings – not the books – were faulty. Once we realize that all reading is problematic, we can try to pitch our reading of one book against our reading of the other, both to test given readings of certain passages and to prompt some new readings. This is give-and-take between two sets of readings. Clearly, medieval religion committed thinkers to consider the religious truth superior, more interesting, *etc.* Yet, they often found philosophy superior, especially those among them who viewed religion as a system of philosophy expressed in common language. Perhaps Maimonides, the great Jewish thinker of the medieval Muslim world, was so popular in Christendom, even though he was scarcely understood, because he said, religion makes it possible to lead the best life possible, the contemplative life, the life devoted to philosophy. Does this give priority to faith or to reason? Maimonides said, this question rests on the false assumption that the two differ, but the true philosophy is in harmony with the true faith. He applied this idea to purge religion of all the unphilosophical elements in it, mainly the erroneous literal readings of metaphors. This is the background to the metaphor of the Two Books. Yet, the metaphor comes to do away with the technique of Maimonides and establish the autonomy of science. It came to free researchers of the need to harmonize the Two Books.

This was no by-product. Copernicus lowered the authority of Antiquity: he said, as the Ancients were in disagreement about astronomy, we must decide for ourselves. Galileo said, Aristotle and Archimedes were in disagreement about gravity, and so we must decide for ourselves. He explicitly preached individual autonomy in all matters but the faith. To maintain the harmony between faith and reason under these new conditions, it is essential that faith should accept a division of labor and not adjudicate on matters that depend on autonomous reason. The fact that faith and reason may adjudicate on the same item invites a rule of priority. There was an established rule: reason prevails, since faith should not be unreasonable. The quarrel between Bellarmino and Galileo was about hegemony, not about priority: to repeat, Bellarmino wanted Galileo to ask for permission to adjudicate. Galileo reluctantly obeyed, but he rebelled.

We may sharpen Galileo 's enlightened attitude by contrasting it with a more pious one. Enlightened theologians interested only in religion – in the Bible and in the aspects of life that fall under its lights – will yield to science, and all too quickly, merely in an effort to spend as little time as possible on scientific considerations, including considerations of the proper reading of the Bible. This greatly differs from contrasting the Two Books in efforts to read them correctly.

In the Middle Ages, thinkers tended to see the Two Books as covering more or less the same ground. In the Renaissance, thinkers tended to see a division of labor between faith and reason. This division of labor was the justification of the Renaissance argument against religious criticism of reason. Bacon put it in a maxim that is the best expression of this division of labor. We ought not seek philosophy in the Bible, he said, nor religion in philosophy. The division of labor is problematic. It became, quickly, the division between factual and moral knowledge. Robert Boyle, a follower of Bacon and a co-founder of the Royal Society of London (1660), gave it the expression that soon became the official view of the modern learned world. This did not settle matters, since, on factual and moral matters alike, the authority of the Bible has eroded. Attempt soon followed to erect scientific ethics, namely, to found ethics on reason.

The break from the medieval view lied in the self-reliance of the Renaissance and the Enlightenment Movement. Attempts to break the code became superfluous, first through the claim that we can read the Book of Nature without looking at ancient writings, and then through efforts to reduce the authority of the letter of the Bible. Attempts then appeared to use new scientific ideas as means for interpreting the Bible. Kepler attempted to explain the events of the Good Friday as caused by a total eclipse. Incidentally, his attempt failed. He resisted religious meddling in science yet practiced the opposite. This is very significant. Next came the claim that the rainbow is a natural phenomenon, not an *aide memoir* for God not to cause another Flood. Soon, many views changed under the impact of new research. The metaphor of the Two Books lost its value and vanished.

By default, the Roman Catholic Church found itself adopting Galileo's view of the way to harmonize faith and reason: both are autonomous. And so, followers of the one need not attend to the other, though they may, and then they can modify understandings in one field to fit those in the other. Reason is unlikely to yield to faith, but faith may yield to reason. This is as it used to be, yet with one huge modification. Reason is now free to modify its ideas in accord with its own light. If faith follows reason, there is the risk that it will take some new idea more seriously than reason does. This is what Bellarmino said is unreasonable to expect of the Church. It is too frivolous to alter the reading of the Bible in the light of ephemeral ideas. Again, we must agree with the Saint, however reluctantly. His refusal to follow the whims of science is honorable. His error was in the idea that the Church has the right to tell people whether a scientific idea is a true explanation or merely a hypothesis whose role is to facilitate computation. The Church, Bellarmino rightly judged, cannot instruct believers to alter their readings according to mere scientific fashion. He refused to consider the possibility that the Church should not adjudicate such matters: most of the universities were Church institution, and he wanted the Church to preserve its intellectual hegemony. Now that this hegemony is lost, the Church has less interest in the question, is some scientific idea solid or ephemeral. Meanwhile, a new option appeared, the idea – of Einstein and Popper – that scientific explanations may be false, yet approximately true. This new option has changed the scene.

We may want to know this anyway. We may want to judge for ourselves what part of science is reliable and what part of it is ephemeral. But we do not know.

Galileo suggested that those who are familiar with mathematics could decide for themselves. This turned out to be too optimistic. And so, not pretending to know for sure what the Book of Nature says, we let it go.

This was manifest already in the works of Bishop George Berkeley, the famous eighteenth-century critic of science. In his book *Visual Language* he says, God speaks to us – instructs us as to what we should do – through both revelation and perception. He did not attempt to correlate the two, much less to harmonize them. He deemed experience and revelation complementary. He did not deem it necessary to double-check one against the other. He took the idea that scientific theories are merely hypotheses for computational use not as a stopgap, as Bellarmino did, but as a general feature of science. Conflict between science and faith became impossible. The need to double check science and faith against each other disappeared. The metaphor of the Two Books had become useless. The superiority of science and the need of theologians to accommodate it are now commonplace even though Berkeley's view of science is not the most popular. The Two Books metaphor was inherently a defense of science from religious attacks. When the attacks subsided the metaphor faded away. The counter-attack of science began early, with Spinoza's biblical criticism if not with Descartes' earlier explanation of the rainbow. Scholars began to reject the idea that the Book of God is inspired. The counter-attack became fashionable. Darwin's followers if not their immediate predecessors, declared the Book of Nature, too, not God's responsibility. Scientists attacked religion in earnest only when they became atheists. With no divine author, there were no divine books. The transition was complete.

The nineteenth century atheists were convinced that time is on their side: as science grows and religion stagnates, interest in religion should wane. This did not happen. Popular attitudes to religion vary greatly in the Uniuted States, in western Europe and in Russia. Cynical use of religion for political ends and of political means to impose religious observance – even in countries where science is a very popular commodity, such as Israel – breed fanaticism. The academic study of religion and of theology then split into the dogmatic and the critical. Critical students notice that whereas the Book of Nature is one, many Bibles are contending. Philosophers still debate natural theology. There, the impressive argument is the one that Robert Boyle christened the physico-theological argument for the existence of God, and that others have called the argument from design. It is the only theological argument that magisterial Kant has allowed (not as proof, though).

NOTE

1. See my "On Explaining the Trial of Galileo", reprinted in my *Science and Society*, 1981, esp. p. 329. See also Michael Segre, "La 'rehabilitaciòn' de Galileo: un punto de vista popperiano", *Galileo*, 13, 1996, 3-7; "Light on the Galileo Case?" *Isis,* 88, 1997, 484-504; "Galileo: a 'rehabilitation' that has never taken place." *Endeavour*, 23, 1999, 20-23; "*Hielt Johannes Paul II. sein Versprechen*?", in M. Segre and E. Knobloch eds., *Der ungebändigte Galilei*, Stuttgart, 2001, 107-111.

Chapter 3.7

SCIENCE AND TECHNOLOGY

1. TECHNOLOGY AS AN AID AND AS A HINDRANCE TO SCIENCE

Consider the ways in which technological developments have stimulated or obstructed scientific change. Some of the ways in which technological progress stimulates scientific progress are all too obvious. The simplest, perhaps, is the improvement of the life of scientists through technological progress. It frees time for education and research. Innumerable technological advances raise the standard of living in myriads of ways, and help science grow in all sorts of manners. These are the indirect ways. Then there are technological advances that open the way to new scientific experiments: the development of ever better tools of observation, the ability to go to new frontiers to look at new places, the development of new computers and computing techniques. These are some of the direct ways. Technological advances also stimulate scientific progress in the most indirect way, in the most round and about way – by arousing the quest for knowledge, by inviting scientific studies of those phenomena that constitute technological advances, such as the success of antibiotics.

Yet there is at least one more way, a still more incomprehensible and round-about way, in which scientific progress may profit from technological progress – the institution of greed: all those who crave more fruits of further technological progress may be induced to institute incentives for scientific progress on the firm or shaky ground that is technologically useful.

This is an impressive list of ways in which technological progress can and often does stimulate scientific progress: with no claim for completing it, we have already listed four items, each of which is a vast cluster. Let us, then, survey the ways in which technological progress can impede scientific progress. It is much more of a challenge. We can assess empirically the extent of the challenge. Researchers can report to their peers that they have a list of ways in which the progress of technology can aid the progress of science but no way in which the progress of technology can hinder the progress of science. They may then ask their peers whether they can help. Experience shows that most people can meet the first half of the challenge with ease but the other only with difficulty and often not at all. Yet, the challenge is easy to meet.

Technology can hinder science in many ways. In the late nineteenth century, Samuel Butler made a suggestion (*Erewhon*, 1872) that many a science-fiction author took up. Humanity might fear technological progress and outlaw it. He did not say this need hinder scientific progress too, but lesser minds, such as religious and political leaders with all sorts of unsavory ideas spoke freely in this vein before and after him; only later writers, such as Aldous Huxley, envisaged the rise of specifically anti-scientific cults. In his opinion, these are inevitable outcomes of disasters brought about by technological means which science made available, such as nuclear war and the industrial destruction of the natural environment. Of course, technologically assisted disasters may hinder scientific progress very simply by destroying universities, laboratories, libraries, printing presses; or by killing scientists and their students and their friends; they may make people busy seeking bread with no leisure for scientific pursuits. Population explosion plus pollution – both highly technologically-assisted – will do just that. I have thus far covered three out of the four clusters. The fourth, the intellectual factor, calls for more deliberation. Of course, we can easily say, one mind finds in technology stimulation to advance science, and another mind finds in it stimulation to desert science. But there is more to this than a mere response to technology. There must – or at least may – be objective, systematic sense in lumping science and technology together. Let me elaborate on this, which I shall label the vision of Horace: *dulce et utile*: the vision of having something both sweet and useful in mental exercises, something that should both delight and instruct. Sir Philip Sidney, Ben Jonson, and Bacon shared it and made it the central idea of the age of Reason. It is still very popular and worthy of study.

2. THE VISION OF HORACE

The idea that sweetness and utility go together is a mere hint, a suggestion, a mere sense of abundance. It translates into an abundance of instances, one of which is, truth is beauty. Another is that knowledge is power. This part of the vision of Horace concerns us here. But we should not exaggerate this part either: it, too, is a mere suggestion, a mere sense of abundance. An idea may become more than a suggestion due to its rational elaboration, or due to its integration into an intellectual framework, or due to rational debates concerning it. Bacon was a great influential thinker – the most influential in the emerging modern world second to none but Newton (says Paul Hazard) – but a systematic thinker he was not, and none of his rational arguments goes beyond one or two hints: there is no single sustained line of reasoning in all of his works. His argument in favor of his slogan, knowledge is power, is partly commonsense, partly transcendental – with a moral aspect to the transcendental. The commonsense argument is powerful and in no need of elaboration: greater knowledge brings better controls. This is unassailable: knowledge can increase control but not decrease it. On second thought – but only on second thought – we can seek counter-examples to this. One metaphor captures them all: do not look down, says the mountaineering guide to the novice. Of course, this is but a metaphor, since the guide wishes to prevent only the physiological ill effect of looking down, not any knowledge; yet, the metaphor stands for known cases of people going on to fight out of ignorance of the likelihood of defeat and thus improbably winning. Bacon must have known this argument, since he answered it, *a propos* of the argument that science is dangerous to society since it may breed atheism. Little knowledge, he conceded, may be dangerous, but not much knowledge.

The proviso to Bacon's claim – that if more knowledge does not lead to more control, then still more knowledge will – makes the claim untestable. It takes us out of commonsense and into metaphysics. There we meet Bacon's transcendental argument, which is even shorter: God has promised humanity dominion over Nature, and we can accomplish this only with the aid of science, since knowledge is power, provided that we take the opportunity and invest hard work in the performance of scientific research.

The vision of Horace comes strongest in Bacon as a vision. The *locus classicus* of the vision is the posthumous utopian *New Atlantis* that, like practically all of Bacon's writings, is but a fragment. In his utopia the vision comes strong from the start: when the marooned narrator and his friends are rescued by the citizens of Atlantis they offer their rescuers guns and similar impressive items, which are refused with a condescending smile. In the New Atlantis there is a secular college just on the outskirts of the capital, yet the college head enters town in a procession like the one that, commentators observe, participated in when James I was crowned. The secular college conducts empirical and technological research. It has great technological achievements. It makes most of them public, but some of them it keeps as state secrets because they are dangerous, and some it even deems too dangerous for that and therefore it keeps them secret even from the state. To date researchers cannot accomplish this feat.

The vision is stupendous. The argument that technology needs science never received a serious examination. If technology needs science, does science need technology? It is not obvious that to be scientific a theory must be useful. Some branches of science do not exhibit any prospective usefulness, yet many people think that this is quite impossible – or at least superfluous. Researchers who seek practical applications for science, says Bacon, are the benefactors of humanity (*New Atlantis*). The expression became current in the Age of Reason. Does science, however, benefit from the advancement of technology? Again, Bacon must have thought of this question, since he answered it, though, in his usual aplomb, he did so implicitly, and in one brief sentence. The mark of science, he said (*Novum Organum*), is its usefulness. That is to say, science includes tests of new theories, and their practical applications comprise such tests. Since technology is the touchstone by which we distinguish genuine science from fake, technology is intellectually significant. This, however, is my reading of Bacon; he offered no elaboration and no comment.

We need not put under scrutiny the idea that applicability is inherent in genuine science: it is not fair. It is but an expression of the vision of Horace writ large. We can just as well put under scrutiny the immortal stupid lines, Beauty is truth, truth beauty – that is all ye know on earth, and all ye need to know. (Obviously, truth can be ugly, and beauty can deceive.) And so, to be a bit more palatable, the vision of Horace needs a modification. The modification should admit that things might go the wrong way, that gain in one aspect may be loss in another. The modification should deny only that this need be so. According to the modified version of the vision of Horace, progress is always and in principle an open option. Barring unforeseeable cataclysms, it will sooner or later be the inheritance of Man, the rightful property of the sons of Noah – because someone will sooner or later take advantage of existing available opportunities and use them for some good ends. Of course, this is highly questionable.

3. TWO VERSIONS OF PROGRESS

For the vision of Horace to be a rational idea rather than a mere poetic vision, its starting point should be the observation of a plain fact: most people live in misery, and wealth of any kind is welcome: there is no need to suppose that wealth in one dimension is accompanied by wealth in another. Perhaps this is even impossible. Philosophers, from Thales and Socrates to the Talmudic scholars and the oriental sages, declared material wealth an unwelcome burden to the spiritually rich. The Talmud even sees here a sense of justice. The Lord, it says, endowed this world to Esau and the world to come to Jacob. The opposite idea, the idea of global injustice, has been recently termed by Robert Merton "the Matthew effect". ("The Matthew Effect in Science: The Reward and Communication Systems of Science", reprinted in his *The Sociology of Science*.) He did not ascribe the view of cosmic injustice to St. Matthew, of course, except in jest. More frustrating was the frustrating problem that Alfarabi faced as he contrasted the universalism of Greece with its immense fortune in line with the vision of Horace of wisdom, beauty and goodness going together. Is this vision not a theory of cosmic injustice?

This question has found its explicit expression in two literatures. The one is the debates concerning modern international politics, particularly in relation to the fact that the rich nations pollute the whole globe to the detriment of the poor ones who are defenseless against this evil. The other is the discussion about cargo cults. Cargo cults are rites performed with the expressed purpose of bringing modern technology to poor south seas islanders, and, by extension, any rites performed to that effect anywhere. But the idea of cosmic injustice must have occurred to Europeans the moment Europe emerged as the richest and wisest of all of earth's regions. Now, the sixteenth and even the seventeenth century still saw the ancient Greeks and the Chinese as both wiser and richer. In the seventeenth century, few writers presented China as a kind of Utopia. The Europeans of the eighteenth century could not uphold such a fiction, and so the question must have arisen there: is Europe graced and elevated to the special status of the wisest, most beautiful and best? In eighteenth-century Europe, one of the most influential social and political thinkers of all times – David Hume – answered this question in his classical essay, "Of Refinement in the Arts". The essay that follows it, "Of the Rise and Progress of the Arts and Sciences", forcefully points out that by arts he means both the mechanical and the liberal arts – in addition to the fine arts – but with the accent on industry and refinement in the mechanical arts.

Very briefly, Hume claims in these two essays that despotic regimes impose poverty and superstition on the majority, and bestow luxury and high-handedness on the ruling minority, thus disqualifying everybody, ruled or ruler, from the development of the refinement and discretion and thoughtfulness natural to free people living in an orderly, well-governed republic of one sort or another. And Hume deemed Britain of his days sufficiently republican in this sense. Hence, proper political conditions give rise to all sorts of flourishing of the culture, including science, technology, and the arts. It hardly needs saying that this is only one aspect of the matter. Two others are missing – the moral and the epistemic. In his book on the principles of morals Hume argues for the vision of Horace, for the combination of the *dulce et utile*, the pleasant and the useful, and as a moral principle. The last item occurs in his studies in the theory of science, for which he now gains much honor, due to the great influence that the first historian of modern philosophy, Hegel, had on the history of this field. In his epistemology, Hume argues against causality; everything real, he insisted, is independent. He thus insures the independence of the diverse ways in which we may get rich. Given a fertile soil, to use his metaphor, we cannot but get rich eventually – in each dimension: economic, intellectual, artistic, and moral.

This, then, is the canon of classical liberalism. Both the American Revolution and the French Revolution declared allegiance to the liberal creed. Not surprisingly, after the failures of the French Revolution this creed lost its popularity with writers of the Reaction to it – beginning with Edmund Burke. The most important reactionary, the most intelligent and influential among them, was doubtlessly Hegel. He attacked liberalism at its core, declaring men unequal in principle – the Master-Slave Relation the principle is called – but it is too complex and elaborate a principle to present here. Rather, we may pay attention to Hegel's strongest argument against classical liberalism. Usually people are unable to judge what is good for them, he said, unable to run their own lives, much less their government. Hence, the liberal explanation of the vision of Horace is an error. Hegel has offered an alternative theory of progress to explain the apparent coalescence of high science, art, technology: victor takes all; the refined talents come to the rich and victorious capital. What remains to explain is the fact that each period has one leading capital. This is only an alleged fact, of course: whatever was the leading capital in ancient or medieval Europe, or even 18th Century Europe, who can doubt that 19th Century Europe had different capitals each excelling in its own way in its own specialties? Yet, the idea of universal progress caught and, due to both nationalism and Marxism, the theory of universal injustice caught, and with it, the idea that a certain remedy will eliminate all injustice. Soon the whole of humanity will consist of equally progressive societies, each of which will see progress in all the arts and all the sciences alike. Those who will not catch up will presumably vanish. This is the iron law of progress: join the forces of tomorrow or perish! Tomorrow the vision of Horace will come true for every society, then, and presumably also for every individual!

When we clean this progressivism of its messy metaphysical chit-chat, does it come up different from liberal progressivism? Yes: it is optimistic in a different way. Liberal progressivism presumes that progress in every dimension is possible under favorable conditions, and that these conditions must eventually obtain. Tom Paine has put it thus: independence being a single simple line, contained within us, it gives the answer without a doubt. And, he added, such is the irresistible nature of truth that all it asks and all it wants is the liberty of appearing. The progressive liberals believed in free enterprise because they considered selfish, enlightened individuals able to solve every social and political problem *ad hoc* – with no need for principle. In the nineteenth century, Malthus, and then other social and political thinkers, discussed problems for whose solution ideas were necessary. Nevertheless, the idea remained popular that all social and political problems are soluble. Let us reconsider the possible impediments that different kinds of progress may raise and see what is called for to insure success.

4. AID AND HINDRANCE AS PRINCIPLES OF ACTION

A brief recapitulation. We have four clusters of modes of influence of technology on science: (a) conditions of life in general, (b) conditions of empirical facilities, (c) specific problems and their modes of attack, as well as (d) inspiring curiosity. Obviously, the influence can be both favorable and unfavorable in all four clusters of possible cases, yet, since we live in an ethos of intense cooperation between science and technology, most people will readily imagine cases of favorable influence that technology may have on science but not of unfavorable ones. It is clear, now, that this suffices to show that the vision of Horace is in contrast with some other visions. Yet, translated into a progressivist philosophy proper, the vision becomes hard to dismiss. To take nineteenth-century progressivism proper – say a variant of Marxian or Spencerian optimism – we may translate it to say this. Technology may impede scientific progress in the sense that a kettle may freeze when placed on the hot stove, but the laws of nature normally force kettles to boil when placed on hot stoves and they likewise force science and technology to give both incentives and helping hands to each other.

To see how much the vision depends on the naïve optimist philosophy that goes with it, we may contrast the progressivism of nineteenth century thinkers – such as Marx or Spencer – with its predecessor, the 18th-century liberal version of progressivism, or, to bring that version up-to-date, its neo-conservative variant, especially American, of the twentieth century. For, this philosophy rests on – puts its trust in – private vices that are public virtues, as Hume explains, using the expression of Machiavelli with acknowledgment. Now, of course, the reference is not to vices proper, but rather to what the superstitious erroneously deem vices, namely, to the caring for one's own interest as best one can. That is to say, whereas the superstitious deem all self-interest vicious, modern liberals trust enlightened self-interest. This is, indeed, why in the scientific ethos the claim that technology may do harm is not very popular even though everyone admits it. One may concede even the claim that technology may harm science itself without allowing this concession to carry much weight. For, the scientific ethos admits that there is a remote possibility that science and technology might clash yet the accent is on remoteness. The scientific ethos insists that in principle science and technology must be on the best of terms.

Technology should not impede science, as they are natural allies. Instances of conflict between them, possible and even actual, it seems, are therefore marginal and easy to rectify. This is an error. There are admittedly no good reasons why technology should impede science, but there are some bad reasons for it, and these often obtain – partly because they are unexpected and so not guarded against. Employers demand of the experts whom they employ to neglect their own studies and serve their employers instead. Technology competes then with science for the efforts of researchers, and science often loses. Perhaps, since science only loses to technology here, the loss is cancelled out; but perhaps the very support industry lends science impedes the procurement of larger and freer public funds for scientific research unimpeded. Whether this is so is no small question, involving as it does science policy questions that belong to political economy – a very controversial and notoriously intricate field. No one seriously claims to know how science policy decisions occur. And studies of such matters sadly suffer neglect due to the prejudice that since science and technology are natural allies the support of industrial research institutes for scientific research is always beneficial. It is not.

Science policy studies comprise a new field. The introduction of arguments from this new field to discussions of scientific and technological progress has disposed of eighteenth-century philosophy. The claim that private vices are public virtues may be true; it is irrelevant to modern science policy studies: a twentieth century neo-conservative would hardly say, directors of public affairs should follow their own self-interest in an enlightened manner. This is a change. It sounds like a small amendment to the classical liberal view, and this is how neo-conservative political philosophers, political economists, ideologues and party politicians, present it these days. The view that they offer is a hybrid. But then neo-conservatism is a hybrid anyhow, merely posing as pure-bred. Since neo-conservatives are all staunch defenders of the cause of scientific and technological progress, since they stress the importance of technological progress for national defense, their philosophy is vague just about the very questions on which it is called to offer fresh answers. It concerns matters that committees, not private citizens, should handle. How can these committees best serve the public? Neo-conservatives have no answer.

To conclude, no one knows how much technology aids and how much it impedes the growth of science. It is a matter both of fact and of principle. It is hard to approach facts without some guiding principle. Classical liberal eighteenth-century is the only guide extant. It says, the enlightened self-interest of the individual is the best guide for social policy. Nobody holds this today. What neo-conservatives hold is a modification of it, yet they do not specify the modification. They agree that national and civil defense are public: they should say what research whose fruits serve defense deserves public support and under what kinds of constraint should it run. Also, and more important, neo-conservatives do not allow the market mechanism to be the sole control mechanism, especially in matters of education and of immigration. There is a need for guiding principles for decisions on such matters.

5. THE BETRAYAL OF THE PHILOSOPHERS

Contemporary rationalist philosophy is largely philosophy of science – of science in general and of the specific sciences, as well as their foundations or metaphysics. A glance at the program of the present international conference of the Philosophy of Science Association will confirm this with ease. The philosophy of science, further, concerns epistemology and methodology. Most philosophers take for granted that research is the acquisition and possession of the best rational belief possible and rational technology is but its adequate application. They are inductivists and empiricists. Others are mostly conventionalists-instrumentalists, or pragmatists, who view science not as rational belief but as technologically adequate mathematical systems stuffed with useful empirical Information. To date most philosophers of science thus write on the subject in general from an eighteenth-century point of view. The paradigm is Carnap. He was a brave socialist who openly declared himself a socialist in the height of the days of anti-socialist hysteria in the country that gave him refuge; he was busy then with his work on induction. His 1950 *magnum opus* (*Logical Foundations of Probability*) contains one rule, formulated five times for the sake of improved precision. In all five versions, however, his rule aims at business people – telling them how to optimize prospects. Clearly, it was not the making of money that occupied his attention. Clearly, he was aware of the problematic aspects of the case; yet he deemed these were hardly challenging: socialists, at least the best of them, shared with the liberal progressivists the view that enlightened self-interest is the foundation of wise politics.

These days, to continue this brief survey, the philosophy of Thomas S. Kuhn is very popular and very prestigious. He was no neo-conservative; he was a plain reactionary. He found impossible and, more importantly, unnecessary, to divide science from technology, since significantly in both cases most research is what he approvingly calls normal: it is pedestrian, routine, and submissive. He analyzed only normal science and technology. Of abnormal science he only said this. It is episodic; each episode opens a new era of a new style of normal science for the rank-and-file to endorse submissively; and styles are not comparable to each other. Thus, a follower of Kuhn has to say, what advances or impedes science and technology must undergo scrutiny in each episode separately, yet science and technology are always one. Kuhn never said why. He laid down the law like a Czar. He thus may be a symptom of the end of an era, the end of the vision of Horace, and the beginning of the age of technocracy. Assume that public bodies will heavily finance all research, all researchers will be public employees, and their work will be mostly normal and thus under public control in accord with public regulations. Then we can trust the short-range interests of public figures in general and of political leaders in particular to guide all research and make it technological – perhaps even on a perverse reading of the optimistic Baconian maxim that usefulness is the mark of genuine science. Of course, technology may serve different ends, and public bodies already largely determine which ends have priority. A simple extrapolation will render Kuhn's vision truly Orwellian.

Kuhn meant well. His demand for the autonomy of the scientific community was well intended. This demand originated with Michael Polanyi, who began his philosophic career by attacking the idea of planned science – once popular amongst radicals and socialists of all sorts – from a distinctly conservative point-of-view. But when he began his advocacy of an autonomous scientific community, before World War II, big science did not exist. Both Edison and Marconi were paradigms of technologists-entrepreneurs; they were the biggest that anyone could imagine at the time, as contemporary fiction easily illustrates. The rise of big science rendered Polanyi's view exceedingly bleak. Kuhn is his chief follower. Unlike Polanyi, Kuhn regrettably ignored the political implications of his philosophy and the risks that it invites.

6. PRACTICAL PROPOSALS

Now this is not a call to return to the vision of Horace; it is a call to reject the vision of Kuhn as but the careless perversion of it, to reject Kuhn's vision regardless of what will replace it. We need the new science of science policy in order to meet the challenge to discuss the aid and impediments from technology to science. We need to investigate which technology under which conditions aids or impedes the growth of which science, and how it does so. The standard hackneyed argument here is that technology is but a tool and so in itself neither good nor evil. This argument will not do. A powder keg is a tool for good or for evil use, but it also is, in itself, dangerous; it is especially dangerous in a blacksmith's workshop, where sparks fly in all directions. We are sitting on a growing powder keg, and the sparks are flying. It is time to wake-up: it is time to find safer ways to harness science and technology to the service of a better control of and of a better integration into our environment – human and natural. For this we should jettison the old idea of rationality as the enlightened self-interest of rational being as well as the old idea of enlightenment as rational belief. Science and technology are social and political institutions that we want to know more about and to control better and to prevent their reckless use.

Big science is only as old as the Bomb. The size of some science-based technologies before the Bomb may have been impressive – not only with some large military and chemical industrial organizations of the mid-nineteenth century, and with Kelvin, Edison, and Marconi. But big science is public. So, the very recent loss that scientists suffered of their private individual freedom as scientists, is understandable. Kuhn was a historian of science, yet he generalized from physics of the second half of the twentieth century in flagrant disregard for the centuries of science as described by Hume. It is a shame that as soon as generals, party hacks, and college administrators led scientists by the nose to servitude, their loss of freedom looked inevitable. This calls for an urgent and extensive study. But at least that much we know: the Bomb that brought much incentive to research also brought must loss. The idea that all change brings mixed blessings is obvious, but the vision of Horace that contradicts it has a strong hold on the imagination. Dr. Jerome Wiesner, then MIT president, repeated this vision with reference to the funds that purchased the American universities for the American military: he said this was the cause of a blooming akin to nothing except the blooming of the Renaissance culture! Only about a decade later the advisability of military aid to research was questioned – and even then on moral grounds alone and with reference to one and only one moral question: was the Vietnam War moral? The voice of protest against that war included some protest against the military involvement in scientific and technological research – at least in the universities and in other civilian public research installations. Yet, this voice of protest, limited as it was, was also extremely weak. Thus, the leading figure in the anti-war protest, Professor Noam Chomsky of MIT, opposed then the proposal to close the research institute which MIT ran with the US Air Force for the exclusive purpose of conducting in it research for military ends. He explained: closing the installation would cause unemployment. This explanation always holds; allowing for it is a declaration of bankruptcy.

This is but a glimpse of some obvious points. I appeal to those able to stay firm in the face of so much glamour, money, and gambling with enormous funds and with millions of lives. Science is so great, that to discover one law of nature, as grand old Democritus said, is more thrilling than to rule an empire. Yet, we should not overlook the negative in science. Knowledge is power, and power corrupts all too easily. We need badly the science of science policy to apply and advance the knowledge and the power of control over the power that knowledge breeds – in order to avoid the corruption uncontrolled power surely brings in its wake. We need, thus, to fortify democracy with weapons to combat technocracy. As a prime proposal, I suggest, we should declare publicly that the aim of the science of science policy should be to help democracy enhance the rationality of the public so as to secure technology as an aid to the growth of science. To this end, we should institute a public policy of investing less in research for military ends and more in the research for the end of democratically controlling both the military and the research that serves it. We have to consider the public support for the development of the science of science policy an item with very high priority on the public agenda.

4. PHILOSOPHY

Chapter 4.1

SCIENCE AND PHILOSOPHY

I. THE CONTRAST BETWEEN SCIENCE AND PHILOSOPHY

What is it to be human? What kind of a thing is the human animal? The answers to these questions are fragmentary and conflicting. They come from diverse sources. Religious teachers, philosophers, moralists, and scientists, all respond to it to some measure and in various ways. It is therefore not very surprising that more often than not religious teachers and moralists offer conflicting views in response to the question, that philosophers and scientists likewise offer conflicting views on it. Such conflicts are not limited to the study of humans – the conflicts between science and metaphysics are much more general. We usually regard the doctrines that religious teachers espouse as metaphysical or as parts of a metaphysics. Indeed, philosophy includes metaphysics or ontology, or the theory of the nature of things, as well as the theory of value, including ethics, or the theory of the principles of morality, as well as theories of philosophical method and of scientific method. Let us ignore all other contrasts and center on the contrast between science and philosophy.

When facing this contrast, traditionally scientists tended to dismiss philosophy as superstitious, as remnants of old prejudices. Alternatively, they were ready to view philosophy, and possibly religious doctrines too, as not doctrines about the world, as not informative, as not statements proper, but as moral teachings, as descriptions of points of view, or as anything except information. Philosophers who were in sympathy with science often advocated this view. Philosophers who were out of sympathy with science usually praised it as mere technology. They denied that it is informative on any level except the most superficial level of appearances, on the practical, technological level. In particular, they either declared that science (or its part that they did not approve of) contains no information about the nature of things, or that the information it contains on such matters is unreliable and thus it is better to ignore it. The example that comes to mind is the view that the chemical elements are the final building blocks of all matter. Some philosophers say, science has never really asserted this view. Others say, it did, but we should ignore it anyway, because it is false. Yet, why should we trust philosophers about the nature of things if we are not at liberty to trust scientists about it?

A school of philosophy has developed that says, the nature of things is not for us to ponder about – perhaps because it does not exist, and perhaps because it lies too deep for experience. We must then be content with the available facts and their organization within science. This is a version of positivist philosophy, so-called, and it goes back to the works of David Hume and of Immanuel Kant.

Another school of philosophy says, science cannot perform its task of organizing observed facts without some general idea of how the facts might come together, an idea of how to regulate research. And, as it happens, regulative ideas come from metaphysics and concern theories of substance, causality, and similar traditional metaphysical items. In brief, philosophy, on this view, offers the metaphysical foundations of science, and building on these foundations regulates research. This philosophy, too, originates with Immanuel Kant, though its development began only about one century after his death.

This is very exciting and very gratifying. Yet it evades the initial difficulty. The difficulty was a matter not of foundations but of conflict. Philosophy and science do disagree about the nature of things. Which should prevail? The philosophers who oppose metaphysics, as well as most scientists concerned with this difficulty, declare outdated all ideas that metaphysicians advocate. The philosophers who support metaphysics against science deny that science offers worthwhile information about the nature of things. (Some philosophers even go so far as to say, certitude belongs to metaphysics, not to science.) Nevertheless, interesting metaphysics and interesting science at times do clash – at least seemingly so. All efforts to look at worthless metaphysics, at the technological function of science, or at cases where science and metaphysics are at peace with each other – all these are not to the point. The point is, at times science and metaphysics do clash, and interestingly so. What shall we do in these cases?

Of course, there is no difficulty in admitting that both science and metaphysics talk about the nature of things, to admit that at times they clash, yet to notice that they perform different tasks. Otherwise, we would hardly be able to distinguish between them. I use the word "metaphysics" as descriptive of the important book by Aristotle on the first principles of physics. This is how Bacon, Descartes, and Boyle understood it in the seventeenth century, as Christian Wolff and Kant understood it in the eighteenth century, and as the modern defenders of metaphysics as a regulative idea for science understood it, including Meyerson, Burtt, and Koyré. According to them, metaphysics offers overviews, intellectual frameworks, general maps, principles of the nature of things. Meta-physics is the first principle of physics, meta-biology, is the first principle of biology, and the same holds for meta-psychology, meta-sociology, *etc.* Metaphysics, or *metaphysica generalis*, is the name for all of them taken together – to mean meta-science. In this sense, meta-science comprises cosmological principles; their relations to natural laws are similar to those of the constitution to the laws of the land. Epistemology and methodology comprise meta-science too, and they function as protocol, with no reference to the content of scientific ideas. Metaphysics offers over-views and science offers deductive explanations. An explanation, when broad enough, may be an over-view. Perhaps a series of related explanations may share an overview. For example, all Newtonian theories, whether about gravity or about electricity, or about anything else, share Newton's view of space, time, and (conservative) force. This way metaphysics regulates research. Yet, the general viewpoint that diverse specific explanations reflect need not cohere. The general viewpoint that a scientific explanation reflects may clash with the general viewpoint that a philosopher suggests. And two explanations may reflect general viewpoints that clash with each other. Clashes create tensions.

What happens then? Researchers may overlook them, at least temporarily. At times, they dismiss the metaphysics that clashes with their science. Yet, time and again something more interesting and fruitful happens. They may reinterpret their scientific explanations to conform to other metaphysics. Such an interpretation may then become a new, independent scientific explanation. There is then room for a crucial experiment between them that might accord with one better than with the other. For example, Newton declared gravity to act instantaneously at a distance. Field theory denies the possibility of action at a distance. A clearer clash can hardly exist. According to field theorists, gravity acts not instantly but very nearly so, probably with the speed of light. The scientific explanation that goes along these lines is Einstein's theory of gravity. Eddington carried out the crucial experiment between Newton and Einstein in his total-eclipse expedition of 1919.

2. CONCERNING SCIENCE

When attempting to apply any general considerations to the human sciences, one encounters the usual difficulty. It is the difficulty to produce a veritable scientific theory within the field, a scientific theory concerning anything specifically human. The first suggestion is usually physiological, and is at once admitted and dismissed as not specifically human. Many thinkers doubt that humans are physiologically different from other animals. Of course, it all depends on what counts as a difference and on what difference is physiological. For example, only humans speak. This is in dispute, as animals communicate too, and simians and birds can learn to use language. But this is piddling. More to the point is the claim that language is not a part of physiology. But perhaps it is, as it is anatomical: the human brains contain speech areas. Moreover, the human erect gait, the dexterous human hand and mouth, even the relatively problematic human birth that relates to the large human brain plus the erect human gait, all these offer a broad field of specifically human physiology.

Supposing this to be so, we can seek scientific theories elsewhere. Say in the behavioral sciences. What is specifically human about human conduct? The theory of animal behavior gained increased prominence in the twentieth century. It is the theory of species-specific modes of conduct, relating to fighting, courting, hunting, and rearing the young, learning from experience, social conduct, and more. These studies exhibit the great difficulty one encounters in seeing order in the world of animal behavior. It is not so difficult to observe certain ritual patterns of behavior, especially of courtship, but it is far from easy to decide whether they are innate or acquired. As it turns out, the answer is a complex of the two: they are partly determined genetically and partly learned. How much all this applies to humans is very hard to say, but at least the fact that most behavior patterns are partly determined genetically and partly learned is very stimulating and efforts to apply all this to humans may be very rewarding. For example, what in the choice between heterosexuality and homosexuality is innate / acquired?

This approach too seems very frustrating. Why, anyway, should we confine to biology our search for scientific theories specific to humans? Why not start with what is so very obviously specific to humans, such as language, politics, customs, art, religion, or science? The very diversity of certain human characteristics is decidedly specifically human. There is evidence that certain bird species have variations in birdsongs that depend on location or on groupings. This diversity is reminiscent of the diversity of human languages, but only very remotely so.

The answer is that the studies of languages, art, religion, and so on, belong to the humanities: they are not empirical. A fashionable complaint is that most of the humanities, far from being empirical, rest on concatenations of metaphysics and methodology. This complaint is odd. After all, we have an enormous store of factual information about societies in diverse parts of the world, their languages, customs, institutions, religions, rituals, *etc.* We have a lot of information about economics and politics and law in diverse parts of the world. The complaint is not, then, that we have no empirical information, but that we have no theories, that we do not know how to use the available information. Efforts to use it to theorize comprise comparative studies – comparative anthropology, economics, government, *etc.* These studies result almost only in contrasts: they exhibit nothing in common to all cultures. This is the odd complaint.

This odd complaint looks oddly obvious. Let me report a familiar empirical observation that I find somewhat significant. It is this. Diverse groups of social philosophers and social scientists spend time complaining about the absence of empirical generalizations in the study of humanity. The conversation is usually dull when it hits this topic. It livens up only a little if someone contests the current view, since this leads to a challenge to contestants to give an instance of an empirical generalization in the field. But the expectation is that contestants will fail at once. Every possible instance that they may mention must be either biological, especially physiological, and so beside the point, or patently false. In the last resort, challengers declare right instances too trite to mention.

Example: all societies have families. Either families are specific structures and so not universal, or they are any groupings for procreation and the generalization is too meager.

This game must stop. Not only does it rest on an erroneous view of what is an empirical generalization; it is an expression of a self-fulfilling malaise and self-pity. People who play this game complain that they do not possess the status of a shaman that researchers in physics or in medicine habitually possess in the modern world. It is the prejudice that the empirical nature of research automatically grants status to those who are engaged in it. This blocks attempts to discuss the empirical nature of theories that is more interesting than status.

The demarcation of empirical generalizations is rather intuitive; not so the demarcation of empirical theory. The only intuitive idea here concerns the logical link between the two. It is this. A theory that has no logical link with empirical generalizations is remote from experience, speculative, metaphysical. The prevalent view is that empirical theories are conclusions from empirical generalizations in some sense of the word "conclusions". This view is obviously either false or too vague for critical discussion. The next popular view is that empirical generalizations verify empirical theories, or, if they do not verify them, then at least that they render them probable. The usual assumption about probability is that an empirical generalization renders a theory probable if we can logically deduce the generalization from the theory, namely, if the theory explains the generalization or predicts the occurrence of instances of it. The usual characterization of both explanation and prediction is deductive: a deduction of facts from a theory is explanation of the known and predictions of the unknown. Predictions allegedly verify the theories that entail them thus rendering them scientific. All this is commonly accepted. It is a blunder – the confusion of tests with passing tests successfully. It causes much heartache and the oversight of much valuable research. The famous psychologist D. O. Hebb reports that many of his research students make predications in their research, fail to verify them, and suffer all sorts of hardship, up to, and including, nervous breakdown. This is simply the refusal to view predictions scientific unless they are true (or correct, namely sufficiently near the truth). The same refusal hardly applies to explanation, since we explain the already observed. Instead of explaining this to his students, he advised them to be cool.

Explanation is possible without prediction. This is a well-known case of *ad hoc* explanations. As long as an explanation has no role to play in any prediction, as long as it is not predictive, to use the jargon, we may suspect that it is *ad hoc*. Even when predictive, we may consider its predictions hardly of interest, and then we may still brand it *ad hoc*. Also, a theory can be predictive without being explanatory. In that case, one might test it, and if the prediction comes true, it will thereby become explanatory. Otherwise, the test will refute it. Usually, before testing it we ask, is there a good reason for trying to test it. If it is explanatory, then testing it may be interesting enough, and so it may call for a test. The test may then come out one way or another.

A theory does not lose its empirical status because the test came out contrary to expectations. Otherwise, we will have to say that Newtonian mechanics is no longer empirical – and this is absurd. Many philosophers have noticed this fact. They should have noticed, thereby, that when they insist on prediction they mean correct prediction and that they should replace the demand for correct predictions by the demand for any prediction, correct or incorrect. Instead, they insist that though Newtonian theory logically entails false predictions, it is nevertheless true. This violates basic logic: false conclusions follow logically only from false assumptions. Denying this is sheer confusion.

It is amazing how much sophistication has gone into the philosophical and popular literature on science in order to entrench this confusion. The wish to be logical while insisting that a falsified theory is nonetheless true is impossible to grant. It rests on the mistaken view that all empirical theories are true.

3. DESIDERATA FOR SCIENCE

The liberating idea about science is that predictive power of a theory is not limited to true predictions. Popper's suggestion that the empirical nature of a theory is its predictive power only if the predictions may come out false. It follows at once that we have many empirical theories in the human sciences. Here are some very important false empirical theories in the human sciences. No social conflict is irresolvable (functionalism). No society changes from within (functionalism). Newer social classes are more progressive than older ones (Marxism). Legal divorce and permissiveness threaten the institution of the family (Catholicism, fundamentalism, communalism). Humanity is impossible without a sense of guilt (Catholicism, fundamentalism, Freudianism).

This is not to endorse Popper's view, since we may want – contrary to Popper – a theory to be more than merely empirically refutable. Popper noticed that we want theories to explain, and said he had taken care of this, since the demand for refutability includes the demand for explanatory power. This is not true. As he has noticed, we may defend a theory against refutations by *ad hoc* measures. And *ad hoc* theories are still explanatory, but they do not initiate predictions. Hence, we may have the choice between a highly explanatory theory and a highly refutable one. Here my recommendation differs from Popper's.

(Example: B. F. Skinner's theory of behavior modification explains any of the experiments that he or his followers have conducted and deemed supportive of their views. The theory explains, but is not refutable. Duhem has claimed that all scientific theory is of such a character. Popper rightly denies this.)

Duhem has suggested that we deny that the alleged refutations of Newtonian mechanics are logical conclusions from it. He suggested that there is a measure of flexibility to the question, what prediction follows from a given theory, what not. We can try to suggest that a prediction follows from the theory, try it out, and confirm our suggestion if and only if the prediction comes true. Many philosophers studied this flexibility. They tried to describe it with the tools of the theory of meaning and inference. As it happens, the general idea is simple, but the details are not easy to come by.

Admittedly, in some cases, the *ad hoc* rescue of a theory from refutations succeeds and then the theory is explanatory but scarcely testable. In other cases, the rescue operation fails and then it empties the theory of both explanatory power and testability. In such cases, the theory may be metaphysical not in the sense that it is unscientific but in the sense that it is a broad general theory – such as Skinner's theory of human beings as behavioral mechanisms of sorts. That theory is neither explanatory nor testable. Many philosophers deem such theories worthless and they – including Popper – deem them unscientific.

This is evidently not fair. Theories such as mechanism offered immensely important regulative ideas for scientific research. I suggest that to be scientific, a theory should be more than explanatory and testable. It should also conform to an interesting metaphysics. We classify theories that do not conform to our metaphysics not as science but either as challenges to scientific research or as technology. The early theories of spectral lines, for example, were unacceptable, and so researchers took them as formulas to explain. Quantum theories explained them in manners that many deemed unacceptable and others deemed acceptable in the light of an utterly new metaphysics. Such cases are marvelous and exciting. They are rare. In most cases, explanatory theories that do not conform to the received metaphysics pose no problem, as we deem them mere technology. Theories of elasticity, for example, are technologically very important and so they regularly undergo severe tests. But since they rest on the assumption that matter is continuous, they are scientifically uninteresting; they are not scientific proper. The same holds for many technologically useful theories about human beings, except that most of these theories are useless even though we continue to apply them, especially in education and in legislation. But that is another story.

What, then, would make a theory about human beings exciting, interesting, valuable, and worthy of our attention? Contrary to almost all writers on the matter, especially those who concern themselves with the human sciences, I recommend that we begin with metaphysics. This, let me say at once, is not a necessary condition. A theory may be very significant without having a metaphysical import, or have a metaphysical import *post hoc* and quite unintentionally. Nevertheless, it is most economical and stimulating to start with a general picture of the matter, of the case of human beings.

There are a few possible options here. We may take an existing metaphysics for our starting point and take it as a putative truth. We may, alternatively, work at it critically and offer a modification or an alternative to it. Yet, once we have a metaphysical system, or better, competing metaphysical systems, we can use them as intellectual frameworks. Current empirical theories and generalizations may sit well within a given framework or not. When in conflict, we may try to modify the one or the other – and we can try it both ways. We may then try to design a test or a crucial experiment. With the reiteration of the process, something interesting should turn up even if each and every test refutes each and every theory under test. For, the reiteration is likely to create series of theories with increasing explanatory power. At present, this is what social science seeks.

4. THE METAPHYSICS AND THE SCIENCES OF HUMANITY

The most dramatic and visible interaction between philosophy and science occurs in the study of social matters, where radical, conservative and reformist philosophies dictate empirical research and alas! even its empirical results.

In the past, all great studies of humanity appeared in terms of certain very simple polarizations that have given these studies both their strength and their weaknesses. At the center of the struggle that engages the *avant-garde* students of humanity is the attempt to free us of these polarizations. To take obvious examples, all traditional theories of humanity assume either that we are all morally good or else that we are all morally bad. Those whose metaphysics have left the world devoid of all value – the mechanists – usually used their metaphysics of nature to defend the metaphysics of a human being as good. Nowadays many students of humanity try to avoid the extremes, either by evading the question, or by giving it a new middle-ground answer: human beings consist of a mixture of good and evil; none of the two prevails (except perhaps – perhaps – in the cases of saints and of the Shoah). Another example is rationality: some declare all human beings perfectly rational, others go to the opposite extreme; only few contemporary writers now argue for the view that human beings are rational in some measure. Moreover, the two sets of opposite views are connected: the view that human beings are good and that they are rational are historically bound together.

The same holds for the socio-political nature of human beings. The school that deems humans good and rational is politically radical and the other is politically conservative. The alternative idea of partial goodness and rationality offers a middle ground also in politics, in the form of liberal democratic reformism. The relative novelty of this view explains the odd discrepancy between the extremism of traditional political theory and the reformism of traditional political practice.

The same holds for the theory of knowledge. The radicals rely on scientific proof. The conservatives rely on knowledge embodied in traditions – of society or of the commonwealth of learning. Einstein, Popper, Bunge, and others inaugurated reformist philosophy of knowledge. It is approximationism.

The fundamental metaphysical question regarding humanity is, do we have souls or are we mere machines? Here, paradoxically, the mechanists, the reductionists, are radicals, whereas the anti-mechanists are conservatives. Here, again the two extreme positions invite an alternative (Bunge).

The fundamental question reductionists and their opponents bring to philosophy concerns society and social institutions. The reductionists not only deem humans machines; they view society as a mere collection of individuals. Conservatives, on the contrary, postulate that in addition to individuals, also societies have souls (sociologism, group minds, mono-psychism). Often enough conservatives shrink from this dangerous step, yet it repeatedly allures them.

Claude Lévi-Strauss has brought to our attention the mode of thought that polarizes views, and he described it as characteristic of magic. It also appears outside magic, though in quite a different way. This invites further examination, criticism, and modification. I have attempted some preliminary steps in that direction in my *Towards a Rational Philosophical Anthropology*, 1963, 1977. I suggest there that the root of western philosophy is in the polarizations that characterize it. The major source of these polarizations, and of their hold on the imagination up to the present, is in ancient Greek metaphysics that equates the following pairs of polarities in the manner that Claude Lévi-Strauss has described:

reality : appearance :: truth : falsity :: logos : mythos :: necessary : arbitrary :: nature : convention :: good : evil.

To confine our discourse to these polarities is objectionable. Science is between reality and appearances, approaching the truth by degrees, employing rationality to increasing measures. Social reform is the use of the better conventions not as arbitrary but as the best substitutes for nature that we have at present. This makes reformism not a Lévi-Strauss-style juggling with the extremes, and not a compromise between the two extremes, but a conscious improvement over old polarizations. The concepts of uncompromising reformism, of an extremist middle of the road, and similar ones, cease to be inconsistent. We can now speak of uncompromising reformism in new ways.

Chapter 4.2

PROGRESS IN SCIENCE AND IN ART

0. ABSTRACT

The view that there is no progress in the arts the way there is in science rests on the observation that whereas newer scientific theories are better approximations to the ideal truth, newer works of art are not necessarily nearer to the ideal beauty. Nonetheless, there is progress in the arts in other respects, relative to other ends that art seeks, both as techniques and as explorations of human inner life – perhaps also as a mode of expression.

1. COMPARING THE ARTS AND THE SCIENCES

A huge literature compares and contrasts the arts with the sciences and with other aspects of human activity. Now any two sets have both similarities and differences: being human activities, they must share much; and to be distinct they must differ. For the exercise to be of any interest, the couple that it compares / contrasts has to exhibit similarities / differences other than other pairs exhibit, and to do so in ways that are not too obvious. Even then, it is hard to know what problems the exercise addresses, though possibly they refute some important theories. If such theories exist, and if they come to solve some significant problems, then their refutations revive important old problems. Otherwise, possibly the exercise leads nowhere in particular.

A few simple facts about the comparison of the arts and the sciences should be taken as given. First, science progresses towards the truth, whatever exactly this means, and any piece of research that somehow advances this cause is science, in contrast with applied science, technology, and art. The fine arts, in contrast with the applied or the practical arts, supposedly produce things beautiful. This supposition is under dispute. Clearly, some artists wish to create beautiful things, and to the extent that they achieve their aim they create works of art that are indeed beautiful. Yet this is not always the case. Artists may mean to create works of art without trying to make them beautiful, and then usually they are not. Even calling them ugly need not be an insult and it will not detract from their value as works of art. And these may be works of art quite generally acclaimed as great. Works of art of dubious quality may gain popularity simply because they are mistaken for great art on the evidence that they are clearly ugly and clearly intentionally so; they may easily reach this quality by mixing the vulgar with the distasteful. Suffice it that a pornographic scene in a play or a film is clearly sexually repellent, that some critics will call it great art. This is very silly, even though it is an understandable reaction to the opposite, equally silly view that what is not beautiful and/or pleasing cannot be art. Examples are the *Quartet to the End of the Days* (Messiaen) and *A Survivor from the Warsaw Ghetto* (Schoenberg). Some deem these works poor art or no art at all. Some performances of these works are as beautiful as possible. This renders them artistically barely tolerable: they were not meant to be. Art, unlike science, is autonomous: it can choose its own goals freely.

All this is less revolutionary than it sounds. After all, already Aristotle noted that tragedy is far from pleasant as it depicts tales of horror. He explained this away, though, by his theory of the catharsis: tragedy floods us with emotion and thus makes us feel clean. This rendered it obligatory to end all tragedy with a catharsis – until Bernard Shaw broke the rule and added a didactic epilogue to his *Saint Joan* that works miracles. Nor is it obligatory to perform a work in line with its intent: it is a celebrated fact that the best way to produce a tragedy may be to treat it like a comedy and *vice versa*. This is so because it becomes a great understatement, and understatements may soften, disguise, or even obstruct its message – and then let it storm out as a *tour de force*. This, however, is not true of energetic performances that prefer overstatements, like those that usually take place in a *commedia dell'arte* and in a fanfare.

What all this shows is that the diversity of art blocks all general descriptions. Each work of art has its aim, and critics should judge it in accord with its aim – unless they judge the aim itself objectionable, of course, in which case they may dismiss the work with no analysis.

Criticism of performances, such as the rejection of beautiful performances of works not meant to be beautiful, is a major item in the present discussion. Anyone who disagrees with the criticism presented here can substitute for it any other criticism of any other objectionable performance of any other piece – as long as they consider it not incompetent but misguided, thereby accepting the idea that right and wrong performances are possible. I ascribe great importance to the very fact that there is art criticism, to the fact that art can be subject to critical debate and critical assessment. There is a very popular view to the contrary; according to it, there is no valid objective rational art criticism. I will not debate the error of this view here for want of time, if not for want of interest. Suffice it to mention that some very able performances of the most pained chamber works of Schubert, are very sweet and thus very distasteful, as any comparison with more honest performances of the same works will lay bare with ease: the improvement of the taste of the subjectivist is a powerful argument against subjectivism in aesthetics. Of course, they are at liberty to deny that their tastes were ever improved and to refuse to improve their tastes further. We are at equal liberty to ignore them.

The comparison between the arts and the sciences is simple. Progress of the sciences is to the truth, and this progress takes place in each step towards this goal. There is no such progress in the arts, either towards the truth or towards beauty, since, even if the arts do approach beauty, it is not true that every step in the development of the arts is nearing beauty. Nor is it true that there is no progress in the arts.

2. PROGRESS IN THE SCIENCES AND IN THE ARTS

Let me mention one more preliminary point, though it is not so obvious, perhaps not obvious at all. The comparison between the arts and the sciences often opens with the admission that art is of no intellectual value. The great philosopher Descartes advocated this view and thinkers of the rationalist persuasion scarcely ever challenged it, and often irrationalism is identifiable as the very idea that art is of intellectual value, that its intellectual value is higher than that of the sciences. (Whereas Hegel spoke of religion, art, and science as if they were on a par, it was Wagner and Heidegger who declared art more significant than science as a vehicle for expressing a higher truth, poetic truth as Heidegger referred to it.) Perhaps the twentieth-century philosopher and archaeologist R. G. Collingwood is the only exception to this, as he said, art, science and religion share a purpose, and in some respect perhaps art is the most successful of the three, yet he still was a rationalist philosopher. (See the conclusion to his *Speculum Mentis* of 1924.) Let me also report in this connection that Albert Einstein, whose dedication to science and to rationality was as exemplary as it was unwavering, expressed gratitude in his intellectual autobiography to a few individuals who had contributed to the background knowledge he used for his fabulous research. Among these he counted Mozart. This fact puzzles commentators: how could music inspire research? This, I conjecture, is why they disregard it. Einstein called Niels Bohr's first breakthrough very musical. He was perhaps alluding to the fact that Bohr drew his inspiration from the great astronomer Kepler, who had claimed to be hearing the music of the spheres. Einstein himself did not study anything remotely related to the harmony of the spheres. Rather, he was speaking of the harmony of the universe.

The point is precisely this: no one can put the idea of the harmony of the universe into adequate words. This is why some people say that art has no cognitive value and other people say the very opposite: there is no way to assess critically the truth or falsity of an idea until it is fully articulated. And so the rationalists say (as Bertrand Russell did), art is devoid of cognitive value and the irrationalists (like Martin Heidegger and Hand-Georg Gadamer) say that it has a greater cognitive value than science. They share the idea that art and science are utterly separate. Art criticism, says I. C. Jarvie, evidently opposes this idea. It has to relate to the aim of art in general or to the specific aim of the specific work of art under examination, given it by the artist who has created it. How, then, is rational art criticism possible? Before we can discuss that question we may wish to ask, what is / are the aim / aims of art? How does art progress?

Progress is improvement. The idea of universal progress is of the improvement or betterment of everything or of every aspect of the world. This is a serious confusion, since to have any improvement one needs a specific purpose. The ability to speak of progress without specific purposes led to the view that such progress is superior to progress towards a given end. This view led to the contrast between art for art's sake and technology, to the identification of science with technology, and then the view of artistic truth, to use Heidegger's locution, as superior to the scientific truth that is merely of technological value. This attitude and this terminology changed the intellectual map. First science went with philosophy. Technology then went with art, as practical art or as applied art. Now technology left art and joined science. Science and technology then became the applied arts, as opposed to the fine arts and the humanities. Scientists and pro-science politicians concurred on this fusion of science and technology, as they found it easier to procure grants for practical projects than for purely scientific ones. The mere search for the truth for its own sake was not too popular with voters. Further, art and science became identical. It became popular to prove this identity by marshaling cases of strong interaction between them (as if the strongly interacting husband and wife are identical). The spokespeople for the arts who were ignorant of science concurred too, and enthusiastically so, as it exempts them from the need to show any interest in science and even allows them to confess proudly their ignorance of it. Still, the applied arts, namely, technology, especially science-based techniques, were nearer to the sciences than to the fine arts. This is so despite the fact that *Tafelmusik* and even church music, for example, originally belonged to the applied arts and became fine art only after they reverted to the concert hall and to the record industry. And this is so despite the fact that architecture and the recording of music, to take two conspicuous examples, are clearly science-based technologies yet they are nearer to the arts than to the sciences.

The discussion here rests on the claim that progress is towards given ends. As scientific progress is towards the truth, so moral progress is towards the good and artistic progress is towards beauty. We do not know if truth, goodness, and beauty exist in themselves, and if they are distinct. Plato said they are identical and many philosophers say they do not exist. Yet these very philosophers often admit universal progress; this is inconsistent, as the concept of universal progress leads to the concept of the perfection of the universe as a whole. Also, it is now agreed, in the light of modern logic, that the idea of perfection as such is very problematic, though we do know what is a perfect fit, say, between certain means and certain ends in certain circumstances. This latter idea seems simple enough. It makes simple the sense of the idea of progress towards any given ends, of progress of whatever we want improved towards the goal of the improvement. This accords very well with our intuitive idea of improvement. Hence, not only is the idea of progress as improvement quite intuitive, it includes an appeal to our intuitions directly as it appeals to our sense or intuition of satisfaction or of increased satisfaction.

This, however, is only one part of the story: it is its conspicuous part. But it is easy to elicit different cases. It is easy to show examples where the feeling that there is progress clashes with some criterion of progress. We may then prefer the criterion. Physicians are often not of the same opinion as their patients as to their progress. Take the sense of great relief that certain patients have and that they consider an indication of improvements; their physicians may very well consider this very sense a symptom of an oncoming collapse. Physicians seldom use the simple intuitions that their patients use. Rather, they use criteria, namely rules for judging ("kritēs" is the Greek for judge), about what is progress what not, and their criteria and intuitions, if they have added intuitions, concern not what they as physicians feel, but what their patients feel as patients ("patient" means sufferer). The criteria are not very good, and they seldom undergo tests as they should. At times we seek criteria. We may ask, how do we judge whether our criteria are good or bad, and whether one criterion is preferable to another? That is, we are in search for some meta-criteria. Some meta-criteria are very simple: we want an instrument to perform certain tasks under certain constraints, and these serve as criteria for improvement. We do not always know how to compare, as an instrument may be better in one respect and not so good in another. But this is how things are. This presents real problems. And these need not relate to the study of progress as such. Consider such questions as, what can best improve the quality of life of the citizens of the United States these days. It may become easier to handle it if the discussion concentrates on a section of the American population. This is particularly so if the discussion concentrates on a section of people who share a serious problem. We need not go further into that matter, as the preliminary discussion here already shows how very vague is the traditional optimistic theory of universal progress that culminated in the end of the nineteenth century.

3. NEW IDEAS OF PROGRESS

Philosophers of science and of art follow these days the ideas of Popper and Ernst Gombrich and their critique of the traditional theory of progress. Popper and Gombrich admitted only scientific progress: comparing the scientific theories of two epochs, the latter are the better; otherwise we say that there was no science in the period in question, such as the Dark Ages. The traditional concept of science is of perfect knowledge. It follows then that there is no progress possible on any matter on which science has made its move, though progress is possible by adding new items and by unifying theories. After Planck, Einstein, and Bohr revised the classical theories of physics, many people felt the need to give up the idea of scientific perfection. This led some, like the famous historian of science Thomas S. Kuhn, to render truth relative. He also declared that in science paradigms (whatever this word means; the Greek word means chief example) are above criticism (as if they were absolutely true). With no concept of absolute truth, the concept of relative truth loses its meaning too; and so, Kuhn could safely decide that one cannot compare the relative truths of classical and of modern physics. This means that there is no absolute progress, no progress in the abstract, not even in physics at large: progress as such is thus in principle absent from science. And all this is so because Kuhn declared the concept of absolute truth meaningless. Popper disagreed to the last. He stated the commonsense view that scientific progress is manifest, and he explained this by the Einsteinian claim that the newer theories are nearer to the truth than the older ones. It is not sufficiently clear by what criterion we should judge a move in science as an improvement. But obviously, assuming that the task of science is to trace God's blueprints of the universe, to use Einstein's metaphor, and then the proximity to the truth is the proximity to the attainment of the task at hand. Kuhn denies that the task of science is the search for the ultimate truth. He thus sees no possibility of overall progress in science.

There is a simpler criterion of scientific progress than the ones that either party employs. When we study science, we do not always begin with the latest theory extant. We often study first older theories for which the newer ones are substitutes. We then learn how these old theories somehow integrate in the newer theories. This is not true of all old theories. We do overlook some of them, as we find them valueless. We study the valuable ones. Supposedly, scientists are familiar with these, not in the exact versions in which they first appeared, but as part of the new theories. More precisely, later theories should absorb the parts of the old theories that are valuable. If not, then we are in trouble. As a matter of fact about scientific research, this fact about science operates as a criterion for scientific progress. The best in its old theories is somehow included in the new theories.

This is not to say that science is merely theories, and that the sole task of science is to approximate the absolute truth. Science may just as well include scientific technology. And if so, then Kuhn is right or at least half-right: technology, even the most scientific of scientific technology, need not be progressive in the sense outlined above. Proof: if we omit from the history of scientific technology any given episode, then there will be no way to retrieve it. Also, the skill, the ability to implement a technique, is lost with scientific progress: the best operators of the best locomotive of today will be scarcely able, will scarcely have the needed skill, to operate an old one. Likewise, sailors do not usually know how to operate large sailboats, and this is why sea scouts learn to use them and some schooners stay in operative shape and operate, both as sport and as means to keep alive the knowledge of operating them. Survival techniques are another obvious example. Popper spoke in this vein of the significance of all traditions, and he said that in this respect the tradition of performing music is in the same boat as that of any other skill, of any other performance. It is intriguing, incidentally, to notice that the tradition of scientific research is a similar case: there is progress in research technology, but without tradition, all progress will not help, and without tradition the skills of research will wane. If this is true, then we can compare scientific theories of progress only within one given tradition. As it happens, Popper stresses, there is only one scientific tradition extant.

The technology of scientific research progresses regularly. It is manifest in many different avenues. Different senses or criteria operate when we judge technology or anything else as progressive or not. We can compare different skills of operating a machine to perform a given task. This permits judging one of them as preferable, so that the move from the one to the other is progressive in the respect for which we test their progress. We may have to be specific, as tasks do not coalesce. Even the different ways to perform one and the same task need not coalesce, not even if we perform the test under the same constraints.

There is no general way to assess progress even when tasks do coalesce. For example, consider the tasks that collectively contribute to the improving of the standard of living or of the quality of life of a given population. Since there are so many ways of considering any such improvement, it is extremely hard to judge which of them contributes more to progress. Such questions trouble people in charge of budgets.

All this justifies Popper's view of artistic progress. He says, there is no progress in the arts. He does not explain but points at counter-examples to the claim that art progresses: he points at old masterpieces: we replace old theories with new ones but old masterpieces are as compelling as ever. This is certainly true, but it is equally true that a newly discovered Rembrandt picture will scarcely surprise Picasso but a Picasso painting would have amazed Rembrandt. This is another paradox. Perhaps it is not; perhaps Rembrandt would be unable to see in Picasso's work more than a child's doodle. After all, many artists think so even to this very day. Be it so. Nevertheless, at least an artist who struggled with an artistic problem can learn from another who has solved it. This, indeed, is why both Rembrandt and Picasso were so very influential. This is at times most obvious and, even when it is not obvious, honest artists repeatedly say that they learn from predecessors and peers. Also, it is a fact, the progress often if not always, evolves in processes of rational deliberations and criticisms. Is there progress in the arts then or not? What theory of artistic progress does Popper's example refute?

Is it possible that art progresses not only in its improvements of techniques but also in development of new aims?

4. THE AIM OF ART BEYOND ENTERTAINMENT

Clearly, what Popper's example refutes is the theory that art has an overriding aim that receives expression in each stage, as is the case of pure or theoretical science. Art, like any other techniques, clearly has many aims. Even the finest of the fine arts have diverse aims, whatever these are, as is clear to anyone who is willing to compare the art of Raphael with that of Breughel. Relative to any given aim, then, there may be progress in the arts the way there is in any other technology, and this should resolve the sense of paradox.

Gombrich first suggested this idea. Though art is not progressive, its technology is, he says. He takes as his paradigm the plastic arts. Their aims may differ, but one of these aims is to create illusions and the techniques of illusion do improve. Of course, the fact that modern commercial art creates with ease better illusions than Renaissance masters does not make today's commercial art greater than Renaissance fine art. This is obvious: art has aims other than mere illusion: and there are many works of excellent illusion that do not count as fine art at all.

The studies of Gombrich are exciting and invite assent, but they suffer from certain narrowness due to the lack of a discussion of the aims of art: the only aim he discusses, and shows progress towards, is the creation of illusions. This is why he can study abstract art as minimal art, not as abstract proper, and it is hard to see what exactly the sounds of music give us the illusion of. But clearly, he will concede at once that all rational action is goal-directed. So those composers have their goals. What are these? They need not be the same for all artists, and certainly Beethoven invented orchestration techniques that Schumann rejected just because he had different aims when he orchestrated. What exactly were their aims as they worked on their orchestrations? I confess I do not know. Perhaps they did not know either. And certainly Gombrich will not say that illusion is the sole end of any painter, yet he does judge contemporary paintings merely as (failed) attempts at creating illusions. Since no one is sufficiently competent to judge the art of one's contemporaries, we need not complain that Gombrich approached the paintings of his contemporaries only through a limited prism, even though as a result his study of them is limited. Popper's attitude to contemporary art is harsher than that of Gombrich: he dismissed all contemporary art, and on the ground that its aim is to outdo traditional art rather than follow its own aims. But what exactly are these? And is it impossible that some aims require more radical attitudes to tradition than others? Popper's assertion is right that utter radicalism is impossible. John Cage summarized it wittily in the way he put his criticism of Schoenberg: twelve tones in a row; which row?

What counts as great fine art defies criteria: we do not know what makes it great, though we do recognize a great work of art when we see one. Gombrich says, we will never discover a criterion that will tell us which work of art is great, but he does not know, and he does not mean that we cannot try and even improve our proposal of such a criterion; he means that we will never succeed. We will never succeed, he says, since when we will succeed the arts will cease to be challenging. Possibly. Perhaps the arts will one day cease to challenge us anyway. This only means that Gombrich requires that art should be challenging, perhaps that it should forever stutter. If there is progress in the techniques that artists employ, as he readily admits, then there are criteria for progress, as there can be progress relative to each and every one of them, even if there will never be any overall criterion. Hence, we can see that Gombrich's discussion is limited to a rejection of one kind of progress, and the endorsement of other, but the list is long and the discussion is still in its cradle. We may view one of the ends of art as the exploration of our inner life. The end would then be the finding of new means for expressing or invoking new experiences. The successful use of these new means rests on the older means being not adequate for the purposes that the new art serves. And in this sense there surely is progress in the arts.

This will be noticed clearly when technology is brought nearer to fine art than to science. For, the aims of the arts, practical, applied or fine, may easily be given externally, whereas the one overriding aim of science is always given: the attainment of the truth. Thus, when after the Council of Trent leaders of the Catholic Church were concerned with the claim that music may drown the words of the mass, Giovanni da Palestrina promised his employers to see to it that the singers will express the words very clearly. And whatever he did, certainly, one way or another, this influenced his music and the music of those whom he influenced.

In the twentieth century art reversed or altered direction. Thus, Gyorgy Ligeti wrote his *Lux Eterna* with the instruction to the singers to omit all consonants and sing only vowels, and Robert Cogan wrote his *Whirld...s 1* with the aim of exploiting every spoken sound to the limit. We cannot say that the art of Palestrina is more beautiful or less beautiful than the art of his successors, but we can say that in our opinion, such as it is, in some sense their works are satisfying and his is not. Otherwise we would constantly prefer to listen to his music rather than to theirs. At times we do just that, so that, clearly, we do not consider the works of his successors superior to his in all respects. This is the same way in which in technology we often prefer to apply old theories, instruments and techniques, rather than the latest. And this surely means that artistic progress of sorts does take place but not of the overall sort.

One final point. Gombrich will not disagree with me, since what I say here is fairly commonsensical and familiar material, especially the claims that art can have different ends and that progress can be attained relative to each of them separately and at a fortunate moment even together. Indeed, he said so himself on diverse occasions, and in diverse and very illuminating ways. The real test of whether there is any disagreement here might center on the question, would he agree to check this against contemporary art and if he would, might it not make him a bit more open to its influences and then would he not be more appreciative of it?

5. POSTSCRIPT

Sir Ernst Gombrich was kind enough to read the above discussion and to respond in a very friendly private letter to the challenge in its conclusion. He mentioned there the very few *avant-garde* artists of his days whose works he liked. Readers of the latest editions of his classic *The Story of Art* may find it surprising that he mentioned but a few, as this book includes insightful remarks on many. He did not enjoy their works, it seems, and he found them hardly challenging. Many people have observed that much of contemporary art is not very exciting. This is true of all periods. The public participates in the artistic process by helping separate the grain from the chaff. As he was a great scholar and an art historian, Gombrich knew that very well. He also knew, however, that some but not all periods excel in the production of great art. And he felt that his times were not too happy – not only politically, but also artistically. Such matters are hard to judge – beyond agreeing that there is no guarantee for success in any human venture, artistic, scientific, or any other.

At least this much is easy to admit, and we should admit it: the idea that great art is challenging is some sort of a criterion for artistic quality (Collingwood) – or at least a challenge. But this criterion is not sufficient: some find today's minimal and distorted art challenging, and others try hard to find something in them and end up shrugging their shoulders. If they are honest and not snobs, they admit it. And Gombrich certainly was both. Only time will tell. As he noted, new idioms have to assimilate before clear judgment of their use is easy to make.

REFERENCES

Agassi, Joseph, 1976. *Science in Flux, Boston Studies in the Philosophy of Science*, 28.

—, 1976. "Spontaneity in the Arts", *Poznań Studies in the Philosophy of the Sciences and the Humanities*, 54-64.

—, 1979. "Art and Science", *Scientia*, 73, 127-140, Reprinted here as Chapter 4.6.

—, 1983. "Technology as Both Art and Science", *Research in Philosophy and Technology*, 6, 55-63. Reprinted here as Chapter 4.3.

—, 1985. *Technology: Philosophical and Social Aspects.*

Collingwood, R. G., 1924. *Speculum Mentis.*

—, 1938. *The Principles of Art.*

Gombrich, E. H., 1950. *The Story of Art.*

—, 1960. *Art and Illusion.*

Jarvie, Ian C., 1961. "Towards an Objective Film Criticism", *Film Quarterly*, 14, Spring, 19-23.

—, 1985. *Thinking About Society, Boston Studies in the Philosophy of Science*, 93.

—, 1967. "The Objectivity of Criticism of the Arts", *Ratio*, 9, 67-83.

Popper, K. R., 1974. *Unended Quest.*

Richmond, Sheldon, 1994. *Aesthetic Criteria: Gombrich and the Philosophies of Science of Popper and Polanyi.*

Chapter 4.3

TECHNOLOGY AS BOTH SCIENCE AND ART

0. PREFACE AND SUMMARY

The word technique comes from the Greek word, *téchnē* , whose Latin cognate is *ars*. As often as we hear of surgery or of acoustic engineering or of any other technique, that it is an art, we also hear that it is a scientific art or technique. It is both art and science, in the sense that some techniques are scientifically attested, some not, as well as in the sense that every item of our contemplation has unique aspects, not given to science, as well as repeatable ones, subject to scientific investigation.

This leaves open the question, is technology as cumulative as science? In a sense, science is indeed cumulative – though not in the traditional sense that most modern philosophers of science use. In the sense in which science is cumulative, technology is not. Even so, a scientific theory of given techniques may succeed in rendering the technology in question cumulative.

Applying this to the science-making technology, namely, to scientific method, would make methodology scientific and thus unite science and technology. Such ventures are not without promise, and at times they may produce exciting results; yet their success must remain partial: unforeseen discoveries and inventions regularly threaten it, and invite renewed efforts at integration. As uniqueness is inexhaustible, unification by repeatable means may forever meet new diversification. Science and art are thus competitors and partners.

1. UNIQUENESS

It is customary to identify the distinction between science and technology at times as that between thought and action, at times as that between the universal, and the particular and at times between the abstract and the concrete. Especially since the problem of induction is the questioning of the validity of scientific generalizations, often philosophers take comfort in the assertion that action is particular and so it is not vulnerable to the same problem. This excuse is poor, since the problem regarding action is not whether it is universal but whether it is rational. To be rational, so the classical view goes, it has to rest on science. And there we are, where we have started. A specific action, a technique, may be particular, but the theory of action, technology, is not: to the extent that a technique determines an act, the act is not unique. It is unique because the technique does not determine it exhaustively.

Philosophers prefer to talk about statements and systems of statements rather than of events; as long as it is not too restricting, we may employ with benefit this practice of discussing statements. Science then splits into general and singular statements. Consider then also the repeatable and the unique. Usually, scientists make universal statements and historians make singular ones. They may deviate from this and shift concerns between the general and the singular, between the scientific and the historical. A thought experiment may illustrate this. If we pick out universal statements from historical texts, and discuss their critical examinations, then their authors will refuse to get involved. They may tell us to discuss them with the scientists among their colleagues. Or they may try to do without these universal statements if they admit that we have reason to examine them. Or they may change occupation for a while and become scientists. The same will hold for examinations of singular statements that historians may pick out of scientific texts. Scientists would rather repeat an experiment than examine any specific report about it; otherwise, they may change occupation for a while and become historians. (Traditional astronomy is the paradigm of this shift back and forth.)

The singular statement can report on a historical event, a historical individual, a single work of art, or a period in the history of art. Possibly, this is how art historians were drawn into the study of singular individuals of genius. Genius is but the name we give to a talent that is both unusual and unique. Now whatever in unusual talent is not unique is repeatable: any technique that a genius uses and that is repeatable may become public and perhaps hundreds and thousands may repeat it. Society may forget inventors of techniques or credit them with their unique inventions either of techniques or of the molds with which to cast and recast them. Hence, it is uniqueness that makes a genius. Indeed, every individual has some (capacity for) genius. Otherwise, we are interchangeable with our neighbors.

To recast the point in linguistic terms, a finite list of characteristics may depict and fully characterizes more than one individual. These individuals are then exact equivalent of each other. Hence, as Plato suggested in his *Parmenides*, individuality is not exhaustible. Leibniz had two proofs of this inexhaustibility. One of them is from God's omnipotence: He is versatile enough to be able to avoid repeating Himself. As this argument depends on the concept of omnipotence that is logically problematic, we should ignore it. The other is from physics: every two individuals must differ, as they must have at least different spatio-temporal coordinates. This holds at most for physical objects, not for minds. Leibniz's view of minds is then questionable, since he refused to place minds in space. (Euler has noted this point and he thus shook up Kant's faith in the Leibnizian system.) Moreover, it is open to question whether space-time coordinates are in any sense characteristic of a physical object; and to pertain to identity they should be. Strangely, students of quantum mechanics have agreed with Leibniz regarding the identity of electrons but not of protons. But this is questionable too.

Statements that refer to space-time coordinates are singular, Popper has observed. For example, all the people here and now are adults. Since classical astronomy up to Newton referred to our earth and to our sun, classical astronomy by this criterion was historical rather than scientific. Interestingly, this shows an unexpected link between science and its philosophy. But we need not go into this either – except to stress one point. The word "science" here refers to generalizing science, not necessarily to empirical science, since some generalizations are metaphysical and not empirical (such as ones concerning the immortality of souls), whereas some historical statements are eminently empirical, including those relating to the study of the earth or of the sun.

Historical studies are distinct from generalizing ones by their attention to singular statements – ones that employ proper names. Each may but need not be empirical – in the sense of being empirically founded or empirically testable, or in any other sense of being empirical. We have, then, physical science as generalizing and the history of some physical systems as historical – with both being empirical in some sense (Rickert; Poincaré).

Where is the place of technique, then? There are repeatable techniques, of producing gadgets, curing people, or reproducing works of art or computer design. There are unrepeatable techniques, such as making works of art, or designing a concert hall or a computer-generated design. The expression "golden hands" that we apply to mechanics and to surgeons refers to something that is eminently unique.

2. REPEATABILITY

Discussions of the uniqueness of the unique convey the feeling that it is problematic whereas the repeatability or reproducibility of the repeatable is plain sailing. Not so. The cases are entirely symmetrical.

The only simple manner in which to introduce repeatability or reproducibility – repeatability for short – is the way that Robert Boyle introduced it in historical fact. He had predecessors, of course, including particularly Bacon (*Parasceve*), Galileo (*Letter to Castelli*), and Descartes (*Discourse*, VI). Robert Boyle introduced his idea in an essay ("On the Unsuccessful Experiment," *Certain Physiological Essays*, 1661); but what really counts is that he introduced it into the rules and regulations of the Royal Society of London. (To be precise, the president proposed the rules; Boyle only seconded the motion. He could not accept the presidency since he never took an oath, and the regulations required that the president should take an oath.)

Repeatability was required of an experiment or an observation before it was entitled to scientific status. Before it was entitled to the characteristic of repeatability, an experiment needed at least two *bona fide* independent witnesses to testify to having observed it on two independent occasions. Observations are entitled to scientific status on two assumptions: that two witnesses could report them in court to the satisfaction of the court and that they were independent.

This is a necessary condition, not a sufficient one; to tighten it one may wish to specify independence. Multiple discovery is surely independent enough, but it is too much to demand. Two observations from the same laboratory are not independent enough. A more precise line is hard to draw. Nor are people usually concerned with this overmuch, since there is always the possibility of more repetitions just to please the skeptic. Some philosophers have even suggested that the more repetitions the better. Scientists have never agreed. Only philosophers have voiced it, and amazingly most of them are still voicing it. Of course, repetitions may be useful in circumstances under which one suspects them to lead to different results. But then, if the result is different, it also has to be repeatable in order to overturn the generalization of the initial result. Otherwise, the refutation of the scientific result will not have scientific status.

And here lies the trouble. The requirement for two witnesses derives from civil courts and that derives from the Mosaic Law. That law demands the independence (not of repetitions but) of witnesses testifying to one event: it demands the acknowledgment of the validity of the testimony of two (or more) independent witnesses – not relatives, disinterested, *etc.* Testimony from two other witnesses to the contrary may cause a stalemate, but not so testimony from two other witnesses refuting the initial witnesses by the testimony that they were elsewhere at the time of the said event. The refuting witnesses then get the upper hand. Why? asks the Talmud. Because so the Law decrees, is the answer. This is not acceptable to the scientific community, which never recognizes such claims: science accepts only generalizations, and so it resolves every ambiguity by demanding fresh repetitions. History, by contrast, refers to unique events, and so it is full of cases of stalemate due to competing testimonies. Any break of a stalemate is then a hypothesis in need of a test.

The role of the repeatability of experiments is to aid the advancement of learning. The demand for repeatability is the sole convention that was never in dispute within the scientific community. The domain of the unrepeatable is parascience. Yet, the repeatability of experiments may concern the community at large. This happens every time it relates to practical matters. Consider a disaster related to anything scientific and / or technological. How many times need a Comet 4 airplane lose its tail in flight before the authorities should revoke its license of airworthiness? More than that, how does a model gain a certificate of airworthiness in the first place? Error in decisions of this kind is unavoidable, both technical and legal. Legislatures reform laws after the repeated appearance of tragic errors. How many? No answer.

Obviously, repeatability or reproducibility relates to reliability in a simple way: the ability to repeat a successful experiment is the ability to ensure success. The ability to reproduce a failure is not much different, since the ability to reproduce is the ability to select and control. What counts as success depends on aims and on the ability to realize them. Repeatability, then, is control, and control is guarantee. This is quite plain. But do we have guarantees? Do we control? Can we ever repeat any experience or any event?

The people in charge, the people whose job it is to control, safeguard, and guarantee, quite repeatably and unfailingly demand credit for success and refuse discredit for failure. They usually explain success as due to control and repeatability. In a bullish mood, they often brag that they can repeat every given event to any specified degree of accuracy (within the limits of quantum theory, since beyond these limits specifications lose all empirical significance), and in a bearish mood, they usually excuse themselves by reference to the impossibility of complete repeatability. One need not be a logician to see that this is inconsistent – and that one can easily evade the inconsistency by noticing that the possibility is only with respect to specific accuracy while the impossibility (that quantum theory guarantees) is with respect to absolute accuracy. The people in charge, thus, evade inconsistency by shifting standards of accuracy.

This evasion is expensive: it blocks the way to finding out which mishaps are avoidable. The question concerns the repeatability of the conditions that make the difference between success and tragedy. What is the rule for the choice of an answer to apply? Harder still, what is the rule for deciding whether that answer is scientific? The scientific status of factual claims depends solely their repetition plus claims for repeatability for them; hence, claims for repeatability are scientific if and only if they are repeatable; and so on. This is an infinite regress. This should not surprise: the discussion pertains to the insoluble practical problem of induction. (This is why Popper declared that scientific status is partly conventional. He added that the convention is scientific if it sides with the critical and against the defensive .)

3. CUMULATIVENESS

Both uniqueness and repeatability are thus problematic, perhaps because our discourse is too abstract: each item has both unique and universal aspects and we do not know where the borderline between the two lies. So be it. The repeatability of scientific experience still is the most important requirement in science. And the repeated refutations of claims for repeatability still are empirical discoveries, and they still lead to restatements of initial claims in qualified versions. These must be a burden to the theory of science. If only repeatable experience is scientific, if claims for repeatability are questionable (and often improved upon), is not the entire venture questionable then?

The refutations of some claims for repeatability lead to their dismissal – for example, the alleged beneficial effects of bloodletting or of cupping – while refutations of others lead to their qualifications – for example, Ohm's law of electric resistance that the law of electric impedance qualifies. Instances of total rejection – bloodletting – are of spurious or pseudo-scientific claims, while instances of rejection subject to modification – Ohm's law – are inherent to science. Popper's demarcation of scientific theory by its refutability thus explains the frequency of scientific modifications and the sedimentation of past successful theories into newer ones. This way, modifications are progressive and the venture is attractive.

All this is both intuitively acceptable and quite *avant-garde* philosophy of science. The philosophy of science that is still most popular amongst both scientists and philosophers presents science as a growing collection – of descriptions of the structure of reality or of mathematical tools. The choice between these two options is subject to a secondary dispute among the majority, who deem science cumulative. Some find it hard to say in which sense science is cumulative; usually they declare both mathematics and empirical knowledge cumulative, and they see the evident expression of this accumulation in technological advancement. But consider a little the extent to which technological advancement is the guarantee that we can solve the problem posed here about reproducibility. It is obvious that only to the extent that technology holds its ground can it offer any guarantee; and this we do not know: technology has dangerous side effects. Suspicion lurks that the side effects may already have started an irreversible process of doom. This would be a loss of control and / or repeatability. Hence, there is no guarantee here; indeed, less than in science.

A new scientific theory should explain the past success of its predecessor by showing that successful observational consequences of the old are good approximations to those of the new. And the explanation of the success of a defunct generalization is similar. The old generalization follows from a new one under the assumption that certain conditions obtain, and that the conditions in question usually obtain in our environment. The defunct generalization holds under some qualifications, then (Newton, *Opticks*, penultimate paragraph): in a qualified version, the old generalization still claims repeatability. Claims for repeatability, however, may be empirical and theoretically unexplained – as is common in all sorts of technology, particularly in medicine. Still, even these have some root in theory. Generalizations that make no sense – such as ones relating people's occupations to the stars they were allegedly born under – we do not notice (at least the learned community does not). Admittedly, at times an idea that sounds crazy, such as that of continental drift, gains popularity, but not before a major change in our intellectual background has taken place. (This is well illustrated in Rachel Laudan's pioneering study of the fate of Wegener's hypothesis of continental drift.)

Technology does not and need not have such intellectual instruments of progress. Consequently, technology does not possess the means by which science keeps alive the memory of its past; technology may forget. We forget how the Egyptians mummified their dead; we forget how the Arabs forged Damask and Wootz steel; and we nearly lost the ability to sail ocean-going sailboats, but sporting organizations have saved it for us. Repeated experiments show how hard it is to revive techniques of mining and of factory operations that were popular no more than two centuries ago.

This is no romantic lament: the past often gave way to a better present. Moreover, scientific theories concerning lost practices may revive them. Hence, a successful scientific theory of technology renders a part of technology as cumulative as science. But we cannot expect technology to solve problems concerning science, including the problem of how to judge repeatability in science and in technology: any such solution should count as scientific.

4. THE ART OF SKATING ON THIN ICE

Our image of the situation largely depends on our Darwinian vision: different societies have different systems of adaptation, and these include intellectual and practical components. As only some societies survive, those do that exhibit a tolerable fit between human and eco-niche. Fit need not be ideal, or else variety would be extremely limited. Hence, the survival of the fittest is in no way evidence that the fittest is perfect, but merely that the fittest ousts its less fit competitors for the eco-niche. This picture may easily mislead, however, as it may seem to suggest that eco-niches are given, whereas they are mere aspects of the system and they alter as their inhabitants do. To go further, prospective inhabitants have to discover their eco-niches (as possible eco-niches: they create and inhabit them). This is obvious yet it plays no role in short-terms studies and this may contribute to their oversight and thus to the oversight of Darwinism.

Darwinism seems to be the clever scientific escape from skepticism. The skeptic may be unimpressed with scientific success, and even ascribe it to mere good luck. The skeptic may even employ science to support skepticism: if Einstein's theory is true, then the enormous empirical success of Newtonian astronomy is in part due to the unexplained fact that most of the mass of the solar system is in its center. The Darwinian response to this is smashing: all odds are against us, and we survive due to our wits; hence, wits do count for something. Science proves its validity – partial validity, to be sure – by allowing the physically weak human species to survive and even inherit the earth. Not the lion but the human inherits the earth.

This argument is popular and its popularity is on the increase. Yet philosophers of science and social anthropologists view it differently. Social anthropologists show that even the most primitive humans are intelligent enough to survive. Hence, Darwinism is as valid as a defense of all culture, primitive and scientific alike. Philosophers of science consider only up-to-date science. They may overrule social anthropologists, by the claim that scientific society can destroy primitive society. Would this enable Darwinism to validate science? Such a validation assumes that the modern is winning over the primitive. On this the verdict is still pending. Philosophers are eager to depict a rosy future; they present this wish in the guise of the need to answer the skeptics, as the problem of induction. Hence, any victory over skepticism that Darwinism may help achieve is Pyrrhic: all victories in the battle for survival are possibly short-lived. They are often very costly and they threaten survival. Every victory that upsets a delicate eco-balance may easily destroy an eco-niche altogether, if not wipe out the whole eco-system. Geology reports such things, and ecologists suggest that they are happening with alarming frequency.

Thus, the argument has not yet made progress. The claims that diverse parties are making may be forceful or vain. We had better scrutinize them as best we can – scientifically. We thus wish to study – as scientifically as possible – not only Darwinism but also the question, how legitimate are its many applications to science? This will be the scientific study of science, the so-called "science of science". The science of science must also be the science of technology. It should include all sorts of theories of how to do things with our brains. A scientific theory of this sort will describe both science and technology. It should account for past science and technology, possibly not for all future developments. As humans can describe animal knowledge so can super-humans describe human science. The hope of the advocates of the science of science is a dream of becoming super human. But possibly, the science of science (and of technology and methodology) will be partial and aid science and technology progress only in limited measures.

The emerging situation might, then, be that of progress through trial-and-error, progress raises hope only as long as the process continues. When stopped, due to a catastrophe or due to a decision, then nothing existing can tell us the outcome. As long as the process continues, however, it attempts to unite science and technology. These attempts are partly successful, and they may have further success. This may call for innovations that pull science and technology asunder. This is simply an unstudied matter, thanks to the glib superstition that science and technology are one anyhow. Yet, despite this superstition, a number of studies are now converging into theories of method that may help scientific progress the way theories help in other fields; the *locus classicus* for this is computer science, which is exciting, as it is, and has been from the start, a hybrid of methodology and engineering. It is the field of super expertise, and so its success invites careful study and assessment. Without such a study, the uncritical view of ideas that this field suggests vaguely – as if they are already achievements of technology – has done much to confuse and has invited attacks that further confuse simple issues.

5. CONCLUSIONS

The goal of uniting science and technology is laudable but (fortunately) not attainable – just like the goal of obtaining scientific knowledge. Everything has both unique and repeatable aspects, including everything we do. When we center on the repeatable in any artifact, we deem it a product of scientific technology; and when we center on the unique in it, we deem it a work of art – particularly because, trivially, we center on the unique only when we value it. At times, it has sentimental value for us; at times, its value is in its beauty. But these two possibilities are not all the available reasons for centering on the unique. For example, the value we find in the golden hands of a surgeon is neither; it rests on the value we place on the lives we hope they save. We call the owners of golden hands artists. This means, we cannot reproduce their skills in schools. (This may alter in time, and today's average medical school produces average surgeons who easily outdo the best surgeons of a generation or two ago – though many of the skills of those past surgeons are now lost, some perhaps for good.) This is a claim of progress – of making repeatable what was once a matter of genius. Any claim of progress rests on the claim of repeatability; and this claim is always questionable and the success of any challenge to it will lead to its revision, and the revised version will again be questionable. Consequently, we cannot guarantee the whole venture, though we can repair a part of it when necessary, as long as the whole venture is a going concern.

The most important corollary to all this is that the whole is unique. Forecasts of the future of the scientific and technological venture as a whole, then, can never be scientific. Yet, although hopeful, the venture faces risks. Both the hopes and the risks should undergo scientific and technological examination. The next most important corollary, perhaps, is this. Attempts to unite science and technology are worthwhile, but their success cannot be complete: every success invites new challenges. These may help keep the venture going – and hopefully with increased success.

Hence technology is, and will remain, both science and art.

Chapter 4.4

ARTIFICIAL INTELLIGENCE

The issue of *Daedalus, Journal of the American Academy of Arts and Sciences*, Winter 1988, is labeled *Artificial Intelligence* – AI, for short. It is always interesting to observe the official position of the intellectual establishment on intellectual matters, especially where much grant allocation is concentrated. AI was initially a program, perhaps a mere vision. It was the idea of a computer able to emulate human intelligence in all respects, including the ability to acquire language and to contribute to art and science. The American research establishment officially adopted the vision. This issue of *Daedalus* announces the demise of this official recognition. In a quiet revolution in the field, partial AI systems – expert systems – are replacing AI proper.

The preface, by Stephen R. Graubard, Editor of *The Academy* and of *Daedalus*, opens with an admission that sets the tone for the whole issue: the label of artificial intelligence has helped create a myth: a duplicate of the human intelligence made in the computer lab! It was "a kind of hubris, ... unbecoming and unnecessary".

That is all. The second paragraph moves from the negative side of a balance sheet to the positive side. The negative side is swiftly over with the admission that the term AI is an unbecoming and unnecessary kind of hubris. Does the balance sheet concern the allocation of grant moneys, modes of speech, philosophy, science, or technology? What is the balance going to look like? Will it illustrate yet again the famous fact that every cloud has a silver lining or will it present one of these rare cases where the spilled water miraculously returns to its bottles as wine? Was the lot of grant money that went to AI waste or intelligent investment? The matter can stand an investigation.

Not surprisingly, already the definition of AI introduces a bias. The initial definition of intelligence was behaviorist and extremely biased. It ignored the inner world. AI is any successful emulation of intelligent behavior, where by definition success is the emulation that fools the expert. To become a little less biased it should pass the test by showing that it can (1) learn a natural language and (2) create art and science (the Turing test, 1950). AI, then, is the research program for creating a computer program that can do these things – or at least the research program for creating a theory of such a computer program (1956-61). The idea is traditional; it antedated computers by centuries. Mechanism views all operations as mechanical, as following some algorithm.

The outcome was terribly clever and exciting and fruitful interdisciplinary study that involved many fields, including philosophy, logic, mathematics, psychology, and neurophysiology. AI entered the picture with efforts to use this outcome to emulate intelligent conduct. The interdisciplinary work was but a preliminary to that. This important point is lost in the reshuffle.

The expert-systems programs now on sale are computer programs that may help specialists to a significant extent. Teams of computer experts create them, each with the help of experts from any one field. Diagnostic expert systems for heart specialists may spot digitalis poisoning in a patient faster than the average specialist. Advocates of AI find vindication in this success – admittedly partial, yet quite significant. This invites the unjust ridicule, on the just argument that expert systems are mere [!] computerized dictionaries of sorts (see pp. 148, 215, 250-1 and 270; see, however, p. 78), whereas at issue is the intelligence of the program not of the programming team.

When then do programs (rather than programmers) exhibit intelligence? Does a chess program exhibit intelligence when it beats its maker in a game of chess? We do not know. As Yehoshua Bar-Hillel has forcefully argued, one can beat a chess program, however intelligent, by making a move so stupid that its makers have overlooked it. Hence, the program cannot distinguish the stupid from the clever. A minute aspect of this will come up later on, in the discussion concerning the lack of commonsense flexibility, of most computer programs (pp. 149, 196-7). This lack of commonsense flexibility is called brittleness .

From the definition to the history of the AI-related interdisciplinary study. Classical associationist psychology described perceptions as occurrences of items from a given list (of elementary items, of sense *data*) in the groupings or associations in which they occur; it describes recording or memory as the residue of perceptions, akin to grooves that passing vehicles leave. This theory has received more refutations than any other theory ever did. It is still alive and kicking. It justifies the repugnant practice of learning by rote. It animates much current research – in AI, philosophy, psychology, and education. Researchers often take it for granted. Pity.

Neurons entered in 1940. In 1947, D. O. Hebb presented neural paths as paths-and-grooves of memory – like scratches repeated on a hard surface. By then, computer memory was available. It is more like the written page than like scratches. Theories that compare the memory of brains and of computers heretically broke away from associationism. They center on retention and ignore retrieval. Plato and Freud viewed only retrieval as problem-ridden. Computer researchers took up these problems later on. The standard problems of memory are still about retention – even in this book. (See p.114, and cf. p.151.)

Already in the earliest days of computers, in 1943, Warren S. McCulloch and Walter H. Pitts had introduced a formal neural network theory to account for memory. It is a network of abstract neurons each operating one of the simple logical operations of conjunction, disjunction or negation, just as in computers. In 1956, a historic meeting took place in Dartmouth College at the invitation of John McCarthy, who introduced there the concept of AI. Allen Newell and Herbert Simon presented there a computer proof of a (trivial) logical theorem.

Marvin Minsky and Seymour Papert met there. They then cooperated on AI research. In 1958 Frank Rosenblatt described a complex of formal neurons that can learn by trial-and-error. He called it a perceptron.

Thus a quiet revolution away from associationism began. In 1961 Minsky presented the program for AI research as algorizing effectively anything recognizable as intelligent. The associationist bias is still manifest in it. Soon a model of associations appeared as a complex of perceptrons – with associations as composite trial-and-error processes, however. David Marr and others inaugurated studies of individual brain-cell functions in the late sixties. He ascribed to different kinds of cortical cells different logical functions from the quasi-associationist neural-networks formal model (see below). In 1969 Minsky and Papert produced a crisis. They showed that the program of presenting the brain as a set of perceptrons is hopeless. Consequently, a new program appeared, connectionism, a breakaway from associationism into the *terra incognita* of systemism that rests between mechanism and classical holism. (Terminology is still unsettled. W. V. Quine has labeled this position "modified holism". Mario Bunge prefers "systemism".) The leading new connectionist essays appeared in 1986 in a best selling volume, *Parallel Distributed Processing* (where the parallel processing is the cooperation of many units, and where the distributed is the not localized – like a holographic memory). The book under review here is a follow-up by friend and foe.

The transition is very curious. Associationism appeared traditionally not as a theory of the way the intellect operates, but as a part of an answer to the bigger question, what is the mind? In the late nineteenth century and in the early twentieth century opposition to it grew (mainly thanks to the idea that perception is of *Gestalten*), but it remained dominant. It began to weaken when Turing recommended replacing the question, what is the mind, with the questions, how does it operate. Answer: the mind operates as a computer. This takes us back to the initial question, answering it with, the mind is a natural computer. Hence, artificial computers that emulate minds are possible. Most of the discussion here centers round technical questions related to Turing – although with an eye on more basic ones.

The book's structure. Its 310 pages contain 14 chapters. The first chapter sets the concern within the AI community and the second chapter sets its historical and philosophical background. Two chapters then explore the concept of AI and five discuss its main issues. The next chapter, its authors claim, transcends these issues, and possibly renders the rest of this volume obsolete. Then comes something that is indisputably substandard, nicely leading to an attack on AI as humbug, followed by a counter-attack. The close is an overview by John McCarthy, the initiator of AI.

Here then is the summary; comments are in square brackets.

1. Seymour Papert, professor of media technology and director of the Learning and Epistemology Group at MIT, "One AI or Many?"

The two schools of AI thought are the old-style programmers who emulate brain processes on the computer in line with the research program that Marvin

Minsky has outlined (see below), and the new-style connectionists, who study brain physiology (see below). [This characterization has the merit of brevity and crispness. It is inaccurate, though: we all try to emulate, hoping to thereby understand.] Each promises full success in the very near future. The success that the initial old-style program promised is gone: all grand-scale promises are gone; expert systems now replace AI. Papert says, he is no party to the dispute. It rests on the category mistake, he adds [erroneously], that feeds on "the quest for universality of mechanism" (p.7). Each side can try to be as universal as possible without limiting the other, as they operate on different levels, Papert suggests [erroneously: levels are not given but determined]. The conflict between the two AI schools, old-style and new-style, was [not intellectual but] financial. They were quarreling over grant moneys (p.7). Minsky and Papert have refuted the connectionists' program. Now the connectionists hope to achieve great success in no time with the hope to employ parallel distributed processing. This is a waste of time and of scarce grant money.

[The paper begins with a plea for pluralism and ends by condemning one of the two schools. The complaint that its fans make exaggerated claims to secure grant money, is evened out by the editor's admission that all AI people do that.]

2. The Dreyfus brothers, Hubert L., the philosopher, and Stuart E., the engineer, "Making a Mind Versus Modeling the Brain: Artificial Intelligence Back at the Branching Point".

This is a history of the dispute over AI as stemming from the older, traditional, philosophical one, over the nature of minds, between the mechanist-reductionists and the anti-reductionist holists. [It is difficult to know what the disagreement is about, as the reference here is merely to two philosophical schools. No single empirical study ever stays exclusively in the confines of the ideas of either. With scarcely any communication across school lines (both in philosophy and in AI studies), an unschooled reader will despair. The Dreyfus brothers deem the messages of the two philosophical schools almost identical. This makes understanding them hard even for the philosophically adept, as is their confusion of systemism with holism.]

What is the disagreement? The answer remains [vague to the last]: it is between competing research programs, the mechanistic-reductionist and the holist. [The rest of the Dreyfus paper is not clear. Catch phrases are of no help here. What views are characteristic of one side and not of the other?] Remarks like, "Minsky and Papert were so intent on eliminating all competition ... while completely ignoring ... " (p. 22) give the tone to the rest of the paper. [They tacitly validate Papert's complaints – as if research without grants is impossible:] "... was discredited along with hundreds of ... research groups ... research money dried up ... had trouble getting his work published ..." (p.24). [When publication counts in competition, it gets hard to publish. Publication should please readers, not grant seekers. How? No answer. Pity.]

3. Robert Sokolowski, philosopher, "Natural and Artificial Intelligence".

A printed page is artificial and seemingly intelligent. Intelligent reading is intentional; the printed page is not. Nor are computers. [This sounds like a criticism; it is a trivial corollary to every theory in the field except behaviorism.]

4. Pamela McCorduck, author, "Artificial Intelligence: an Aperçu".

A readable text is intelligent. So is a program generating or transforming it. This holds for computer art and for expert-systems. This throws light on notions of art [and of expertise]. [In general, whatever is open to formal description is open to reproduction by rote and to delegation to machines. Art goes beyond that. This paper differs from its predecessor; their disagreement is tacit. Pity.]

5. Jack D. Cowan and David H. Sharp, mathematical biologist and theoretical physicist, "Neural Nets and Artificial Intelligence".

This chapter is on the rise of artificial neural nets. (Terminology: "hard-wiring" is programs that are built in the computer and "soft-wiring" is programs that are added by software; "top-down" is deductions from given hypotheses in search for new information, and "bottom up" is search for new hypotheses to explain given information.) "There is still a very long way to go before any kind of truly intelligent robot can be produced" (p.114). Technical problems aside, can intelligence lack all intentionality? If not, then, how can computers think intentionally? [This is the mechanistic version of Darwinism: selection leaves only those systems that behave in seemingly purposeful manners.] Computer technology selects. Can it "telescope billion years of evolution ... into a few decades of neural-net and AI research ... ? Until we understand how ideas and intentions" inhabit "the human brain, rapid progress is unlikely. On the other hand, developments ... We predict that the top-down approach of conventional AI and the bottom-up approach of neo-connectionism will eventually join to produce real progress in ... experimental epistemology, the study of how knowledge is embodied in brains and may be embodied in machines."

[It is hard to assess all this, as the division is largely arbitrary between programs built into the computer and the ones fed into it afterwards, between hard and soft wiring. Nor is the difference between the two schools merely methodological. Possibly, they differ on orders of priority too.]

6. Jacob T. Schwartz, mathematician, "The New Connectionism: Developing Relationships Between Neuroscience and Artificial Intelligence".

Unlike computers, neurons work in parallel and may have local and non-local functions. The clue to brain function theory is that "mental (especially sensory) processes seem to be of very restricted 'depth', in the sense that not many successive elementary neural reactions are required to form the higher level reactions that the brain generates. There is simply no time ..." The time it takes a brain to perform one basic operation is comparable to the time it takes it to perform some complex operations. [This is not much of a clue, since a switch may be most simple or very complex. This makes associationism *passé*, as it permits only one sort of simple switch.] Another clue is the way the brain maps visual and tactile sensations. At first, it retains them in simple geometric images and then it transforms them [to God knows what]. Still another clue is from embryology. In the embryo, brain cells send potential synapses. The random manner in which this occurs suggests a high degree of non-specificity of brain cells (they perform diverse functions depending on their environment), although the (morphological, biochemical, and other) differences between the kinds of neurons in the brain, and even their numbers, suggest that they have specific functions. This apparent specificity invites study. In any case, "biological systems are not wired precisely enough to support this extremely delicate style of information processing" that computers possess (p. 132). The theory of the cerebellum as a set of simple mechanisms capable of learning by conditioned reflex is of help, but all "theories of the origin of neural functions" based in theories of learning are still useless. For, "we know hardly anything yet about the actual locus or mechanism of other memory storage within the brain and even less about the way memories are modified to accomplish abstract learning" (p.134). [The author is not alone in considering the theory of conditioned reflex attractive while admitting that it is useless. This is fascinating. The author notes deviations from the conditioned reflex theory in the process of selection and the modification of information.] The basis to the project, the ideas of thresholds, excitation, and inhibition, should undergo a critical reexamination (p.136).

All analogy between computer and brains is conjecture (p.137). Computer simulation may help neuroscience, neuroscience can hardly help design better computers (p.136). There is an exception to this already. Does the brain function as analog or as digital? [The distinction between analog and digital is problematic. We call analog a processes that a continuous function describes; stepwise functions describe digital ones. The question is vague, as matter is atomic: the answer to it depends on degrees of accuracy of measurements.] Analog computers are less stable and accurate than digital ones, but in some respects more brain-like, and so they may be preferable for "the processing of streams of incoming sensory information like audio information or moving images" (p. 139). "Consequently, there is reason to hope that analog networks can process sensory data in a manner that will profit from ..." [There is trouble with the quotation here. Its logic should demand that it should go thus: there is hope that computer data processing research will profit from studies of neuroscience. The quotation gets somehow lost in detail (p.139).] The last two pages express hope that the two branches of AI will one day unite.

7. George N. Reeke, Jr., and Gerald M. Edelman, both biologists, "Real Brains and Artificial Intelligence".

Two errors obscured the similarity between the goals of AI and of neuroscience. They comprise epistemological conclusions from two views. One is the view of Turing and Church of computers as universal problem-solving machines (namely, the taking of idealized cases as if they are real, p. 148). The other view is of the brain as a collection of units that exchange chemical signals (namely, the mechanistic view rather than the systemic one). "As biologists seeking to understand the nearly dogmatic neglect" of biology by old-style AI researchers, they ask, what are the AI researchers' goals and methods (p.144)?

The success of old-style AI, the successful application of physics and engineering to computers, rests on taking some categories and some information about them as basic and as not given to analysis. This leads to the proposal to consider the possibility that perception and intellectual processes are all algorithmic, computer-like in principle. [The authors identify old-style AI with programming and conclude that it is the view of all intelligence as programmable and of intentions as irrelevant.] This is not evolutionary, as it prevents explaining the rise of characteristics – such as categories in the brain – as the rise of adaptive mechanisms. Biology "might contribute to further progress in AI" (p.145) when AI researchers will try to accommodate evolution and describe machines that emulate not only human intelligence but also its evolution.

In 1961, Marvin Minsky outlined the old-style AI program as that of searching for effective procedures for search, pattern recognition, planning, and induction. This includes some objectionable "epistemological assumptions" (p.146). [Of course: finding effective procedures for scientific progress is a traditional dream (inductivism). If Professor Minsky will find such effective procedures, then he will already have made a tremendous contribution. When he will implement it, he will do away with the need for researchers! The traditional view of scientific research as algoristic – as producible by a strict formula – is not in the least evolutionary. (The same holds for the theories of associations and conditioned reflexes and so on. Their enormous attraction despite their obvious faults is thus explicable as the attraction of inductivism.) Many authors, chief among them is, perhaps, Popper, have observed this in the twentieth century. How is that to effect AI research remains to be seen.] The assumptions of old-style programmers are limited. Their algorithms do not take care of the limitations that they involve and therefore they are "brittle" or inflexible. When they transcend their limits, they "crack". The best solution to the problem of brittleness is only a less brittle program; AI is thus a mere ideal and not the best (p. 149 ff.). [See end of this review.]

Parallel computation is so complex that the only way to learn what a programmed parallel machine can do is to let it run; each machine is unique and thus not really programmable (p.152). [Its results are not repeatable and hence unscientific.] Current computer models of the brain contain too many specific unrealistic assumptions (p.153). Computers are passive. Programmers determine in advance their rules and procedures (p.154), so that they cannot adapt (p.155). [This smacks of Lamarckism, as much of this book does; it considers evolution as due to individual adaptation rather than to selection among mutants.] Live neural systems obey no strict hierarchy. So, no single neuron is indispensable. There is an enormous diversity of kinds of neurons, especially in the more evolved species. "Only patterns of response over many neurons can have functional significance" (p.156), and produce responses that excel not in accuracy, speed or efficiency, but in possessing overlapping wide-range "repertoires" of functions. These are more suitable for unprogrammed systems in unfamiliar hostile environments. The authors have constructed automatic "selective recognition systems" that "address some of the problems of the standard AI paradigm by avoiding preestablished categories and programming altogether" (p.161). [What is "the standard AI paradigm" and what are its problems they do not say.] Each program makes a computer in the system simulate a neuron, and tells it nothing about functions or about programs. One resultant automaton could "act upon the environment to form a complete autonomous behavior" (p.161). [These results are unscientific too.]

[Here is the place to turn back to Papert's critique (p.11). "Although its models use biological metaphors, they do not depend on technical findings in biology any more then they do on modern supercomputers." This invites comments from opponents, especially from those who dismiss the new-style connectionist work. Understandably, the authors describe their automata too sketchily to permit assessment. Do the parallel computers really start with no program? Or do they modify given programs? Is their learning to categorize not also by modifications? Papert's essay is of no help here: his attitude is cavalier and associationist. Modification and emergence differ greatly, and this demands study. To that end, see the authors' contribution to *Proc. Am. Phil. Soc.*, 134, 1990, 36-73, and compare their statements here with their conclusion there, that is much more clearly programmatic.]

8. W. Daniel Hillis, the inventor of connectionism (as his MIT doctoral project!), "Intelligence as an Emergence Behavior; or, The Song of Eden".

We understand neither emergence nor intelligence, yet the idea of an emergent intelligence is attractive as a possibility of "constructing intelligence without first understanding it" and without yielding to the traditional mechanistic view of us as machines. The shift from sequential to parallel processors "is not a deep philosophical shift, but it is of great practical importance, since it is now possible to study large emergent systems experimentally" (p.176). [Very nice.]

Human song is older than speech. Songs are parasites. They survived by the specialization of the moods that they expressed: their survival value was in their usefulness to the community as means of communication and thus as levers for intelligence. What was the size of the storage of the brains of the first humans? Living memory is distributive – non-localized, as in a hologram – and so, its size is hard to assess. The author assesses our storage capacity as surprisingly small. He cannot assess the value of sensory-motor functions for the growth of intelligence.

Little understanding is necessary for construction. This is why constructing systems, including emergent artificial intelligence, is now in style. [Let me quote here a whole passage, as I find it too sketchy for a critical assessment.]

"I have recently been using an evolutionary simulation to evolve programs to sort numbers. In this system, the genetic material of each simulated individual is … a program specifying a pattern of comparisons and exchanges. The probability of an individual survival in the system is dependent on the efficacy and the accuracy of this program in sorting numbers. Surviving individuals produce offspring by sexual combination of their genetic material with occasional random mutations. After tens of thousands of generations, a population of hundreds of thousands of such individuals will evolve very efficient programs for sorting. Although I wrote the program for the simulation that produces the program, I do not understand the detail.... If the simulation had not produced working programs, I would have had very little idea about how to fix it" (p.188).

"The result would be not so much an artificial intelligence, but rather a human intelligence sustained within an artificial mind." "Of course, I understand that this is just a dream..." (p.189).

[This is thought-provoking. The reification of songs is challenging. Is a bit of information a thing? The author does use a questionable analogy, as he speaks of the evolutionary value of the survival of a cell and of the intelligence of genes of the simulated population. Ideas may be mere metaphors unless they become testable theories.]

9. David L. Waltz, computer scientist, "The Prospects of Building Truly Intelligent Machines".

Old-style algoristically programmed AI machines obey obsolete associationism. As the idea that all learning is by trial-and-error replaced it (p.195), a crucial question cropped up: "Given the immense range of possible situations a truly intelligent system could find itself in, and the vast number of possible actions, how could the system ever manage to search out appropriate goals and actions?" (p. 196) [This gives the false impression that the new style offers a solution to this question. Pity.] New-style parallel distributed machines learn to modify programs with respect to a goal. [This is a dismissal of the crucial questions, what generates goals and how?] They have associative recall and tolerate faults. Old-style algoristically programmed AI machines do all this much more slowly than the new-style ones (p.199). They learn each item unambiguously. This is logical [!], but not the human way (p.201). Hybrid machines of old logical reasoning and new associative-memory learning are more adequate (pp. 201-202). [Notice the false implicit equation of algoristic and logical thinking, as well as the recognition of error as illogical. This is perverse.]

New-style machines work in "a process much more like lookup than search", in a process of looking up "items ... more like representations of specific or stereotypic episodes and objects than like rules and facts" (p.197). A new-style "associative memory" diagnostic program prescribes lookups of records of previously diagnosed patients to select ones with symptoms most similar to those of the patient now under examination (p.198). Similarity between patterns is a statistical function averaging over a measure of distance definable between every pair of items. The statistics may iron out faults, yet at the cost of uncertainty [especially since real lists of symptoms are usually much too short and inaccurate], so that the results of such systems invite checking by old methods (p.200). [In simulations this may be so. In live cases, experts must check the computer's results. All this is irrelevant to the promised revolutionary learning psychology. Disappointingly, the author employs Minsky's utterly unrealistic presentation of society as a set of freely interacting individuals (p. 201).] "Researchers have identified perhaps a dozen distinctly different learning methods." [This is very exciting, except that the author says nothing at all about any of them except to name one of these, the psychologically least interesting. Pity.] The author mentions a learning method whose input and output are given in advance (p.204). Its task is to modify a program by trial-and-error and end with one whose output is right.

"The central problem ... in the connectionist" new-style system is the "credit assignment problem". Considering as units the individual neuron-like elements, what rewards and punishment should they receive? What system should allot it to them? [Rewards are relative to goals. The system supposedly creates its own goals, so it should thereby determine rewards. It cannot. It seems able to do so as it confuses the source of the rewards and the punishments with their receiver. The central old problem would not go away.] The management of the "static part" of the problem is through testing units upon their activities [in the old-style]. When they act correctly, they strengthen their connections and *vice versa* (p.205). The "temporal part" of the problem – the problem relating to delayed rewards – is more difficult as the system must remember its past states, to analyze and judge them, and then allot rewards and punishments. [These are metaphors for the strengthening and weakening of the connections of the individual neuronlike items upon success or failure to perform adequately by set criteria. Taken literally, the metaphors are hilarious. Memory appears here in the defunct associationist manner. The rest of the essay, perhaps most of it, is left out here. It defies summary. The author's presentation of elementary learning psychology is too diffuse.]

10. Anya Hurlbert, MD, AI researcher, and Tomaso Poggio, brain and cognitive scientist and researcher in computational vision, both in MIT, "Making Machines (and Artificial Intelligence) See".

"Why ... have we balked at calling vision intelligence?" If AI is to be interactive, it must include robotics, namely, "the study of how to join perception with action. ... In its beginning AI research ... excluded both vision and motor control from the realm of intelligence" (p.214).

Old-style AI follows the Newell-Simon hypothesis that intelligence is "a physical symbol system". Hence, humans are computers (p.215). By contrast, new-style AI is *Gestalt*ist. (P.214; cp. p.43, last note, on the influence of the post-*Gestalt*ist J. J. Gibson.) "Leaps of intuition and instant insights are at one extreme, ordinary perceptual skills such as speech recognition at the other: these are the powers of the mind that traditional AI is hard to put to model. ... Evolution has spent millennia perfecting such unconscious talents" (p.217). Machine vision is a synthesis of the best in AI, new style and old: "the computational approach". It "describes exactly what information a system receives and what information it puts out, and seeks a computation that will transform the input into the output" within the recognized constraints of the normal living visual system. "Machine vision has turned the search for constraints into a science of the natural world" (p.218).

Most of the essay is a competent summary of the Marr-Poggio theory of vision (pp. 218-230). [Whether it is too brief for the uninitiated or sufficiently detailed and succinct, it certainly is as short as possible. Rather, a report is missing of what is specific to that theory. It is that it ascribes a central role to approximate solutions to problems. Trial-and-error is not the same as approximation. A textbook trial-and-error system is that of seeking a needle in a haystack. Approximationism is a trial-and-error system that uses earlier trials as means for raising degrees of precision. One of the most impressive experiments in vision theory is Bela Julesz' refutation of Gibson's theory of stereovision as the result of comparisons of images of contours in two eyes. Julesz made a static stereogram of randomly distributed dots. By Gibson's theory, but not in fact, the absence of contours there should prevent seeing depth. Presumably, dots on the two images appear first as close enough to seem identical, and this enables observers to read small local variations as depth. Here the two steps are inherent in the process of intelligent surmise: unlike textbooks trial-and-error, approximations use superseded stages. An integral part of the final recognition is the partial achievements of the first step and its correction in the second.

This aspect of pattern-recognition theory is disagreeable: different paths may take an observer to different intelligent surmises of the same input. It is an inherent aspect of the situation; it is not specific to machines or to vision; it happens in medical diagnostic practice. This explains the proposal to repeat a diagnostic process *de novo* (N. Laor and J. Agassi, *Diagnosis*, pp. 152, 203, 218). Associationism takes every process separately: it is unrepeatable and so one cannot repeat it *de novo*. Farewell to associationism.

Back to the Marr-Poggio theory. The contrast between associations, textbook trial-and-error, and approximations is essential for the presentation of its achievements as achievements. This is sadly absent from the Marr-Poggio paper and from Marr's book. (See my "The Wisdom of the Eye", *J. Soc. & Biol. Struct.,* 10, 1987, 408-13.)] The penultimate section of the essay contrasts the three approaches, algoristic old-style, connectionist new-style, and computational revolutionary-style. The third is that of the "true believer in levels of understanding". [These levels are levels of approximation. See p. 224.]

The heart of the matter is "the single question: What is the final goal of the enterprise? ... is the goal ... to build intelligent machines? to understand how the brain is put together? to describe the structure and powers of intelligence as a free-floating entity, tied to neither brain nor machine?" (P.232.) Connectionists seek a model of the brain, seeing the resemblance between humans and machines as something abstract (p.233). Traditional AI people construct machines and use brain physiology only to that end; on the computational level they stay on the lower, algorithmic level [where any algorithm that does the job will do, in indifference to the total picture]. By contrast, as connectionists create machines, they ignore all computations and forget that "many of the networks work only because" they contain "the necessary computational analysis" in the style of the other school (p.233).

The final passage of the essay offers an example. The conduct of a fly shares flexibility of goals with that of a driver in stress. [No. It is the flexibility of means. The goal is fixed: escape. The means are flexible just because the goal is fixed, and they alter rapidly in response to rapidly altering conditions.] "Finding the right representation is what computational theory does. ... Machine vision shares the dream of building a machine that can learn ... . Will we be satisfied with simply building machines that can learn? ... We humans should not forget that those who aim to build intelligent machines have the whole future to disprove their starting hypothesis: that intelligence can be reproduced on a machine." [This is a serious error. The old AI theory is programmatic and thus metaphysical. It would be confirmed by success but not be refuted by failure, since "the whole future" is at its disposal for trying again. These sophisticated researchers' ignorance of Popper's seminal writings is regrettable.] "... future more sophisticated machines ... might look fondly back at the days when machine vision, which combines all levels of understanding human intelligence, brought their parents together." [Nice touch.]

11. Sherry Turkle, MIT sociologist, "Artificial Intelligence and Psychoanalysis: A New Alliance".

"If psychoanalysis is in trouble, artificial intelligence may be able to help ... one of the ways computers influence psychological thinking is ... that ... computers provide science of mind with a kind of theoretical legitimation that I call sustaining myth" (pp. 241-242). [I find all this incomprehensible.] AI chess programs have refuted behaviorism (pp. 242-4) and the idea that humans possess autonomous selves, since AI chess players are "more like Freud than Skinner" (pp. 244-6, see also p. 261 ff.). [I do not see how.] Old-style AI initiator Minsky follows psychoanalysis, says Turkle (p.246; see also pp. 259-60 and n. 17). [Minsky disagrees.] "The two AIs, rule-driven and emergent, logical and biological in their aesthetic, fuel very different fantasies of how to build mind from machine." Old-style "information processing put AI in a distant relationship to psychoanalysis" and new style connectionist emergentism parallels the move from Freud to Melanie Klein in taking seriously object-relations. Thus, "when the stuff of AI is expanded to include ... active and interactive inner agents, there is a starting place for a new dialogue between the psychoanalytic and the computer culture" (p.248). [This is mumbo-jumbo.]

12 and 13. Two philosophers cross swords: Hilary Putnam, "Much Ado About Not Very Much" and Daniel C. Dennett, "When Philosophers Encounter Artificial Intelligence".

Putnam: AI research is parasitic on computer science and its advocates mislead the public. Natural languages and science are not algoristic.

Dennett: Putnam mentions troubles and polarizes positions to all-or-nothing, in order to make the trouble tip the balance to the nothing. AI shows that memory problems involve storage and retrieval. [This is a howler: librarians know that anyway.] AI people test theories. The philosopher Putnam dislikes tests and AI.

[Dennett's paper is defensive, *ad hominem*, and untypical of his standards. He evades the challenge. Putnam thus wins though his judgment is facile too. Admittedly, AI was hubris and deception and a failure. Yet, even as a failure it led to some interesting and useful results. Whether expert systems are AI or computer programs is a matter not of definition but of history.]

14. John McCarthy, doyen of AI, "Mathematical Logic in Artificial Intelligence".

McCarthy presents the standard computer as comprising three levels of sophistication. The first is simply that of executing commands. The second is that of operating with sentences (pp. 299-300). "The third level uses first-order logic as well as logical deduction" (p.300). "Examples ... used commercially are 'expert-system shells'", namely, "computer programs that create generic expert systems" (p.301).

This level is not practical, as there are no programs to induce intended inferences from commonsense to sufficiently many simple everyday corollaries from given statements (p.298). There is some progress in this direction, and there still remains a fourth level, wholly in the future (pp. 301-2). It is learning. "The present way of 'teaching' computers programs amounts to education by brain surgery." The hope is that computers will be able to learn as we do.

The difficulty is this. "The fourth-level systems require extentions to mathematical logic" (p.302). Traditional logic is monotonic: adding to the premises of a valid inference leaves it valid. But "some important human commonsense reasoning is not monotonic" (p.303). Many researchers in the field share this [erroneous] view. Example. The inference from "he was thirsty" to "he drank the water" is valid. Now extending the premise to say, "he was thirsty and he learned that the water was polluted", we receive the opposite valid conclusion: "he did not drink the water". Some people try to save logic by the addition of rules of probability implicit in the context, but these are very doubtful (p.303). McCarthy agrees with Quine that first-order logic should suffice for all formalization of ordinary discourse. Also that there is no need to refer to ideas, as we can scarcely say of two individuals that they share an idea. He nevertheless sees a need for a special logic (pp. 303-5), one that will permit "jumping to conclusions on the basis of insufficient evidence" (p.307). He pleads for "incrementalism, or modesty" (p. 307) that will permit AI researchers the privilege of trial-and-error, which we all have anyway. The example for trial-and-error he gives (p. 308), however, seems to be associationist and refuted: a baby first says "mother" thinking it is a singular and then learns it is a universal. Babies do not know this distinction. The essay – and with it the book – ends with the hope that AI will help the evolution of some sort of meta-epistemology akin to traditional meta-mathematics.

[This is the heart of McCarthy's general outline of the current situation. I do not know how to respond to it. I find it magnificent, broad, enlightening and simple overview. Yet, on matters of basic logic it is confused. This embarrasses me. I cannot complain. It is we, the philosophical community, who are at fault. But a person of McCarthy's stature should know what standard logic is and what the best in logic is. AI needs natural deduction theories – preferably Popper-style. (See Bibliography in P.A. Schilpp, ed., *The Philosophy of Karl Popper*.)

Ordinary reasonable thinking is monotonic. The examples McCarthy takes are not of violation of logic but of statements that describe the contexts of his examples. They are implicit yet obvious all the same. We correct them easily and naturally when we alter the context. For example, the premise in "he was thirsty, so he drank the water" has to be explicit: "he was thirsty and assumed that the water was drinkable, so he drank the water" and then one may contrast it with "he was thirsty and he learned that the water was polluted, so he did not drink the water". This is logical. Indeed, logic began as the theory of dialectic, and dialectic is the art of making explicit context statements and criticizing them and then hopefully giving up the ones we refute. "Jumping to conclusions on insufficient evidence" (p.307) is thus the norm – even though most writers on knowledge and on probability aim at overlooking this. McCarthy seeks a kind of meta-epistemology that is already here. It is the theory of conjectures and refutations, of progress by trial-and-error. Popper has presented it. The philosophical establishment is desperately overlooking it for half-a-century.

Cognitive psychologists are in no better shape. *Science*, 30 October 1987, Volume 238, presents a paper on teaching reasoning. Four famous cognitive psychologists wrote it to show that most people, including samples of science students, reason with highly brittle – inflexible – algorithms. They present alternatives similar to the ones that McCarthy presents for programming students with modes of reasoning. They simply ignore trial-and-error.

The incremental attitude of McCarthy will easily show that the fourth level of computers (all-purpose multi-lingual meta-epistemological ones) that he says is wholly in the future, does exist, partly, already now, since computers can spot programmers' mistakes and since they can easily learn to spot contradictions. And there is a great demand for fourth-level machines. There are partial expert systems diagnostic software programs on the market, and they invite coordination and merging into a comprehensive computer-assisted diagnostic service. It can have a partial success only on the fourth level. The service should also include competing expert-systems. Commonsense is therefore essential for it. A combination of formal logic with commonsense requires a better understanding of natural deduction. Interchangeable possible contexts – para-texts – for a given text, and rules for altering contexts are required and are available, at least partly, in current expert systems, but not systematically at all (see N. Laor and J. Agassi, *Diagnosis*, 1990, Ch. 2). Sets of contexts may require a meta-text that would embed some general metaphysical assumptions (see my "The Nature of Scientific Problems and their Roots in Metaphysics" in my *Science in Flux*, 1975) and some technologically significant blanket suppositions (see my *Technology*, 1985, 59-64). A first step would be programming a meta-program for a brittle program to attempt tentatively different programs – different possible contexts – for the mending of brittleness for a while.

Constructing a fully mechanized intelligence is impossible. Also, it is immoral to use one, as responsibility for action is always human. But there is the possibility to do so in limited contexts. Dead languages and dead art are open to algorism, and researchers can use computers in their researches only because they formalize some of their procedures (in some brittle ways). In the very early days of computers (in the early 'fifties) Yehoshua Bar-Hillel argued that fully automatic machine translation is impossible, yet all the same he tried to formalize natural languages – taking it as a project not open to full success but worthwhile anyway. It is, of course, impossible to feed a computer different levels of computation without the extensive use of a meta-linguistic program. As the computational approach does this already anyway, it seems clear that there is a prejudice against loading the meta-language and against second order logic, which prejudice spills over to computer science and to AI. But, clearly, inexact and undeveloped and perhaps as objectionable or redundant as Quine and McCarthy say it is (p. 306), how come so many individuals are willing to emulate the human cognitive process even when it supposedly goes against logic and yet decline the use of hierarchies of languages and second-order logic and alternative frames to play with? Even non-scientists and non-logicians do that. Why not view the "expert-systems shells" as brittle second order logic? Why not broaden the study and develop ideas that can help school children to learn and then help programmers too? The argument from redundancy is the least effective: our lives are full of redundancies and computers have a huge share of them in any case.

AI as the Golem myth is neither promising nor interesting. As human-machine (wet-dry, C/Fe) interactive systems it can excite. (See my "Heuristic Computer-Assisted, not Computerized: Comments on Simon's Project", *J. Epist. & Soc. Stud. Sci. &Tech.*, 6, 1992, 15-18, and Simon's reply there.)]

Chapter 4.5

PHILOSOPHY WITHOUT SCIENCE

0. THE ART OF AUTOBIOGRAPHY

It takes training to open up to the enchantment of pondering the world through different people's eyes and it takes training to benefit from this experience. This is the benefit from reading a diary, a correspondence, and even an account of the table talk of a public figure. Biographies – especially autobiographies – facilitate this practice. They may be good reading as art or as stories, especially stories of great deeds seen through the eyes of a participant. They may also display role models. Bertrand Russell has noted the importance of the role models in Plutarch's famous *Lives*. Somerset Maugham said the same of Kipling's fiction. Intellectual biographies may offer all this and more. They may satisfy the wish to learn about lives of great thinkers. (Outstanding examples here are Ruth Borchard on John Stuart Mill and Constance Reid on David Hilbert.) Their correspondences or autobiographies may supply details about admirable ideas. (Outstanding examples here are many correspondences – between the Mozarts and between the Van Goghs, between Malthus and Ricardo and between Kelvin and Maxwell – as well as autobiographies, such as those of Mill, R. G. Collingwood, and Buckminster Fuller.)

Lives of thinkers may be instructive also as to the life of reason. Learning about outlooks of some thinkers may help understand their intellectual attitudes. (Freud's and Russell's autobiographies are useful in this respect.) Learning about people's intellectual backgrounds may help understand their ideas. (The intellectual autobiographies of Einstein, Planck, and Schrödinger stand out here.) Learning about their interests and their growth may help too. (The autobiographies of Darwin and of Kandinsky stand out here.) All this may be enormously valuable. Information about inner struggles may also be of interest, and this requires description of the character of the struggling individual as background. (A comparison of the life of Beethoven with that of Schubert is intriguing. The mental agony that the struggle of Faraday involved has not yet drawn enough attention; those of van Gogh, Hugo Wolf, and Virginia Woolf are legendary.) This holds for all those whose unusual characteristics have helped them in their quests. (Faraday, Edison, and Jack London were amazing individuals.) The ability to participate in debates and the readiness to do so – or its absence – may have played a significant role in the intellectual life of some great minds. This is of particular interest in the case of people whose life stories reflected significant facts better than their own views on these facts – if, for example, they added to great disputes while their view of science as inductive made them belittle all disputes. (This holds for the dazzling autobiographies of Ben Franklin and Joseph Priestley.)

This contrast between a person's conduct and opinions is quite general. Collingwood recommended keeping the records that we use, as we may have misread them and so we should allow our successors to correct us. His own autobiography is a classic in many important ways, one of which is its display of his own misreading of his own records: he did not see the contrast between his conduct and opinions. It is no accident that his greatest contributions to scholarship, his philosophy of archeology and his logic of questions, find their best articulations in his autobiography. He recommended writing in the dialectical method; probably he never noticed that he seldom followed his own recommendation. Now, the idea that retaining the records may help correct their misreading may suggest that a diary is the best form of an autobiography. Not so. We cannot avoid choosing, he observed, and thus possible misreading of our records. So, we should not destroy our diaries, but use them selectively in writing autobiographies. He thus came close to endorsing fallibilism, which suited his temperament, as he expressed it in his famous essay on Croce. But otherwise, he was captive of the traditional view of science as certitude. This explains much of the contrast between his conduct and opinions. We may improve our readings of lives of intellectuals by asking, were they fallibilists? Or did they think that in some ways they were above error? How did this influence their vision and conduct? These questions improve our reading of texts – even of texts that have no answers to them.

A case in point is the recent autobiography of Ted Honderich, *Philosopher: A Kind of Life* (2001). It says too little about its author's views about his fallibility. He is a minor luminary on the scene of British analytic philosophy, and so he shares this school's view of commonsense as more reliable than science – perhaps as infallible. The book sells well. It has had, if not *succès de circonstance*, then *succès de scandale*, as it discloses sordid details. His love of wine, women, and song, has drawn more attention than his ascent on the academic totem pole. Yet, the academic side of his story merits more attention.

1. WINE, WOMEN, AND SONG; ACADEMIC POLITICS

Wine fills pages here but plays a marginal role. "My inclination to wine, a lot of it, had increased." (228-9) "My usual daily consumption was two-thirds of a bottle of white wine in the evening." (249) "I got drunk, and fell into what was supposed to be amiable … badinage with the guest of honour. His final eruption was Vesuvian. … I delivered an abject note of apology. … As you will gather, wine was in my life." (232) He drove under the influence and once he had his license revoked for a year. (274) Sex is marginal in the same way. "I … keep quiet about the number of my lovers, and how far I am ahead of Russell and lag behind Freddy" – Professor Sir Alfred Ayer to you and me. He describes his position as missionary in all matters sexual except loyalty. (28) (127) He makes light of loyalty in matters sexual, not in matters professional. He is loyal to colleagues, especially Ayer, whom he deems his mentor. Appearances aside, he demotes sex. Of love – of any sort – he says much too little, and still less about songs and of partying. He describes himself as at times depressed and at times a jolly reveler. He is not only a distinguished professor and an amateur politician, but also a proud member of the Hampstead Set. He finds this worth noting; some of his readers may not.

Surprisingly, Honderich reports no knife-in-the-back. One minor unpleasant case concerns conduct less open than it should be among colleagues. (312) Another concerns a sneak attack. (321) His report about stench – "*odium academicum*" (306) – concerns muck that is on open display. In comparison to my own experience, it is tame. So is a remark that the college provost made to Honderich upon his retirement: "dealing with a department of philosophers was like herding cats", he said. Better cats than hyenas.

Still, not all is well. The trouble lies with loyalty. Honderich says he cares for under-dogs. Yet, he avoids condemning the injustices of the system. He got powerful academic and editorial jobs through camaraderie, as he frankly explains. He was on his way to success well before he graduated. In professional philosophy, social skill is still the main route to success. It looks fine to him but just ugly to me. He seems ambivalent and envious. Are Oxbridge graduates the best? "It would have taken a great faith in a certain conspiracy theory to resist a certain proposition – that Oxford and also Cambridge graduates contained, along with all the noodles, a somewhat higher proportion of the able or exceptional than the graduates of other institutions." (140) He observes that "there was more to the story" of their success than skills. Class and connections contribute to it. Oxbridge Philosophy is "the best mutual-support and approval-conferring society in academic life". "Some had very low foreheads, but almost all got a hearing." (148)

Honderich got his hearing too; not always, but often: he is not so high on the totem pole. The most famous theory of justice today is that of Harvard big cheese John Rawls. He says, justice is fairness, with no adequate response to Anatole France's famous dictum, the law in its majesty forbids sleeping under the bridge equally to the poor and the rich. Honderich has refuted Rawls with a simple argument: Rawls supports all law, even racist or Communist, as he demands only its even-handed application. (209-13) The learned public ignores this. The established regularly stonewall good ideas for political reasons – even if they value them, as they show when they finally endorse them with fanfare soon after the demise of their originators. They were unjust to Honderich. (He shares this complaint with a few other critics of Rawls. The criticism is obvious, and the established repress it only because they are its targets. Had Rawls applied his ideas to his own case, he would have protested.) Overall, however, Honderich got a fare share of academic glory. He gratefully recognizes this fact.

2. THE BOOK

The book's popular success owes nothing to its author's academic input, whatever it is. It is in part due to its being the story of a lad from a far corner who won fame and fortune in the metropolis. He became an eminent philosophy professor, a leading editor, a media intellectual, and a member of the Hampstead set. The book is an ornate personal diary. Its mood is reminiscent of that of some famous diaries. So, future social historians may like it. Its style is awkward and high, at odds with its spicy tale. It is episodic, full of *aperçus* and asides and heaps of pointless details, and laundry lists aplenty. It reports precise addresses long vacated. Two court cases use up pages and pages, although two paragraphs would do amply. If anything in them is relevant, it is the grumble about some abuse of the state welfare services. Since Margaret Thatcher is Honderich's *bête noire*, he might have noticed that her argument includes objections to such abuses. They helped lend her convincing power. But he does not concede an inch – to her or to anyone else. This book praises impartiality and displays relentless one-sidedness.

Honderich is frank, though. He reports odious matters, private and public. They did not impede his success. Nor did his reports about them. Otherwise, he would hardly have sent this book to the press. He presents himself as a brave heretic; he is not. He poses as an "apostle of violence" (297) and as a sharp critic of democracy. Reporting on a conservative victory in some British general elections, he says, it was sham. (339) Sadly, this sort of frivolity is popular fare. Above all in academe. Political information is usually more dependable the less credulous its intended recipients (presumably) are. The better-educated democracies have leaders who lie less blatantly and less frequently than elsewhere. This holds seldom for the intellectuals who play amateur politics. They describe politics in black and white pictures that deceive only the innocent. Honderich's politics is all stark black and white. So is his political philosophy.

He displays contempt for rival opinions. In the opening of *The Jewish Wars*, Josephus Flavius said what such contempt amounts to. Those who discredit the Jewish soldiers, he observed, inadvertently discredit the Roman soldiers too, as they outnumbered them and yet they subdued them only after three hard years of bitter combat. Criticism bespeaks respect, said Plato. Hence, added Popper, credit for innovations should go to different parties in some critical debates.

Honderich's political actions may be all to the good. If he defends the cause of the under-dog, then power to his elbow. If he is better at it because he uses his position as an eminent philosophy professor, then so much the better. (324-5, 336) This is politics, however, not the sort of serious critical thinking that invites critical examination. Such thinking must be appreciative, or else all sense of value is lost. Honderich repeatedly dismisses conservative thought as muddled. He is in treble error here. His dismissal is unenlightening. Some muddled people are not conservative. And, some conservatives are not muddled. Some of them were truly bright. Remember Churchill. Thatcher was no Churchill, Honderich may rightly parry. He would thereby unwittingly admit that the poverty of her thinking is not a symptom of her conservatism alone.

3. ANALYTIC PHILOSOPHY

Honderich belongs to mainstream English-speaking philosophy. It is all analysis. This word does not mean what it meant to Newton, for example. Today analysis means clarification. Analytic philosophers allegedly clarify – concepts, words, even ideas. They judge their own success using their own native intuitions, which they consider healthy and normal. No criterion. No analysis (clarification) of normal native intuitions. The outcome of an analysis is allegedly obvious to its author. (Analytic philosophers often find each other not sufficiently clear.) Their guru was Wittgenstein, who forbade philosophy new insights. His prohibition clashes with my intuition, I dare say. My dissent should not disqualify me, as Honderich dissents too. (26, 316) And in dissent, he does more than clarify. What more? I cannot tell. Nor can I see deviations from Wittgenstein. The application of quantum theory to the brain-mind problem ended in a "dismal murk". (279) Honderich prefers his own hypothesis. This hypothesis is so clear, that it is "a heaven of light". (279) This happens often: many people brag that they are clear. With no criterion of clarity, it is my word against Honderich's.

Here is his hypothesis. Whatever is physical takes some space. Subjective events also take some space – in the subject's head. (278) But, he adds, and this is the crunch, his

> "nice new thought that mental events could be physical ... was not to be confused with the mad Identity Theory that mental events have only neural properties .... My happy discovery about the nature of consciousness was not the disaster of neutralizing the mind. To say that mental events were physical was not to say that they were the particular kind of physical events that were neural. ... They were some other kind of physical thing." (278-9)

This passage displays a remarkable confusion – about all the basic paraphernalia of philosophy: events, things, qualities, and relations. More significant is the confusion that it displays between explanation schemas and satisfactory explanations. One can find older and clearer versions of the same idea in the writings of Willard Quine and of Mario Bunge. They happen to disagree with one another in fine detail. It is the details that count. Honderich says that he can advocate determinism without giving a "sharper picture" of the situation, and "leave that for further reflection another day". (172) Yet, it is exactly the "sharper picture" that may signify; without the detail, such a disquisition is but a thin web of aimless speculations. He seems at times to recognize this, but only at times. It is especially in the matter of the relevance of science to culture in general that determinism comes up. And Honderich has no interest in science.

He has tried hard to be clear: he followed the standard rules common among analytic philosophers. Bertrand Russell said of them that they are too parochial and in conflict with science. These same rules, let me add, also decide positions on the totem pole. It is no surprise, then, that Honderich has "not taken the philosophical world by storm" – much to his regret. (28) By the expression "the philosophical world" he refers to the tribe. The tribe deems irrelevant to philosophy any clash between science and commonsense. Honderich concurs: he is a member. (8-9) He wished to impress the tribe's leaders and has failed. For them he was never sufficiently an insider.

Stagnation is not an initial obligation there. It becomes an obligation through adherence to a static view of common sense. Having no theory for common sense, and taking it as a criterion of rationality, they cannot let it grow. Consider Honderich's study of time. (23, 42, 193-4) Intuition does not go with Einstein. Too bad for Einstein. Intuition does go with Newton. Sadly for him, Honderich ignores him nonetheless. This makes sense. For, had he followed Newton, then a scientifically minded reviewer might brand him old-fashioned. He need not fear this, however. Commonsense is here mercifully inexact. And so, he is safe as long as he has enough common sense to reject the hostility to science that Paul Feyerabend advocated. (414) But if intuition is the judge and if science may clash with it, then possibly we want some rules for handling such clashes. Analytic philosophers ignore this need assiduously. Fortunately for them, this need is here not too urgent. Common sense here is the outcome of ages of internalization of science. In cultures without science, common sense is magical. Eventually the analytic school will have to recognize this.

The stagnation of the analytic school is that of its rules – of its analytic method. It had a hold on generations of its adherents long after the demise of its rationale. This is intriguing. This book hints at an explanation. Publication pressure makes this method very useful. It helps churn out with ease reasonably intelligent papers. Honderich tells us how he works. (194 *ff.*) His readers might emulate him with ease. The basis for a paper hardly needs more than a usual grounding in philosophy, good general education, and some unease that it should air and put to rest. Science is irrelevant, even when the analysis stems from science. High-powered logic is irrelevant, even if the analysis stems from logic. Religion is irrelevant, particularly if the analysis stems from theology. Useful for such analyses is courage to advance a thesis that colleagues may want to advance but dare not. Honderich has it. This recipe will not yield an earth shaking paper. Had he published one, then justice would demand that the tribe should make him a leader. He admits he has no earth shaking paper, although only the leaders can judge this, and their judgments are erratic. For my part, I would not judge this matter, maybe it's because he is a Londoner: were he an Oxonian or a Cantabrian, he might have reached the top of his totem pole.

The role of Wittgenstein's output is that of pulp, dull pulp everywhere. This role is to oust the better stuff. Students reared on it need great courage plus greater curiosity to make them surf libraries in search of the better material that their professors deliberately hide from them.

4. ANALYTIC POLITICAL PHILOSOPHY

Not all of Honderich's ideas are divorced from all practice. Some of them are moral and political, and some of these include proposals. Most of these are familiar to the British public. One unusual idea of his is a conjunction of two familiar ones that do not sit comfortably together: all legally sanctioned violence is evil and some illegal violence is good. (His example is terrorism.) Against legally sanctioned violence he says, let one who has not sinned cast the first stone. He has a point: public conduct often rests on shaky morality and on untested information, he rightly observes. (The wording is mine; I have replaced his journalistic idiom and omitted his analytic jumble.) He finally comes close to the sound idea – best articulated in Popper's *The Open Society and Its Enemies* – that violence against tyranny is permissible if earnest efforts to overcome it by other means have failed. (297) He expresses pride in having such a surprising idea. Not so. The idea is not surprising; unlike his defense of terrorism. Guerrilla warfare may at times be justified. Never terror against civilian populations (195, 196, 215, 217, 231, 297; see for details Walter Laqueur, *The New Terrorism*, 1999). Honderich defends the IRA, especially in its early days. Perhaps he forgets that it started its activities by attacking not soldiers but the Catholic prostitutes whom they visited. Had the local Catholic clergy had the courage and the wisdom to protect the civil rights of these poor souls, they might have prevented much needless bloodshed. Honderich justifies terrorism with the *tu quoqe* argument. Whitehall is a bunch of terrorists too, he declares. (195) This is a famous argument. Since utter cleanliness is obviously impossible, it says, let us wallow in the sewage. No; thank you.

We do not know what role penal systems play, Honderich rightly observes. Supposedly, they deter. He discusses retribution, not deterrence. No matter. He says rightly, there is too little evidence about the efficiency of the penal system, no matter how we judge it. Bernard Shaw said (Preface to *Major Barbara*), the English penal system is utterly useless. We do not know if this is true. We do know, however, something about the correlation between penalty and moral standard as well as between penalty and socio-economic standards. Usually, the more socio-economically advanced a society is, the more lenient are its penal code and practice. We should not extrapolate this to anarchism, though. We should be careful about any reform, including that of the penal code. We should also not procrastinate where the law incurs much needless pain. What balance should we strike between the two urgent needs, for action and for prudence? We do not know. It is a shame that advocates of attending to one of these needs ignore the other. Some penal law must stay. Otherwise, only good will would secure the peace, and it is insufficient as political control.

Honderich mentions some paradoxes involved in the very idea of democracy. His discussions (178-9, 184) prove his ignorance of Popper's *The Open Society and Its Enemies* that is full of wonderful discussions of them. Also, Honderich pours scorn on political inactivity, thus showing ignorance of the claim that diverse thinkers repeatedly assert: the right to political indifference is essential for democracy. Only Communism imposes political activity on ordinary citizens. Honderich's sympathy with Cuba and with Iran is not only care for the underdog. Few aching populations suffer from states as brutal as Iran. The democratic solution for political indifference that makes sense is the proposal that governments should try to enhance the means for encouraging citizens to partake in local and national political activities of diverse kinds and on diverse levels. We need activities that may yield encouraging results.

5. PHILOSOPHY WITHOUT SCIENCE

Honderich also displays ignorance of Popper's philosophy of science. He criticizes him for his claim that he has solved the problem of induction. As Hume has proved, he rightly observes, this problem is insoluble. (14) As is well known, Popper embraced Hume's proof. His solution to the problem is his assertion that science has no use for induction. If Honderich wants to criticize him, he should confront this assertion. Instead, he is content with his preference of commonsense over science.

Honderich shows contempt for Popper and Eccles on the mind-brain problem. He dismisses their book as beneath critical attention. The book covers vast areas of scientific information – psychological, biological, and more. He either ignores them or dismisses them as irrelevant. After all, science and analytic philosophy are worlds apart; nothing in the one is relevant to the other. So much the worse for analytic philosophy.

He recognizes the need for a concept of truth. Not a hint on Alfred Tarski's classical theory of truth. Honderich praises one or two Oxford pundits to whom he ascribes some minor corollaries of that theory as if they were innovations. (140) He proudly offers an improvement on them. His report on it is too cursory for me to grasp, especially since I do not know what its import is. It is his pet bias, he explains. All philosophers suffer from bias, he continues, due to "commitments" that "grip" them in childhood. (141) This is true, but it is no excuse for clinging to bias. He tells us that he opened up to philosophy when he read Ayer's best-selling book and that he remains faithful to him. (79, 387) This loyalty is not for me to judge. Ayer's book was the rage in my youth, but I now seldom find mention of it, as its author tacitly but clearly renounced it. In it he had backed Wittgenstein's identification of sense with science. Later he shared Wittgenstein's uncompromising reliance on common sense.

Wittgenstein's early work is all science and nothing but. His later works ignore science except for very few incidental remarks. If science is not divine, then it must be fiendish. So, in the study of psychology he and his fans ignore empirical psychology, where J. J. Gibson has radically altered the scene. So, they are way behind. This is a pity.

As it happened, I was a student of Popper's in London a little before Honderich came to study under Ayer – only to find that his hero had left for Oxford. I had read as much of Ayer's output as I could, due to his immense reputation then. I went to his seminar in University College, where Honderich soon entered to spend most of his adult life. Ayer's discourse puzzled me. I now see that he was describing the world of everyday experience as he saw it, taking no recourse to science or mathematics. In one such seminar session, he spoke of direction in space and in time, with deliberate indifference to topology. As a science graduate, I saw no point in it and left. A dictum of James Clerk Maxwell excited me then: yesterday's science is today's commonsense. Ayer's *The Problem of Knowledge* appeared soon. I did not understand it then. Now I see: it examined the rationality of everyday beliefs with no recourse to science. Why did he not say so? He knew the import of science on everyday life. Why did he choose to ignore it? I still do not know.

Honderich goes further than Ayer. He ignores science even when examining items from scientific philosophy. As so many questions are still unanswered, science does not cover all questions. Nor does commonsense, nor does anything else. Commonsense endorses causality, but it also endorses free will. Hence, it does not adjudicate on the question, is determinism true? Honderich does. His main philosophical claim is his defense of the (Spinozist or Schopenhauerian) kind of determinism that shares with commonsense the recognition of the freedom of will as the ability to choose, to make a decision. As he is a member of the analytic school, he declares his view commonsense. He dismisses quantum mechanics as irrelevant to it. His reference to Einstein as an ally is faulty. Einstein did not defend determinism, least of all as commonsense. He found it challenging as a scientific research program. Also, he found it appealing as a metaphysics. Honderich does not advocate any scientific research program as he is not interested in scientific research. He does not advocate any metaphysics, possibly because it is fallible. He says that determinism is compatible with observed freedom of choice and with observed randomness. This is a conjecture. He tries to prove it on grounds of common sense. This is absurd. Alfred Landé, Popper, and Mario Bunge have used common sense to criticize it. But they do not belong to the analytic school. As Honderich has the assurance of common sense that he is right, he need not discuss their ideas.

The determinism that captivated many thinkers' imagination is known as scientific determinism. It says, the laws of nature plus initial conditions determine every event. Nonscientific determinism is poorer. It says, every event has a cause and every different cause has a different effect. This makes sense if any two events (in different space-time regions) are definitely inherently identical (identical as causes), and then and only then their effects are inherently identical too. This requires a criterion as to which events that differ in space-time coordination are identical in all other aspects. Such a criterion is a natural classification of events. Knowledge of natural classifications requires a solution to Hume's problem of induction. Honderich finds Popper ignorant of its being insoluble, we remember. The boot is on the other foot.

His error is in his view of common sense as intuitive and so as unrelated to the problem of induction. But the world of science is the same as that of commonsense. Honderich must know about Buridan's ass, and he must know that in our vast universe some occasion is likely in which many bodies collide and then the forces of nature cannot decide how to direct each of them. This should raise doubt in his mind, even without the proof (by Henri Poincaré) that Newton's laws do not determine such cases.

6. COMMONSENSE, ALAS, ALL TOO COMMONSENSE

The ordinary commonsense to which the analytic school appeals is not ordinary. As they know, commonsense is supple and fallible. Hence, appeals to it do not settle all disputes: it is not as a universal decision generator as they insinuate.

Is Honderich ignorant of this? If not, could he not have commented on it in his long autobiography? Every sentence in it that he declares clear and commonsense seems to me fuzzy and question begging. I am here after comprehension, not after agreement. Comprehension must come first, of course. Do I understand him rightly? He says, different effects must have different causes. (407) He refers to Einstein's determinism with approval. But this determinism was scientific, not commonsense. Einstein was never certain about it. Honderich seems to suggest that analysis offers assurance here. (406, 412) This is the positivist idea of William James and Charles Saunders Peirce. Now, Pierce was an indeterminist; not Honderich. Let me do him the courtesy of appreciating his saying at one point, "my confidence seemed to decline". (411, 415) This is commonsense: we should never be too sure.

I should withdraw this at once. I do not know what common sense is. Perhaps it is less doubt and more balance. Some express it thus: common sense is healthy doubt, not any doubt, not the universal doubt of Descartes. Wittgenstein said, to ask a philosophical question is sick. He invented a lovely anecdote about two people listening to two other people talking. One of the observers says to the other, these are not lunatics, they are philosophers. R. D. Laing said the same: he said, Descartes' expression of doubt might lead a modern psychiatrist to doubt his sanity. If this were the sum total of the legacy of Wittgenstein, then I might be ready to endorse it. But he never said this. He never investigated the difference between healthy and sick questions, healthy and sick doubts. He said once, he did not object to scientific doubt, only to philosophical doubt. Even this he never said clearly. Nor did he ask, what is the difference between science and philosophy. He did say, to ask a philosophical question is sick. He could not explain, as he had no logic of questions at his disposal.

Perhaps all this is beside the point. Perhaps the point is this. It is alas sheer commonsense to yield to the demand of peers to accept a gospel on the mere say-so of a leading thinker in order to get an academic job, and only then to try to improve matters. Is this what Honderich told himself when he first learned that Ayer's first book was not the peak of perfection? If so, then he is smarter than he credits himself – perhaps because even he has to observe some limit to frankness. His seniors may tolerate tales of drinking bouts and of carnal indulgence, yet frown at displays of intellectual opportunism. I salute those who do frown at intellectual opportunism. They must be sufficiently numerous to make a difference. But perhaps this discussion still takes place in the wrong framework. So, let me try again.

Sociologists distinguish between ascribed and achieved status. Ascribed status belongs to the nobility. One gains it by choosing the right parents. Is talent achieved or ascribed? Wrong question. It is not a status at all. Struggling artists know this. The aim of their struggle is to achieve status for their talents. To become an item of status, talent must gain recognition. What then is the status of a recognized talent? When I was a science student, I noticed that most of my peers took it for granted that they must work hard to let their talent catch the eye of their seniors and that the rest would take care of itself. They viewed talent as given and hard work as the sole cause for recognition for it. I still find this fact enormously significant. The view of my peers, naïve and erroneous though it obviously was, is still very common. The first to criticize this was

Michael Polanyi. He observed that most people who have won recognition as talents – in the arts and in the sciences – were once apprentices to recognized masters. Joseph Ben David used this observation to explain the existence of centers of excellence – in the arts and in the sciences. Let me call this fact "the transmission of excellence".

Honderich notices, we have seen, that the transmission of excellence is of different sorts. Possibly, excellent teachers nurture excellent pupils. Possibly, he reluctantly admits, this is not always so. He regrets that he is not Oxford but only University College, London. And he suggests he equals his Oxford peers. I see little merit in this comparison. Can he see that possibly his and their alleged excellence is pathetic? No. It is contrary to common sense. Let us remember, then, that universities were not always excellent, that Oxford philosophy professors today deem pathetic the output of their predecessors – with amazingly few exceptions. Science professors hold similar views. This is why the scientific revolution is the product of amateur learned societies and not of universities. This is why few of the great modern philosophers were academics. Today leading universities are centers of science, but not as a matter of course. Who is to judge, then? We do not know. We still live in a free society. We still have the right to dissent from the consensus. Common sense says, this dissent is risky. It also says, at times this risk is worth taking. Professor Honderich is retired; he sits on his laurels. Am I naïve to suggest that he may still surprise me? I said similar things in other reviews and was disappointed. I refuse to lose hope.

7. CONCLUSION

In a Coda to his book Honderich applies its own analytic philosophy to it. The outcome is not gripping. I commend it all the same. I commend the first chapter too; it is a pen-self-portrait. As such, it is rather unusual and of interest, especially for those concerned with the academy. As to the book as a whole, it is hard to read or even to skim. To those who have managed to do so, I recommend rereading the first chapter. Next comes the story of the author's early life. It is readable and worth reading.

The story as a whole is sad. Honderich is intelligent, industrious, and ambitious. (294, 297) He boasts a broad education that he has acquired the hard way, and he is worldly. He suffers from chronic depression and he overcomes it through constant struggle to make a difference. This deserves high appreciation. Regrettably, however, the outcome is poor – simply because analytic philosophy is. He never questions his affiliation to it. He criticizes naïve realism, but he is a naïve analyst. His silent disregard for science is his downfall. His loud ignorance of Continental philosophy is harmless. (384) Still, his moral and political interests should have made him cast his net a wee bit wider. He brags about his not having read Marx. (12, 242) He should know better, as he offers some observations about him. They are adequate, but he did not check them. (250) He could do so by reading criticisms of Marx by Russell, Popper, or Gellner. He is a victim of gossip. His peers dismiss them – Russell as antediluvian, Popper as ignorant, and Gellner as a "merely sociological adversary of analytic philosophy". (14, 291) They dismiss wonderful critics on flimsy excuses since they know that they are right. Their self-assuredness rests on their faith in common sense. But self-assuredness is never commonsense. Honderich could be less assured of the correctness of the analytic creed; this would have made him a different person.

His autobiography is one of the few positive contributions that analytic thinking has prompted. It is a monument to the idleness of a popular school of thought, and to its author's self-exposure that is outstandingly frank with no lack in self-esteem. His WebPages display harsh responses to critics. His self-exposure does not render him a soft target. I hope that he will respond to this review and to my challenge to him to examine his creed, to study criticisms of his basic tenets. Philosophy is criticism of basics if it is anything at all. The basic assumption of Wittgenstein is that there can be no (articulated) basic assumptions. This assumption is its own refutation. His followers need to free themselves of it – by a clear understanding and by open, brave admissions. If this will end that school, then so be it. Honderich is brave enough to meet this kind of challenge. So it stands, awaiting his response.

To the extent that this book is an intellectual autobiography, it stands out as frank self-exposure in self-esteem. Regrettably, its intellectual value is slight. Its narrative is of interest – the way Honderich became an intellectual despite narrow background, the way in which he struggled with difficulties on his way, and his rise to position. It is the kind of story of the growth of a character that we find in fiction and in semi fiction. (The fictional classic of Goethe has set the mold. The semi fictional jewels are Samuel Butler's *The Way of All Flesh* and Somerset Maugham's *Of Human Bondage*.) It seldom appears as an autobiography proper. On this, too, the book stands out, except that regrettably the character of the book's hero is not too impressive. As a work of art, the book is too much of a diary. Its better parts are the self-portrait and the story of an early life – including details on a childhood background, a search for identity, and a first love. Other detailed parts have the flavor of life in London at the time, in college, in the Hampstead set, and on the outskirts of British politics. Alas, much of this is lost on me, as it is too snobbish for my taste – or too smug, I cannot say.

Philosophy has struggled for millennia with the problem of rationality. The problem has many facets, regarding the rationality of action, thought, science, and lifestyles. On all its facets, the problem has one traditional solution, and it is too demanding. Philosophical autobiographies that have addressed it seriously have failed. Others philosophical autobiographies are shallow. I found no evidence that this book's author was concerned about it. Even when discussing problems that traditionally belong to scientific philosophy, he handles them flippantly, suggesting that neither science nor rationality has ever troubled him. I find this surprising. And so, unlike the autobiographies of Mill and of Collingwood, it displays for me no philosophy. I find this frustrating. It is a report on many serious philosophical efforts, and on little to show for them. Sad, but hopefully instructive. Reading it was hard work, but it was worth it.

Chapter 4.6

SCIENCE AND ART

0. BETWEEN SCIENCE AND ART

Much interesting information is at hand regarding the cooperation between the arts and the sciences. One domain may borrow an achievement or a technique from another. There is the use of photography for artistic purposes, even the use of microphotography, and even the use of microphotographs. And there is the scientific and technological employment of artistic achievements, not only of perspective and of psychological insights to diverse phenomena, but also of sensitivities that artists have discovered. Science and art inspire each other. The artist Beardsley has inspired the development of new techniques of photography. Acoustic innovations have suggested electronic music of all sorts.

This is a rich field that I will not discuss. Let me say, rather, that the very cooperation between science and art is very thrilling – particularly to those who deem the two domains distinct. We all enjoy the cooperation of poetry and music, of paintings and music, of the theatre and music, of the theatre and poetry. Somehow, we feel, the cooperation of science and art differs from the cooperation of art with art, however natural or problematic. (We surely find more natural to mix poetry with music than paintings with music.) The reason is that we feel that science and art are so significantly different as to make their cooperation much more intellectually exciting, challenging, and thought-provoking than the cooperation of diverse arts.

The distinction between science and art seems all too clear. Art is emotional, science factual; art is subjective, science objective; art is pleasant, science is useful. So goes the common distinction. It implicitly merges scientific technology with science, as against art – as a matter of course. Yet, it is not clear why it is so. Quite possibly the opposite is true. It is hard to see why installing an air-conditioner is a matter of objective technology, yet using it is subjective. And admitting that its use is objective renders the use of a record of Beatles' music objective no less. The standard answer to this objection is that the Beatles are not everybody's cup of tea. But then neither is air-conditioning.

We divide technology into an art and a science. We call art whatever requires a knack and is a matter of trial-and-error. We call a science whatever we can do by rote with assured success. This is how we understand the debate as to whether medicine is art or science, for example. Science is allegedly sure and this is all to the good. But what is sure is also dull. Art is the doubtful, but thereby also the thrilling, given not to every practitioner but only to one with golden hands.

Not quite. The humdrum of the conveyor-belt is not science; it is terrible and barely excusable. The bungling of a medic or of an acoustic engineer is not art, reasonable though it may be. There is much more in common between the bungling of a Newton and those of a Beethoven. Their zeal is comparable, as is their quest for the perfect and their experimentation. They both tried to remove the merely subjective and the accidental as much as possible and they even hoped against all common sense to eliminate it altogether.

Nevertheless, science and art are inherently distinct. They can unite or merge perhaps, but only as a remote, unattainable ideal. Let me take the major aspect of the difference, to show how unbridgeable it is, how useless are efforts to close the gap between scientific and artistic activity. The aim of art is the attaining of beauty, whereas the aim of science is the attaining of the truth. And so, much as we may narrow the gap between the two, all efforts to close it must fail.

1. BEAUTY AND TRUTH

Perhaps the simplest way to close the gap is simply to consider how either art or science may betray its aim. If both are to serve the state, for example, then they could merge. If science and art were to aspire to what the Nazi philosopher Martin Heidegger called "poetic truth", and had they placed the value of poetic truth above that of scientific or artistic truth, then they could merge. Science would then produce conveyor-belts and art would then settle on them.

Heidegger's proposal is thus for science and art to betray their ends and serve the state. Those who view the value of both science and art as measurable by the yardstick of utility – private or social – may have sense of neither art nor science and can thus shortcut them and talk of the utility of Newton or of Beethoven. Alternatively, these people may have an independent sense of art and of science whose value they could discuss without reference to their utility measure – even if, for one reason or another utility happens to equal intrinsic value. In the end, acknowledging the intrinsic value of art and the intrinsic value of science is a moral matter, a matter of a primary, honest, straightforward attitude.

The word 'straightforward' plays an important role here. With sophistication we can get anywhere. Perhaps there is but one final and ideal end. If so, then beauty, the final end of art, is identical with truth, the final end of science. It is no accident that Plato's most satisfactory exposition of his theory of ideas is in his *Symposium*, that classic study of beauty, where he presents them as true, beautiful, and good. But artistic creation aims at beauty, not at truth, and scientific research still is the search for truth, not for beauty. Hence, though the final end for art and for science may be the same, perfection, their immediate ends quite obviously diverge.

2. REASON AND EMOTION

The giants of the high Renaissance were Platonists and tended to blur the distinction between science and art, truth and beauty. They did so in many ways that fill the literature of the history of Renaissance art, science, and ethos. The idea of the distinction between science and art comes from the idea of the purity of science, from the view that science is good only as long as it stays aloof. This idea belongs to Bacon who blamed older philosophers for mixing science and religion, science and metaphysics, science and politics. Bacon himself did not speak of art or of aesthetics. Descartes accepted the division between the arts and the sciences. The Royal Society of London, and with it the whole tradition of the Enlightenment Movement, followed Bacon and Descartes. The separation of art from science raised the problem: what role do the arts play in the life of a culture? Descartes hinted at an answer, and the English poet John Dryden, Fellow of the Royal Society of London, said it explicitly: art should serve science by making it agreeable to the public; art is a mere ancillary to science; it has no role of its own. Not accidentally, then, eighteenth century poetry was full of the scientific spirit and of scientific texts in verse. Similarly, John Constable, the famous early nineteenth century English painter who was the forerunner of impressionism, viewed himself a naturalist as he painted from the hand of nature. His pictures reproduce images as one sees them, through an untutored eye, a naïve eye, or an innocent eye – not what some theory would tell one to see. The impressionist painters said the same. The naïve observer sees things as they really are, said Bacon, uncolored by prejudice; and impressionists attempted to illustrate this. William Whewell, a contemporary of Constable, developed a new theory of science, anchored in the idea that there is no innocent eye: perception is always guided by ideas.

In *Art and Illusion*, E. H. Gombrich has discussed at great length the idea of the untutored eye and the role it has played in the works of Constable – who declared himself a scientist – and his followers. He has also discussed there the empirical evidence for the impossibility of seeing pure facts. This evidence accumulated since Helmholtz. J. J. Gibson and R. L. Gregory have gathered it. The fascination and drama of such matters derives from the fact that Bacon's influence is still tremendous. It is strong enough to cause discomfort to the believers in the theory of pure facts and its corollary of science as isolated.

Many people find the mixing of science and art disturbing, and for an obvious, good reason. The place of wishful thinking is in art, not in science, since science is the search for the truth. This reasonable idea rests on the unreasonable idea that art is emotion alone and science is reason alone, that reason and emotion dampen each other as they commingle. By this view, *qua* artists, artists must have brains made of straw, that they must think with their hearts – just as *qua* scientists, scientist must have a heart made of stone, that they must feel with their brains. Thus, a very reasonable and commonsense view gets its popular expression couched in a string of confused and irrational suggestions. Their convincing power deserves examination.

3. THE REQUIREMENT FOR A SHARP SEPARATION

The mixing of science and art is disturbing. People expect them to stay separate in as strict a manner as possible. Now, a requirement for a strict separation rests on a fear that the slightest overlap may lead to harmful results. The harmful result is usually the outcome of a total confusion of the two. Somehow, the feeling is that the moment art has the slightest foothold in science, science loses its claim to objectivity without any further ado; that, likewise, the moment art is cerebral to any extent whatsoever, it loses its right to a claim on beauty. Thus, for example, when P. A. M. Dirac claimed that he used aesthetic arguments in order to retain his celebrated equation against empirical evidence that contradicted it, he gave the impression that science lost all claim for its aspiration for truth. And we hear that the cerebral approach of Schoenberg has deprived music of its status as art.

This is an error. It is admittedly the case that Dirac used aesthetic arguments, yet only as encouragement, as giving him some hope that his equation will prove defensible against known initial criticism. When such hopes do not come true, ugly facts do destroy beautiful theories. Though Dirac's equation did withstand the initial objections and thus attained its prominence, it did not withstand all objections. Dirac himself expressed dissatisfaction with it, or in general with quantum theory as he knew it. Hence, clearly, though aesthetic considerations did play a role, even a significant role, in the history of science – of quantum theory in this case – the decisive considerations are not aesthetic but those that pertain to the truth.

The same holds for cerebral considerations in art. Young artists often display aversion towards the acquisition of techniques that require cold calculations. Often painters overcome such aversion by making a rather silly distinction between real or pure and commercial or applied art. They allow themselves to acquire cerebral techniques for applied art, and then they manage to smuggle this knowledge (preferably transformed and altered to their tastes) into allegedly pure art. I have encountered many budding authors who struggle bitterly because they permeate with the conviction that a novel should grow spontaneously, organically, one page naturally following another. They regard with disdain the very thought of planning a novel by cerebral, analytic means. To write a synopsis of a novel, a table of contents, an estimated length of each chapter – this seems easy and pointless, not to say commercial. So, they refuse to try it. And so they suffer hopeless frustration. Only such people, and the popularity of the prejudices that frustrate them, can make the public gasp at the publication of Dostoevsky's notebooks. They are highly analytic, entirely cerebral. They record plans, experiments, suggestions. They refer not to their author's own emotions but to his reason. Yet, this very characteristic that we should expect *a priori*, is what did cause a stir.

4. THE ERRORS BEHIND THE SHARP SEPARATION

The sharp separation has a deep reflection in our present way of life, and the mere observation that art has a cognitive side and science an aesthetic side will not eradicate it. Consider old works of art and original versions of scientific theories. All of them require tradition, training, definite approaches. We do have a tradition of the one and not of the other: we do learn to appreciate the arts historically, but not the sciences. Admittedly, we do learn certain ancient doctrines, and some classical scientific doctrines, such as those of Archimedes, Newton, and Lavoisier. This is so by some sort of default: we do not know how to avoid teaching them, and we teach them in some highly modified versions, akin to Lamb's *Stories of Shakespeare* and Hamilton Harty's version of Händel's *Royal Water Music*, not to mention cinematic vulgarizations. These do serve their purposes. They are substitutes for the real things, and with luck introductions to them. Nevertheless, science connoisseurs, unlike art connoisseurs, have no access to the real thing, to the originals of past scientific creative genius, unless they undergo training as historians of science.

Similarly, the cognitive side of art is more accessible to scientists than the aesthetic side of science is to artists. Lord Snow first noted this in his enormously successful *The Two Cultures*. In this talk, as well as in its follow up, Snow found it necessary to blame artists for their insensitivity, through willful ignorance, he suggested, to the aesthetic side of science. He only grudgingly acknowledged that scientists are often ignorant of art, and he admitted their ignorance of any history, of art or of science, only reluctantly. He deemed the scientific temperament progressive, and thus, in the ethics of politics it is on the right side (which is left), whereas artists are natural Luddites. This is just a degenerate version of Dryden's view of art as a handmaid of science: art should serve both science and progressive politics, but, alas, it tends to go its own devious way.

The traditional inability to take seriously the aesthetic side of science comes from the view of science as purely objective, as purely impersonal, as utterly devoid of all connection with other aspects of the culture except through the economic side of things. Even the life of Madame Curie by her daughter shows no significant link between the character of that remarkable woman and her remarkable research. Even Helmholtz's researches – that so obviously reflect his personality, his conservatism, his philosophic taste, his taste for music – appear in reports on his scientific work not as personal but as a part of the vast impersonal tapestry that is science.

The best way to show that the claims of science for utter objectivity and for the absolute truth demand its utter aesthetic neutrality, is to see why some authors about science allow themselves to deny that it is aesthetically neutral. Philosophers of the conventionalist school, Duhem and Poincaré in particular, insist on the aesthetic quality of science. They can allow themselves this because they forego all claims for scientific objectivity. They deny that scientific theories in any way reflect the structure of the universe, the nature of things.

The way in which the cognitive side of art and the aesthetic side of science come to fore, is mainly through the more modern philosophy of science, the one that Popper calls the critical view, and that we may call gradualism. It recognizes a personal, a subjective, element in science, but it declares, nonetheless, the aim of science to be objective and impersonal. According to gradualism, science undergoes a gradual process of *objectivization*. Every product of science in time may appear as a mere approximation to the truth. Scientific theories may appear from different angles and at different times as different approximations. The same holds for the history of the cognitive side of the arts, that is part and parcel of the history of science and technology proper. At times, this is all too obvious, as when we compare the diverse tools of artists.

5. OBJECTIVIZATION IN ART AND IN SCIENCE

Distinct as science and art remain, they may have significantly more in common than hitherto indicated, quite apart from the interactions between the arts and the sciences and from the fact that the arts have a cognitive aspect and the sciences an aesthetic aspect. These are the processes of objectivization. Let us begin with the processes themselves: with the artistic process, the act of artistic creation, and with the scientific process, the act of scientific research. We can contrast the art with the product, the process with the achievement it leads to. Somehow, we can compare the achievement with the struggle and evaluate them separately. We greatly appreciate the struggle of those who brought what we possess into our possession. This is why it takes training to learn to appreciate works of art of old masters: we have to learn to appreciate their struggle when we contemplate its product, to see the process in the product.

This last point is a cornerstone of the philosophy of art of the celebrated Gombrich who is, significantly, a disciple of the philosopher of science, Popper, the chief exponent of gradualism.

Much as I endorse this idea of Gombrich, however, I cannot endorse it in all detail. Considering only truth, but not beauty, an ideal towards which to progress, he recommends the objectivization only of the cognitive part of art, of its beauty. He views the struggle of the artist as purely cognitive. Though he sees the cognitive achievement of an old master as much inferior to what cognitive tools a mediocre artist has today, he declares the very struggle for that achievement essential to the old master's greatness. (I cannot offer an exact quote from Gombrich to substantiate this reading. If I have misread him, I have to declare him elusive: I have tried to read him sympathetically and carefully, and I have compared notes with others, *etc.*) I disagree with his view as I find that many artists have made important artistic innovations that they themselves put to use not very forcefully though their less inventive contemporaries who were of greater artistic power exploited very effectively. Gombrich notices that some great masters of the past were arch-conservative and some of the greatest innovators were poor artists. He explains well why people who break from old techniques in impressive ways need not be impressive artists, since they may have nothing interesting to replace them with. But he cannot explain why great innovation and great art, much as they go hand in hand, do not always go quite together. The question then forcefully arises, what is the struggle of the artist who is a great artist yet not a great innovator of techniques? What else does a great artist innovate other than techniques, other than the cognitive or cerebral part of art? I shall return to this question soon; here my point is that Gombrich forces the question.

The analogue of this discussion as far as the history of science is concerned is also rather new but it is much less problematic. Considering the cognitive aspect of science – as productive of valuable informative ideas – we may well imagine researchers struggling with cognitive problems: is this or that idea true? What explains this or that? How do two theories cohere? Now the history of science as the history of a struggle is a new field. Thus far discussion centers on personal, moral, and aesthetic aspects of intellectual struggles, not on their non-cognitive aspects. Comprehending the non-cognitive aspects of such struggles is essential for their comprehension, yet the measure of their achievements is certainly cognitive. The process of science is that of objectivization – largely the increased approximation to the truth – and thus largely a cognitive process, though it necessarily contains non-cognitive aspects. If we wish to retain the parallel, we would perhaps try to say in complete parallel to the above: the process of art is the process of artistic objectivization, where objectivization is the increased approximation to beauty, and is thus largely aesthetic though also necessarily containing non-aesthetic aspects.

This is Platonic: it rests on the assumption of the existence of both ideals: of truth and of beauty. It is the fundamental bias that the present essay discusses. It is problematic. Its problematic aspect, however, appears most forcefully in the philosophy of mathematics, and we require a solution for the problem more urgently in that field than here. So here we may put it aside for a while. Later on we may borrow any extant solution and try it for size. To return to our discussion, then, we might ask, are truth and beauty identical? And if they are, does this not collapse the arts and the sciences into a unity? The answer is clearly, no. Our stress is on the struggle, on the process, and the processes of art and of science differ even if in the end they turn out to be working towards the same ideal. This, however, does not remove the present central problem: what is artistic objectivization if not progress of their cognitive aspects? What is aesthetic objectivization? What exactly is the approach to the ideal of beauty? How can we reconcile viewing beauty as the value of a classical work of art as manifest in its process and viewing it as manifest in its end? After all, the beauty aimed at differs from the beauty achieved. Is there progress in art, and if so, what does it consist in?

Suppose that both science and art undergo processes of objectivization, of progress towards their final aims. Though we are not clear about the final aim of science, the absolute truth, we know it better than we know the final aim of art, the absolute beauty, and we know more what is approximation to the ideal truth means than what is approximation to the ideal beauty. This is why art as such is more intriguing than science as such, even though a scientific theory may be more intriguing than any work of art, and why a problem in science is more intriguing than any problem in art, including one in a thriller.

6. FORM AND CONTENT

I tend to view the history of art, like the history of science, as a process of objectivization. And I owe this attitude to Robert Cogan, the contemporary *avant-garde* composer of New England Conservatory. There are sets of traditions. Each tradition has the personal imprint of its giants. The tradition absorbs only a part of this personal imprint, the part that it objectifies. It irons out the rest of it and eliminates it in the process. Trying to move towards newer levels of higher objectivity, later contributors contribute to this process of objectivization of earlier parts and replace them with newer material. This material, too, is partly subjective partly exploratory. The question still is, since art does not pertain to truth, what can it objectivize and how? What in art is objective anyway? I find this question hard to handle for want of tools due to the following fact.

In the Western tradition it is customary to identify the form of art with the objective and the content of art with the subjective. This need not square with observations about the cognitive and the aesthetic, though it easily looks this way, merely because all too often we tend to identify the two distinctions as a matter of course. The distinctions form/content, objective/subjective, and cognitive/aesthetic look identical but they are not. The aesthetic aspect of a painting by Bosch or Breughel, or of a novel by Dostoevsky or Zola, cannot possibly be in their content, as that is intentionally unaesthetic. Therefore, those who identify the aesthetic with content will have to declare them ugly and prefer to them the pre-Raphaelites and Peter Pan, and all other chocolate-box works one could mention off-hand. One might protest that chocolate-box works are not beautiful or not art. But there are counter-examples to this, if not Raphael, then Renoir and Modigliani and Chaplin.

Even those who assume that art is subjective, is emotional, is content, is partial to form. This invites clarification of the question, is art content or form? Advocates of the theory of art as form are formalists, and their opponents are expressionists. Both parties to the dispute on the matter agree that there is no content without form except unexpressed and inarticulate feelings; they likewise agree that there is no form without content except perhaps in the world of ideas that is incommunicable as well. Every communicable item, then, not only art, has both form and content. (The common view of logic as purely form is a slight, tolerable exaggeration. The very vehicles of logic comprise its content.) Formalists, for example Schoenberg, say, pay attention to form, and content will take care of itself; or, pay attention to form, and choose content as you wish. Expressionists like Wagner, will not declare form insignificant or capable of taking care of itself; they will merely declare form ancillary, and strictly so. Thus, if the recipient is more aware of the form than of the content, then the transmission of the item of art is cerebral. Expressionists view that item inferior. Superior art, they say, requires both significant emotional content and the form proper to the task of transmitting or conveying it. Ideally, then, art should be the immediate transmission of strong feelings from the artists to their publics.

This applies well to the problem of the sense in which we can objectivize art. Expressionists view form as cognitive and content as aesthetic. This allows only one way to objectify beauty – to objectify content. A way to do this, then, is to develop emotional capacities. It is not the only way. To objectify must mean here also to make more formal, more cerebral. That also is not the only way. There is room for widening the range of feelings that a culture offers. This has to do with the role of art. The psychological novel is but an obvious example; the garden scene in Mozart's *Don Giovanni* is a subtler example. Though formalists deny that art is content, they may still look for both its aesthetic and its cognitive aspects that are capable of development and objectification. But it is still not clear how.

7. HOW CAN WE OBJECTIVIZE EMOTIONS?

Let us explore the possibility that emotional content can be made objective in the sense that a culture offers, objectively, a range of emotions through its artistic achievements. These emotions are possible subjects for study and evaluation, then, through the study and evaluation of the culture's arts. For example, Beethoven has expressed anger in music, thereby giving both anger and music newer and wider scope. Similarly, Wagner gave in music the expression of a collective, of individual helpless, delicious dissolving into the tribe. I am speaking here of emotional content, not of psychology. For, the content of psychology is statements that are true or false, whereas emotional contents are just there, in the world of facts. There is then a possibility to approve or disapprove of arousing these emotions; there is an intellectual possibility of weighing arguments in favor or against submitting to the emotional slough of a Wagner opera. Patients of psychotherapy know (or should know) that they have such options as to wish or not to wish to like someone or some feeling, and perhaps do something about it. (These are second-order preferences.) We may appreciate a person or an attitude and wish to endorse it or not to endorse it. Therefore, it seems at first, the content of Wagner's operas, or of the paintings of Delacroix (in contrast to his drawings and sketches), or of Baudelaire's poems, or of Rilke's, is more important than Schoenberg said it is. That content is of great value, though a negative one and an objectionable one. Even the art of Hermann Hesse, or Rachmaninov, or Käte Kollwitz, being an expression of self-pity, is of a rather negative value, though less than that of Wagner.

But no. Much as I agree with these assessments, I object both to the tone and to the generalization of the previous paragraph. As to the tone, it is functionalist, such as Tolstoy's theory of art that is a generalization or an extension of Dryden's. Dryden saw art as valuable as an ancillary to science. Tolstoy viewed art as possibly useful for diverse worthy causes. He approved particularly of folk-art, as it integrates in folk-life. He viewed function as accentuating aspects of art, like wedding song and dance or stylized funeral wailings. The view of art as merely ancillary, as at its best when it is frankly applied art, is the denial of the contemplative view of it as an object of contemplation (as Collingwood has put it). The view of art as autonomous opposes functionalism. The present problem – in what sense can art become objective – makes no sense without admitting the autonomy of art. We may try to turn the tables and see if we do not thereby get better results. Whereas functionalism in art makes aesthetics a part of ethics, aestheticism in morality makes ethics a part of aesthetics. Both are objectionable, and we have to take for granted here the autonomy of both ethics and aesthetics. Let us take a hint from aestheticism. Let us say, though our moral code does not require magnanimity, surely magnanimity is very beautiful, as Victor Hugo showed in *Les Misérables* (and Kurosawa emulated). Though our moral code does not forbid self-pity, surely it is ugly. Harpo justly ridiculed it, in *A Day at the Races,* in his hilarious performance of Rachmaninov's famous c-sharp minor *étude* on a piano as it literally goes to pieces.

But no. This will not do either. The sentiment the previous paragraph expresses is wrong again, and the generalization it offers is the same as before it. The sentiment of aestheticism is somewhat decadent and objectionable both morally and aesthetically. Many counter-examples refute this generalization. The misanthropic novels of Dostoevsky, the vulgar and reactionary movie *Gonga-Din* that excited even Berthold Brecht *malgré lui*, Degas' portrait of his parents and Grant Wood's *American Gothic*, they all excel in malice and petti ness, and are nevertheless justly celebrated. As such examples, they are all too obvious and all too obviously devastating. Even tribalism can become something out of this world in the music of Charles Ives, and Marcel Proust and many of blues singers have made self-pity an object of aesthetic contemplation.

And so, within the compass of classical aesthetics, on its division between form and content, the present study remains frustrated. We have a clear impression of progress in aesthetics through objectivization, yet we can barely say what it is. Someone may object to the claim that objectivization in the art is obvious or that we even know what it is. We do not. This is the major frustration expressed here. But it is obvious all the same: anger in music as opposed to Beethoven's anger, neurotic helplessness as opposed to that of Dostoevsky's or Chekhov's heroes, a sense of worth as opposed to Goethe's *Faust*. Likewise, perspective as opposed to that of Brunelleschi or Alberti or Leonardo. Tonality, atonality – not necessarily Schoenberg's, catharsis, not necessarily of this or that dramatist – even that of Bach's Mass in b minor will do. We know that it is there. We see lesser artists learn it from their betters. What exactly?

8. BETWEEN CONTENT AND FORM

The classical aesthetic approach divides content from form and leads to the question, does the beauty of a work of art lie in its content or in its form? This approach has a long history and for many historical reasons – good, bad, and indifferent. Yet, many discussions for a long time past, couched in terms of that approach, made it burst at its seams. The idea that certain contents fit certain forms and not others has its expression in the use of the hackneyed metaphor of the *New Testament* regarding old wine and new bottles. Without adjudicating on how convincing is Albert Schweitzer's correlation of the religious content of Bach's preludes and fugues with their form, we may notice that he was using a hackneyed idea. The sensitivity of artists finds its expression not only in their sensitivity to content, but also in their feel for the need of art for a new form, a new medium, a new technique of expression – in word, sound, or color.

The cognitive aspect of art is not purely formal. Beethoven's ability to express anger was an innovation, both cognitive process and technical. His creation of new kinds of sound, new textures, created new problems for the art forms he used, the piano sonata or the orchestral symphony. He solved his problems by creating new forms, or more precisely by modifying existing ones. The same goes for a religious sentiment. It may be easy enough to emulate Bach, or Bruckner, and express even today religious sentiments in modern compositions that serve as background to some Hollywood movies. When Schoenberg wished to express it, say, in the choral bars at the end of his *Survivor from Warsaw Ghetto*, he could use neither tonal nor serial sequences. So he created a new type of melody that is neither, though, admittedly, akin to the one used in the violin sonata of Roger Sessions, for totally different purposes to different effects.

Here I would be bold and say, perhaps again in agreement with Schoenberg's recommendation to take care of the form and not of content: artists struggle to solve new problems because they have new contents, new experiences, which plainly do not fit old forms. They invent, then, new solutions that are new forms capable of adequate expression of new feelings. I do not quite feel at ease with this: I feel that artists may widen the extant range of feelings while exploring them as artists. Art need not be only the expression of already given feelings; it may be the creation of new feelings too, as well as the new evocation of feelings, or their new combinations: even old feelings may transform as they enter new forms – or even new media – in new inventive ways. But I do not know if there is a clear disagreement here. In the classical theory all was clear enough, where content was emotional and subjective, form cerebral and objective, and beauty residing in one or the other – possibly in both, but separately nonetheless. It is no longer so. As the composer Roger Sessions and the philosopher Milič Čapek stress, it is a widespread prejudice that the cerebral part of art is open to articulation in words no less than in painting, music, or poetry. Certain ideas, certain images, can better come across in a picture than in a thousand words, as old Confucius said – unless the words belong to a poem.

There is an anecdote, certainly spurious, of the reply Beethoven gave to the question, what did a sonata of his mean that he had just finished playing. In reply he played it all over again. This anecdote is spurious as it is expressionist and he was a formalist. The Expressionist approach invites contempt for mediocrity. But contempt is seldom just. (A superb photo in the Warsaw Jewish Museum depicts a Jew looking with pity at the Nazi soldier who humiliates him by cutting off his side-locks.) The formal approach invites the use of formal analysis as means for finding partial answers to questions regarding the meaning of given works of art. We may view a picture as a canvas divided into parts along some broad outlines. We may try to reveal the form of a symphony. But all this is very limited. It is quite possible that an idea finds a better articulation in art, such as an answer to the question, what do you think is the character of a friend, or what is New York like. If Sessions and Čapek are right on this, then it is hard to say any longer where content ends and form begins, and certainly it is impossible to locate the aesthetic exclusively in either. It may very well be that the aesthetic is what widens our range of feelings, that the aesthetic is some fit between form and content, that we thus have layers of feelings, aesthetics and other, building up alternately in bootstrap fashion.

All this has a consequence for art appreciation and for art criticism: every work of representative art has both an abstract and a representative aspect, no matter how realistic or abstract it is.

9. CULTURE AS AN ART

In his book on *Movies and Society*, Ian Jarvie joins many others in viewing as a work of art anything well designed. He concludes that one may – but need not – live one's life as a work of art. This is rare as a conclusion, but common as an observation. Literature is full of descriptions of people who tried to live their lives as works of art to varying degrees of success. I wish here to go further and see whether it is possible to objectify this conclusion: all sorts of factors may enter the design of a person and even of a culture. (And passing a character to the culture is what objectivization amounts to.)

This is hackneyed to the extent that it raises suspicion – and rightly so. We are over-used today to correlate the diverse arts and sciences and philosophies of a given period, and it is now fashionable to call baroque and rococo not only all the art styles current in these periods, but even their sciences and philosophies. This can be merely a matter of coy expression: we may call baroque anything that happened during a certain span of time in Western Europe when baroque architecture flourished. The history of such expressions goes to Hegel's theory of the spirit of the age (the *Zeitgeist*). He said it enters everything current in that age that is worthwhile. If so, then the mere name for the spirit of an age should not suffice. We should seek the same in all the achievement of a given age. If we can find it, then the effort should enrich our feel for, and experience of, the diverse activities of the age, by exhibiting its unity. And here readers are rightly suspicious that authors use names of styles as promises for the harder thing, for seeking and for conveying the sense of unity of an age.

There is some valuable truth in Hegel's idea of the spirit of an age. However remote two activities of an age are, when one considers in perspective enough information and some contrast with other ages, one may – just may – come up with some interesting resemblances between different activities then. However, Hegel's philosophy is disastrous in its excesses. Can we say that every age has a unity to it? I doubt it. There are diverse degrees of integration, and by some lucky accident a culture may achieve a high degree of integration. For this different traditions must bloom. This is seldom the case. The music of the high Renaissance is an example. Hegelians may say, there was a delay in the rise of great music, but when it came its spirit was the same as that of the plastic art of the high Renaissance. This is an admission of error put in confusing verbiage.

One tradition, artistic, scientific, political, whatever, may be better or worse integrated in the culture than another tradition. This final point is so trite that I wonder why I have to make it. In his famous *The Sleepwalkers*, Arthur Koestler has declared that the Pythagoreans represented an integrated culture where science and art were one, and that humanity hardly ever saw such balanced integration. Bruce Mazlish and Jacob Bronowski saw the Renaissance as such a period. They showed this not by analyzing the diverse parts of the culture but by reporting that it had integration as its ideal: they were Platonists. Plato's theory of Ideas, in its very a-historical declaration of the unity of Truth, Beauty, and Goodness, disregarded the possibility of closing the observed gap between them. Hegel has claimed that his relativism introduces a historical dimension into philosophy. Every achievement, he said, is not better or worse, but always the best for its time. The strength of his view was in his seeing much of the limitations of traditional philosophy and in his allowing us to transcend these. His weakness is the confusion that he caused, since he criticized and he declared all criticism wrong. We can see limitations in the achievements of old ideas, we can appreciate the struggles that went into them, we need not confuse them with us, or their views with ours: we have since learned to keep apart yet sympathize, to sympathize and criticize. We can critically assess different degrees of integration of cultures.

The study of degrees of integration will be significant if it will be open to critical assessment. For this we have to discuss art and science in a manner that is objective to some extent. For this we still need a good idea of objectivization.

Chapter 4.7

THE INNER WORLD

0. ABSTRACT

Standard current philosophy is stuck. Dividing the world and our experience of it into two compartments, the inner and the outer, it holds fast to the defunct thesis that information about the outer world reaches us as information about the inner world. Refutations of this thesis do not reduce its popularity. Perhaps exposing the deceptive character of the arguments in its favor might. This discussion is politically important, since the thesis in question was initially individualist and progressive and it has become romantic and reactionary.

1. THE GRAND POLARIZATION

Current philosophy comprises almost exclusively the two big philosophical schools of the twentieth century: the analysts and the phenomenologists, alternatively known as English-speaking and Continental. (1) The analysts are more or less pro-science, and the phenomenologists are more or less hostile to it. Also, the latter are reactionary, perhaps the former too. In my youth, the two schools were frank adversaries. Things have changed, and now, probably recognizing the poverty of their output through the decades and feeling fenced in, they are looking for a rapprochement – still with some sense of embarrassment. Sir Michael Dummett wrote about their joint ancestry, but this hardly counts: Cain and Abel had it too. They are seeking similarities between their respective doctrines, but this is beside the point, since both used to stress that their peculiarities lay in approaches and in methods, not in doctrines.

The two schools are similar, though, especially now, since leading analysts have relinquished the quest for clarity that used to be their trademark. And they have replaced their platitudes about everyday experiences with platitudes about life and death. Both schools rely on intuition. Since intuitions of people not familiar with science may easily clash with it, both schools retreat into the inner world, in the vain hope for safety there from clashes with science. (2) So, they join traditional philosophy and divide the world into the inner and the outer, and hold fast to the defunct thesis that information about the outer world reaches us as information about the inner world. This is subjectivism, or methodological subjectivism or transcendental subjectivism. (The word "methodological" means, with no ontological decision – without asserting that solipsism is true or that it is false. The word "transcendental" means, avoidance of reference to the outer world.) Husserl, the founder of phenomenology, put science aside, and claimed that clinging to subjectivism will bridge between the inside and the outside. (3) Wittgenstein, the source of the new analytic techniques, discussed only the inner world. Both Husserl and Wittgenstein invented methods of inquiry that rely on inner experiences, and so they isolate it by fiat and thus they polarize the inner and the outer worlds. We need not discuss them here.

On polarizations Claude Lévi-Strauss has the last word. (4) We owe him his insight on mythical thinking. It comprises intuitive extensions of tensions between similarities and differences between diverse polarizations. He writes

a : b :: c : d :: e : f, *etc.*

The colon signifies polarities and the double colon signifies a parallel. It reads, a is the opposite of b and the polarization of a to b resembles that of c to d. The present case is this:

inner : outer :: subjective : objective :: emotional : cerebral :: concrete : abstract

Lévi-Strauss has suggested also that science is a myth system. As his illustrations for this show, he meant not science but its function, or its cultural ambience. Already Bernard Shaw noted this (5), saying using it as a myth hurts our culture, not our science. He was in error: it hurts both.

The most important and the most popular aspect of current subjectivism is the idea that the most important and most urgent task of philosophy is to show the way out of the inner world into the outer world: philosophy must start inside, but staying there is insane. It is insane in many senses, including the clinical sense, as R. D. Laing has noted (6) and as John Wisdom explained (7). Carnap, who was during his active life a leading member of the analytic school and the leading philosopher of science, called his view "methodological solipsism", namely, the idea that philosophical research – perhaps all research – must begin with the assumption that the first-person singular exists and no other assumption. (8) He never changed his mind on this, he explained, but he dropped the label "methodological solipsism", as he found that too many people misread it to designate solipsism, the crazy view that only I exist. The misunderstanding was his, not theirs: they simply took it for granted that anyone who begins with the inner world only and seeks a bridge to the outer world is doomed to failure and thus to imprisonment in one's inner world. Thus, methodological solipsism brings in solipsism as such. The philosophy of Martin Buber, by contrast, begins with the assumption that at least two people exist who can converse with each other. Wittgenstein took over this idea and added that philosophers must take for granted not only the other but also the community and its language. Buber, however, found interpersonal empathy prior to language, and for reasons too obvious to need articulation. (9) He said, certainty is redundant. Wittgenstein never ceased looking for certainty. Since the reason for starting with the inner world is that it allegedly guarantees certainty, and since he never gave up the wish for it (as his last book shows) (10), he could not give up the alleged philosophical method of starting with the inner world. All his criticism of the philosophical method did not make him give it up.

2. THE FORCE OF THE POLARIZATION

It is hard to know why it is so hard to give up this thesis. Beginner philosophy students notoriously find very odd the texts that expound it and that their teachers make them read. To begin with, they hardly know what these texts are about. This fact is notorious. Wittgenstein's disciples often use it against philosophy, yet they still teach these texts and do not explain their oddity and wait for it to disappear. For, it does disappear: after a course of study of the history of modern philosophy, students may find the same texts intuitively agreeable, and then they try to discuss them critically. All too often they accept subjectivism as a matter of course, and then they ask, what can we do about it and how can we find a bridge from the inner world to the world out there.

There is one simple argument that renders subjectivism convincing. It is from the thesis of privileged access: only I have access to my inner world. (11) Even if this does not endow our inner world with certainty, perhaps it is the nearest to it that we can ever get; having privileged access to our inner experiences we allegedly know them best. Wittgenstein had one new idea about this. Suppose direct access to other people's inner worlds was possible – by telepathy or by neural linkage akin to the way we link personal computers. Would this change the picture? If not, then privileged access is of no philosophical import. What is important, said Wittgenstein, is to be right, and right we are about our inner world. On this, Wittgenstein and Husserl were in agreement: we know our inner worlds infallibly. As long as philosophers seek certitude (or foundations), they must ascribe privileged access a significant role: it is a part of all established philosophical methods and it vitiates them all.

The thesis that privileged access is certain has met with systematic refutations in some very convincing experiences. They are on display in perception theory laboratories in almost all psychology departments in almost all large universities. Sensations of colors are the most popular example of inner experience; both Locke and Wittgenstein used them as rubberstamp examples. Many experiments deceive the eye. (The most striking of these is the spread effect comprising a juxtaposition of a blot of paint α on the background of paint β and a blot of β on background α. They look like four colors, not two.)

Since some inner experiences mislead, the quest for certainty leads naturally to a search for criteria for reliable inner experiences. The best criterion was that of Descartes: clarity assures certitude. This may invite a criterion for clarity. We have none. So, we can only exhibit examples for it. If they turn out to be false, we may replace them. This is a tacit identification, not to say confusion, of clarity and truth. Given a clear true statement, an alternative to it may feel equally clear, so that feeling clarity is no criterion. (Descartes said, it was clear that one solid object cannot penetrate another. Leibniz said, we can imagine clearly ghosts penetrating solid objects.)

It is easy to refute the justifications of traditional philosophical methods. The best example is that of Popper. (12) He recommended instead an older method, the historical, or the dialectic, or the historical-critical method. Its users express concern in cosmology. He explains the source of the view of the inner world as primary in its link to the view of meaning and thus also to the view of the analysis of meanings as primary. It was to no avail: the analytic school is still going strong.

Can the picture change? Perhaps, but then the change will be slow. It is hard for philosophers to change, especially if the change may require the mastering of new techniques and new knowledge. Philosophers need some preliminary discussion and some coordination before they will be ready to invest effort. This requires a public recognition of a crisis. Only acknowledged leaders can do that. Until they do, rapid change is not on the agenda.

What can change that? The best mover is curiosity: most of the output in philosophy today is painfully boring. Unfortunately, some interested parties demand that philosophy should stagnate. Their interest is in public relations and they claim that progress is and must be sure and to that end it must be slow – so slow as to be imperceptible to the lay public. Their interest in public relations stems from the wish of the scientific establishment to employ philosophers as public relations agents of science, as its defenders. This renders much philosophy of science shallow advertisements. The establishment of the political segment of the academy wants philosophy to stay reactionary. This keeps much philosophy unclear as long as reactionary politics is not too popular. The academic establishment wishes to maintain the fiction that scholarship must be boring, and keep political philosophy separate from the philosophy of science.

All this, however, is unimportant. At issue here is the power of the theory of privileged access. As things are we all feel that we have a better grip of our inner world than of the world out there. This is an error that we should refute.

3. AGAINST THE GRAND POLARIZATION

Let us ignore the desperate effort to find certitude and the idea that our grip on the inner world is better than on the outer one. Let us attend instead to the following two arguments for it. First, we are not familiar with sensations that we have not experienced. This argument is irrelevant: it applies to all experiences, inner and outer: it rests on the assumption that the inner is primary, which is what it comes to show. Second: access to the outer world is through the inner world; not only is access to the inner world privileged: access to the outer world is not immediate. This is Carnap's defense of methodological solipsism. Its refutation is the final attack. It is smashing – until adherents will come up with a response to it. It is this.

We do not know what the inner world is. Our notion of its diverse items and their characteristics is a theory. This theory is amazingly convincing despite all of its many refutations. The greatest expositor of this theory is David Hume, who gave it its final, definitive, and canonic presentation. This greatest critic of philosophy was amazingly gullible here. His criticism regarding our inner experience of personal identity, of the identity of things "out there", and of causality, too, came to refute not them but their claim for certitude. He never doubted the law of causality and the law of association that allegedly leads to it, for example. And he never doubted the primacy of the sensations that John Locke has listed. Today, no perception theorist accepts this list. Indeed, it is so odd, that to understand it we need some background knowledge. Let me offer one instant, perhaps the most important one. The idea that science rests on experience rests on the idea that some experiences are utterly reliable. Which ones? Bacon said, those of people who are not biased. For example, he himself, he said. And as he saw the sun's motion, he rejected the Copernican hypothesis. Locke reconciled Bacon and Copernicus by denying that we see motion. Rather, we surmise motion from observations of locations, he said. This is false. Even animals see motion: freezing prevents detection.

Locke was aware of sense illusions, and he said this was no refutation of his view that some experiences are utterly reliable. This led to the study of sensations, and the more the study developed the more remote certitude became. Today, perception theorists agree that we learn to perceive, and that the traditional list of primary perceptions suffers from both excess and deficiency.

This refutation need not be fatal: we may wish to rectify the list. We may wish to find out which nerves carry which sensations in order to determine what are our true primary perceptions, and we may wish to construct science out of sensations afresh. Older efforts in this direction, such as that of Laplace and of Mach, were admirable failures, and now we may wish to do better.

Even were this possible, however, it would not do: to perform the task we must connect sensations. Traditionally the connection was associationism, the idea that a constant conjunction of two sensations leads to the expectation of its repetition. This doctrine is insufficient: memory is not only retention and storage, but also retrieval. If we do remember all of our associations and all of their components, we see the world as a maddening jumble (Borges, "Funes the Memorious"). Of all the thinkers who have studied memory, Plato and Freud said, retention and storage are perfect and only retrieval is problematic. Most others, Locke and Hume included, said the opposite. Brain studies and computer science have taught us how mistaken they all were. Moreover, it is clear that we do not notice every association. We are born with some preferences for some perceptions and some associations, those that our imprinting mechanism determines from birth. Moreover, we modify our learning apparatus early. Here not only do we build expectations, but also we learn from disappointments that they need modification. This is trial-and-error of sorts. It is anyway a very different kettle of fish.

All this does not refute the idea that our inner experiences are primary and our experiences of the world come second. Yet it offers already two options and presents the traditional one as inferior to the new one. J. J. Gibson developed the new theory as a new perception theory. He said we need ask not how we form a picture of the external world from our inner world but rather, how does the external world shape our inner world. Because it clearly does, both phylogenetically and ontogenetically (both as a race and as individuals).

All this is very powerful, but it does not yet impinge on the strong sense of conviction about the primacy of the inner world over the outer world. And as long as a strong sense of conviction persists, some may prefer to follow it or at least postpone decision until we can understand its persuasiveness. Well, we need not wait: a strong, convincing experience in favor of the primacy of the outer world explains the situation better. It is this. We use information about the outer world to correct our inner experiences: we use thermometers to rectify our sensations of heat and cold, for a conspicuous example, and we measure wavelengths and play with lights and colors to correct our color perceptions. We can alter the molds of our inner experiences, thereby influencing their results. But the strongest argument about the matter is abstract – much more abstract. We often doubt the adequacy of our experience and of our conduct. We can then run a reality test. If we doubt the propriety of our conduct, then we can imagine the situation in question and imagine a role reversal: if the other person did that to me, would I judge it fair? If yes, then my conduct was fair. When this powerful method does not deliver the goods, there is a stronger one: we can ask others. This accords with Buber's assertion that the very possibility of interpersonal contacts logically precedes the inner experience of them. This, says Gibson, is how we judge the correctness of our experiences. He goes further and explains hallucinations: we all hallucinate, but we usually disregard our hallucinations, or else we check them against others. We may illustrate this by the conduct of paranoiacs: they have lost the will and the ability to undergo a reality test. (13) This replaces the dichotomy between the objective and the subjective by the inter-subjective that is in between. Popper has ascribed this idea to Kant – probably too generously.

4. POLITICAL CONSEQUENCES

There is a catch here, and this is the source of the whole of the present discussion, though it surfaces only at the very end of the discussion no matter how it is structured. It is this. Asking others, we normally ask for the conventional answer to our question. Philosophy began with the distinction between truth by nature and truth by convention, and the preference for the truth by nature over the truth by convention. To follow this preference, we detach ourselves from others and try to see things in as unbiased a manner as we can. This led Descartes to his monumental, brave, world-shaking thought experiment. It has failed and most philosophy still lives on its debris instead of seeking new ways. In our fear that others will feed us the conventional answers and the desire to be autonomous, we lost the ability to cooperate. The catch is this: to cooperate properly we need means that we can acquire only through cooperation.

The thesis of the primacy of the inner world came to support autonomy and thus liberalism in two ways. First, my knowledge depends on my own experience and reasoning, not on what I learn from my teachers. Second, my preferences are dictated by my personal tastes, not by others. Both Locke and Hume stressed that we should leave it to individuals to decide what they should try to achieve. Now this sounds reasonable when ordinary needs are on the agenda, the ones that the welfare state should guarantee. But we do not really know what our preferences are, and reactionary philosophers, beginning with Hegel, declared that some dark motives hide in the recesses of the mind and try to take over and give vent to highly anti-democratic sentiments. The view of the primacy of the inner world thus can easily support anti-democracy. The romantic philosophy came to transform the liberal variant of the theory into an anti-liberal version. Until today, many art critics do not notice the anti-democratic character of the version that they support as they do not connect art with politics, and perhaps with some justice, but still, others may do so with a quite unhealthy outcome.

The new, reformist philosophy thus also relinquishes the dichotomy between conservatism (truth by convention) and radicalism (truth by nature): we should reform tradition, not reject it wholesale. The radical rejection of all received opinion leads to the loss of the ability to pass the reality test, and no reality test imposes the endorsement of tradition: suffice it to be aware of deviations from it (which awareness the paranoiac has lost; see Fried and Agassi (13)).

Thus, the new philosophy also relinquishes the dichotomy between conservative and radical politics. These extremes are contrary to the democratic intuition that is parliamentary and so reformist. The thesis about the primacy of the inner world backs the autonomy of the individual and avoids paternalism. It is thus radical and more democratic than authoritarian. But it is problematic. The diversity of inner worlds is largely socially determined, of course, particularly in matters of tastes and dispositions. And these spill over to opinions. Thus, one who likes people and tends to trust them tends to endorse views that justify this and to view them as the foundations of democracy. Similarly, those who are disposed the other way tend to become authoritarian. Neither attitude is right. People deserve trust not because they are trustworthy but because they deserve the benefit of doubt – because they need trust to improve. Also, no matter how untrustworthy we all are, this does not render any authority trustworthy. So, democracy and openness are the order of the day. (14)

The radicalism of Bacon and Descartes led to the view of the primacy of the inner world. They are admirable as they wished to set the spirit free of a stifling tradition. But they went too far in seeking certitude there. Their view that the inner world is primary and that full autonomy must precede all research is an expression of this excess. The courage and logical force of their move is what is so captivating and looks so convincing despite so many counter arguments. But they could not guarantee full autonomy, and so their readers became dependent on the promise of certitude and of its origins in the inner worlds. The insistence on autonomy is what made classical philosophy individualist and progressive. As autonomy was lost, the inner world could and did dictate ideas and feelings that led to the worst political disasters humanity has ever known.

We should relinquish the traditional philosophical dichotomy between truth by nature and truth by convention. What we call science is not the truth by nature, which we still seek, and not the truth by convention, which we wish to improve upon; it is in between, a surrogate for truth by nature. So we have to improve upon it by a constant struggle. This is the best version of fallibilism that we have thus far. It is enormously common sense – or it should be. Once we consider this, the convincing power of the traditional views about the utter (or at least maximal) reliability of the inner world vanishes like the morning mist.

NOTES

1. The existentialists count as a stream within the phenomenological school. The other schools of philosophy are mostly older, as their use of the label "neo" indicates: neo-Thomists (Catholics), neo-Kantians, neo-Marxists (including the Frankfurt school, some of whom refer to themselves as Marxians), and more. (The analysts also call themselves neo-positivists and neo-empiricists.) And then there are the critical realists (mostly Popper and his fans). I do not know what the older schools say about the inner world.
2. See Nathaniel Laor and J. Agassi, "How Ignoring Repeatability Leads to Magic", *Philosophy of the Social Sciences*, 30, 2000, 528-86.
3. Edmund Husserl, *Cartesian Meditations*, 1960.
4. Claude Lévi-Strauss, *The Savage Mind*, 1966.
5. Bernard Shaw, *Saint Joan*, Preface.
6. R. D. Laing, *The Divided Self*, 1960, first chapter.
7. John O. Wisdom, *Philosophy and its Place in Our Culture*, 1975, 154.
8. See my "Conditions for Interpersonal Communication", *Methodology and Science*, 26, 1993, 8-17.
9. Martin Buber, "The Word that Is Spoken", *The Knowledge of Man*, 1965, 109-119, esp., 111.
10. Ludwig Wittgenstein, *On Certainty*, 1969.
11. See my "Privileged Access", reprinted in my *Science in Flux*, 1975.
12. Karl Popper, *The Logic of Scientific Discovery*, Preface to the 1959 English edition.
13. Yehuda Fried and J. Agassi, *Paranoia: A Study in Diagnosis*, 1976.
14. Karl Popper, "Public Opinion and Liberal Principles", republished in his *Conjectures and Refutations*, 1963; John Watkins, "Epistemology and Politics", republished in J. Agassi and I. C. Jarvie, *Rationality: The Critical View*, 1987.

5. RESPONSIBILITY

Chapter 5.1

VALIDATION

0. THE PROBLEM OF VALIDATION

Scientific hypotheses possess a high status: they are valid. (The word "valid" comes from the Latin for strong.) How do scientific hypotheses acquire validity? This is the problem of validation of scientific hypotheses. It is standard fare: it is the major item of discussion in many philosophical texts especially those devoted to the philosophy of science, be it a textbook or a history or a monograph, in which a valiant philosopher tries afresh to tackle it. Some of the most powerful minds ever have tried to tackle it, yet to no avail. This is hardly surprising, as there is a famous proof that it is insoluble. It is simple: every validation is the application of a criterion that also needs validation. Diverse variants of the proof appeared repeatedly ever since Antiquity. The modern tradition presents it in the variant that David Hume has offered, and as Hume's criticism of induction. Most philosophers today still declare it the most central problem in the philosophy of science, perhaps in philosophy generally: how can we answer Hume's criticism?

I spent years in the effort to comprehend the problem and to find its import. Why is it so important to validate? I still do not know. Validation seems to do not with something abstract like science and its merits, but with something very personal, the profound sense of insecurity that is psychological rather than philosophical, like the fear of falling off the bed during sleep or of the brakes on a motorcar not properly functioning. This is how I understand the import of the problem. My understanding may be an error. Let me report how I came to it. Hume's criticism leads to idealism, namely, to the denial of commonsense realism. As it happens, in my adolescence I had the profound feeling that the admission of idealism is just intolerable. And I repeatedly meet people who feel this way and cannot free themselves of this feeling. There is a prevailing feeling of a need for a disproof of idealism, and its absence makes one feel awful.

A sister of Wittgenstein reported that philosophy hit him – presumably, it was Hume's criticism – in his late adolescence, and that he never recovered. Perhaps. At least we know that he spent his last days on earth struggling with the problem, writing a brief study that appeared under the title of *On Certainty*. And no, he did not refute Hume. He tried to explain there why there is nonetheless no need to worry. I do not know why he said we should not worry. But let me report my experience with some who do worry.

1. VALIDITY AS RELIABILITY

The beginning of my academic career was in London, a few decades ago. I was a lecturer in philosophy under the tutelage of my revered teacher Karl Popper. He was then a local celebrity. Many people wanted to acquaint themselves with his ideas without investing much effort in the process. To satisfy this wish I gave popular public lectures. Popper's leading idea is that Hume's criticism of induction is valid. The past cannot justify expectations about the future. So this was the first thing I tried to explain to my popular audience. As the style of my lectures was interactive, the discussion got stuck at once. My audience expressed a sense of shock and incredulity and would not let go until I would answer Hume's criticism. They asked me insistently questions like, do you mean that pressing my car's brakes pedal is no reason for me to expect it to stop?

The question was insistent and painful. I found this very revealing: it looked like an expression of a worry about the brakes of cars. This worry is reasonable. Too many times brakes of cars fail. It is not exactly the possibility of a failure of a car's brakes that is frightening, of course, but the possibility of this happening while traveling fast on a highway, since under such conditions the failure generates disaster. So I assumed that behind the *façade* of philosophical abstraction there is the genuine problem of the reliability of some hypotheses. Is the hypothesis reliable that my brakes are reliable? Which hypotheses are reliable?

This is a general human concern. Already ancient prophets accused some of their peers of providing deceptive reassurances instead of ideas they deemed really reliable. Differences between ancient and modern conditions should not obscure what they have in common, namely, the wish for a secure future. Regrettably, today philosophers tend to hide genuine human concerns behind abstract remote ones. Perhaps it is hard to avoid both displaying and hiding it.

The aim of these lectures was not to allay fears about road accidents but to present Popper's philosophy. I had to begin with his endorsement of Hume's criticism of induction. Most philosophers in England at the time – the late fifties it was – were trying hard to diffuse Hume's criticism, alleging that it flies in the face of common sense. They bravely ignored the simple, obvious fact that common sense proclaims nothing certain except for death and taxes. The future is unknown and not even knowable. Life is constantly full of surprises, though also of surmises about the future, only few of which come true. They are partial, short-term, and often frustrated. This commonsense observation is besides the point at issue, which is the problem of the validation of science. Our reliance on science fully accords with commonsense: the reliability of our surmises increases with the help of science. The question here is, how?

The presumption here is that the reliability of science is common sense. Most English philosophers ignored this presumption then. As it couples science and common sense, most philosophers of science ignored it too. Bertrand Russell repeatedly presented the problem of induction this way; but he also put it differently, as relating to the reliability of the principle of induction. A. J. Ayer put it almost this way, but not quite. I was not capable of handling it better. My audience asked me the traditional question: how can expectations be reliable? This question is unanswerable, since behind it stands a false presumption. Contrary to it, most expectations are unreliable. A better alternative to this question is, how does science increase the reliability of some expectations. This question is much better, and it is common sense. I will return to it later on.

Repeatedly I faced questions like, do you say that when I press the brakes pedal on my car there is no reason for me to expect that it will stop? They are hardly questions to answer. If not pleas for help, they are expressions of a reasonable worry. They also display some poverty of comprehension or of the imagination. Properly understood, Hume's criticism of induction leaves completely undecided what should happen after pressing a car's brakes pedal. It should prompt a question not just about the car coming to a halt or not, but about any sequence of events, including, say, the possibility that the driver will then be somewhere on the moon. I tried to explain this to my audience. It left them convinced that I was way out – on the moon, perhaps. My lectures did not fail totally, however, if only because I came from the university to present ideas of a famous professor. But if we do not know what to expect from pressing a car's brakes pedal, how can we know what to expect of a famous professor, not to mention his unruly young assistant? The only thing that was certain about my lectures is that they raised a furor. They were challenging, a kind of intellectual exertion, not useful but diverting. This frustrated me no end.

Questions like the one about a car's brakes did concern me, however. They are instances of a famous ancient problem, known as the problem of rational choice of action. How should one choose a course of action? What is the criterion for rational action? What is the rational criterion for action? The problem of rational action can be put in the most practical manner, as no one doubts that some actions are more foolish than others, and perhaps some are the wisest, the optimal actions. Is there a general criterion for improving conduct, and if so what is it? Can one find such a criterion and use it in order to optimize actions or, alternatively, at least render them less inadequate? Surely, we may learn to improve our conduct. How?

The most popular answer to the problem of choice is inductivism. It rests on a few presuppositions. First, action rests on belief. The rationality of action then rests on the rationality of belief. Second, science is a set of rational beliefs. This explains the concern of inductivist philosophers of science with the problem of rational belief: what belief is rational? The answer is, one that science supports. This is what my audiences expected me to say, and as I frustrated them, they sometimes tried to help me say so. But Popper explicitly refused to say so, and representing him I had no choice. Now that I no longer represent him, should I say, reliable conduct rests on science? If I will, would it put me over the worry? Would it now put Hume's criticism to rest? How?

A funny thing happened in the last paragraph. Far from putting the matter to rest, the answer forced out of me worsens things. If I reached the right answer, then I did so by a shortcut: rational action rests on rational belief; rational belief rests on science; hence, rational action rests on science. No need to believe – in science or in anything else. This is annoying. Can one endorse science blindly? Should one not first believe in science? Can one be a convert to science? How do we render beliefs rational? These questions are terrifying. We need not tackle them. This is fortunate, since they are unanswerable – for a few strong reasons.

The idea that science is set of beliefs is disturbing: it is no doctrine and no religion substitute. It is a search. Though I knew little about Popper's ideas when I was engaged in expounding them wherever I received invitations to do so, I liked the title of the book about science he had authored in German; in English the title reads "the logic of research". (Unfortunately, in English translation it has a different title.) I also enjoyed his ridicule of the view of science as a supreme authority. Science with a capital S, he used to call it in class.

Belief is not the outcome of decision; it is not open to manipulation. (Diverse thinkers have observed this, including Robert Boyle, Benedict Spinoza, C. S. Peirce, Bertrand Russell, and George Orwell.) The search for an answer to Hume's criticism rests on a few errors. One: scientific ideas are objects of belief. Two: belief can be rational. Three: action rests solely on the belief. This is obviously false, as Bernard Shaw observed (*Doctor's Dilemma,* Preface), and Eliot Friedson reported; (*Profession of Medicine*, 1970, 168-9): patients often obey physicians, and assume that recipes rest on science when physicians act in self-interest or trust others or act on mere expediency.

Some people follow instructions of physicians without trusting them. Often action follows no instruction, no understanding, and no belief. Often decision takes place under pressure. The worst is, people often act against their better judgments. Being tired of fighting everyone who has their welfare at heart, or in order to give others the benefit of the doubt, or in preference for the more prudent course of action over the correct one, or in deference to seniors, or because the law makes demands that are not quite right but prevails anyway, they often yield, at times too readily.

Popper rejected the inductivist answer to the problem of choice and spent much time and effort arguing against it. What was his alternative?

I asked Popper all sorts of difficult questions when they troubled me. I regularly asked him to talk on the problem of rational choice. He had two very different answers to it. One of them I heard him assert in class, which I will refer to as his fleeting answer, and which is disappointing. The other he published. It goes well with his general outlook and is very intriguing. The fleeting answer is this. We try hard to refute our conjectures (such as the ones that present the brakes on cars as reliable). A failure of attempts to refute a conjecture corroborates it. We then endorse it. What else can we do? I am reporting his fleeting answer as literally as I can. It ended repeatedly with a rhetorical question: what else can we do?

It is unbecoming to answer a rhetorical question, but I did so anyway. We can reject them, I said. We can refuse to believe even the best, most corroborated conjecture. Nothing can stop us from doing so. (All democracies specifically design laws in order to safeguard the right to accept or reject any idea, let me add.) Forcing oneself to endorse any conjecture is futile. Realism allegedly imposes the endorsement of corroborated conjectures, but then realism is uncommon, and whether it is good or bad is open to debate. My popular audiences took it for granted, most people I know do, that realism is sensible. I wonder. I do better than most people I know without being much of a realist. Colleagues say that I am a realist, proving by my sane conduct that I consider Hume's criticism of induction invalid, at least when I do not think about it. But then these same colleagues, when they do not think about it, deny that I am a realist; they say so whenever I offer them some specific advice. They dismiss it, they explain, because I am no realist.

These colleagues respond angrily to this, saying it is casuistic. The realism that refutes Hume's criticism of induction, they say, is different. It emerges from our walking on God's earth and breathing fresh air and eating our daily bread, they say. As Hume's criticism does not alter conduct, it is void.

This last inference is invalid and its premise is false. Hume's criticism is valid whether it leads to some specific action or not. Validity need not induce change in daily conduct. At most it increases its rationality. What rule makes conduct rational? What application does that rule have? What application can it have? Does Hume's criticism of induction invalidate all such rules? Does it destroy rationality? The advocates of the criticism known as Hume's, Sextus Empiricus, Pierre Bayle, as well as Hume himself, have all endorsed common rules of conduct. Their critics say, their criticism is invalid, as it does not alter conduct. Not so: they showed that criticism need not alter conduct.

Popper's fleeting answer, we remember, is that we have no choice but to endorse the verdict of experience, voiding Hume's criticism of impact on daily life. This is an error: Hume's criticism has influenced Einstein, and Einstein did exert a great impact on daily life. And it has more applications and should have still more impact on daily life. The paradigm is the dictum that, since we are never certain, we should never enact the death penalty. More generally, Hume's criticism encourages granting people the benefit of doubt. This is imperative, as it is imperative to be optimistic, to give the good Lord Himself the same benefit, as it were, though in our troubled times optimism and divine benevolence are unrealistic and hard for empirical evidence to support.

This is too sophisticated, and so it is suspect. In one profession, however, Hume's criticism of induction is significant, and regularly so. That profession is scientific research. As a physics student, I read Einstein's autobiography where he explains this in a simple and personal manner, with a grateful reference to Hume. I was deeply impressed. I learned from him that it is not easy but very useful to criticize firmly established ideas. This is what made me reject Popper's fleeting answer.

Corroboration can systematically mislead. In detective stories, when suspects too obviously appear guilty, the tough realistic detective says, no, this is too obvious to be true: there must be a catch here. And usually there is, at least in detective novels that remain challenging. Such catches occur in science too. Science reporters conceal them. Over-eager to convince, often they thoughtlessly conceal some difficulties, even well known ones. They are then surprised that it took great minds to make discoveries that they describe as very obvious.

Bernard Shaw noted (*Don Juan in Hell*) that the reasonable try to adjust to their environment, and only the unreasonable try to adjust their environment to themselves. Hence, he concluded, we owe improvements to the unreasonable. This sounds like a paradox. Like so many things that wise people say, it sounds more like a witticism than the obvious truth that it is. For example, those who recognize the scary unreliability of the brakes on their cars are reasonable. Those who try to improve them are admirable. Assume that inductivism is reasonable, that it succeeds as a means for the selection of the best objects for belief. Rebelling against it in search of something better is then unreasonable but still admirable.

My own reason for saying one can refuse a well-corroborated hypothesis, has nothing to do with the fear that the brakes will fail. It gains its rationale from the history of science. The most firmly established conjectures, Popper always stressed, inhabit science. The most firmly established among them is Newton's theory of gravity. Einstein presented an alternative to it and paved the road to its refutation by devising a crucial experiment. Over half-a-century earlier Michael Faraday tried to refute it, failed, and refused to accept the verdict of experience. He prepared the ground for Einstein's enterprise. If Faraday could do so, then it may make sense to do so. His own reason, incidentally, was metaphysical: he could not believe in action at a distance.

2. VALIDITY AS RIGHT CHOICE

What is rational belief? Popper's published answer to this question is intriguing. His fleeting answer was, we cannot but endorse the most firmly established views; he also vehemently denied that he had ever said it. Perhaps he was not clear in his mind, or perhaps he was in two minds. He repeatedly affirmed that doubt about any assertion about the world is always permissible, however undeniable that assertion looks, and he said explicitly that the inability to imagine anything is no argument, as it only shows how limited the imagination can be. And he also always said, it suffices to know what not to believe; the obligation to endorse the right conjecture is unreasonably excessive. Yet, this leaves a pressing problem unanswered. The practical question remains. When should I rely on the brakes of my car? The answer depends on the theory of choice. At times Popper said, try to find fault with the best conjecture around; more you cannot do. This way he possibly endorsed the view that (rational) action rests on the (rational) choice of beliefs. This answer to the problem of rational choice of action raises the problem of rational belief, and Popper tried all his life to avoid answering it, since in his view its proper answer is partial. It is only in the negative. Do not believe a refuted hypothesis. His view does not offer a positive answer. It does not say what hypothesis we should believe in. On this he was clear, especially when he said, if we must have a positive theory of rational belief, then the best candidate for an answer is the following hypothesis. The most credible hypothesis around is the most highly corroborated one, the one that best stood tests devised to refute it. But it was amply clear that even the identification of corroboration with credibility he considered excessive.

Popper thus rejected his fleeting answer, the view that we cannot but endorse the best-corroborated hypothesis around. Following Russell, Popper denounced it as the preference of instinct over reason. Hume stuck to his inductivism, and when he could not justify it rationally, he justified it pragmatically. He said, we cannot help but trust past experience as the best guide for the future. Both Russell and Popper took this to be the beginning of the irrationalism that followed the Age of Reason. Perhaps this was unfair. The disappointment in reason resulted not from a philosophical trouble but from a political one, namely, the failure of the French Revolution that expressed the ideals of the Age of Reason and led to a reign of terror.

The view that we cannot but endorse the best-corroborated hypothesis around is hardly applicable to real life situations such as the matter of reliability of car brakes. We can do all sorts of things about brakes, as experience clearly shows. In particular, reliable or unreliable as brakes are, people concerned with the matter make good sense when they prefer cars whose design includes more safety devices. The simplest safety devices are auxiliary brakes that come automatically into action in emergency. And so these have entered the standard design of cars. Their construction costs are marginal, and European manufacturers had them installed for some time before their American competitors did. These hesitated, since doing so might draw attention to the fallibility of the original brakes of some cars, and perhaps even involve manufacturers in costly lawsuits. Yet, competition required improvement. So, the American manufacturers inserted emergency auxiliary brakes in unison, and they did so by first lobbying in Washington to legislate the requirement to insert the added brakes. Since this was a very practical solution to a very practical problem, we may compare it to the inductivist solution.

Inductivists say, experience can and should justify forecasts. Reasonable actions, then, should rest on experience. How? The reasonable should expect the future to resemble the past. This sounds very sound, but only because it is sufficiently vague. We cannot possibly take it literally. Surely, my youth is dead and hidden in the past. Suppose it means, let us repeat actions that have led to success and avoid repetitions of those that led to failure. This should not rule out progress. Inductivists suggest that clinging to experience is scientific and so it is truly progressive. This is a serious error: science does not progress by induction. The readiness to try out new ideas, and to apply them only if they stand up to tests, seems a better candidate. This sounds like induction. It is not, especially since researchers try and fail and try and try again, refusing to admit defeat. Bacon said, they cling to their ideas; Einstein said, they are opportunists. Bacon opposed trying unlikely new ideas; Einstein favored it.

As this discussion is abstract, let us apply it to experience. Inductivism has no proviso for class actions as it takes each action individually, urging each individual to act in accord with the view best supported by all the facts available to that individual. A coordinated action by many individuals is not the same as class action, even though the two kinds of action share the idea of personal choice. What is it, then? By the inductivist standard, class action is the sum total of actions of many individuals. For this to obtain, each of them must possess more knowledge than most of us can possibly have, and perform many complex calculations that are too difficult, if not impossible, even for experts. Modern decision theory adds that individuals differ in their strategies and some are more prone to take risks than others. Prudence always suggests procrastination when possible, and often in the form of a search for additional information before taking any action. Notoriously, people who want to be sure that their actions are sound suffer from paralysis and cannot act at all.

Conduct is decided by additional differences between individuals – physical, psychological, and social. The same individual in the same physical circumstances will behave differently in different institutional settings. Views on the conduct of individual drivers and their rationality may be controversial, but the rationality of the American car manufacturers is usually transparent, and it clearly deviates from the policy that inductivists recommend, since they operate in their official capacity and according to the law. It is unreasonable for any of them to act alone, so that no matter how reasonable the idea behind some possible action is, performing it was unreasonable unless it was in unison.

This argument conflicts with the claim that the rationale of cooperation is individual belief in its future success. It therefore enables us to generate many examples to refute the idea that scientifically attested belief alone guides rational action in a given situation. In particular, the law forbids marketing innovations without license, no matter how well science establishes them. Some licensing procedures are strict and clear. Safety laws regulate cars, and more so planes. Before a plane receives a certificate of airworthiness, it has to undergo certain strict procedures, including, in this order, calculations, simulations, wind tunnel tests, and test-flights. It must pass each test before it qualifies for the next. Suppose a test pilot has a rational conviction about the validity of the design of a new plane and is ready to forego the early tests. Insurers will not allow this. They will not underwrite the test flight before a wind tunnel test, and so the manufacturer will not allow it. Suppose, on the contrary, that a plane is ready for a test flight but the pilot refuses on some grounds. Then another test pilot will take over. Inductivists may respond to this by saying that legally sanctioned procedures have undergone inductive validation. Not so. Laws are faulty, and they often meet with challenges. Faulty laws are valid in the sense that they are obligatory, even if they are not good. These will undergo reform, we hope, but, for now, contrary to the inductive canons, we must comply with them. The same holds for many new cures. The use of a new cure without license may land a practitioner in jail regardless of clinical evidence. This law is reasonable, but not always. For example, forbidding the use of alleged cures for AIDS on safety grounds makes no sense for patients who have no time to spare and nothing to lose. The law is still binding in these cases until we reform it. Sometimes defects lead to hasty reforms. Are such procedures valid? This is the second level of the problem of induction.

The law of the land develops by the method of trial-and-error, not by the reliance on hypotheses based on any pool of given experience. Experience may later validate it or we may find it harmful. How then do we validate it? Inductivists say, the validation of a hypothesis is its accord with the pool of given experience, whatever this pool may be. Popper has suggested a view about validation that better accords with experience, as well as with the law of the land. The validation of a hypothesis is its corroboration – the accord of that hypothesis with the facts that accrue as the result of severe tests, which are the failures of attempts to refute it. The harder and the more ingenious the attempts to refute it, the more corroborated it is when these fail. A factual observation that accords with Popper's view and clashes with inductivism is as follows. Any challenge to any scientific theory causes the suspension of all of its past validations. The challengers then devise new attempts to refute it. The challenge to the tests of Newton's theory is the observation that they took place in weak gravitation fields. This led to tests performed in a strong one. Similarly, the challenge to a statistical test is a claim that it is biased, namely that its sample was not representative. The description of a bias paves the road to a new test whose sampling method aims at its avoidance.

Innovation has to undergo prescribed tests before its implementation. According to Popper, the tests are attempts to refute it. This sounds odd. But why not test hypotheses as we test other products? When tests of products fail to meet the legally prescribed standards of severity, we complain and force producers to test their products properly, namely continue trying ever harder to find all their faults. The legally prescribed test procedures are usually not good enough. They help a little, but then even the market mechanism does that. The market mechanism, too, works by trial-and-error, by tests that aim at refutations, not by any application of any inductive canons: as no one knows what these canons are, it is impossible to implement them.

Popper's view is thus very general; it is superior to inductivism, since tests abound in science, in technology, and elsewhere, and all tests are attempts to refute. His view still is inaccurate, however, since the logic of science differs from that of technology. Science hounds error so as to have a more accurate image of the world; so the aim of tests in science is to refute, and so corroboration is just failed tests. The end of technology is practical. So, the aim of tests in technology is to fail to refute; there corroboration is a condition for success. In technology, the primary aim is to refute the forecast that the product will not stand up to expectations of reliability and of absence of side effects. And the law specifies the list of forbidden side effects. Tests of scientific theories may have a secondary technological aim: to use corroborations as means to find new useful practices (such as the use of radioactive elements for tracing in biochemistry and in biotechnology).

Popper's philosophy is problem-oriented. As it became more so, it grew increasingly easy to follow. Just consider the question, why and how do tests take place wherever they do? In technology, the answer is simple. The law requires tests. Now why does the law require tests? The answer is still simple. This is the best-known way to prevent unwelcome side effects. Newer and better ways to do that lead legislators to improve the law as fast as possible. Yet improving the law is also a technology, and it is also not the best possible, but it is the best available, and finding a better way to improve laws leads to a suggestion for a constitutional reform. This answers the question, why technology conducts tests, and it explains how Popper could move so easily from methodology to politics.

Why and how do we conduct tests in science? Scientific tests are the means by which we hope to improve scientific knowledge. It is a matter of ingenuity and ability of individual researchers. They try their best, since testing is in their interest as researchers. Those who want to improve upon extant theories will do well to try to refute them. An alternative method is to seek improved variants in order to devise crucial tests between the original and the variant – and refute at least one of them. There is no general reason to prefer one method to the other. Researchers try both.

Most researchers are still inductivists, and so they repeatedly suggest that the aim of research is to validate hypotheses. They do not say how they validate: they do not know. Popper does. He observes that it is the result of attempts to refute hypotheses, that what they view as validation is the corroborations that happen only as tests fail. This is thought provoking, as it suggests that researchers do not know what methods they follow, and that they may learn this and thus improve their mastery over their methods.

3. VALIDITY AS REASONABILITY

Inductivists are still unbending. The expectation is unreasonable that we will be on the moon when we push a brake pedal. The reasonable expectation is that we will still be in our cars. They say, here experience guides expectations. They suggest that this is the whole truth. It is not. Rather, beliefs about the laws of nature guide some expectations. The laws of nature prevent transportation to the moon with no vehicle. Now this is obviously true, yet it repeatedly meets with great displeasure: it is not the expected answer. We expect validity, namely assurance, whereas our views about the laws of nature may be erroneous. This is always so. Validation is no assurance. Some things that once we considered possible we now consider impossible, and *vice versa.* (Transmutation of metals is a good example. We viewed it as possible and then as impossible, and now it is a fact.) The expected answer should provide validity and carry the stamp of science. Hence, the desire to validate is the desire to avoid error, to be divine. The demand from science to validate is the view of it as divine, Science with a capital S, as Popper used to say in class, we remember.

This leads to an angry response from both popular audiences and philosophers of science. They view this as a gross, damaging exaggeration. In response to it, they lower the target of science from certitude to high probability. This response is common, though Hume demolished it centuries ago. It is an unsupported conjecture at best. If it is unsupported, then it is simpler to hold scientific theories as unsupported conjectures than to support them by the unsupported conjecture that they are probable. If it is probable, then the question reappears: what is the status of this probability? And the reiteration that it is probable undergoes repeated reiteration, and then the probability in question diminishes repeatedly and thus it approaches zero. If it is certain, then, again humans possess absolute knowledge. So, at best the claim for probability leads to the same impasse as the claim for certainty.

How is it that sophisticated philosophers of science ignore such a simple reasoning as the previous paragraph displays? I do not know but the next paragraph offers a possible answer.

What are the laws of nature? We do not know. But we look for them. How do we do that? The answer is simple. We observe regularities (the four seasons). First, we deem these regularities laws of nature. So, we explain away the exceptions to them (droughts). We then take the exceptions seriously, and try to find the regularities that control them. These should secure both the observed regularities and the deviations from them by reference to some conditions – initial conditions so-called. Repeated exceptions to regularities are also laws. Vital regularities (the seasons) depend on initial conditions (especially the tilt of the Earth) more than we had suspected. The story moves away from common sense. Even doubt about tomorrow's sunrise, that once came up only in fantastic fairy tales became real. Philosophers of science seek assurances that the sun will rise tomorrow, as science raises ever-newer doubts about it, such as the possibility that the sun will explode one day.

We want to know on what initial conditions our lives depend. We make conjectural statements of the laws of nature, deduce from these conditions under which the observed regularities will be regularly violated, and test these. This picture of the theoretical sciences is satisfactory, but it is no guide to action, and it gives no assurance, no validation. How, then do we validate? Common sense often seeks validation, and efforts to obtain a clear idea of this come up with two *a priori* unsatisfactory results, one that only science validates, and the other that nothing does. The truth is clearly in between: we do repeatedly observe qualified validity and we are desperately trying to capture it. Attempts to equate validity with probability again identify validity with science. So let us try again.

By inductivism, validation makes an idea scientific. Example: science supports and thus validates the expectation that the sun will rise tomorrow. Our concern is thus not only with the efficacy of car-brakes but also with tomorrow's sunrise. Stories about silly people worried about tomorrow's sunrise display the view that skepticism is not serious, and yet doubt gnaws all the same.

This illustrates again a violent deviation from common sense. The traditional question is, how is validation possible? We should replace it, says Popper, with the question, how is increased validation possible? We may likewise alter the question of reliability: not, how is science reliable, but how does it increase reliability? This question, to begin with, moves the discussion from full validity and reliability to their degrees, and from science in general to extant technology. The new questions avoid the extreme expectation from the answer: there is no perfection. The new questions, then, rest on the observed regularity that technology does increase validity and reliability, but within reason, not absolutely. For example, the introduction of irrigation reduces the dependence of agriculture on the unreliable weather.

This example raises matters thus far overlooked. Common technologies like irrigation are pre-scientific. Tests are proper when aimed at failing conjectures, and so, surely, irrigation did pass proper tests. Hence, tests may properly apply also in pre-scientific or extra-scientific contexts. Science is demarcated not by its test procedures alone, but also in the regular tests and the regular efforts to produce new explanations to put to test, in regular efforts to make test procedures and new tests, in the construction of laboratories, and in the development of the modern laboratory life-style.

The example of irrigation illustrates the limitation of validity. We all agree that irrigation increases the ground's salinity to intolerable levels. Excessive irrigation turned the Garden of Eden into the desert of Iraq. At the time of my story, Popper made this point in class: new solutions raise new problems. He published it later.

This point is not new. Knowledge of the existence of undesirable side effects of technology appears in ancient myths, beginning with the one about the expulsion from the Garden of Eden, the one about the strife between Cain and Abel (the farmer and the shepherd) and the one on the golden touch of King Midas that destroys all life. The prophets of scientific technology always expressed the hope that whatever side effects technology may have, the growth of science will help overcome them. Thus, sophisticated methods for the reclaiming of polluted arable land evolve, and with it the idea that scientific technology can fight all sorts of pollution and other side effects of technology as well. This is the view that problems tend to shrink. It is shaky. Technology does not put an end to pollution. Hopefully, people may do so if they set their minds to devise new technologies to that end.

Solving problems invites investments of time and effort with no guarantee for returns. The idea that new technologies will overcome all side effects is an expression of a hope. We can never be sure, and we can never even have grounds for hopes. (Even inductivists have to concede this, since the discussion here is of unknown effects.) The implementation of new technologies to overcome side effects of older technologies is often a matter for legislation, since usually pollution damages the public. (Economists call involving others without their consent external effects. Usually external effects invite redress. The market mechanism does not take care of it. Thus, damage from passive smoking found its redress in law courts. Advocates of the market mechanism as a cure-all oppose legislation that defends consumers, and so they are dangerous.) As legislation is also a technology, it also takes time and it also has side effects. The hope, however, is that the combination of scientific technology plus democratic legislation will alleviate matters rather than aggravate them. But the idea that side effects become ever smaller rather than ever bigger is still but a hope, perhaps a refuted one. Some remarks on this hope will bring the present discourse to its end.

4. VALIDITY AS SECURITY

However satisfactory the validation of a technology is, it has side effects, and so will the technology that should overcome them. What does all this amount to? As to safety, that of the situation in general and that of my car, will it improve or deteriorate?

This question concerns the system as a whole. The whole system does not become more reliable by rendering any item in it reliable. On the contrary. To insure a part is to raise its dependence on other parts, thus making the whole less reliable. All insurance follows the same model. It applies to expected damage with an unknown exact location, and it does no more than spread the loss. Some of us are sure to get sick, but mostly not anyone in particular, and so the simplest insurance is that of creating a job pool, of agreeing to work for those who fall ill. This is the oldest system of insurance extant. It exists in all villages. There is no insurance against unexpected damage and against damage whose cost we cannot spread. Thus, when all the villagers are sick, the system fails. There is no insurance against epidemics. (An epidemic is any trouble involving too many people for the insurance system to cope with – sickness, war, or the collapse of the solar system.) Hopefully, an insurance system does not make life less stable, since hopefully it has no known ill side effects. But in the case of a national system, excess insurance may lead to a collapse of the system and then the subsystems become autonomous (Herbert Spencer; Hughlins Jackson). The aim of insurance is to raise the stability of insured sub-systems and thus that of the system as a whole – but only up to a point: beyond that, it may very well cause a collapse. At times, it may even raise the risk it comes to meet, as does the motorcar insurance that reduces the incentives for good driving and for good car maintenance.

In Spaceship Earth, things are scary. However beneficial the insurance of any sub-system is, and however improved the possible benefit due to the tremendous advances of technology, the stability of the terrestrial system as a whole, of Spaceship Earth, is for the first time under threat from within, and science is a major contributor to the new threat. This is the outcome of the combined effect of Auschwitz and Hiroshima. The former shows that even the most civilized, the technologically most advanced society, can become barbarian beyond imagination. The latter enables the technologically equipped to destroy Earth. Their combination is a recipe for extinction.

We began with the discussion of the worry about the reliability of the brakes on a car, moved to discuss the validity of the theories behind its design and then to insurance, ending with the threat to human survival. Anti-intellectuals send us to look into the brakes and examine them for reliability. Intellectuals raise the general question of reliability. The terrible conclusion appears unavoidable then, that no brakes system is infallible, even though the insertion of emergency auxiliary brakes improves matters somewhat. The search for foolproof brakes, then, is a fatal error. It rests on the view of science as foolproof. We have nothing external to rely on, only the internal system of insurance, and it does not work against epidemics. (Insurance against a local epidemic spreads damage to wider populations. But the limit to that is the limit to all useful insurance.)

So, let me try for the last time, very briefly, to present validity in basic terms. The world is insecure, but secure enough for the survival of the species. This is the basic insight we owe to Darwin, and this is why his contribution always had a tremendous philosophical import. The world, he showed, is for the time being safe enough to have escaped extinction, but not as safe as we want. We want to raise our security. How? This is a reasonable variant of the problem of validation. We do so by exploration. Like puppies, humans sniff around. Humans also try to characterize their environment. Other than proper names and definite descriptions, parts of speech are universal: they are there on the supposition that what they denote is shared, possessed by different things. This makes even language, let alone science, universal. It applies everywhere. Science is also universal in a stronger sense: it is the same for all societies. This leads to the idea that all humans are alike in a most significant sense. We are all rational – if we are – in the same way. The name for this idea was "the brotherhood of men", except that this title is unfair to both women and children; it is better to speak of the siblinghood of humanity.

The siblinghood of humanity is a vision that preaches toleration. The vision has promised perfection, and thus it prompted intolerance. As Robespierre has put it, why should I tolerate opposition when I know that I am right? Bertrand Russell found this an intolerable fanaticism. In his *Skeptical Essays* he said, in principle only that which we know we may teach. He had studied logic in search of what is knowable, and concluded in astonishment that only logic and mathematics are.

And so, we are back at the start. How do we decide what we should teach? How are we to decide what is valid? What is validation? Russell, who was skeptical all his life, found this last question central because, he said, there must be a difference between science and insanity. Hume wrote a moving letter regarding his failure to obtain an appointment as professor of philosophy in his hometown Edinburgh. This hurt him, but the ground given for disqualifying him hurt him more. It was the view of him as a skeptic. He protested: skepticism is facile. Discrimination and action somehow reflect serious thinking. How?

5. CONCLUSION: VALIDITY AS RESPONSIBILITY

The classical question still stands, but not the classical answer to it. That answer rests on presuppositions that are too simple-minded. We do not act on our opinions or beliefs alone. Validity is not the same as reliability. And reliability only partly determines action. Hence, the choice of action is a matter of responsibility, and the classical view cannot account for responsibility, even though it is a commonsense concept. Strangely, responsibility is almost entirely absent from philosophy and even from social science. The first to have given it serious attention was the sociologist Max Weber, early in the twentieth century. The philosopher who first studied it is Popper, in his *The Open Society and Its Enemies* of 1945. It still has not yet elicited much response on this point.

What action is responsible? The answer is not general but specific. We have to examine matters case by case. There is, however, something common to all cases of responsibility. Responsible people do their best and they can then only hope. At times the law prescribes what this best is. The law is also not good enough. It does not apply to every case, and in most cases to which it applies it is very partial, as it excludes some options but it hardly ever limits action to one option. As even the best laws we have are imperfect, they cannot avert all disaster. And so, when disaster strikes we try new means to handle the situation. And when we fail, we institute inquests. We ask there, did the people in charge act responsibly? Did they do the expected? If yes, then it is wrong to blame them for the disaster. If they do more than is required by law and custom, then we may declare them heroes. This is an admission that what we require is not all that is possible, and so it is possible to learn from our heroes how to improve matters. It is therefore an integral part of Popper's teaching that heroism is never obligatory, that only openness to criticism is.

Recognizing our limitations is basic. It is wonderful to be open and to encourage reform and prospective reformists and rebels, and to encounter unexpected heroes. This is possible because unknown conditions always limit validation. This is a mixed blessing – as responsibility complements validation. Philosophers who spend much effort in search for unconditional validation may be suffering from the desire that validation should be relief from responsibility. Even were it logically possible, it would be morally and socially neither possible nor desirable: life without responsibility is both unpleasant and impossible. Philosophers should cease seeking it and discuss responsibility instead.

Chapter 5.2

THE POLITICS OF SCIENCE

ABSTRACT

The myth that the politics of science is non-existent is dangerous as it prevents the important and urgently needed institution of some democratic control of it within the commonwealth of learning. Feyerabend's attack on science makes sense only when understood in this way.

1. PRELIMINARIES

There is almost no literature on the politics of science proper. The reason is the generally received view that there should be none. In clear opposition to this view, I propose that politics is unavoidable and therefore it should be the subject of open discussion. This will be a democratization of the politics of science. The first step towards this is the exposure of the existence of this politics and a survey of its problems on the way to setting a proper agenda for solving them democratically.

Let me backtrack, then.

There is almost no literature on the politics of science proper. The little that exists is incidental. It is usually incidental to stories – novels, autobiographies, or even biographies – and at times it is incidental to adjacent studies – the sociology of science, the participation of scholars and of institutions of learning in national politics, and, to a very small measure, studies of the history of science. The field of science policy, in particular, is utterly devoid of the politics of science, and for a good reason. Science policy takes for granted the fruits of science – extant or projected – and considers their role in national politics, whereas the politics of science studies the political aspect of the very process of the growth of science. To conclude this survey, let me observe that most writings on the politics of science are not only incidental, but also, largely, the expression of gratuitous moral indignation that results from the view that the politics of science should not exist. This view is a simple-minded error. Also, it is also not to the point. Whether we like it or not, the politics of science is there, and exempting it from rational criticism is an invitation for irrationalism.

Here then are some empirical findings. When I report – to intellectuals, scientists, philosophers, and university administrators – that I wish to discuss the politics of science, they ask me if I mean science policy. When I say, no, I wish to discuss the politics within the commonwealth of learning, they respond, saying, there is no such thing. When I say, I have ample empirical evidence to the contrary, they say, there should be no such thing. When I retort that there should be no more war, they say, war is a part of political life, willy-nilly, but academic politics does not belong to science. They say the same about politics in learned societies, and about intrigues and power struggles there. They say, scientists may be politicians, but not as scientists. There is no room, they say, for politics in science.

Science and politics cooperate in every situation within the domain of science that is also case of discrimination – by religious affiliation, skin color, or gender. The rule that there should be no politics in science (or in any other intellectual and artistic activity, incidentally) rests on the right demand to restrict judgment of a product to its quality and to let people act as their own judges of quality. This ousts religious judgment as it rests on some authority. Democratic society always allowed for some kind of discrimination, as long as it applied with no favoritism. As an example, we may take the distasteful denunciation of books and of movies that rests on an authoritative Catholic code. It is not objectionable as long as it is voluntary and shows no preference for Catholic authors or moviemakers over non-Catholic ones. This is usually the case in the United States. The same goes for a book or a candidate for an academic appointment: universities with religious base hire co-religionists in preference to others. This becomes objectionable only when as a consequence academic standards suffer too much neglect. But, obviously, the university system as a whole allowed for gross discrimination by religious affiliation, skin color, or gender. Despite much improvement, matters are still not fully settled. In particular, women are still regularly subject to discrimination. This has an adverse effect on research.

Thus, the claim that there is no politics of science has the status of an obvious empirical fact, yet pointing at its empirical refutations invites moral indignation. This signals a surreptitious change in the claim that the politics of science does not exist. Under fire, it becomes a moral demand; it ceases to be factual.

This is a very general phenomenon. It is defensiveness. It often appears in religion and in politics. The fact that it is present in science too is a major fact of the field of the politics of science. I will speak of it as of the presence of *the apologetic symptom*. This symptom clashes with the principle it comes to defend. The principle is that of impartiality. Defense in its name shows partiality to science by an apologetic attitude to it. Scientists have nothing to be apologetic about. They are apologetic because the scientific establishment or its political leadership is apologetic and the scientific establishment is apologetic in its effort to justify its very existence by an obviously false hypothesis. That hypothesis identifies establishment with the leadership. (The concern here is with the intellectual or scientific leadership, but this is a mere example.) This way we debase culture in general and science in particular. The false equation of the establishment with the leadership automatically justifies the fact that for decades the establishment delayed awarding recognition to such scientists as

Barbara McClintock. The equation of the establishment with the leadership is an error: the latter has recognized her much earlier.

But I anticipate myself. Let me then return to the apologetic symptom and give my initial response to it. Let me report my initial response to the simple-minded theory and to the gratuitous moral indignation that constitutes the major part of it. My initial response is moral: I tend to meet the moral indignation of scientists with a moral indignation of my own. Doubtless, it is equally gratuitous. The simple-minded view does seem to be dogmatic, pig-headed, and harmful. Nonetheless, I should not meet it with indignation. It does not become a philosopher, and it is a standard defense of the *status quo*. Traditional Chinese mandarins said that everyone could become a mandarin. They did not deny that all mandarins came from mandarin families, but they still insisted that in principle everyone could qualify. This is the way top dogs tell under dogs that there is no *status quo* and no top dog, that the road to the top is open to the able. It is the way top dogs tell under dogs that they are to blame for being under dogs. Advocates of the simple-minded theory that there is no politics of science rely on *naiveté* and ignorance. It is self-serving for top dogs and unjustly painful to under dogs. Only indifference allows saying that we all get our just deserts.

My second response to the simple-minded view is empirical. The politics of science exists. Some top dogs are incompetent and they maltreat some competent peers. Here is a simple test. Take any period in the history of any science and prepare a list of a few top dogs there. Then go to the people who have spent their lives in that period and ask them, who would they nominate for the same list. Oh, of course, these people are dead, but, when documents exist, historical methods enable us to ask dead people some questions and elicit reasonable answers from their literary remains (at times even if they never met these questions in their own lifetimes). As an example, historians of astronomy would list among the top dogs in astronomy in the first half of the seventeenth century Kepler, Galileo, and Borelli. Their contemporaries, however, would not include Kepler. They would include Galileo, but as second to Lothario Sarsi or some other individual who was then very powerful and esteemed as a great astronomer but is now hardly known. We list as leading nineteenth-century electricians no action-at-a-distance theoretician except Ampère; the contemporary list included only action-at-a-distance theoreticians and no field theoretician. At present, allegedly leading scientists are more likely to be members of Ivy League universities than excellent scientists, though very often these too are good and sometimes even very good. The excellent, the truly leading scientists, are people who may but need not come from the Ivy League. These facts pertain to my claim that there is no science without politics.

My third and last response to the simple-minded view is theoretical. The claim that science and politics do not mix is not empirical. It is, we remember, the principle that they cannot mix. It is the principle of impartiality. Now, what forces science to stay impartial? The traditional answer is this. Science is inductive, induction is the basing of beliefs rationally upon facts, and given a state of a science, inductive logic prescribes unanimity. Schools and factions are thus unscientific. This answer is conclusive, but it rests on an error. Inductive logic is a myth and in different ways. First, it should preclude the misguided choice not of data but of theories in the light of data. Thus, for example, when the possibility of contraception by the use of hormones appeared on the horizon in the thirties of the twentieth century, researchers shied away from the pursuit of data in that direction for about two decades. Thus, also, researchers do not pursue as vigorously as they can data regarding the causes of addiction and the ill effect of laws that proscribe drugs. The same is true of data leading to the Wolfenden Report and to the subsequent legalization of deviant sex between consenting adults. These data were available long before researchers collected them and used them to create the general picture that led Wolfenden's parliamentary committee to its liberalizing recommendations. Popper has observed that even if inductive logic forbids the disregard for some unpleasant data, it does not forbid the shying away from the search for them.

Scientific practice is not inductive. There is no unanimity in science. Each school decides which data it finds unpleasant, and it then attempts to explain them away. Though inductive logicians oppose this practice, they nevertheless participate in it – directly and indirectly – by making it sound insignificant. They gloss over the fact that scientific schools and factions exist and contribute to the growth of science in many ways. This claim is more plausible than the claims of the inductive logicians. As scientific schools are a part of the scientific scene, science is inherently political. Its politics was reasonably democratic before it became big business. It is time to rescue the democracy of science before the damage becomes too extensive.

End of preliminaries. I will now elaborate on the moral, empirical, and philosophical responses I have thus far outlined. I will conclude with some proposals for research that should bear fruits, both interesting and useful.

2. THE ETHICS OF THE MATTER

We all have ideal images of all sorts of people, politicians, researchers, or taxi drivers. As Ian Jarvie has noticed, Hollywood had to develop the ideal New York taxi driver for its own purpose. The Hollywood ideal image was close enough to the real thing to elicit attempts to emulate it. This is understandable, helpful, and innocuous. A follower of Jean-Paul Sartre may apply here Sartre's critique of the waiter who tries to play the ideal image of the waiter, as if to say, I am not really a waiter, I only play the part. He expressed improper contempt for the poor waiter. He ignored the simple fact that playing a part along some familiar guidelines usually makes life less difficult.

We may take it as standard sociology, along Max Weber's ideal-type methodology, that the ideal researcher plays a role, as the ideal waiter, taxi driver, or bureaucrat does. It is useful, at least as a starting point. Even if we do not agree with Weber, we may initially endorse his proposal to develop an ideal bureaucrat or taxi driver. The most important part of an ideal role is the standard of conduct and its (alleged) expression of an ideology, of views and values that those who play that role endorse. It is easy to describe the views and the values of researchers. They believe in science and in scientific method. They believe that all scientists agree about science and about its content and that scientists do not disagree among themselves. They do disagree – about religion, politics, departmental affairs, at times even about scientific conjectures. But, they believe, where scientific method applies, controversy quickly peters out. When this is not the case, they add, scientists should speak (not as scientists but) as private citizens, preachers, or politicians.

This view finds repeated expressions and meets with regular approval and as quite unproblematic. Particularly researchers find it quite unproblematic, and they nod their heads fairly solemnly in agreement, speak softly when explaining their grounds for agreement, and quite generally give off an air of polite and carefully considered commitment, firm and humane. The ideal demeanor – of waiters, taxi drivers, and researchers – is as appealing to the public as the ideas that accompany that demeanor. Indeed, there is a certain coherence to the image that modern researchers project. They know that knowledge is power, and that power obliges. They know that their opinions are the best and they smile in a friendly, understanding manner at dissenters, since they see where dissenters go wrong, not being as familiar with the facts of the matter as they should be, and not having thought matters out as well as they should. Given the opportunity, researchers could and gladly would straighten dissenters out. They do not resent the absence of opportunity; they merely regret it.

The one thing wrong with this projected image is that it is projected. Jean-Paul Sartre resented the projected image of the waiter because of that. He was unkind. Waiters then had to be servile. This hurt their self-respect. So, they took the servility as part of the job and refused the humiliation. Sartre rightly saw the accent that waiters put on impeccable performance as self-distancing. His claim that it is bad faith is thus objectionable. The case of researchers, oddly, is the reverse: they accentuate the perfection of the image they project in order to identify with it. If they are not perfect fools, then they know that they are not perfect researchers. Trying hard to fit that role, they cannot examine the ideal role and its impact on others. When one draws their attention to these facts, they express puzzlement. The ideal, they say, is of rationality. We all know we are not fully rational, they add, but we are trying hard to be. Displaying the ideal to others for emulating it seems to them above reproach, though it is highly objectionable: pretending to be ideal, they cause harm, especially to students. Even pretending that we always try hard to become ideal is harmful. It forces on students either the view that they are much inferior to their teachers, or, alternatively, the dreadful observation that their teachers, too, have feet of clay. And training them to try to be ideal is not training them to live in an imperfect world and contain imperfection. When they mature and face their imperfection, and find it terrifying, they do not know what to do. They make desperate efforts to cope. These efforts sweep them into the intrigues that they find all around. They do not have to realize what they do. In confusion, they may simply follow gossip. But gossip is manipulative, and the manipulators are intrigue-mongers. Granted, of necessity intrigue-mongers in science are much more intelligent, knowledgeable, and respectful towards the truth, than their counterparts in business, politics, and the church. But even when gossip reflects recognition of the right things, it is not honorable, and following it is not rational. There is too much of it. Saying that it does not count is harmful, especially to the young. They blame themselves for their inability to attain, and lose and become unable to criticize the ideal as well as the investment of efforts to attain it.

3. SCIENCE AS A SOCIAL PHENOMENON

I know that a play is any good, says a theatre critic in a Bernard Shaw play (*Fanny's First Play*), if its author is good; otherwise I cannot know if it is good. This elicits laughter. That playwrights who write good plays are good is all too obvious, but so is the fact that occasionally accredited playwrights write poor plays. Critics who declare a play good because it comes from a pen of an accredited playwright are incompetent cowards. Their admission of it in Shaw's play exposes them also as fools. Shaw alerts the audience, whom incompetent critics so often bamboozle. He wanted to educate them. He said explicitly as often as he could, he told jokes only as vehicles for his own political opinions, as he deemed them sufficiently important to bring to the theatre. He agreed with the view of the critic that accredited playwrights are more likely to produce masterpieces than unknown ones. He exposed critics who act on this view as parasites and as harmful. He did not resent this fact. He simply found it very important to advertise it. Every band-wagon, he said, has its joiners from among the rabble and the mixed multitude. And critics who join the bandwagons of accredited playwrights were but one instance of this general fact. Christians, socialists, physicians, and scientists, he said on various occasions, are either genuine or, more often, bandwagon joiners. We cannot, therefore, trust that critics – be they Christian, socialist, medical, or learned – use their judgments, unless we learn to discriminate the genuine from the bandwagon joiner. As official critics are probably bandwagon joiners, we must act as our own critics.

This analysis, as well as the lesson from it, is a bit trite and too detailed. Shaw would not have been offended by this judgment. His task was educational, and if education requires to make a trite point, especially if it is a public matter and so a matter of political education, then he could liven it up with a joke. He always presented humorously the trite points he felt a need to drive home.

About one generation later, Robert K. Merton, a leading American sociologist credited with the invention of the sociology of science, presented an aspect of the bandwagon effect as the Matthew Effect (mentioned above). Public platforms are available more easily to accredited scientists than to others. Merton's observation receives public presentations on many occasions, and, unlike that of Shaw, these cause no laughter. Nobody observes the fact that journal referees are as inept and parasitic and harmful as the critics who boost the famous at the expense of the young newcomer. This is terrible.

I should defend journal referees. They are volunteers. Occasionally they do recognize good products of newcomers and bad ones from accredited sources. Yet anyone who knows a bit about these things can report that praise for contributions of novices is often due to personal support from some old pros – and that many unsupported first fruits had to wait for decades for accreditation. (The paradigms here are Doctor Seuss and Barbara McClintock.) Historians of science describe with pious horror cases in which it took generations for good ideas to win recognition. They refuse to ask whether the same holds for today's scientific ideas, whether our referees are the best available.

The important point is not to complain but to record the facts. Presumably, referees are excellent. At times, they are. They are more often about as good as theatre or literary critics: they often do follow the Matthew effect. The role of this effect is to keep the establishment in the position of power, in which they can determine for a few years who is the best. And they decide that they themselves are the best and they give themselves a few years to learn who are the best candidates for replacing them, to induct them into their ranks, and then to use this as evidence that you cannot keep a good scientist down. But at least they try, and this secures their power.

The scientific establishment also has duties. It performs them better than other establishments do and deserves gratitude for this. Yet, it suffers from the apologetic symptom, and it is a symptom of a sense of guilt – for the inability to keep the image of the calm rational scientist and to listen to the empirical facts presented here. This makes the established too culpable in their own eyes. Their sense of guilt leads to the conduct that Walter Kaufmann has described in his *Without Guilt and Justice* as the attempt to improve by trying harder, by repeating old errors with a vengeance (rather than by self-appraisal and self-criticism).

Let me bring this discussion to the politics of science by an observation about the politics of the Matthew effect. The effect is very popular. Ample empirical evidence supports it. It appears somehow as sociological, not as political. This is amazing. After all, what is more obvious than the political aspect of the Matthew effect?

What gives any scientist power? Influence on others; holding office in a learned society; the ability to influence appointments to lucrative jobs; the ability to get jobs to pupils and to allies; the ability to secure for an unknown writer publication in a prestigious periodical. Everyone knows all that. We all try to get a powerful colleague to recommend us for a job, for a grant, for an acceptance by an editor. The Matthew effect, thus, is the means of maintaining the stability of power in a society in which there is no use for the normal means of the perpetuation of stability, such as money and parentage and political connections.

Rather than parentage, we have discipleship. Can I become a disciple of powerful scientists? How? Polanyi says, by becoming their student: apprentices become disciples. He spoke not of power, though, but of leadership, meaning power. He ascribed to leaders excellence and originality. Yet, most powerful scientists are not original, especially since original scientists are usually too busy to cultivate power. Can I become the student of powerful scientists? No, says Polanyi: I am too old for that. Can I, as a colleague, follow a powerful scientist? Of course, says Polanyi: we all do that, and this is the same as the fact that the powerful are at the helm.

Sigmund Freud said, your fate is set by the time you have acquired your parents' characteristics (through efforts to emulate them). Polanyi substitutes teachers for parents. Moreover, both spoke not of real individuals but of ideals projected to us (not expert but master; not father but father-ideal). Now children may innocently view their parents as ideals; not graduate students. The view of them as children is quite remarkable, yet it is Polanyi's view, and it is becoming increasingly popular, in recognition of the fact that most of the powerful scientists are pupils of previous powerful scientists. Polanyi justified this by declaring that powerful scientists are powerful because they are original. Not always. They are often mere bandwagon joiners, as facts easily illustrate. Originality, anyway, is rare.

Polanyi wins acclaim, especially through the writings of the popularizers of his ideas, chiefly Kuhn, simply because he wrote in a vacuum: no one wrote of the leadership in science before he did – because of the popular prejudice that science is inductive and so utterly rational and so in no need for leaders. Polanyi's rejection of this popular prejudice enabled him to present science as a closed society of sorts, as one with leaders who impose their views. He was in error: science is not a closed society. It can be more open, to be sure, but this requires the recognition of the influence of the scientific leadership – for better and for worse.

4. SCIENCE AND THE GENERAL PUBLIC

Polanyi simplified matters. He has emphasized this, so that his view is much more sophisticated and reasonable than that of apologists of science such as, say, Kuhn. In Polanyi's view, science is autonomous; it is not democratic, but it functions well within democracy when left to its own devices. He knew that the scientific community does not ignore outside prestige; the prestige of physics rests on excellence judged not intellectually but militarily. The market in mathematics that computers have generated has greatly raised the prestige of mathematics; the excellent Claude Shannon is a *persona grata* because Bell Telephone says so, not, alas, because we all understand and value his lovely ideas: most of those who sing his praise do not.

Polanyi expressed desire for a better approximation to the ideal of an autonomous scientific community. But he was a realist. Even within a church, prestige is at times regrettably due to outside factors, such as family ties, wealth, and national origins. Yet, for the sake of the freedom of science, Polanyi had hoped for less interaction between science and society, between the inner politics of science and national politics. One may advocate this same separation not only in the name of science, and not necessarily for the sake of the autonomy of science and against the encroachment of outside influences, but even against the imperialism of science. Essentially, these two campaigns have the same immediate end of the isolation of science from other aspects of public life.

Despite the scorn and ridicule that the ideas of Paul K. Feyerabend regularly encounter, his popularity has been constantly on the increase. However just the scorn is, the resilience of his views under attack deserves attention. I propose that his views are popular because of his just complaint that science exhibits intolerance that reaches levels that should be a cause for concern. We may join

Feyerabend and refuse to ask simple relevant questions. What exactly is science? Who exactly is a scientist? We may endorse with him the naïve views: that a scientist is anyone publicly known as one, and that a leading scientist is anyone influential among scientists and known to the general public as a leading scientist. We may join Feyerabend and refuse to listen to arguments in favor of any militant conduct, and also refuse to consider the negative overtone of words such as 'charlatan', 'pseudo-scientist', 'prejudice', and 'superstition'. We may then observe the conduct of scientists towards others. Then, he said, their aggressiveness will not elude us. Consider an astronomer telling a commercial publisher not to publish a pseudoscientific book on astronomy, asserting that it is harmful and succeeding in forcing this opinion on that commercial publisher. What makes this different from the conduct of any other pressure group, say religious leaders forcing a publisher to avoid publishing Tom Paine or Albert Schweitzer? Of course, what makes the difference is that science is progressive, correct, valid, *etc.* But religious people deem religion the way scientists deem science. Hence, religious and scientific conduct may be intolerant, and recently science has been increasingly intolerant towards folk medicine, religion, *etc.* In Feyerabend's opinion, science is harsh to all of its competitors, and to every activity that may enter the curriculum, or compete for research grants, since it competes with science for places in the curriculum and for grant moneys. Hence, we must check the tyranny of science before it goes too far and causes permanent damage. Cares for pluralism, be it pro-science or not, must bring about some effort to control the aggression of science.

This is a possible reading of Feyerabend. It makes him popular as the champion of all sorts of underdogs: non-science, pseudo-science, para-science, and alternatives to science. The philosophical literature is full of analyses of his claims and of proof of their ludicrous excesses; his admirers gladly indulge him his excesses. Let us, then, accept his complaint, and cut his philosophy down to size. How, then, should we curb the aggressive conduct of science?

Feyerabend has a proposal: separate state and science. Let me explain this proposal in my own way. Consider the question, is religion inherently tolerant? It is problematic, since the religious and the anti-religious may differ about it along party lines. It is not a very happy question. A much better question is, how can we prevent religion from acting intolerantly? Those who think religion is tolerant will not oppose curbing its intolerance. Those who think religion is regrettably intolerant may be glad to curb its intolerance. And let us ignore the advocates of religious intolerance. The separation of state and church has done the trick. Since science is aggressive towards increasing areas that were traditionally in the domain of religion and folk-lore, it may seem quite reasonable to effect a similar separation so as to restore a sense of balance.

Yet, Feyerabend's proposal invites some adjustment. The separation of state and church does not imply any separation of politics and religion. Politicians may act out of a religious conviction. And religious leaders cannot ignore politics. The early leaders of the American civil-rights movement were clergymen of different denominations. The Pope has the right to declare his convictions on political matters and to ask his congregations to exert political pressure to boost religion. The separation between state and church is thus a separation of authorities, of sets of power, not of sets of ideas or of activities. Such a separation is even unhealthy. As different authorities, state and church represent different populations, nation and congregation, and these overlap, thus ensuring no separation of politics and religion. State and church represent, organize, and rule these different publics: nation and congregation. The nation has its nationality (the citizens have their citizenship), and the congregation has its creed and rituals. The authorities are separate in the formal, technical sense, and interact via the overlap of nation and congregation, of national and religious affiliations.

We need to examine the parallel between two separations, of state and church and of state and science. Consider the triad nation, nationality, and state. Compare it to the triad congregation, creed, and church. What is the parallel scientific triad? The first item, the human grouping, should be scientists, or the scientifically educated, or the scientifically inclined. The second item, the characterization, in parallel to nationality and creed, should be science. This is quite unorthodox, since traditionally philosophers of science declare science to be a body of rationally binding doctrine. Many scholars suggest that what is a creed is not at all clear, and this more than legitimizes Feyerabend's parallel refusal to discuss the question, what is science. The third item is more troublesome. What within the sphere of science is the parallel to state and church? The starting point of the present discussion is the assumption that nothing is.

Let us deviate from this, and assume that science does have an institutionalized authority that exercises power – at times quite unjustly. Anyone who is not willing to admit this claim is excessively naïve or, worse, simplistic, not to say ill-willed. Admitting this is endorsing Feyerabend's view cut-down to size. Allowing Feyerabend his exaggerations, then, as mere rhetorical devices, we have no quarrel with him. And he admits that his exaggerations constitute a mere rhetorical device. So let us take him to complain merely about injustices performed by the authorities of science as acts of intolerance. Such injustice is indisputably possible, and legislation should block it. His advice, then, is to do so by separating those authorities from the state.

Among the natural candidates or this are universities, scientific societies, and national centers for scientific research and development, perhaps also suppliers of atomic bombs to the military, suppliers of nuclear plants to agencies in charge of the supply of energy, the medical corps in the armed forces, and their likes. Medical schools are powerful political institutions. Since they grant diplomas that are prerequisites for the joining of the medical corps of the armed forces, they have strong ties with the military. Is this healthy? No. Is this avoidable? Yes.

Is Feyerabend right, then? No. To the extent that the authorities of science are separate from science, we have to control them and prevent them from misusing their power – on the condition that they may actively participate in public affairs no less than the Pope. Thus, clearly, it is a mistake not to put to public scrutiny the decisions that heads of national medical associations make for the nation, as to what criteria exactly should candidates for medical schools satisfy and how many of them may qualify each year. It would be an improvement if not the medical guild and not the government were the public body that grants medical degrees. That public body will employ medical personnel, of course, but this will be great progress, since that medical personnel will have to represent the public interest, not that of the medical guild and not the government. This way the abuse of power will diminish through legislation. This concerns not only the medical guild, but also the legal and the accounting ones, though they scarcely come under the subject of science.

The separation will benefit both sides, of course. When British universities became excessively dependent on government funding, British legislators demanded the establishment of a University Grants Commission whose major function was the prevention of governmental intervention in university affairs. In the United States, authorities used the Research and Development funds almost openly as means to bribe university administrations to enlist their greed in the cause of forcing professors to work for the military – though also to enlist the same greed as means to fight segregation and discrimination in the academy. At the time, saying all this aloud elicited hostile responses. Now that it is past history, it is common knowledge. Yet, though Feyerabend was not interested in the cause of science, he would have a bigger rhetorical advantage explaining the benefit of the separation of state and scientific authorities than by talking irresponsibly as he did.

Medical officers are military authorities, not scientific, even when they are also engaged in research on the side. Medical research unit administrators, however ignorant in matters scientific they may be (and all too often they are), comprise scientific authorities proper. The complaint, regularly made, that only second-rate or superannuated scientists become administrators of scientific institutions, is irrelevant to Feyerabend's case. So is the claim that many powerful heads of research teams are also not very familiar with science.

The scientific authority and the political authority thus fuse in two ways. They fuse when the state legitimizes science, particularly scientific research. And they fuse when the authority of science legitimizes the state. This, indeed, ought to be precluded. How, exactly, is hard to say. When cardinals crowned kings and kings designate cardinals, intolerance was not in check. When college presidents appoint premiers and premiers appoint college presidents, Feyerabend's proposal should lead to a change. Though in no modern society is this the case, and in traditional societies in which it is the case science barely exists, let us not be complacent. It is better not to dismiss the complaint of Feyerabend. It throws some light, for example, on the organization of science in the Soviet Union, even though there the state interfered with research, and researchers did not interfere in state affairs until in the very last phase of that regime.

As long as we deal with Feyerabend's complaint, it is hard not to indulge him his exaggerations and deem them mere rhetorical devices. Yet, when we translate his proposal to a clear language, we may detail it into many proposals to try out one institutional device or another, and we may judge each detailed proposal on its merit, so that there is hardly room for exaggeration in the debate about them. So, rather than pour scorn (in the present case, on Feyerabend), it is better to render messages clear and try to implement each of them to the extent that we would judge it democratically advisable.

5. TOWARDS AN INSTITUTIONAL ANALYSIS OF SCIENCE

Consider the division between science and non-science. It may be between ideas, sub-cultures, social strata, and institutions. When we take science as social rather than as theoretical, we demarcate it in quite different a fashion than we did until now. For example, consider the university as an institution; to what extent is it scientific and to what extent is it educational? Should we view the action of recruiting people to the scientific community science? What part of scientific education is recruitment? And we can examine in similar manners learned societies, NASA, grant-giving foundations, and more.

I began with the inductivist ideology. That ideology deems researchers separate individuals and allows for no more room for recruiting of researchers than that of chess-players, moviegoers, and cat-lovers: they are researchers whenever they show concern with science. (Modern sociology studies even the recruitment to movie going, but let us ignore this.) In the wake of Sir Francis Bacon and Robert Boyle, the inductivist ideology considers the researchers amateurs or members of a college who have very few duties and no financial problems. A philosopher, said Boyle, must have a brain as well as a purse. They both considered important the problem of information flow and they both set the standards for scientific publications that led to the rise of scientific periodical publications in the mid-seventeenth century. The standard included the demand that readers of an experiment should repeat it for themselves. We know that this is how educationist Dr Joseph Priestley became a scientist in the mid-eighteenth century as he wrote science textbooks. Priestley also agreed with Boyle about the need for a purse. This was still the practice with Michael Faraday, in the first half of the nineteenth century. It died out, however, when he was still active. Some of the reports about his experiments came from pens of people who, doubtless, had not repeated them. (They got them wrong.) During his time, science became a profession, though not nearly as lucrative as it became after the Bomb, and the word "amateur" became synonymous with "dilettante" although the greatest scientists in history were amateurs (including the young Einstein). Faraday was a professional scientist, yet he insisted on living by the standards of the amateur scientific culture. (See my *Faraday as a Natural Philosopher*.) Inductivists assume that individual researchers possess all the available information, understand all the extant theories, and make up their own minds at leisure. There is no room here for scientific leadership or for scientific politics. When inductivists applied this model to national politics proper, they found no room for political leadership either – only for political delegation of personal decision. And they found little room for reasonable disagreement that would stretch over generations, as it evidently has done in real political life.

The application to political life proper of the theory that denies the need for leadership had varied complex adverse consequences. The politics of science is but a special case of this. It also had beneficial consequences, as it contributed to the growth of liberalism and to the enhancement of the autonomy of the individual. Yet, it also left the field fallow and enabled leaders to act under too little democratic control with too little sense of responsibility. Most importantly, it allowed any leadership and any institution to continue under its own steam. Thus, in particular, in Europe aristocrats played the role of political leadership despite all democratization and liberalization up to World War I and even later. (And when their hegemony waned, they did not hesitate to wage a bloody war.) Old institutions continued to exist, as did their patterns of leadership. Universities and professional bodies superficially adapted their medieval patterns of guilds to modern democracy.

Consider the problem of recruitment to scientific institutions. Very little is known as to when a person is committed to it, except that at the age of 30 or so, hardly anyone changes careers and at the age of 15 or so everyone may have dreams and feel committed but hardly is committed. What, then, happens in these crucial years? Where does commitment settle for good? How does discrimination discourage little girls, say, from taking interest in science?

We know more about the professions organized in guilds than of others, because the guild determines many factors *a priori*. These are medicine, law, and accounting. (The last one is a latecomer.) We do not know how many physicians or lawyers become scientists and why. But we do know that the number of individuals who become physicians much depends on the leadership of the profession. Deans of medical schools, who are very powerful, decide these things together. Are these deans researchers? Do they cater for the recruitment for medical research? There is much complaint that not enough M.D.s are active in research, that most medical researchers are Ph.D.s. Is that true? Are not research teams staffed by Ph.D.s but headed by M.D.s?

Who is a scientist? Kuhn says, normal scientists are small-scale researchers, employees in the science-based industries who need science degrees for qualification, and whose jobs he considered research. They do work for their livelihood as scientists, and are, then, normal scientists, yet they are not normal *à la* Kuhn, as they scarcely perform research. Consider, then, science professors. They are scientists, and some of them, not all, are engaged in research. Professors in community colleges usually are not; professors in Ivy League universities usually are. And so it goes. With so much sociology of science around, we do not even know who is a scientist and who is a scientific researcher and exactly by what criteria. This fact is truly amazing.

Who determines such things? To make an empirical observation, the scientific leadership determines them. Who are the scientific leaders? Again, we do not know the answer in principle, but we do know it in fact. Why Einstein was not a leader the way Niels Bohr was? This was a matter of self-selection. What is the cause of self-selection I do not know, except that it may be of very different sorts, including a sense of responsibility, personal ambition, self-assessment as a good organizer or as a burnt-out researcher.

How, then, do leaders decide? *Ad hoc*, but not entirely arbitrarily. Why not by a rule? Because the accepted rule is inductive, and thus inoperative. Why not arbitrarily? Because there is a limit to the degree to which a leader can be high-handed, arbitrary, oblivious of facts and impervious to criticism. Moreover, since science is open to criticism, the scientific leadership is less high-handed than the leaderships of education, art, business, religion, or national politics.

So much for recruitment for leadership. The first step to recruitment is the registration in a college for undergraduate studies. As long as college degree is essential for becoming a scientist, science is at a risk of sclerosis, ossification, petrification. The Matthew effect is only the tip of an iceberg. It has specific ill effects through the use of the leading scientific press. When an Ivy League periodical has a racist essay appear in it, racism gains respectability. This happened when the Jensen Report appeared in *Harvard Educational Review*. How should we learn not to be so gullible as to swallow rubbish just because it has appeared in the Ivy League press? How should we learn to view the Ivy League colleges critically? It is unbelievable how little research goes into young people's applications for studies in a college. Distinguished alumni of a distinguished college often make their offspring follow their footsteps. Otherwise, the choice of colleges rests on the strength of accidental grades of candidates and of the advice of the most unsuitable people around – the uncritical intellectual snobs. A little research in this direction may be enlightening and also prevent a lot of unnecessary disappointment.

Recruitment to graduate studies is a veritable disaster, and the only reason that it eludes notice is that there are no reasonable standards against which to check results. Anyone who wishes to set standards – or scales – of success of graduate education will meet with hostility from the educational establishment. Yes, this is an empirical finding too, on attempts made over many years and in many places.

Oxbridge and Ivy League nabobs, self-selected adjudicators on excellence and its guardians, control recruitment and job placement, the nub of political power.

There is much more to the politics of science, including the political power of those who know how to find funding for diverse valuable academic activities. They work with journal editors, conference organizers, and controllers of access of scientists to mass media, legislators, and more. To repeat, these raise many background questions that as yet are hardly studied. Also, clearly, there are almost no longitudinal studies of current policies of editors, of grant-giving committees, and the like. We can study the contributions in journals whose declared policy is to publish only uncontroversial material. We can ask how fruitful this or that controversy was. We will find that considered controversy is the leaven of progress. We should replace the current institutions of science that rest on hostility to controversy, with institutions that rest on the canons of proper scientific conduct, controversy, and public debate. The reassessment of researches that led to Nobel Prize awards has hardly begun. The case of Einstein stands out: the Nobel Committee delayed awarding him the Prize, and finally awarded it for an uncontroversial empirical finding. He gently refused to go to Stockholm and lecture to the Academy there as was required. (See appendix below.)

To conclude, until recently, the canons of conduct of the commonwealth of learning have served it admirably well. This is no longer so. We should study empirically their desirable and undesirable effects and try to reform them. The inductive philosophy that has helped establish the commonwealth of learning rested on the ideas that researchers are amateurs and that scientific information is easily available. It is no longer adequate. We need most urgently standards and values regarding scientific schools and their controversies. What scientific politics wants is, in a word, democratization.

APPENDIX: EINSTEIN'S NOBEL PRIZE

Einstein's biographer, Philipp Frank, skips his absence from Stockholm (*Einstein: His Life and Times*, 1948, 254). Abraham Pais reports (*Subtle Is the Lord: The science and the Life of Albert Einstein*, 1982, 503) that Einstein learned about the desirability of his presence in Europe to receive the prize before he started his trip to the Far East. He adds, "The news of the award must have reached him while he was *en route* to Japan. I do not know, however, when and where he received word. The travel diary he kept during that journey makes no mention of this event." Albrecht Fölsing reports (*Albert Einstein: A biography,* [1993] 1997, 535) Einstein having said, he could not alter his plans.

Frank reports (254) that Einstein lectured in Göttingen, Sweden, in the presence of the king of Sweden. He seems to have tried to avoid rocking the boat. Pais reports (527), "1923: July. E. gives a lecture on relativity in Göttingen in acknowledgment of his Nobel Prize." Fölsing reports (541) that, unusually, instead of the Nobel lecture in Stockholm, Einstein lectured to the Scandinavian Scientists' Convention in Göttingen, and, unusually, he was allowed to choose his topic. It was relativity, clearly, in silent protest.

Initially, the German ambassador to Sweden represented him and received for him the traditional medal and scroll. Fölsing reports (540) that he wanted them sent to the Swiss legation in Berlin. "The Nobel Foundation found a tactful way out", he adds: the Swedish ambassador to Berlin presented it to him (with the consent of the German ambassador).

Einstein had a choice, exercised it, and preferred not to explain. The explanation for this reticence is clear: the award demanded an expression of gratitude, not a complaint, yet he had a complaint too severe to overlook. Unforgivable Philipp Lenard, by contrast, devoted his 1906 Nobel lecture (for 1905) to a concealment of Einstein's work, and later he launched more than a mere complaint about Einstein and about the Nobel Committee. Ever so sedate Einstein lost his temper during an altercation with Lenard in a public meeting.

Chapter 5.3

SCIENCE AS A PUBLIC ENTERPRISE

ABSTRACT

Science has changed radically after World War II. Big science has replaced traditional science, which was little, and had little public concern. Today big science has new, exciting tasks. This should not be at the cost of a loss of little science. There is a need to revive and protect it, in order to insure the continued well-being of science as a whole. For this a new system is required, since the old one transformed into big science and is thus no longer adequate. Similarly, there is a need for new scientific education. All this invites much careful planning. The starting point of the venture should be the rejection of scientism, of the view that science has a monopoly over rationality. This will invite little science and concern with the quality of daily life.

0. PREFACE: SCIENCE AND EDUCATION BEFORE AND AFTER WORLD WAR II

Science has changed radically after World War II. Big science has replaced traditional science, which was little and had little public concern. There is a story, possibly true, that illustrates forcefully the difference between contemporary big science and traditional, small science – science as it used to be before World War II. The leading chemist James Bryant Conant, later the celebrated Harvard president and father of the science education program and the history of science program there, it is told, asked his military friends if he could be of use to the armed forces. After inquiry, the response was negative: the military had already a chemist. This, of course, is a joke: soon the military could not hire enough chemists. The Manhattan Project, the nuclear weapon research under the military, inaugurated a new era, and there is still almost no study of the effect of the new situation on the character of public education. Researchers take it for granted that big science boosted big science-education, and that is all to the good. It mostly is. But it also had a price, and it invites investigation. The neglect of it rested on the reasonable supposition that the price was sufficiently low. Low or high, prudence requires its investigation, and, if need be, attempt at a recovery of the loss. Unless we know what the loss is, we cannot tell. This requires a study of the difference between traditional, little science and contemporary, big science. Some of the cost has become so obvious that it invites no investigation. In the wake of the Manhattan Project came the new era, with its risk to human existence from the pollution of the environment, the proliferation of nuclear weapons, the population explosion and the poverty of the underdeveloped world (the four P's). I will leave out this matter, as I have covered it in a book (*Technology: Philosophical and Social Aspects*, 1985). Solving these problems by giving up both science and technology is unthinkable. The positive impact of the tremendous growth of science is great, especially and primarily on education. Yet, both research and education have radically altered. Big science has new, exciting tasks. The planning of education increasingly gears towards securing sufficient scientific cadres to insure the proper replacement of the old guard of research and of high-tech, and for the proper placement of the new high-tech jobs that the well-being of the national economy requires. A comparison of the two systems of research and of education, the traditional small-scale and the contemporary, large-scale, are ever more necessary. We hardly know what impact the transition from the one to the other had on education and on the quality of life in general. Big science should not replace traditional science and traditional scientific education, at least not without deliberation. Let me sketch a brief outline of this task.

1. THE LIFE AND TIMES OF LITTLE SCIENCE

Officially, modern science began at the time when Europe was less advanced than China – socially, politically and technologically – in 1660, with the founding of the Royal Society of London. The rules and regulations of that society soon became scientific etiquette. The Society was a club of gentlemen with a vision. The vision was the improvement of human understanding and, with it, the improvement of the human condition. It was officially indifferent to religion and to politics, and it insisted that it had no ideology at all, only respect for natural phenomena and for the effort to understand them and to apply that understanding for the betterment of the human lot. As to supernatural phenomena, they refused to deal with them in their capacity as fellows of the Society. They assumed that the improvement of the human condition will have great political import, but they kept this out of discussion. Today, they may look like typically narrow researchers. This is an error. They were not professionals but amateurs, and they were not narrow at all, as their interest covered all science. Their ideology was that of individual autonomy. So, they considered all knowledge strictly a matter of the individual mind, and so each of them wanted to be a universal scholar, a true polymath, in order to be able to judge every idea. Of course, this is impossible. But the belief in it was sincere and potent. They saw every individual as able to achieve autonomy and general knowledge and contribute to the growth of human knowledge and thus to human welfare.

Thus, without debate, scientism became the received opinion: the enlightened, rational individual must believe all scientific ideas, and disregards all other ideas. Researchers became the most open society ever. Scientific doctrine served them as dogma. In the nineteenth century, they became a relatively closed society. In the twentieth century, they acquired an incredibly shiny, hard veneer. The expression of their ideology became a combination of two vulgar and dangerous ideas about science, which, as it happens, are mutually incompatible. They are rational belief and normal science. Rational belief theory says, the theories that the community of researchers endorses are the most rational and so credence in them is obligatory not on the authority of anyone but on the authority of facts and reason. Normal science theory or the paradigm theory of science, describes scientific individuals as highly technically qualified, hard-working professional researchers who abide by strict authoritarian discipline. They believe what their leaders tell them to believe. The two theories are in conflict: the one bases belief on rationality, the other on the authority of leaders. If leaders back wrong ideas, what should experts do?

The two theories share the rejection of controversy, especially within science. Science always involved controversy and evolved through engagement in it. This historical fact is one of the many arguments that disprove scientism. Leaders ignore all this – in both philosophy and science. In the scientific community today, authority overrides criticism. Leaders try to suppress everything politically damaging. This is a new characteristic. (The pioneers in this were the early twentieth century physicist Werner Heisenberg and philosopher-physicist Moritz Schlick.) Unless things improve officially, this will obliterate official science. It will also obliterate science, if we do not fight the official identification of science with official science. Time is short.

My aim here is not to refute and not to advocate a reform but to present the difference between traditional, small-scale science, and contemporary big science. Science education is the field where the difference is most obvious, since induction and training new cadres is essential for any lasting enterprise.

The cutoff date is the time of the American and French Revolutions. The education of a researcher before then was no different from the education of a cinema-goer, or a reader of novels or of comics. The induction to scientific activity was as easy: one would subscribe to a scientific society, receive its journals, in which new experiments were described, and repeat them. This is how, for example, Ben Franklin and Joseph Priestley became researchers and contributed immensely to the advancement of science.

2. THE GROWTH OF SCIENCE

Research is unthinkable without laboratories and specialized training. The growth of these two aspects of science is interesting.

First, laboratories. With the exception of the astronomical observatories, the laboratories of classical little science were typically garden shacks, attics, or basements. Before the year 1800 two university lecturers gained fame because they laced their lectures with some simple experimental illustrations. There was no public funding of science before 1800, with notable exceptions. There were remunerations for the Astronomer Royal plus a few assistants, and for members of the French Academy. I remember how surprising I found reading a protest that a leading twentieth-century historian of literature had launched against

Cavendish, the great eighteenth-century English researcher who was filthy rich and single, against his not having bequeathed a penny to science.

In my youth, I wanted to do something about the lack of understanding of the conduct of Cavendish. I wrote about the attitude of the traditional scientific community to money. I quoted sources; I illustrated it with the case of Robert Boyle, the discoverer of Boyle's Law and the father of the Royal Society of London. He was the richest man in England. He bequeathed his fortune to charity and to missionary work. He left nothing to science or to the Royal Society except for his scientific collection. He even excluded the collection of gems from that – evidently because of its high financial worth. He refused to finance the Society despite pressure from friends. Some of them must have thought he was a miser. They procured for him land that the government had confiscated in Ireland. He accepted the land and still refused to give money to science. (He said he had made no request for this procurement.) In a long detailed paper I explained why. I argued against some of Boyle's biographers in some detail, explaining that his motive was not avaricious. Opposing the view of Descartes on the essence of the mind as reason, Boyle said, it was reason plus the moral sentiment. He said, natural religion suffices for the reasonable, but for the rest the Lord has mercifully given a second chance in the form of revealed religion. And charity belongs to sentiment, not to reason.

I could not get the paper published; it is still unpublished. This is an example of the uncritical attitude of contemporary scientific society. This alerted me to the possible pitfalls of the present attitude to money of the scientific establishment. Socrates was proud of his poverty, as were generations of rabbis throughout history, and as was the great philosopher Benedict Spinoza, who was Boyle's contemporary. The Socratic doctrine of poverty is outdated: high remuneration for intellectuals is essential, especially these days, when research inevitably involves great investment. But the result of this is sad. Criticizing a lucrative research project in the professional research press, directly or otherwise, is almost impossible; I have first-hand experience of this. Control over the suppression of ideas out of financial concerns, especially in matters intellectual, is becoming an urgent need. Until the middle of the twentieth century, public laboratories and university laboratories did not exist, nor did censorship. There was hardly any big science and there was no need for a control mechanism over it. Now we have to stop the robber barons who, masked as administrators, have started to vandalize universities and research institutes.

Let me outline briefly the role of the general educational system of the time of traditional small-scale science, particularly the universities. Most people at the time were illiterate; most of the rest went to church schools for very brief periods. Until the end of the eighteenth century, most schools were church anyway, including the universities. They were parts of religious organizations, with the exception of a few private schools and city universities (free universities). The level of scholarship was unspeakably low. Professors did not view themselves as researchers. Some professors did engage in research, but not as professors. James Clerk Maxwell, of the famed Maxwell's equations for electromagnetism, who lived in the late nineteenth century, who was busy in research, had an academic career that suffered from his neglect of activities such as fox-hunting. This was in England. In other places, the process started earlier. After the cutoff date, the university systems quickly secularized. They became increasingly scientifically oriented. This process took generations, and some university leaders are still very uneasy about science, even though they will scarcely say so out loud. The first scientific laboratory in England was the Cavendish Laboratory in Cambridge University, inaugurated in 1870. The professor in charge, Maxwell, spoke in his lecture that inaugurated that laboratory in the traditional vein. Fortunately, he said, the lab was so poor that students had to construct their instruments. An American industrialist founded The Royal Institution of London in 1800 for the education of the illiterate poor. It became an institution for popular scientific lectures and for research. The first new university in England was the University of London, founded in 1830, as its charter says, for Non-Conformists, Jews, and the poor. Before World War I, there were in the whole of England less than twenty professional research workers, and this number includes the Astronomer Royal and his few assistants, whose positions were instituted already in the 1660s.

The picture here drawn is misleading, though I have borrowed it from the literature of the time. The truth is that however poor and anti-science the universities were in the period before cutoff date, however poor they were till the advent of big science, however poor some of them still are, they provided and still provide a significant base for education. Comparison of modern industrial society with modern underdeveloped societies makes this painfully obvious.

The distortion of the picture is more distinct in the matter of professional science. The early nineteenth century saw the rise of a new class: employees whose scientific training was useful for their work. These were chemists, and soon also engineers. They are the professional members of the middle classes that sprang into being in the nineteenth century and are of increasing size and significance ever since. There were almost no professional researchers until World War II. Most of the individuals we call today professional researchers are either not researchers or researchers who perform no research at all, like ever so many science professors across the land. The same scientifically trained people, who traditionally did not count as researchers, now count as such, largely because in grant applications all collaborators in a research project count as researchers. Applied science never left the scene, and during the industrial revolution, around 1800, things began to change, even though much of that revolution was scarcely in any debt to applied science. Already Newtonian mechanics had some applications, especially in navigation, and so naval academies used some of his formulas. So did military academies, whose scientific concern was mainly, but not exclusively, with fortifications. Their concern after the industrial revolution expanded to cover mainly weaponry, of course. The industrial revolution introduced professional activities rooted in scientific knowledge almost as soon as it got going. Individuals of low birth and high aspirations soon worked in industry as experts of sorts, especially in chemical analysis. Others, especially in the United States of America, but also in Germany, were professionally engaged in invention, including the development of the pharmaceutical industry.

Still others joined Justus von Liebig in his launching of the agricultural revolution in the mid-nineteenth century. These individuals would count today as scientific employees, or, to use a fashionable term, as normal researchers. English experts designed and supervised the construction of the French railways. Soon German experts took over and even migrated to England to run their investments there. One famous German industrialist who settled in England was Friedrich Engels, the lifelong collaborator and friend of Karl Marx. On the Continent, the universities soon became centers of technological activities. In England, the universities were much slower to develop, and only in the twentieth century the advent of the new, red-brick universities created openings for research and education. Not so in the nineteenth century: science-based jobs opened, and their occupants effected two great revolutions in the Royal Society of London, around 1830 and around 1870, so that it ceased to be a gentlemen's club. Today it is, of course, one of the most prestigious clubs. Finally, the diverse medical schools and organizations would today count as scientific, but in traditional England physicians and surgeons were more concerned with their status as gentlemen than with their scientific status. Their colleagues elsewhere did not fare much better. Only the success of Pasteur and Lister in their struggle for new standards in hygiene opened the medical schools to science proper.

Since the most important difference between traditional and contemporary science is that between amateur and professional, it is not enough to speak of scientific training in terms of the curriculum. One has to attend to the status that training confers on its graduates. I should therefore say something about status.

3. STATUS ACHIEVED AND ASCRIBED

The rise of the professional class, especially the academically trained, is a complex matter. Before the cutoff date, specialized productions of goods and services, including music, involved largely professional families and church schools. This altered radically afterwards. Specialized secular schools proliferated. Today much of the medical profession, perhaps the majority, runs in families, but medicine still is a major means of social upward mobility. For practically all members of the medical profession, training in specialized schools is obligatory. Access to medical education is not quite free, yet the guardians of this access swear that the only criterion of admission is proper qualification. The same goes for the intellectual leadership of the western world in general. Leaders are allegedly the best intellects around. The theory of the paradigms, we remember, declares this the very essence of science. Its leaders are graduates of the most reputed schools and their seniors handpick them. The most reputed schools are allegedly the best, they allegedly recruit the best candidates, and that senior leaders are allegedly best qualified to select their successors. Be it so. It is the way prophets and gurus select their heirs. But then prophets and gurus do not wield power; academic leaders do. Democratic procedure requires some control here. There is almost none.

The political power of the intellectual leaders raises envy and this prompts the pursuit of excellence, to use the current euphemism. It is the pursuit of accolades, of what the establishment recognizes as excellent. What today's establishment deems excellent is better than what those of non-scientific societies did. But it is not as good as what the society of researchers deemed when it was rather aloof from society at large. The intellectual leadership is not as inept as some other achieving *élite*s, say, as the military *élites*, but this will not persist. In principle, the selection mechanism of intellectual leaders is not different from that of the military. All leaders claim that their presence as leaders is proof of their merit.

Democracy requires control over power. This leads to discussions on the question of equality in intellectual matters. The matter of intellectual equality is obvious: we are intellectually unequal. Equality is possibly a matter of academic politics, and in politics the question is whether the intellectual leadership is democratic enough in matter of power politics, regarding colleagues and students. Regarding colleagues, the question is, do the leaders do what they can to prevent discrimination and unjust penalties for deviant thinking. Regarding students, educational equality is not intellectual equality. Today's ideal is of freedom of opportunities in and through education. A political-religious sect has developed it in the early nineteenth century – the Saint-Simonists. It was unthinkable before the rise of universal education, the grandest revolution ever.

The intention behind total equality of opportunity in and through education was to realize another new ideal concerning social status. The early nineteenth century saw a great revolution in the concept of social justice. It was very tempting to try to create a society with no social differentiation at all, with no social status at all. The communes in the nineteenth and in the twentieth centuries, the Israeli Kibbutz, and other, less famous groups, attempted to create such societies. So were the socialist revolutions in Europe. It was a revelation that the Kibbutz failed to free itself of social status even though its members share property. And in the light of this experience a new idea appeared. If status is unavoidable, at least let it be a matter of achievement rather than of ascription, as the new terminology goes. If we must have social status, at least let it be fair. Let everyone work for it and let those achieve status who have shown merit.

There was an antecedent to this. The medieval Church was a most important institution in the western world, and its ranks were in principle open to all. However different things may have been in practice, at least the Church recruitment system acted through its educational system. Local priests were in charge of the first stage of recruitment. They would send bright boys to college, to be novices of sorts. In almost all societies there is some social mobility of one sort or another, due partly to luck partly to skill, and skill usually comes from training. The main difference between traditional society and the early industrial society is here. Due to industrialization, social change was rapid. It involved new opportunities and new ideas of social justice. These evolved into the relatively recent idea of equality of opportunity through education.

Many anthropologists take for granted, following Marshal Sahlins, that very poor preliterate traditional societies have no ascribed status. This is obviously a modification of the theory of the famous nineteenth-century pioneering anthropologist, Lewis Henry Morgan, that Engels and others endorsed uncritically, that primitive societies are classless. It is doubtful. Admittedly, almost every class society (not cast society) has some means for equalizing existing class differences. They are not prominent in education. In most societies, learning is sacred, and so the individuals who partake in it are primarily shamans, namely, physician-priests, much akin to the members of the Levi or the Cohen tribe as described in the Bible, especially in *Leviticus*. They had status and position ascribed to them by birth, yet this is no refutation of the theory in question, since the Bible says that the priestly lineage begins with the advent of the written law. In the early biblical literature another sort of leaders are prominent, the judges, especially as described in the biblical *Book of Judges*. They had to earn their status. Inasmuch as the Bible is a source of social history, it describes the next stage of social change as the increased affluence and the rise of social barriers – a phenomenon that the prophets forcefully condemned. But our concern here is with education, not with social justice. With the expansion of knowledge, the well-to-do are the first to gain access to it. Next to the priestly scholars, the better off are naturally the first to benefit from better education. This is not the whole story. In protestant England, traditionally the top clergy were mostly second sons of the landed aristocrats. And other stories are more complicated. Nevertheless, in most traditional societies, scholars are usually priests in high position. The exceptions are the researchers of traditional, little science, of science in the period that began in the scientific revolution of the seventeenth century and ended late in the nineteenth century.

4. THE MIX OF ACHIEVED AND ASCRIBED STATUS IN REAL LIFE

The differentiation between achieved and ascribed status is a twentieth-century innovation. Traditional sociologists noticed usually ascribed status, even while noting that most societies have some social mobility. Status is always a combination of the ascribed and the achieved, and the more social mobility, the more it is achieved. Even in primitive societies with hereditary priesthood, where education was available only in priestly families, even there not all who could become priests took advantage of the option. The same goes for other skills, such as music and medicine.

Anthropologists suggest that in almost every society some knowledge and some skill serve less the love of learning or of the application of skill and more the desire to mark class-distinctions. This is no news to anyone familiar with Bernard Shaw's *Pygmalion* or even with its musical version *My Fair Lady*. The expectation from European aristocrats was that they should possess smatterings of knowledge. The expectation from Chinese Mandarins was that they should possess scholarship proper. Allegedly, they achieved their high status by investment of much scholarly work, though everyone knew otherwise. Scholarship stayed in families and some very worthy scholars failed the exam that supposedly marked achievement.

Other than noble aristocracy and poor multitude, traditional society had an aristocracy of low-birth and commerce and trade of all sorts. Records ignored them, or enshrined the rise of some of them to the ranks of aristocrats in myth and fables. (It is a sobering thought that in Britain today, unlike in traditional Britain, membership in the House of Lords is an achieved position.) Merchants and trades people are the most outstanding non-aristocratic well-to-do parts of traditional society. The word "trade" is a synonym for a skill not suitable for aristocrats. The skills for aristocrats are few. They need high proficiency in leadership – in war and in peace. So, their domain is the martial arts and the law. Even in the law the ideal in the upper classes was what in the United States is known as a gentleman's C, namely, the demand that gentlemen do not excel in learning, as their livelihood does not depend on it.

Merchants and traders, then, depend on their skills and labors for their livelihood. They are marginal in traditional society, and nineteenth-century English novels are full of expressions of contempt for them, especially when they aspire to enter the aristocracy. Other merchant and manufacturing and artisan groups constitute societies unto themselves, commercial societies, or commercial enclaves within some non-commercial ones, such as the Indian Pharisees and the British Jews. Even Gypsies are nomads with skills of one sort or another. They do not usually enter the context of learning because their members who excelled educationally left them. In other skilled societies and sub-societies, learning is at times so important that it becomes the mark of these societies and their lifestyles. Traditional society ignored commerce and trade as much as possible. We still consider the western early modern societies as traditional, yet there commerce and trade won respect due to new developments. Their social history and economic history fill volumes, like Max Weber's popular study of early capitalism. (Unlike rabbis and unlike wealthy Jews, prominent secular scholars were marginal in Jewish community.) The modern world, where classical science could grow, included middle classes whose members were sufficiently well educated to be able to pick up scientific education individually. There was no social status in it.

5. THE BACONIAN VISION

Though the society in which science grew had very poor education by modern standards, it offered at least some literacy to a growing portion of the population. Thus, though the moderately educated today know more medicine than the best physicians of the nineteenth century, at least these physicians were literate and to some extent they checked their superstitions by science. As all professionals who applied some science in their daily work, they had no particular interest in science, let alone in research. Only amateur clubs, whose memberships were voluntary, equated science with research. They conducted research outside society, with no remuneration and almost no personal benefit.

Not quite: the general scientific ethos of the day had a profound impact on society in general, on religion and on politics and on the arts. The missing link in the story is the ideology of science, which was then the motive power of progress and is now a dangerous myth that hides the ugliness in the contemporary scientific community. It is the apotheosis of science, its presentation as a substitute religion of sorts. All societies except those that have benefited from the industrial revolution, there are traditional. Learning there is marginal, part magical-ritual, part leisurely pursuit. A new vision changed that. Its best expression is in a formula of the early modern philosopher René Descartes: I think therefore I am: my thinking makes me something, a somebody. Descartes did not mean the slogan this way. Nor did he invent the vision. Ignoring some impressive predecessors (such as Thomas More), we should name Bacon as its originator. He described a science-centered technological society. The Royal Society of London endorsed his ideas and became the paradigm to the whole scientific world. No matter why Bacon was so influential, his idea of science and of the place of science in society determined the official view of science as a public enterprise. his philosophy of science and its social aspects has effected all discussions of science as a public enterprise.

Bacon viewed the growth of science as independent of any factor except the good will of researchers, and its influence on society as the source of powerful technology and the seed of affluence. These ideas are no longer popular. These days we view society and science as influencing each other. The social dimension of science enters the picture in a way that was unthinkable before World War II, and this raises a plethora of problems, particularly in matters educational.

Bacon's vision is incredibly naïve. He presented science as facts that generate theories. They comprise observations and experiments that amateurs can devise and perform with ease. Ideas evolve by themselves out of experimental results. The only ingredient necessary is the good will of researchers. They are curious individuals who are motivated by no more than the desire to know and to benefit humanity with the fruits of their research. To settle matters clearly, Bacon disqualified (as scarcely able to perform interesting or useful research) all those who wish to apply their research to ill purposes or even to gratify personal ambitions. He thereby demolished the myth of Dr. Faustus without mentioning it. This was gratifying, and prevented the perpetration of the image of science as sinister and dangerous – until Mary Shelly revived it two centuries later. It also rendered Bacon's theory of science irrefutable as he ascribed deviations from his description to impure motives. His theory is prescriptive, though the promise that its faultless followers will find the Golden Fleece is descriptive (though still irrefutable, as no one is really faultless).

Bacon's vision is charming. Yet, the excuse he offered for the failure of his method – impure hearts – is still in use, as a means for dismissing all critics of science. The change in the social status of science changed the picture. The powerful scientific class and its powerful lobby make the sociology of science interesting and the benefits and ills of science-based industry make science policy vital for the life of the nation and even of humanity at large. This puts aside Bacon's excuse.

6. A REALIST VIEW OF SCIENCE

Science got out of hand. It was a great surprise. In retrospect, it is surprising that it was surprising. No matter how noble the enterprise, since it has become a means of livelihood for a whole social class, it is a social matter: not a mere individual search for the truth, but a part of the fabric of society, with all the good and ill of society. Let us agree that by definition science cannot be bad, since it is pure and unadulterated knowledge. And let us agree that knowledge cannot but be good, for both its advocates and its consumers. There still are so many troubles concerning it. Take simple information, the like of which investigative journalists handle. Consider the question of the malicious use of the truth. Is the malicious use of scientific truths possible? Perhaps one can insist that scientific knowledge is abstract, and so its malicious use is impossible. This is easy to refute by the observation that the abstract has concrete applications. But as science competes with received opinion, especially received religious doctrine, its misuse is easy. The attitude of the scientific world towards religion was that of neglect and toleration, and so it scarcely invites criticism. Yet, the neglect seemed to be nothing short of contempt. The success of science and its disregard of its social and intellectual background gave the impression that intellectually only science counts, that science equals rationality, and that then everything involving thinking, even to the slightest, could not escape the judgment of science for better or for worse. Bacon has preached or at least implied this attitude. It is scientism, and it is still popular. It rests on the pretense that science uses purely internal criteria of proof. Twentieth-century physics has banished this pretense. The alleged perfection of science permits its isolation by putting it on a pedestal. And the pretense of science for perfection makes the hope to develop a social science utopian, since the recommendations of social researchers will then be perfect. In litigations and in public investigations of all sorts, contesting parties invite researchers to testify for their causes and against the causes of their opponents. This suffices as grounds for the dismissal of the ethos of perfect science.

The ills of science require exposure without hostility to science and without fear. Some people say that more is required, that the commonwealth of learning should admit the justice of the criticism of science, when it is just, and that it should take steps to rectify wrongs. This is reasonable suggestion, but it is excessive, and insisting on it dooms the project to failure. To speak of the commonwealth of learning is to speak of its leadership, and the leadership of science is powerful in social and in political terms so much that Robert J. Oppenheimer said that now science knows sin. Before World War II, researchers were not that powerful. It is too much to expect the leaderships of science to be responsible to the extent of putting the public interest above the interest of their peers. Individual researchers may and do rise to that level of decency, and we should discuss them and their possible contributions. Perhaps some of them will contribute to the study of the scientific leaders and to the possible means of raising their level of responsibility. But we cannot demand this, only hope for it and encourage those who are willing to do something about it.

7. THE INTERNAL AND EXTERNAL CONTROLS OF SCIENCE

The first serious study of the sociology of science is that of Karl R Popper, who published in 1935 a classic study of scientific method (translated in 1959 as *The Logic of Scientific Discovery*). His interest was then philosophical, not sociological, but his attack on the Baconian philosophy led him to view the individual researcher in the context of the public of researchers, and he spoke there of the attitude to criticism and of the public of researchers as those who welcome and foster criticism. Later, in his classic *The Open Society and Its Enemies* of 1945, he spoke of the need to defend the institutions that defend criticism. He spoke as if these institutions exist. In reality they exist only in civil society, not in the scientific clubs and organizations. In the early nineteenth century a young English chemist by the name of Davy criticized the most famous and established chemist of the time, the Frenchman Lavoisier. He could not publish his ideas in the leading French periodicals even though they were very important (and his criticism proved just). When he found his way to a minor French publication, the scientific leaders of France threatened the editor that they intended to send the police to prevent the printing of the article. The police did not come: they had other urgent business, as anyone familiar with the cartoons of Goya and of Daumier should know. I have published some documents related to this incident and met with censure of colleagues.

Even the institutions of science that do defend critics against abuse, such as university committees, operate within the law of the land, and only tolerably well. There are some institutions for the dispensation of justice within the commonwealth of learning too, such as peer review. I have explained elsewhere why these do not work at all. My point here concerns the rather idealized picture of the openness of the scientific community to criticism as Popper drew it. If it ever was true, it is true no longer. There is here an interesting development: The standards of science are lowered, but they became increasingly acceptable in general society wherever they are applicable, so that the standards of society at large are rising in certain areas – alas, only in the modern world, and even there the threat of irrationalism is still considerable. This is perhaps never avoidable, or at least not as long as science education remains as authoritarian as it still regrettably is.

There is progress in science education too, due to the monkey trial, for example. This is stupendous: today's modern youth cannot help but bump into science, natural and social, during their education process. But teaching is still authoritarian. The standard biological theories are full of holes and receive criticism of some of the finest scientific minds, yet schools present them as if they were perfect, particularly when teachers do not understand what they say. The creationist complaint against the dogmatism of biological science teachers is just. The complaint had no force because the dogmatism of creationists is worse. Nevertheless, the introduction of science into the core curriculum is largely to the good, as knowledge is superior to ignorance, even when absorbed dogmatically: hopefully its benefit exceeds the damage due to dogmatism. (See my "Science in Schools" (a discussion note), *Science, Technology Human Values*, 8, 1983, 66-7.)

Things are not so in higher education, where science comprises the whole curriculum for science students. This is due to understandable pressures, but it is dangerous all the same. The prime example is medical education in the United States. It followed the magnificent reform that Abraham Flexner instituted early in the twentieth century. He demanded that students should receive some general education before concentrating on medicine. He instituted the rule, still operative there, that medical schools should confine admission to college graduates. Yet, unofficially medical-schools requirements make them stuff their timetables with courses in what they consider essential premedical subjects. Even if this is a myth, it prevents American undergraduates wishing to enter medical schools from taking luxury cultural courses for their first degrees.

It is useful not to lose the sense of proportion. The current system of higher education is wonderful in ever so many respects. A nation that would limit rather than expand it will lose economically and culturally. But this is not to overlook the absence of education for little science, for education for science as a part of our culture, as the intellectual love of God. There is no need to explain the significance of the possible service to science in general of the pursuit of scientific truth independently of everything else. Even science fiction, not philosophically the most sophisticated literature, explains it well. Nor is it necessary to observe that happily some individuals will perform this task no matter what. But, clearly, institutions are necessary for individuals whose pursuit of science takes precedence over all other interests. The gentlemen's clubs the like of the Royal Society played a supremely important role in this respect, even though they were far from perfect. We need such organizations, organizations that do not confer status on their members but help them find the interesting ideas even before the scientific establishment finds courage to admit them. And the first thing such institutions should develop is irreverence and education for the criticism of the very best ideas around. This should include the view that possibly science is great not the way scientism says it is but the way that responsible, critically-minded attention to one's own curiosity is. So, the philosophy of science and of science education may be the first target of this irreverence, as becomes them.

Will our society ever give rise to institutions of learning that confer no status despite its competitive character? Some such institutions already exist. The highly prestigious club of the anonymous contributors to the mathematical corpus published under the fictitious mane of Nicolas Bourbaki is not the right model, admirable though it is. Can there be incentives and remunerations to prompt some sense of proportion towards incentives and remunerations? After all, a sense of proportion is something to value for itself, is it not? Why else do so many people frequent community colleges and consume popular science?

Chapter 5.4

SCIENCE, AND COMMONSENSE

Our understanding of both science and commonsense is deficient. This is obvious from disagreements about them. We disagree about both the scientific and the commonsense character of both scientific and commonsense ideas. The status of many items in both science and commonsense regularly meet with new challenges and undergoes change. We reject many ideas that were once commonsense. (Victorian ideas about masculinity and femininity serve as rubberstamp example.) We may limit both to what we commonly endorse. We often consider ideas commonsense if and only if they are standard, if most people with healthy normal understanding endorse them with no hesitation. Yet, at times we deem some people eminently blessed with commonsense, inconsistently with the idea of commonsense as standard: these people often come up with ideas that become standard. By what criterion?

Is science commonsensical? This question is interesting for reasons that are all too commonsensical, so that they hardly need specification. Or, are they? Since we do not quite know what either science or commonsense is, how can we judge? Since neither science nor commonsense is infallible, we may ascribe some erroneous idea to both. If we do, where does the error belong? Is science commonsensical? Is commonsense scientific? We all agree that some science and commonsense are in agreement. We all agree that they influence each other. Hardly anyone will go so far, however, as to declare them co-extensive. Clearly, science goes beyond commonsense. This is so even though today's science will hopefully become the commonsense of tomorrow. The rubberstamp examples are the Copernican hypothesis and Darwinism.

What passes for commonsense here and now is not commonsense for all eternity. There is no such thing. We may leave to sociologists the empirical study of what passes here and now for commonsense. We may wish to study commonsense in the abstract. Whatever we consider today commonsense, we may want to know certain things about it. We may ask, what makes commonsense what it is? The answer suggested above is this. It is what gains immediate consensus as intuitively acceptable. What makes it so? Partly it clearly is common experience. This is also clearly not the whole story, since different cultures differ despite the experience that they share. Wittgenstein said, people follow rules of speech, and these rules dictate to some extent what they say. This is true but not satisfactory. Some descriptions are open to translation from one language to another in accord with some commonsense rules, yet the rules and content of commonsense differ from place to place, and from one language to another – as manifest in difficulties that translators experience. This shows a gulf in commonsense that rules cannot bridge. But the best proof of this is that here and now some of us endorse some magic as commonsense and others laugh it off. Language embeds magical expressions, and Heaven forbid that using them should amount to endorsing magic. Wittgenstein agreed. He did not see this as evidence for the extreme limits of the role of rules of speech in the formation of common sense. None is blind as those who will not see.

Our question was, what makes commonsense what it is. We have rejected one answer to it, that rules of speech determine commonsense. Perhaps this is not Wittgenstein's opinion. We may leave this to experts on his ideas and proceed with our question, what makes commonsense what it is? More important, perhaps, is the question, how can we transcend or expand commonsense? Some say, we should know the answer to the first question in order to be better able to answer the second. Is that true? Not quite. We saw that the correction of errors – logical as well as factual – may open the door for some views to enter commonsense. So, for a while perhaps we may examine the second question while ignoring the first.

This is too fast. The correction of error only opens the door to new commonsense ideas. In itself, it neither transcends commonsense nor expands it. What does? Even if we do find examples for what does, we may do better if we answer the first question first. What is the commonsense answer here?

If we do have an answer to the first question, it may facilitate the study of the second. Yet, we may get stuck in the study of the first question and never reach the second. This has happened repeatedly. When one question has another as a preliminary and scholars stay with the preliminary while they forget the one that it is preliminary to, then they are scholastics or Talmudists. Feeling not ready to take the bull by the horns, scholastics take preliminary or outdated questions.

Let us stay with the first question for a while. How does an idea enter common sense? This question has interested many thinkers as they sought to bring back to science its durability and they had hoped to find durability in common sense. This is an obvious error: we are all fallible, and within their joint area, science is less erroneous than common sense. Hence, we should modify our question. We may ask, how does an idea enter common sense, and we may ask, how should it enter common sense? Behind the question stands the supposition that common sense has some quality that singles it out. What is this quality?

Commonsense is volatile as it is a kind of consensus. If we all share an idea as unproblematic, we consider it a part of common sense. Hence, when we offer a new idea that clashes with common sense, we need not despair or attack common sense. All we need do is contrast the new idea with the old and show its superiority. How? First, we may try to criticize the old idea. If we fail, we may devise a crucial experiment between the two. A crucial experiment may go this way or that way or no way at all. Thus, it may, but need not, decide an issue. The idea of crucial experiments comes from science and is now utterly common sense. Some philosophers oppose it, as it is imperfect, forgetting that we no longer cling to the demand for perfection. Common sense is merely commonly received opinions, yet with a hefty dose of a (common) sense of proportion.

We have here a beginning of an answer to the question, how does an idea enter common sense? It often does so by replacement. We have a common sense answer, we refute it, and we seek another. If we find one, then we test it, and if it turns out to be unproblematic, it becomes commonsense all by itself in accord with the above suggestion about the nature of commonsense as unproblematic and generally assented to. Is this the right way? Who knows? Is there a better way? Bertrand Russell suggested what he thought was a better way.

Russell's *Skeptical Essays* is a marvelous book. In it he made some terrific commonsense suggestions. It is important that he was full of common sense. He tried to entrench it, and he failed. Some commentators notice his failure and declare his philosophy utterly contrary to common sense. There is sense in this judgment, but we should reject it. We all have the right, said Somerset Maugham, for judgments that rest on our very best output. The best of Russell was his tremendous sense concerning doubt. At times, doubt makes excellent sense. But if it paralyzes, then it is unhealthy. The idea that science should be perfect is an example for an idea that leads to the kind of doubt that is unhealthy. In short, there is no need to raise doubt for its own sake.

Russell deemed it common sense that whenever science and common sense clash, common sense should give way. This is not to say that he bowed to any authority – of science or of common sense. He knew no authority. On the contrary, he was fully aware of the regular occurrence of disputes – within science and within common sense and between the two. Common sense may judge some disputes reasonable, some not, and most of them it will suspend judgment about. As commonsense, Russell suggested, we should abide by the advice of experts if and only if they agree among themselves. This is dangerous. People may feign agreement in order to gain the status of experts. Russell was aware of this. A famous philosopher who overlooked this possibility is Thomas Kuhn, who advocated endorsing the consensus of experts unconditionally as authoritative. He would have admitted that he was inflating authority, were he not speaking for the experts of his day, especially in physics, as in that field there is no conspiracy of experts. He said he was not speaking about the fields of social studies. Nonetheless, he may have increased the authority of experts in these fields too. What does the authority of experts rest on? Commonsense can come to the rescue again: there is no full assurance ever, but partial assurances may be reasonable. We use them regularly.

This is a cause for concern. All assurances operate on the principle of insurance: the principle is of distributing risk to avoid the collapse of any part of the system within which the insurance obtains. This is possible as long as the faults against which the insurance obtains are within the system. There is no insurance of the system as a whole. This is why systems tend to become sub-systems of larger ones. But there is a natural limit to this: the whole globe is not insurable. (Things would probably be different were colonizing other planets possible. Right now this is not a live option.)

This is why bureaucrats and others who wish to cover their asses are staunch conservatives: they know that their actions are not perfect, but they insist on staying within the system. Innovation is an invitation for risking the whole system. It is important, nonetheless, to be innovative, as this gives an opportunity to avert dangers that the system may face, and to expand common sense. For, when the system itself is in danger, commonsense is useless, and the system is at the mercy of the innovator. The more risk carried from individual cases to the whole system, the more the system itself is vulnerable. And, of course, as insurance is merely distributing risk to avoid collapse, no one can possibly insure the entire human race. It is not capable of insurance. There is no insurance against global epidemics. The optimism of classical rationalist philosophy was unreasonable.

Our current system is increasingly the human race, or Spaceship Earth, to use the terrific expression of Buckminster Fuller. It is only half a century since we have the ability to destroy humanity. And as every system has some functionaries whose task is to watch the system, humanity has to appoint such functionaries. And these have not only to watch the safety of the system. They also have to study the new subject matter that the new situation has given rise to. And the first move is for us all to acknowledge our ignorance of the matter and the way we should avert the real risks. This way we will open the road to the study of the matter and to implementations of some results. This is sheer common sense.

Commonsense has an advantage over the classical philosophy of science in that it recognizes its own fallibility. The philosophy of science has improved this way when it learned that some controversies are reasonable. On this we may, perhaps, speak with the hope of attaining general assent. We cannot put ourselves above error, in science or in commonsense. Nevertheless, both are means for avoiding avoidable errors. These errors are irresponsible to commit. One of them is the idea that we can put ourselves above error. Both science and commonsense deem certain ideas unproblematic. Both recommend taking them at face value, but only until someone shows that they are problematic. Good critics do so regularly. In daily affairs, we see them as blessed with uncommon common sense – especially but not only if they offer alternatives to the received ideas that they criticize. The seeming competition between science and common sense is neither, and comparing them is not too exciting. Their mutual criticism is has the advantages of both and contributes to both.

The interaction is not symmetric, though. Scientific thinking is extremist; commonsense is compromising. A sense of proportion governs the interaction, and it belongs to commonsense. We should open a discussion with a problem, seek its default extremist solutions, consider obvious criticisms against them, and proceed from there. To this end we must see the value of false answers and the greater value of some approximate answers – at least as long as the truth is hidden. This is the benefit from Popper's new theory of rationality, especially in Jarvie's modification. This makes central the following problem of demarcation: what makes the difference between rational and irrational debates?

Chapter 5.5

RATIONALIZING POLITICS

0. INTELLECTUALS IN POLITICS

The discussion regarding the political role of the intellectuals (the *intelligentsia*) is enduring. George Orwell said, intellectuals often favor violent systems, including Fascism and Stalinism. Julien Benda wrote a classic lay sermon, *Le trahison des clerkes, The Betrayal of the Intellectuals* (1926). The commonest current view is that at best intellectuals do not care sufficiently for the quality of current politics. Let us observe those intellectuals who do care, then. Some of them are outstanding watchdogs defending public morality. The general view is that they should contribute by their expertise. This is an error. Improving politics need not involve experts in government and there is little hope that their involvement would rationalize politics. Current views regarding the role of experts are in conflict. This enables governments to misuse their service. Hence, the right question is not who can expertly make government more rational; it is, what will make governments wish to behave more rationally. The most difficult and paradoxical problem in this context is, should a politician not pass qualifying exams like physicians and lawyers and accountants. If yes, which, and if not, why not. Though this question is familiar, it is seldom under discussion. There is almost no debate between advocates of the two leading ideologies, democracy and technocracy. Most intellectuals are indifferent to the conflict between these two political philosophies. Is this indifference legitimate? The popular answer is that it is not, that intellectuals are obliged to partake in the improvement of politics. This opinion is false and dangerous.

I. CONFLICTING MYTHOLOGIES LÉVI-STRAUSS-STYLE

The ideal of a philosopher-king is a powerful myth. Ancient Greek custom, Gilbert Murray relates, sent troubled elders to consult their heroes, namely, their most ancient of ancestors, at their gravesides. They ascribed to their ancestors profound wisdom. So did the Hebrews. The appeal to age as the seat of wisdom is ever-present. When the myth of the wise ruler underwent secularization, it transformed into the hope for rational politics. Rational politics was then confused with the secularized Kingdom of God, with Utopia. Plato, Aristotle, Confucius, and countless others, hoped for rational politics. Plato said, since politics is a goal-directed activity, the best way to serve it rationally is to study its goal. Confucius never thought of this, and Aristotle objected to it. He said, Plato's image of the ideal state suffers from excess and from shortage of detail.

The traditional debate in Western philosophy continued to concern the Ideal State, the City of God, *Utopia*. It got stuck there as it got stuck with the ancient Greek view that rationality is all or nothing, not a matter of degree. Viewing rationality as graded, we can leave the debate about Utopia and deny its immediate corollary, common to radicals and traditionalists, that politics is utopian (radical) or expedient (tradition-bound) and nothing else. Radicals attend to long-term ends, conservatives attend to short-term ends. Only the view of rationality as graded allows for the obvious fact that wise politics belong to the middle range.

Claude Lévi-Strauss has (innocently) contributed to the waning of the Greek all-or-nothing view. He said, mythical thinking goes in pairs of opposite polarities. Each pole is intolerable but they serve well enough as blends, as mediations in degrees. We may take politics as an example. It wavers between the traditionalist attention to the immediate and the radical attention to the final, and quite regularly this belittles the mediate, where valuable politics dwell. The pole opposite the (Confucian) myth of the benevolent king is the myth of the golden age, of the pre-civilized benign anarchy. The pole opposite the myth of the philosopher-king is the myth of kingship as given, be it a dynasty or tradition-minded opportunistic cooks who follow hunches and throw now and then into the pot whatever comes handy and seems proper. To this cluster belongs the myth – the true story elevated to the status of a story with a profound moral – of the failure of Plato's attempts to rationalize the process by entering the kitchen.

In accord with Lévi-Strauss, tension dwells between the desire to commit the wise to politics and the desire to appoint a determined strong leader. The lost in the forest feel two conflicting urges, to stop searching, sit down, take stock, and think, or to make a strong and perhaps senseless decision as to the direction to take, and stick with it to the end. A mixture resolves tension, says Lévi-Strauss, or else mediation does. At times it is fatal. It was the standard nonetheless; the model was the medieval rift between state and church. When church political power faded, the quest for the City of God resurged. Radicals entered the political arena to fight conservatives. These had no response then but through clericalism without faith the City of God. This was a new type of religion. It culminated with nineteenth century reactionary clericalism. The tension then had to move from its older forms to parliamentarian ones. The older forms were diverse, mixed, and confused; the modern ones were sharp. Mixing extremes became then too obvious and thus unacceptable. So the parties invited mediation. It appeared in the form of political science, international relations, and national and international bodies whose functions were to design and implement middle range plans. Failure accelerated conflict and terrible results ensued.

There still is hardly any theory of the middle range. In the long run, we are all dead, protested John Maynard Keynes against economists who had in mind only long run plans in addition to the normal, immediate ones. They had no middle range plans as an alternative to the one that he was proposing. The theory of legislation still appeals either by expediency, in accord with conventionalism, or to utopianism, in accord with radicalism. And there are almost no others. Attempts to justify political science met with the problem of rationality. Rationality, by the received opinion, equals scientific truth. There are only two traditional views about it, the same old polarization. Hence, no mediation by appeal to rationality seems possible, especially since political science, philosophy, and religion, all take political wisdom as salvation, as Utopia. Some argue that utopianism is dead. Others argue that it is the only hope. These are the poles. The truth is in between (Jarvie) Wise politicians call in wise advisers only in times of trouble. They prefer to appear as wise. Also, they do not know who is wise and they fear the wise: they may use their stay in court and take over the throne while they are at it, and secure permanent position on it.

The polarization of political thought to conservative and radical persists. So efforts to attenuate them are welcome. The attenuation of conservative thought is due to Michael Oakeshott. His traditionalism is more palatable than the traditional one as it is thoroughly modern English and opposes clericalism. This made his preaching for muddling through both reasonable and democratic. The old dichotomy between truth by nature and truth by convention was the radical version of the archaic dichotomy between nature and culture that, Lévi-Strauss says, prevails all myth-systems. Oakeshott oddly viewed his conservatism more or less the same as Popper's reformism. From the traditional viewpoint that radicals and traditionalists share, all anti-radicals are traditionalists, especially if they are anti-clerical to boot. Oakeshott could not object to the current practice of governments consulting experts. Being traditionalist and English, he had to respect tradition on this. More crucially, it is the prerogative of politicians to invite experts or not, and to follow their counsel or not. By Oakeshott's traditionalism, politicians should have received traditional education in a democratic society; and then they need not be experts.

Many political scientists and most economists are still radicals, but they attenuate their radicalism in diverse ways, thereby becoming more reasonable but less able to offer a comprehensive view. In particular, economic experts, who are so often radicals, display a mixture of radicalism and conservatism. They allow for government intervention in the system, despite their liberal opposition to it. The middle position, the reformism that aims at middle-range ends and attempts to rationalize politics and reject the practice of both grand-scale rational politics and limited action to meet expediency, the middle position is often the mere mixing of extreme positions, not very logically yet reasonably successfully all the same. Alternatively, they consciously try to avoid extremes and forge a new philosophy. These are the thinkers of the future, the rationalizers of the theory of rationality.

2. THE USE OF EXPERTS FOR CONFLICTING ENDS

The conflict of ideas and tendencies in the political arena testifies to the power of ideas but also to our ability to learn a thing or two about politics. For the sake of progress, the public and the politicians should learn this: democracy needs informed politicians and an informed public, though it defends the right to ignore politics too. Hence, it should produce incentives for public participation in politics.

Rulers may seek experts who endorse official views. Even if all experts are dissenters, some of them will serve and express official views in expectation of rewards. Marx said, we all think in line with our self-interests. Karl Mannheim said, free-floating intellectuals who are political refugees with no roots (as he himself was) are exceptions. But we need not be cynical. The ability to discriminate between experts may help reduce their misuse. For this the need is for an informed public. Assuming with Lévi-Strauss that in normal situations we mix logical extremes, it behooves governments to employ extremist experts from both extremes and implement their advice alternatively. And if each political party sticks logically with only one set of experts, then the logic of the philosophy of Lévi-Strauss forces the nation to force rulers to compromise – by electing mixed governments and legislatures, or by changing their votes now and then. Experts say that this is capricious and thus irrational, and they recommend strict adherence to logic and to expert advice. This is dangerous. Even in the most civilized countries experts-run organizations of expert public servants, such as medical or legal associations, have constantly acted in accord with (what they assumed was) the interest of their members, rather than in the public interest. Experts are anyway not likely to be enthusiastic supporters of democracy.

The possibility of informed publics with uninformed governments seems remote but it is not. Societies of experts, such as medical and legal association, learned societies, universities, and their likes, are examples for this. Members at the helm there are some of the least informed, due to a proper division of labor: the well-informed prefer to go on studying and the uninformed look for refuge in administration. This need not be so, and at times is not. But obviously there is no reason to assume that the better informed will naturally reach the helm. This situation is often satisfactory, although not when robber barons become presidents of universities and run them like corporations. This phenomenon is but a recent development, hopefully merely a passing phase.

This is but an elaboration of Popper's suggestion that democracy is the ability to overthrow governments peacefully. He added that this should suffice. In this he expressed an opinion that is too optimistic. Keeping dismissing inept governments and seeing no progress in the process, the public may easily despair of democracy (Judith Buber Agassi). Governments need a modicum of success. When the level of informed rationality, whether of the members of the government or of the public at large, is too low, then the effectiveness of governments is much too limited. (What the public in question exactly is and what its power may be, obviously depend on education, on the regime, on its institutional setup, and more.) This means that the only way to rationalize politics is that of educating the public. The severest limitation to the speed of rationalizing politics in a given country is the limit to the speed of rationalizing that country's educational system. And here we bump into a vicious circle: the political and the educational systems support each other's conservatism. This comes very closely to Oakeshott's view, since his philosophy of education values not the transmission of expertise but the transmission of the best of the country's traditions, especially social and political. Asking what tradition is best, we receive the traditional answer to it. And every tradition tends to recommend itself.

The myth of the philosopher-king survives in democratic settings as the popular idea that philosopher-kings are democratically inclined experts who consult democratically inclined expert advisors. Still within the democratic settings, but at the opposite pole, lies the idea of the expert. Yet, the rules of democracy make the same obligations on elected rulers as on all other citizens. Here then is the clash between the rule of the citizens and that of philosopher premiers. The same democratic rules bind both. Nevertheless, hopefully premiers can do better than the average. By what virtue are they superior? By virtue of being better-trained in government? By virtue of being political scientists? If so, what rules regarding expertise fit democracy? Should rules be the same for expert and for inexpert? If yes, why do we demand formal training from a physician? If not, why not demand formal training from those who run for office? Is it democratic to demand that a premier should have a degree, just the way we demand it of a physician and of a lawyer? This question is less important than that of public control. As Popper said, it matters whether governors are wise or foolish, but it matters more whether we can dismiss them once we have realized that we have elected them by mistake. This is magnificent, but insufficient. We wish to raise the likelihood that we elect wise rulers. How if at all can we do that? By avoiding older errors, to begin with. This is not so obvious, as the question is, how do we prevent such repetitions?

3. THE RISE OF TECHNOCRACY

The current widespread interest in the role of intellectuals in politics rests on a new phenomenon that enhances an older one. The growth of the popularity of technocracy and / or meritocracy is a general phenomenon that goes back at least to the rise of modern technology, of the modern secular universities and of the social sciences proper. They have contributed to the rise of expert positions in governments, as advisers and as employees, in administrations and in legislatures. These days expertise often spells academic excellence.

Attitudes to experts and expertise are ambivalent, especially in relation to positions of power. When power is in the hands of non-experts, we complain of arbitrariness; when it is in the hands of experts, we feel trapped. How would we feel were it is in the hands of an all-knowing computer? Already in the early nineteenth century Laplace fostered a dream and expressed in his book on probability. He envisaged computers replacing judges and calculate precisely the probability of guilt or innocence of accused individuals. The leading science-fiction writer Isaac Asimov described (*I, Robot*) an all-pervasive benign robot taking over world government to everyone's helpless consent. It seems the most rational thing to do, and yet. (In the very end of his *Foundation* series a robot takes over galactic government and there is no one to complain.) Like it or not, robots do have great advantages over us. They are disinterested. As science fiction writers, East and West, keep telling us, they are capable of endless self-sacrifice (simply for want of any self). Intellectuals are traditional bastions of criticism of power, of opposition to power, of attempts to curb power, but when they are effective they are subject to both threats and rewards (Don K. Price. Already Confucius found that his high position was a bribe. He resigned as he found it was no position but mere means for curbing his criticism. He overcame temptation, but it was not easy. For a robot it is.

If robots are untouchable, they are also not compassionate. And there is the rub. Untouchable, heartless experts are no better than robots. Even if we forget that power corrupts and assume that experts stay honest, we may doubt their concern. Empirical evidence suggests that higher levels of expertise go with lower levels of patience and of compassion. Perhaps they are too busy developing their expertise to have time for compassion, perhaps they escape from embarrassing compassion into the den of narrow specialty.

The competition is between democrats and technocrats, between champions of the common people and advocates of efficiency, between humanists and pragmatists. Both parties are losing the debate, and the fate of democracy is thus in the balance. Hence the sense of great urgency and the ambivalence in the debates on experts, efficiency, and automated bureaucracy. The humanists lose because they do not pay enough attention to efficiency, though it includes the efficiency of saving lives and of making life worth living (Mario Bunge). The technocrats lose as they mistake efficient means for ends. They deny the charge. They say, the ends are clear enough, but the means are not: the spirit is willing but the flesh is weak – the physical, technical and organizational flesh. So, they seldom discuss the question, what ends are most urgent. They say, these matters are trivial. In truth priority is highly problematic; we scarcely know how to begin to study it. They say, the spirit is willing but the flesh is weak. In truth their spirit is dead.

Since both views are erroneous, we need a third, perhaps a *via media*. But, instead, perhaps as a reasonable stop-gap, we use their mixture Lévi-Strauss-style. Except that here we exceed the limit of his theory. Those who cook myths and blend them in proportions appropriate for the pragmatic predicament, Lévi-Strauss's celebrated *bricoleurs*, are shamans, witchdoctors, gurus, and street-corner philosophers. In the modern world they are small-time politicians who view themselves as necessary parts of the cosmic order. They constantly exaggerate, at times one way, at times the other. Powerful leaders of democracies, where humanism is still a political force, at times move under public pressure and then they try to reduce arbitrariness and obfuscation by consulting experts. At other times, they resist public pressures and try to reduce them by calling in anyone who can restore the faith in the current government. This is a most remarkable affair.

Efforts to restore faith in politicians, to bridge credibility gaps, are a phenomenon that historians know well and theoreticians have hardly studied. We still find puzzling episodes such as those in which in Renaissance England Thomas à Beckett and Thomas More played central roles and paid with their heads. One reason for this is that rulers acted in defiance of public opinion. I have mentioned the appointment of Confucius to a fictitious advisory position –

as bait, the report goes, but not what for. It was not only to take off the heat of his criticism but also to gain for it the credibility that it could not gain by silencing him otherwise. The role the authorities expected him to play is presumably somewhat similar to the one that Yevtushenko played in Soviet Russia at its last days. He was a court-jester playing the mock-critic mock-poet. Those who praised him are now happy to forget.

The paucity of open criticism in Russia, until recently at least, is the cause of its bureaucratic clumsiness. Solzhenitsyn ridiculed this in the midst of his heartbreaking descriptions of Soviet cruelties (*First Circle* and *Gulag II*), as well as its attempt at technocracy in the face of a drive of its dissidents for democracy. Erik P. Hoffman says, "In general, Soviet writings in the scientific management of society focus on the *methods*, not on the *purposes* of management" (*Problems of Communism*, 26, May-June 1977, p. 60). There, he added, "decisions are merely efforts to give expression to and to implement the general priorities, policies and values determined in the authoritative 'ideological' (i.e., political) sphere." One may think this an appropriate division of labor between experts – between experts on means and experts on ends. This is the crux of the problem, then. One expert should judge social ends, another social means. But, as Roy Medvedev, one of the leading dissidents of the Soviet Union, tells us, the social ends of the political leaders of Russia was to avoid and prevent all discussion of ends (*On Socialist Democracy*, 1975, Chapter One).

Oddly, the radical individualist movement exhibits the same effort to avoid debating social ends and to center on sheer technocracy, viewing its role only relative to means. In the peak of its revolutionary fervor, in the eighteenth century, the movement rejected all social ends. All it admitted was individual ends, entirely individually determinable, and the right to convene in order to coordinate actions. The most ardent advocates of this attitude today are the Chicago school of economists, especially Milton Friedman. Yet, he has endorsed compulsory education in a mild form. His fans justify this on selfish grounds: your failing to educate your children causes me some disutility. This argument is a shamefaced admission of the limitation of traditional individualistic philosophy. Contrary to it, we devise social policies to exercise choice of social ends, such as having citizens educated. But how much of the public resources should we put into education? This question is still not under suitable debate. We are too much under the spell of technocratic philosophy.

The American Association for the Advancement of Science began a trend when it proposed to reduce the number of admissions to some university programs, in response to the decline of the market's ability to absorb new experts. By classical liberal political economy, the system should put no artificial restrictions on the choices of potential candidates. To deviate from this we need a better theory and new criteria for a planned social policy. In their absence, and in the almost total absence of social plans, professional interests replace the social interest. Experts plan accordingly; they no longer leave ends to others, not even to experts in diverse fields. (To do so invites the super-expertise of the philosopher king.)

4. EXPERTISE AS CONSENSUS

The reliance on experts, said Shaw, is the mark of modern credulity. How do we distinguish the reliable expert from the one who pretends to be knowledgeable? We can pitch alleged experts against each other. But then loyalty makes them close ranks. So we keep them ignorant of the advice of peers and compare their pronouncements. But then they develop a consensus – both as to who is expert and as to what is the right advice for them to offer. They form guilds, follow leaders, and fake expertise in areas that have none. If they do not have a consensus, then they fake one; see the manual for psychotherapists, the DSM. Still, there is progress. All medicine was indisputably foul until the end of the nineteenth century. Not all physicians were charlatans, but hardly any spoke against the phony consensus. Florence Nightingale needed both science and courage on her side to fight the establishment around the middle of that century. And even that was no guarantee for success. Her immediate predecessor, Semmelweis, had both science and courage on his side and he was defeated. Joseph Lister and Louis Pasteur then fought and won a total victory. We should try to apply the lesson from that battle to other fields.

We have thus bumped into the new problem of demarcation of science, Popper's version of it in 1935, when he endorsed the status of Einstein and rejected the claim for this status of Marx and of Freud. I am loathe to ascribe the view presented below (in this paragraph) to Popper himself, since I have heard him clearly, at times enthusiastically, endorse it and at other times no less enthusiastically, and much more emphatically, reject it. Yet, for my part, I see no other point in the problem of demarcation than the question of the authority of science. Whatever this authority may be, a claim for scientific status is but the claim for that authority. Marx and Freud, each of them had his own claim for the scientific status of his ideas. It was in both cases the demand for the unanimous endorsement of their ideas by all and sundry. They also based their claims on the assertion that their ideas are certain. Though Popper rejected certainty and all its surrogates in a most revolutionary and consistent and unqualified move, I cannot shake off the impression I have from his classic book and his later autobiographic writings, that he saw the claim for scientific status as the demand for unanimous endorsement – even though in a new and more reasonable stance. It was still the endorsement of an idea – though no longer once and for ever, and coupled with the endorsement of attempts to overthrow it. This new endorsement Popper awarded Einstein and denied Marx and Freud.

New ideas invite efforts, says Popper, to try to find out whether they are open to criticism. If they are, researchers should try to criticize them. Even if they withstand well all severe criticism, the onslaught on them should not cease. Still, as long as they withstand the onslaught, it is permissible to uphold them. This is permitting the already permissible. But there is more to it. There is the demand to teach ideas. This is important but quite different: some false ideas enter the textbooks and some true ones do not. There is, finally, the permission to apply a theory: applications demand success in passing prior tests in accord with legally prescribed rules.

This is my reading of Popper; it may be false. Nevertheless, let me base a sociological observation on it. My reading or misreading is the same as the officially accepted one – both in the sense that experts ascribe this view to Popper (young and old as well, contrary to his express statements), and in the sense that both disciples and bystanders now usually – there are a few exceptions – consider this view true. For, they deem faith in scientific ideas in some sense obligatory.

Suppose the lay public shows no understanding of some field. They may still consider unanimity a necessary but not a sufficient condition for it to be scientific. But even when there is a *bona fide* claim for scientific status, even within a body of expert specialists, there may be dissent. There may always be the old fogies who refuse to march with the times, the embittered failures, the cranks who have managed to stay inside, and the rebels without causes. We cannot say, on the strength of the mere existence of dissent, that there is no ground for unanimity – especially when we are too ignorant to dismiss dissent as cranky.

This is a practical problem. It merits a practical solution. Popper has observed that the inexpert non-specialist may well be in position to assess critical standards. If this is agreed, it becomes easy to modify the demand for unanimity to take care of established dissenters without expelling them. This is not to say that critical appraisal require unanimity. The commonwealth of learning demands unanimity only in the admission of repeated factual reports. There are, besides, the high critical standard of astronomy and the low critical standard of astrology. No unanimity in astrology will make it respectable, and dissent in physics never caused disrespect. There is a limit to this: economists greatly differ from sociologists and this does not gain sufficient airing. They say, it is irresponsible to cross lines of specialties, as one should stay with one's expertise. This is a poor excuse. Experts must offer some judgments outside their specializations. They offer thereby valuable opportunities for outside scrutiny of their opinions.

Let us consider seriously for a while the vulgar reading of Popper's view that makes it obligatory *pro tem* to endorse a highly corroborated theory, namely, a refutable theory that has weathered criticism impressively. It is essential that assessments of such situations – that Popper has called corroboration – are open to non-specialists, even though they cannot follow the specialist studies in detail. Otherwise, the obligatory endorsement is blind. Were the vulgar reading of Popper irrational, it would be quite improper to endorse it even for the sake of the present discussion. Experts are then specialists capable of exercising their critical faculties *vis-à-vis* their own specialization, and in a manner open to public inspection. When they corroborate a theory, by-and-large they agree about the high degree of corroboration they grant it. They then accept it with a unanimity that they impart to the lay public. The lay public is able to examine the situation within the specialization, even though only cursorily.

So much for the received opinion on the credibility of any scientific idea as tentative-yet-obligatory-while-it-lasts. Let us grant this for the time being.

An aside for philosophically-minded readers. Popper's theory of corroboration is problematic. Here it appears in its vulgar version. It yields a conclusion that the vulgar deem the best solution to the problem of induction. The wish to solve that problem is the wish to rescue the authority of science from the skeptic critical assault in order to restore rationality. It is superior to the extant alternative current solutions. The inductivist solution is the recommendation that we should accept the hypothesis that has the highest degree of probability in the light of the total available evidence. The conventionalist-instrumentalist school does not admit any authority to scientific opinion as it views hypotheses as mere tools for prediction. It is an admission of defeat. Popper said, the failure to overthrow a theory despite its being *a priori* highly vulnerable to empirical criticism is enormously convincing. According to the vulgar reading of his texts, this is his solution to the problem. (By the same reading, the scientific consensus concerns convincing theories.) The idea that science prescribes the right conviction is vulgar, as science clearly concerns opening of horizons, not fixing them.

This holds particularly well concerning the acceptability of expert opinions – as long as the experts are critically minded. We are now prepared to return to the central question of this discussion. Should politicians be experts on social affairs?

5. THE ARGUMENT AGAINST TECHNOCRACY

The only viable competitor to democracy is technocracy. To refute the idea that technocracy is the right regime then is to defend the idea that democracy is. More one cannot achieve, as all contemporary *avant-garde* democratic thinkers concede. They rightly emphasize that this claim is the true spirit of modern democracy. It is not a good regime, observed Sir Winston Churchill wryly, but it is the best available.

As technocrats center on the efficiency of means, a change of ends causes them trouble. So they prefer to leave ends alone. They do make one exception, though. They often speak of stability as having top priority for all efficient technocrats. This clashes with their efforts to take over stable societies and raise their efficiency: they should fear destabilization. Quite generally, efficiency is not the best means for stability. (The most stable societies in human history, ancient Egypt, China, and India, were inefficient. Experts may consider their irrigation systems the cause of their stability. The cases of the Euphrates-Tigris basin and of Peru refute this.) All this is gratuitous: we may grant that some societies need urgently some measure of stability as first aid, and we use then expert advice. But what should we do in other cases? It is silly to value stability (or the free market, or anything else) as a cure for all of our ills.

The chief and most devastating argument against technocracy is simple and decisive. The only possible measure of efficiency is relative to a given end. With no end in mind, the very concept of efficiency is meaningless. This argument often raises incredulous responses. It is hard to view technocracy as so very silly. Also, democracy and technocracy are mutually exclusive, and those who endorse democracy as a matter of course often refuse to see in technocracy a threat to democracy. Admittedly, the concept of democracy is so loose as to allow lots of variation, including variation on the role of experts within the bounds of democracy. What roles can experts play in democracy much depends on the ends that democracy permits. Can experts attend to undemocratic ends? Do they? No. They risk the democracy when they form efficient ways to determine ends; they then establish ends that exclude democracy. They use undemocratic means to those ends. The blanket title for such means is paternalism. This is the idea that Father knows best, that Father knows what is good for you. Established religions are often paternalist. Communism began as liberal and ended paternalist.

A deep paradox may be lurking here. (Jean-Paul Sartre called it the absurd.) Possibly, the venture of humanity as a whole is self-defeating because when humanity will reach the ceiling of its potentiality all challenge will disappear. It is doubtful that we can reach a ceiling. (Sartre spoke of perfection, which is worse on many counts.) Our immediate aim is meanwhile to help people become autonomous. And we have to do so by education, by means of psychology and of sociology and of economics and of political science and so on. This is what democratization needs. It is just the process rendering a higher level of political autonomy attainable for the many. And, experience shows, the best means for democratization is the education for democracy by training in democracy. An increasing number of thinkers today rightly identify democracy as the process of democratization. We do need experts to serve the diverse ends which democracy may determine; democracy can best use experts for efficient democratization.

This will not render democracy stable. It is reasonable to suppose that the ends of defending democracy and of striving for stability are in conflict. At least history offers evidence pointing at this conflict. Stable regimes were not democratic and democracies that went for stability collapsed. The greatest technocratic argument against democracy is – logically, it must be – a matter of efficiency. People do not know what they want; but that is the least of their trouble. Providing undemocratically what people say they wish, is more efficient than doing so democratically: democratic decisions are compromises that fall between stools. Democrats must differ on this basic technocratic thesis on many counts. Decisions that fall between stools are valuable as lessons in democracy. Democratic compromises are better than efficient bureaucratic edicts. Democratic debate often leads to the correction of our greatest errors when still in planning stages, to the evolution of new ideas, to cases where all parties to the debate withdraw their initial proposals in favor of new, exciting ones that they forge together. Parkinson tirelessly points at an important fact: today democracy and efficiency go together. (The argument is not watertight. Nazi Germany could develop efficiently, perhaps. Even if Carl Zuckmayer is right and it could not, since suspicion and wickedness made it decay from within, the matter is still open.) What is efficient in democracy is its flexibility, its ability to absorb new ideas and attempt to implement them, and, most of all, its ability to boost expressions of respect and appreciation for what is the best in humanity.

Technocracy is different. It is shy of new ideas and amenable to streamlining old ones. Technocrats have a predilection for stability. This is no accident: they are political conservatives who hope for little innovation: there are no experts in the technique of expanding horizons, and so it does not count. Reforming the efficiency of a system without changing its aims also precludes radical reform of means, as ends and means always interact, often in unforeseeable ways. It is the best defense of the *status quo*, as it is inherently uncritical and opposed to expanding horizons.

6. THE INTELLIGENT DEMOCRAT

Democratization is open to diverse interpretations. In all of them, the paradox of democracy looms large. It has accompanied democracy since its dawn, as Popper has explained. It holds quite generally: the electorate may choose to banish democracy. Some think we ought to block this, to allow only *bona fide* democrats stand for election. This will only aggravate matters. But the paradox may also work inadvertently. Due to a lack of either interest or efficiency, or due to some other sort of error, the electorate may elect democratically people indifferent or even hostile to it. This is easier to cope with, though not easy and with no proof of success. It is, says Popper, the most significant characteristic of democracy that it allows for the relatively easy dismissal of rulers elected by mistake. Democratization is largely the improvement upon extant mechanisms for correcting the errors of the electorate. Otherwise, democratization may require a bloody revolt, says Popper (including tyrannicide), which is permissible only as the last resort (*The Open Society and Its Enemies*, ii, 151).

Tyrants these days pretend that they rule democratically. What makes a tyranny look like democracy? Or a meritocracy? Or an aristocracy? The cost of candidature for office amounts to limiting it to the wealthy or to those who can trade favors with the wealthy, making the system much less egalitarian than it seems. This holds for posts of mandarins in traditional China and of senators in the United States. It makes the Chinese system much less egalitarian than westerners first thought it was; now they deem it aristocratic. Similarly, though to a much lesser degree, of course, concern today with the plutocratic character of the Senate of the United States may lead to improvement.

What constraints are necessary for a democratic system? Many experts agree that democracy requires that a large portion of the population should be literate, or at the very least have access to radios. They tend to consider it of a supreme, perhaps crucial, import. Experts agree, and here even unanimously, that candidates for office must have some education. (Zapata, for example, was a populist and perhaps democratically inclined, but even his most ardent admirers do not think he could function as a modern democratic ruler.)

Should a democratic constitution demand of candidates for office to be literate? No one thinks so. The question is practically pointless, even were it of some theoretical interest. Voters prefer candidates for office with some education, and often they do prefer those who possess some technical efficiency. Should we require this of them? There can be no objection to the legislation, in some democracy or another, of such requirement of candidates for office for some training. But it is not essential for democracy, as it leaves open to democratic decision the question of what qualities should candidates for office have. This short argument is weighty. It does not resolve the paradox of democracy, however: the paradox as such is insoluble, which means that there is no guarantee for the survival of democracy. Democratic choice may oppose democratic values. This is always possible, and warning against it is always right, since survival is always the first item on the agenda of every regime. It is for the steering committee of each democracy to decide how much time and effort to allocate to the question, how democratic are our aims. The shameful conduct of the notorious House Committee for un-American activities in the United States shows that the question may be urgent. We never know. We can only stay alert and hope, and this is what we should do. Ultimately, we must acknowledge the shortcomings of democracy – in general and in its diverse manifestations. It still is the best we have, but it is more of a cluster of regimes than a regime proper. The choice still is wide open.

Chapter 5.6

THE SIBLINGHOOD OF HUMANITY

The most important way in which science relates to culture is *via* its basic axiom. This axiom is that science is a public enterprise, that it is open to all: it has no esoteric parts. To take a striking example, as an institution, science does not recognize the fruits of classified research, no matter how good they may be. To gain recognition as scientific, an item of information or an idea must be on public display. This justifies the label of the scientific culture "the commonwealth of learning", "the republic of science", or, broadly, "the republic of letters". (That republic includes every cultural item except the obscure – from the mysteries of theology to trade secrets.) The essential openness of science rests on the idea that we are all brothers and sisters. By tradition, the title for this feature is "the Brotherhood of Man". "The siblinghood of humanity" seems to me a better title. It is in line with the general recent upgrading of the English language.

The concept of the siblinghood of humanity is the most powerful in the whole of human history. It appears everywhere and is central almost nowhere. In the whole of known history, it has occupied a relatively high place only in cultures that value science – in parts of ancient Greece and of the modern West. Since the foundation of the United Nations Organization, this idea partakes in all human affairs on all levels – intellectual, moral, and political – significantly, though not forcefully enough. As it rests on the idea of the universal availability and power of scientific truth, it is the religion of science. All humans are siblings in their rationality. Human reason is one.

(Early in the twentieth century, some scientific leaders supported the Soviet Union, more out of the impression that the regime was in principle responsive to science, than out of political conviction. When the Soviet Union advanced the idea of proletarian science, these leaders withdrew their support for it. The idea that science is specific to social class conflicts with the idea of the siblinghood of humanity and thus with science itself.)

Science is inherently (institutionally) egalitarian and so it is politically very potent. The view that it is of a mere pragmatic value (namely, that it is of no cultural value), then, is a thinly veiled expression of hostility to the very idea of egalitarianism that the scientific ethos depends on, and thus to democracy; it rests on the hostility to democracy as such. Usually, this hostility appears within philosophy, sociology, and studies of culture, as well as within the arts (especially in novels, in plays, in movies, and, most virulently, in pretentious, high literary criticism). At times, the expressions of hostility to science comprise forceful and interesting endeavors (with the exception of the reactionary art criticism and literary criticism, as these are consistently dull and pompous – including the works of T. S. Eliot, that first-rate poet and third-rate literary critic). Indifference to science is legitimate though regrettable. (Similarly, ignorance of the cultural value of science is not as objectionable as the denial that it exists, which rests on the combination of ignorance and pomposity.) The hostility to science is objectionable, since it is political at heart, and since its politics is reactionary. Prudence requires attention to the political implications of any hostility, no matter how remote from politics it seems. (Note, however, that exposing an idea as reactionary is not dismissing it outright. Some reactionary ideas are valuable.)

Opposite to the notion of scientific truth stands relativism. It limits the social standing of science and of scientific truth to engineering – in recognition of the practical success that science enjoys. Relativists never study this success; they never attempt to explain it; they take it for granted. They never discuss the useless parts of science. They assume that there is no need to discuss science in order to pursue their aim. This aim is to show that scientific truth has no cultural value – religious, artistic, social, or political. Sometimes they say so outright. Any influence that science exerts, it exerts only indirectly, they say, through its utility. Diverse versions of relativism are rampant these days, gaining in popularity. Today's trendy versions are pragmatism and post-modernism, but the variations between all the versions of relativism matter little – even according to relativists. The appeal of the newer versions of relativism is in their tacit softening of both of its harsh components, the rejection of the idea of the siblinghood of humanity and the rejection of the universal availability of scientific truth. They seem to replace these ideas with some tempered versions of them, with the intent of admitting their force while making light of them – by ignoring them whenever possible rather than openly opposing them. The new versions are the admission that all societies are equal in that each may view itself as the very best, and in relegating science to the community of expert professional scientists. All relativists advocate the supposition that there is no harm in ignorance of science and in indifference to it.

There are two strong arguments in favor of relativism. First, scientific truth is not the absolute truth that until the revolution in physics the scientific tradition has claimed it to be. Second, scientists are experts whose knowledge is esoteric. Both of these arguments are correct but the use that relativists make of them is obviously spurious and even extravagantly so. The repeated revisions of scientific ideas are no license for equating it with Voodoo, as relativists do. And advocates of science have to make constant efforts to bring the newest and most difficult scientific ideas to the public domain. Popular science is vital for both science and the culture at large.

Science developed with the growth of the scientific culture. The scientific culture rests on the scientific worldview. This worldview belongs to a great philosophical, political, social, and cultural movement – the Enlightenment

Movement. It gave rise to opposition in the form of many reactionary movements. To this day, the major battle-line runs between Enlightenment and Reaction. The chief weapon of the Reaction is obscurantism, the hostility to reason. The forces of darkness argue, if this is the right verb, from the fact that humans live within some sort of groups – families, herds or colonies – like many other animals. Biology supports this argument in many ways. Sociology supports it by the claims that many of the people who live in modern societies are lonely and that loneliness is painful and disabling. The jargon word for this painful, disabling loneliness is "alienation". Reactionary writers bewail alienation – not out of sympathy, as they ignore poverty and misery, but as bait: a society that is a close-knit family has none of it. Not so: there is no society without loneliness.

Here is a task for advocates of enlightenment: to cope with these attacks. This task is educational. Some view this task as philosophical, political, social, or cultural. Not so. Their view goes against the right of citizens to ignore reason. The task is impossible where reason is missing; and where reason prevails, it is too easy. It is easy to parry the accusation that modern society is the cause of loneliness. Loneliness prevails in all societies, so that at the very least only detailed empirical studies can tell us in what kind of society this ill hurts most. The reactionaries do not spell out their argument, and not even their aim. If they wish to fight loneliness, then we should all join them and together seek the best and cheapest cure. If they insist that the only way to fight loneliness is by wiping out modernity, then few will support them – at least not without some further empirical study.

The relativists intend to advocate some middle position. (Martin Heidegger and Hans-Georg Gadamer, among others, seem to be saying so clearly enough. But it is hard to be sure, as these are esoteric writers, who never made sufficiently clearly decisive statements. Gadamer has ridiculed in his autobiography the relatively clear style of some philosophical works written in German, as "almost English".) They share the idea of Dr. Sun Yat-sen, who was after a socio-political structure that would accept science as technology and reject science as an item of culture. The socio-political structure that these people seem to be advocating, then, is traditional but with modern technology. This is impossible, as technology strongly interacts with its social framework. The rubberstamp example here is a report on the outcome of the introduction of steel axes into stone-age Australian aborigine tribes. Reportedly, it had an adverse effect on their socio-political structure and so it was injurious. This does not impress the reactionaries. They are willing to tolerate some social change, as long as it leaves them in their relatively secure high position. And it will, as long as their society will endorse their high esteem for some items of culture, but not for science. They have no assurance, of course, that their society will ever fully entrench the system of values that the safety of their high position in society requires. Their ideas are mere dreams. Yet, for the time being, publishing them does keep the dreamers in high positions.

There are famous dreams of a society without science (obscurantism) and there are famous dreams of a society with nothing but science (scientism). They are all dangerous – for science and for culture at large. Culture without science is possible, but it cannot survive the advent of science without efforts to suppress it with much cruelty. Science cannot survive without the scientific outlook, and the scientific outlook imports popular science, and popular science boosts the idea of the siblinghood of humanity. The siblinghood of humanity boosts the critical spirit of science, even if only because science may help fight hunger and pestilence and hopefully war too. And this makes modern culture what it is.

Yet, some reactionary critique of the Enlightenment Movement is valuable. That movement had an erroneous idea of families, herds, or colonies. Members of the Enlightenment Movement repeatedly said that they recognized these. But they recognize them merely as allegedly expressions of individual free decisions; they did not recognize them as expressions of the deep-seated need to belong. They even dismissed this need as unbecoming to the civilized. Yet, the need precedes the decision and is stronger and more widespread. The very idea of humans as social animals invites a new look. Current science divides all truth to truth by nature and truth by convention. The one is universal: the other is particular. The one is binding; the other is arbitrary. Now, this is exaggeration. The idea of science as free of all convention, namely, naturalism, is an abstraction (that scientism advocates and) that is impossible to practice. The leading nineteenth-century reactionary, Georg Friedrich Wilhelm Hegel, gave it a deathblow. “The universal man”, he said, “that is to say, the dead man, … “. The culturally most important single particular item is language, of course. (This is why reactionaries such as Hegel and as Heidegger declare their mother tongue obviously the best and the most philosophical.) In response, some very able thinkers tried to create a universal language, such as Esperanto, and such as the language that today logicians use regularly. (Modern logic plays tremendous social role in the communication-highways in computer science and technology, and thus in science in general. This is a lovely example of the role of the scientific outlook within science and within technology.) Even were efforts to create a truly universal language successful for some purposes, they would not take over the role of natural languages as cultural tools, especially not for poetry. This much has gained general recognition. But reactionaries also assign to the mother tongue an unacceptable role. They want it to bind their nation into one family to the exclusion of strangers.

Members of any specific primary herd are usually blood relations. Developed human societies reduce the role of blood. Even when the development is small, this effect is clear. What binds a human family, herd, or colony, is less their common ancestry and more their culture, says Gellner rightly. The famous legal historian Sir Henry Maine noted this almost two centuries ago. In a complementary move, the famous science-fiction writer Kurt Vonnegut declared a century later that blood is quite incidental to a family. He viewed shared values as much more important ingredients. How then do shared values bind family members together?

(Tribalists always insist on the primacy of blood. It is common to primitive nationalism and to Nazis. A slightly less primitive version of nationalism deems religion a prime uniting factor. Equating nation with congregation has led to conflicts in Northern Ireland, the Balkans, the Middle East, the India subcontinent, and South East Asia. Indeed, with few exceptions, all trouble spots on the global scene are such. Some liberal writers (including Popper) have argued against nationalism and preached the siblinghood of humanity instead. This is a mistake. Just as preaching the abolition of the nuclear family or of the extended family is a mistake, so is the preaching of the abolition of the nation. We must learn to operate with families on diverse levels, and take memberships in them in as liberal a manner as possible. That is to say, we want them to be as open as possible. The demand for travel documents has blocked the most liberal way of changing nationality, and it is becoming increasingly difficult to migrate. So, we should try to liberalize migration. This is becoming increasingly hard. The family used to be very closed, and reactionaries still speak against divorce – rather than examine its causes – as if it were the chief impediment to happy family life.)

Perhaps shared values do not always bind. Few cultures admit that they do. Cultures, however, always bind – by systems of preferences, positive and negative, of likes and dislikes. Even the most intimate and immediate preferences, such as tastes in food and sex, not to mention art, are initially socially determined; they are alterable, but not with ease. And the distaste for some alien preferences is as immediate and as given to influence by some social factors. Advocates of culture tend to veil this fact. They usually stress the positive, since it takes more effort to master than the negative. Dislikes come easier, and often they come as mere by-products. Familiarity with a culture renders it easy to identify others, those who do not share it. (Poor pronunciation splits only the ears of the cultivated; good pronunciation offends only the uncouth. Civilized people are tolerant of all affectation.) And this may suffice for dislikes, even intense dislikes. A culture may (and should) include the value of welcoming strangers, and of tolerating them all unless they are nasty, perhaps. But implementing these values is an uphill battle.

Many scholars preach hate. What should we do in response to their preaching? The Enlightenment movement had a simple answer, as naïve and optimistic as it was noble: ignore them. A little light, they said, dispels much of the darkness. Cases in which darkness won and light lost have sadly refuted this idea – especially the case of the rise of Nazism, since it took place within one of the most enlightened societies ever, yet it brought about the Shoah. So, some action against the venomous may be in order. What action? Perhaps we should merely learn to isolate them. As they often hide their venom behind sweet talk, we should learn to spot venom fast. This task is a bore, important as at times it admittedly is. To learn to spot hate messages under sugar-coatings is uninteresting. It becomes even very easy, once we learn to approach the task without reciprocating hate. Yet, we should not waste a minute on this task until we find empirically that it is vital enough to demand attention. Experience shows that only some hate-messages are dangerous; most hate-messages are innocuous. This is why we have survived so much of it. (The matter obviously greatly depends on context: all things being equal, the morally better immune have less to fear hate.) So, empirical study should reveal more about the dangers of hate-messages. Thus far, we have almost nothing of this kind. And so, some marginal people who want to draw attention know for sure that they will succeed doing so if they perform some act of public display of hate. We do not know under what conditions, what messages of hate require treatment and why. So, let us leave the matter until some empirical information is available.

All hate literature discriminates – openly or tacitly – between "us" and "them". (Each particular family, herd, or colony, has its own jargon words for these terms, and they keep replacing them, to stay "politically correct".) This tacit discrimination often finds its expression in the mere reference to "us" with the tacit exclusion of "them". Often easily available information shows that the point of speaking favorably of "us" is the discrimination against "them". It is very easy to beware of the venom hidden in such locutions. Seemingly, inclusive expressions, such as emphasis on unity and togetherness, also often express discriminatory exclusion. (Exclusions that are not discriminatory are usually outspoken.) For a while, the hidden venom may be mild. This is particularly dangerous, since the excluded comprise growing targets: are first they are the enemy, and then some distant non-members, and then the non-members in our midst (the Jews), and finally members who dare to disagree with official doctrines. Overall, messages of love that express hate are transparent enough. Less transparent are messages that pose as empirical findings. Suppose some empirical findings prove "us" inherently superior to "them" ("we are better off because we are better"). Rather than criticize such findings (they scarcely deserve criticism, so silly they obviously are), we may examine their use. If they come to support our cause, then we may decline this support. Suppose that they are disinterested scientific findings. Then (in line with Popper's *The Open Society and Its Enemies*) we should suggest the following. Any evidence that we are inherently superior to them should entitle them to more government funds, by way of compensation. This suggestion will lead to a radical change in the result of such comparative researches, thereby exposing its crass bias, its lack of scientific interest and scientific disinterest.

Perhaps I am missing the point. The inferiority of the inferior exempts "us" from taking care of them, since they are not our kith and kin. If so, then before the researchers began their empirical research, they have already agreed that "they" are different from "us". (All information about the I. Q. of blacks rests on a stupid or vicious assumption. It is that the quality of black or Afro-American is natural rather than conventional. Did anyone examine the relative frequency of high I. Q. among tall people? Or among owners of pot bellies? Or of its distribution among owners of the different blood-types?) This may invite an appeal to the siblinghood of humanity. Let us forego this move. Let us not deny that humans do live in families, herds, or colonies. Does this absolve humans from caring about other humans? Not quite, but for the time being we must first attend to our kin. How important for this debate, then, is it that humans live in families, herds, or colonies? Does this idea negate the idea of the siblinghood of humanity? If so, how? Does my need for a family require of me to say that my family is better than yours? Why? Bertrand Russell says, nationalism is silly, because it rests on the opinion of nationalists that their people are the best. His characterization fits only some nationalists, not all of them. And so, we may examine empirically the question, who? And then, can we help them improve? How?

Any way we look at it, the need for the nuclear family does not exclude needs of others, though they are not of the same order of priorities. This leaves intact the idea that we are all one big family. The wiser among the reactionaries admit this, but only in the abstract. The idea that families come in many sizes and shapes is ubiquitous. What about it is controversial? What about it is troublesome? The trouble is not in that my family is smaller than the whole of the human family. The trouble is in part due to the scarcity of resources, since the use of resources can and should be subject to regulations in accord with priorities. The trouble is more in the question of loyalty. It rests on the false received view that I must be loyal first to my smallest family, and then to the larger one, and so on up to the loyalty to humanity at large.

This idea has a strong hold. It is valid with no constraints as a part of the legal systems of some very primitive social structures. For example, consider feud as the major system of administering justice. If we refuse to revert to the practice of feuds, then we may also oppose war as a means for settling disputes. This is no solution of problems: it is the right way of posing them: conflicts abound, and there are conflicts between loyalties to different levels when these are in conflicts. There is no universal solution to these problems. We must take them piecemeal. Hence, the idea of cosmopolitanism is on the right lines but it solves no problem. However laudable an attitude it may be, it is a false doctrine.

How do we internalize such matters – or any matters at all? Internalization sounds difficult if not utterly impossible because it usually takes place in early childhood, usually before one can understand what one internalizes, and years before the age of the development of the ability to develop a critical faculty. Can there be internalization past puberty?

Internalization is very strong. Our most immediate response patterns, the ones that we take as most intimate and spontaneous and as parts of our own identities, are usually results of long periods of internalization. Tastes in food and in sex are powerful testimonies to that. And yet we all experience alterations of our most internalized response patterns. The classical example is the change in tastes in food that a young person undergoes upon finding new associates or a new sex-mate. Perhaps we all undergo such changes, as we mature and move away from the relatively closed societies in which we live, or in which we think we live, in order to enter ones that are more open. The most powerful move is the removal of internalized items that are objectionable, and however this is done, the first move is the comprehension of the fact that one was indoctrinated with false and harmful ideas. As familiar experiences indicate, this process may be insufficient – though when it is sufficient, it is amazingly effective. People often say, they got it in their brains but not (yet) in their hearts. What else is needed is hard to summarize. As far as the feeling for the siblinghood of humanity is concerned, it is in part very easy to come by, as we are all born with mechanisms that dispose us well towards other members of our species. (See L. Eisenberg, "The *Human* Nature of Human Nature, *Science*, 176, 14 April 1972,

123-8.) At times this inborn response pattern is inhibited. Diverse factors inhibit it, the strongest of which is the internalization of some tribalist idea, of some idea of loyalty to ones family. What we must remember is that the family in any but the very narrowest sense is an artifact. We can view the family as an abstract entity: the tribe and more so the nation is such a family. The fact is, we do feel kinship of sorts with members of the nation even though we know only very few of them. This should be instructive. What is necessary in addition is the realization, not to say internalization, of two obvious facts: that conflict on all levels is unavoidable, and that settling conflicts by force is primitive and inefficient. Hence, the siblinghood of humanity is more of a program than a doctrine: it constantly invites work on problems that it raises.

Chapter 5.7

FOR PUBLIC RESPONSIBILITY FOR SPACESHIP EARTH

0. ABSTRACT

The present global political situation is serious. It desperately invites public awareness and concern. Local solutions to global problems are of no use except when they can raise public awareness sufficiently to expand and become global. Public awareness should raise a mass-movement that may stimulate. The first thing that it should stimulate is more research. This requires that the leaders of the movement should acknowledge our ignorance of operationally feasible solutions. The movement has to seek a comprehensive view through public critical discussions and efficient means for the dissemination of minimal scientific literacy. The movement must be educational and democratic and it must encourage individual autonomy. It should develop ideas in a combination of imaginative propaganda and proper research. Philosophy can contribute significantly to this by assisting the rise of a comprehensive view. Academic research should attend to practical problems. This already happens to some extent, but in a manner that is not sufficiently comprehensive. The key is grassroots education that should enable participants in it to compel appropriate institutions to discuss global problems and enact legislation towards their global solutions.

1. PHILOSOPHY ON SPACESHIP EARTH

The very concept of Spaceship Earth that Buckminster Fuller has introduced offers a strong image of the interdependence of humans on each other for their very survival.

Our initial problem regarding our joint survival concerns problem-orientation. It is much better to offer solutions to serious problems than to discuss policies of attacking them. But in the absence of any possible solution, it is reasonable to discuss clues to it, that is to say, ways of looking for a solution. Current global problems are very great and no one has any clue as to any solution to them. So, it is better to discuss the problem-situation in general than to wait for someone to come up with a solution. How, then, can we contribute to relieving the current major global political crisis? We should accept two presuppositions. First, survival is always on the top of any political agenda. Second, today human survival is in grave danger. Let me mention once with no discussion this danger of the destruction of life on earth. It is due to the Proliferation of nuclear weapons, the Pollution of the environment at large, the increased economic gulf due to the increased Poverty of the poor nations, and the Population explosion (the four P's). These dangers reinforce each other and there is no saying how much time we have before the threat of extinction becomes irreversible.

The survival of humanity should be on top of every political agenda, global, international, national, and party-political. Clearly, it is not. Some brave, concerned, and dedicated philosophers, such as Arne Naess, have relinquished their ordinary philosophical interests and activities, in order to devote the little they can do to face the central problems of human survival. They have my profound admiration, but not my consent: there is no good reason to follow their footsteps. Even though the concern of these brave and dedicated philosophers is correct, their exclusive dedication may be excessive: possibly the service of philosophers as philosophers may be more useful then their service as political activists and propagandists.

The situation is odd: an urgent requirement for action meets with inactivity. Some activists suggest that the cause of this inactivity is indifference, and the cause of the indifference is insensitivity. They therefore bombard the media with information that is hard to be insensitive to, information that should raise compassion. It does not help. They escalate actions instead of asking, perhaps the cause of the inaction is different. It is: we need a better assessment of the risks, and we need some simple, viable plan for action. Here philosophers can help by offering some severe intellectual discipline.

Here is the place to say something about the general educational value of intellectual discipline, of certain traits of philosophy. Regrettably, they are not sufficiently generally applied. Through the ages, philosophy repeatedly introduced clarity and precision by training to examine problems and get them as much in focus as possible before studying specific solutions to them. This is extremely valuable in all cases, no less in the case of the discussion of the grave matter of human survival. I will overlook it here, as there is little to say about it and most philosophers admit it even if they do not sufficiently put it into practice. Some irrationalist philosophers say that we have no time for the niceties of intellectual discipline. They are too irresponsible to care and they will not reconsider their positions. So, let me address only those who value intellectual discipline. Not willing to preach to converts let me go on without spending time on these matters.

Many of the philosophers who have little patience for intellectual discipline do care for humanity, and they often draw attention to the fact that many highly disciplined thinkers are cold fish. We need not go into that. Suffice it to observe that both compassion and discipline are essential for efficiency, and that even if we are most efficient in our effort to save humanity, success is far from assured. Perhaps the task is hopeless, but we have to try: we do not know what we may contribute if we do. For all we know, we may make a great differenced.

2. THE NEED FOR A PHILOSOPHICAL SYNTHESIS

As to the need for synthesis, it is uncontested. There is a prejudice that analytic philosophers oppose it. Not so. They have never expressed the frivolous view that there is no need for a comprehensive view of the world and of our place in it. Their hostility was never to synthesis but only to poor syntheses, especially to philosophical systems, as their syntheses are metaphysical and in oblivion of science or of common sense. And they usually directed hostility against religion (or religion-substitutes). Let me ignore all this.

The traditional synthetic side of philosophy includes more than some search for a comprehensive view of the world and of our place in it. Some traditional philosophers have supposed that things require a broad synoptic vision of the kind that science lacks, due to its understandable specialization, perhaps, or due to the caution that may be proper in science but that may stifle all the same. These philosophers have tried very hard to develop bold speculations. Few individuals can see themselves qualified for such tasks without suffering from illusions of grandeur. In the 20th century, at least two individuals clearly qualified, Albert Einstein and Bertrand Russell, and we should still try to absorb as much from their teachings as is still relevant and useful today. Yet, clearly they did not have sufficient answers to today's urgent problems, especially since, as times move so fast, new problems accumulate and old ones may deepen, and so it is unreasonable to hope that their teachings should suffice. Still, the end of the cold war has spread a new air of optimism. Optimism is as lovely as the fresh wind of the spring, and it may stimulate the growth of new thinking. It may also lead to lethargy and stagnation. Here we see how important it is to acquire a new synoptic view of the situation. How serious is the current situation? Is the risk of proliferation of nuclear weapons bigger or smaller now that the Soviet Union is no longer and its nuclear arsenal is under much less control? I will not air such problems here. This paragraph came only to illustrate the great need for a more synoptic view. Perhaps this need is all too obvious.

Science is not sufficient for the development of a comprehensive view. This idea has angered many scientists and many analytic philosophers. This is easy to understand. All sweeping overviews are naturally speculative, not too serious, and at times even not too responsible. Speculative philosophers were often ignorant of details of contemporary science. Some of them are contemptuous of science. Ignoring today's science, they unknowingly accommodate yesterday's science. Nevertheless, though their ignorance is indefensible, their courage is commendable. So is their feeling of the urgent need for the synoptic ideas that science does not offer. Their critics admit this need. There is always a dearth of synoptic ideas. Ever since the rise of the scientific tradition, empiricist philosophy was popular with scientific researchers precisely because these researchers are suspicious of speculations. They often suggest very strongly that the safe way to comprehensive ideas is through small, serious researches devoid of illusions of grandeur. Still, what should we do in the meantime to placate our quest for a comprehensive view of the world and of our place in it?

Ernst Mach repeatedly denied that he had a philosophy. When he had to say what his comprehensive view of the world was, he said, the totality of science.

This answer is still popular. W. V. Quine advocated a variant of it. Unlike Mach, he did not boast universal knowledge of science. He simply said, much more humbly, he was willing to receive from science information about the world at large. Unfortunately, scientific information is detailed, hardly ever comprehensive. Comprehensive scientific ideas are extremely rare in the whole history of science. In particular, when it comes to questions of global politics, every decent researcher will admit vast ignorance. Little knowledge is available about the most basic general facts. Little agreement is there about the way to proceed with research. The classical hostility of science to speculations rests on classical empiricist theory of knowledge. This theory was hugely successful despite all doubts and all criticism. Today the picture is different. Things have changed, due to Hume's critique, to logic, and to the revolution in physics. Classical empiricism is less popular these days, even though most analytic philosophers of science cling to it.

This puts in a quandary those who feel the pressure and urgency of global problems and who seek a comprehensive outlook as a tool for approaching them. Hardly anyone can claim even minimal credentials for this task. So, we wait for a qualified person to perform the task, and consequently it will remain in a state of neglect. Perhaps things are not as urgent as they seem to be, even though it is becoming increasingly difficult to be sanguine about the future. But the matter is not only of assessment of the urgency. It is also a matter of responsibility. If the responsible will ignore the task, then the irresponsible will perform it. The task is of planning some global policy. The irresponsible might offer some self-serving poor plan for global policy, and it may have no competitor. If only one plan of action is on the agenda, in times of decision it wins regardless of all objections to it.

3. RESPONSIBILITY AND DEMOCRACY

Clearly, most responsible people are not too qualified. They should present as best and as clearly as they can the problems they wish to discuss and the backgrounds to these problems, in efforts to engage in them as many individuals as possible – preferably including qualified ones, but, still, as many individuals as possible. And these should all be as critically minded about it as they find possible. Facing serious global problems that demand urgent attention, with no criteria for the right way to go about them, requires at least truthfulness about shortcomings. This is essential for efforts to pull resources together. We know of no other way. In particular, we must not fear that criticism will demoralize, and we must not wait for some great thinker to emerge and solve our problems.

All this calls for policies, both educational and procedural. They should grow locally, on small scales and large. After all, this is what the followers of Arne Naess are doing anyway. It remains for me to argue that it is better to do so while staying in the academy than in the political party or the amorphous public institutions or in the street.

What should the general framework be like, within which action should take place? It should include diverse positive attitudes to science, to democracy and to individual autonomy. These are far from sufficient, but they are essential.

The proper attitude to science requires not imposing unanimity in its name. The observational basis of science comes nearest to commanding unanimity: scientific information is information well examined. It is for researchers to explain it, if their concern is theoretical, and to apply it, if their concern is practical. This is a challenge that should win recognition. Applied science needs coordination badly, and nowadays it needs global coordination even more badly. We cannot control pollution in one country: the shipping of toxic waste from rich countries to poor ones is going to hurt us all. And the practice of control of population growth locally, only in parts of the world, increases the population imbalance. The task is to comprehend the facts and to see to the creation of a movement towards global coordination of control.

Unfortunately, for decades people with much good will have made propaganda for the right causes by exaggerating the reliability of information and even fabricating scary information in order to mobilize the public. This is irresponsible on many counts. I will not discuss it here, except to say, the only asset of movements that care for global responsibility is their credibility. Carelessness about it is expensive negligence.

The movement has to educate for respect for science and for democracy. For this some scientific literacy and some familiarity with the democratic process are essential. The best place for training for minimal scientific literacy is the academy, especially since this training has to take place while offering competing ideas about what action exactly is advisable and how to reach as wide a public as possible. And scientific literacy must develop together with a proper scientific attitude, with respect for the autonomy of the researcher.

A proper discussion of democracy is too extensive to develop here. My aim here is to offer a reminder that whenever a huge and urgent task is at hand the proper democratic procedure seems exceptionally frustrating and so increasingly many people begin to hope that some experts will do things more quickly and efficiently, especially if we manage to amalgamate teams of scientific and managerial experts. This idea is deadly. It exempts the majority from all political responsibility and divests experts with uncontrolled power. It is the very opposite of the education for democratic control that is so much in need. Democratic education requires the training for coordination, and this requires practice, like swimming, so that the best democratic education is in training. This should lead to the democratization of schools, and on all levels, and the effort to democratize schools will make some authoritarian experts, especially in science education, scream blue murder. We have to know that standing in their way is part of the assignment of seeking a solution to an urgent global crisis.

The matter of autonomy is too obvious or too obscure. It is not clear what autonomy is. This much we all agree: it is a combination of self-reliance with cooperation. This cannot be an end in itself. We may acquire this mode of conduct and learn to improve it while attempting to achieve some goals that are both individual and collective. For, it rests on a mixture of egoism and altruism.

Philosophers have not noticed this simple fact, as they are busy trying to show that altruism is or is not really egoism.

Scientific literacy, grass-root democracy, and individual autonomy, are basic features that may hopefully help forge a forceful synoptic idea. Let me mention just one small item to illustrate this. Today a new social philosophy is afoot: communalism. Like many buzzwords, it has a meaning that is not clear what it is. Some of the philosophers who speak in its name oppose individual autonomy; others play it down. It is important to confront these other philosophers and ask them to say clearly, is their communalism for autonomy, against it, or indifferent to it? We have to respond to this new or renewed idea, and to do this properly we must know how it stands vis-à-vis individual autonomy. This may be difficult: in the face of the popularity of democracy many anti-democrats prefer not to speak bluntly against it. We have to develop sensitivity to this and we must learn to ask people who propose new ideas to say clearly and in detail what their view of democracy is.

Global politics is the politics of the whole humanity (as distinct from international relations, that is the politics of more than one nation). We have to try to put global politics on the map. Local power seekers will not necessarily find this idea thrilling, and then we have to win them over or neutralize them. It is easy to reject the idea in a cynical mood and say that people will always place local over global politics. This cynicism is a serious factor, but those who abide by it as a matter of course overlook counter-examples, such as the rise of nationalism that expressed a readiness of local feudal potentates to give way to central authority, often very voluntarily, on the understanding that it was in their own interest to give in a little. World security is in everyone's interest. This is really not such a difficult idea. But we should be careful and ask, under what conditions is such yielding of local power to central authorities locally beneficial and how will this be operative. And certainly the current move towards the globalization of world economy is too painful for the poor countries and the rich countries are too eager to impose it, which is unwise even though it may be for a good cause.

We see here why it is difficult to mobilize people for this great cause: the process involves too many unsolved problems. The right move is first and foremost to give vent to the fears concerning them, and to discuss them openly on global platforms. We should always remember that no one wants to be the only volunteer for the cause that can offer remunerations only if it gains momentum. This is true of all great movements, yet some of these succeeded. The analysis of their success may be crucial. The success need not be random. The movement can engineer it. Consider, for example, this question: why do people participate in harmful activities like the transfer of toxic waste from rich countries to poor ones. The people involved in the act may desist, but to no avail: they will make a sacrifice, others will take their place, and no one will profit from the sacrifice. The situation will drastically alter were such transfer illegal. Why is it not? This invites investigation and proper treatment.

To conclude, this discussion is deliberately trite, as its aim is to summarize what everyone concerned with the future of humanity must agree about, until someone comes up with a smashing revolutionary idea. Not having a new solution, I did not aim to offer more than a mere hint at a comprehensive view that may stand behind some future solutions to severe current problems, a hint that comes from the commonest of common stock. Nonetheless, the hints present us with tremendous intellectual and practical challenges. They too are not new. After all, there are now university departments whose task is to meet it – departments for ecology and for global politics. It is the putting of all this together without exaggerating its force and while stressing the great need that may – just may – give the movement its push towards becoming a democratic and scientific grass-roots movement.

BIBLIOGRAPHY

Agassi, J., 1985. *Technology: Philosophical and Social Aspects.*
—— 1990. "Global Responsibility", *J. Applied Phil.*, 7, 217-221.
Popper, Karl R., 1945. *The Open Society and Its Enemies.*
——, 1961. *Conjectures and Refutations*, Chapters 4, 17, and 19.

RECOMMENDED READING

Agassi, Joseph, *Technology: Philosophical and Social Aspects.*
Agassi, Joseph, *Towards a Rational Philosophical Anthropology.*
Becker, Carl, *The Heavenly City of the Eighteenth century philosophers.*
Berlin, Isaiah, *The Divorce between Science and Culture.*
Bronowski, Jacob, *Science and Human Values.*
Budworth, David, *Public Science, Private Views.*
Burtt, E. A. *The Metaphysical Foundations of Modern Physical Science.*
Chamisso, Adelbert, *Peter Schlemiel.*
Disraeli, Isaac, *Calamities and Quarrels in the Royal Society.*
Einstein, Albert, *Ideas and Opinions.*
Evans-Pritchard, E. E., *Lectures on Social Anthropology.*
Fehl, Noah E., *Science and Culture.*
Feyerabend, Paul K., *Knowledge without Foundations.*
Forster, E. M., "The Machine Stops".
Forster, E. M., *Two Cheers for Democracy.*
Freud, Sigmund, *Civilization and Its Discontent.*
Gellner, Ernest, *The Devil in Modern Philosophy.*
Gombrich, Ernst, *Kunst und Fortschritt: Wirkung und Wandlung einer Idee.*
Halévy, Élie, *The Growth of Philosophical Radicalism.*
Hatfield, H. S., *The Inventor and His World.*
Heine, Heinrich, *Religion and Philosophy in Germany.*
Holton, Gerald J., Editor, *Science and Culture: A Study of Cohesive and Disjunctive Forces.*
Hume, David, *Essays, Social and Political.*
Jarvie, I. C., *Concept and Society.*
Jones, R. F., *Ancients and Moderns.*
Jungk, Robert, *Brighter than a Thousand Suns.*
Koertge, Noretta, editor, *A House Built on Sand: Exposing Postmodernist Myths about Science.*
Koestler, Arthur, *The Lotus and the Robot.*
Koyré, Alexandre, *From the Closed World to an Open Universe.*
Medawar, Peter, *The Future of Man.*
Merton, Robert K., *On the Shoulders of Giants.*

Nourrisson, Jean-Felix, *Philosophie de la nature: Bacon, Boyle, Toland, Buffon.*
Oakeshott, Michael, *Rationalism in Politics.*
Poincaré, Henri, *The Value of Science.*
Pope, Alexander, *An Essay on Man.*
Popper, Karl R., *In Search of a Better World.*
Popper, Karl R., *The Open Society and Its Enemies.*
Price, Derek J. de Solla, *Science Since Babylon.*
Russell, Bertrand, *A Free Man's Worship and Other Essays.*
Russell, Bertrand, *Skeptical Essays.*
Russell, Bertrand, *The Impact of Science on Society.*
Russell, Bertrand, *The Scientific Outlook.*
Russell, Bertrand, *Unpopular Essays.*
Schrödinger, Erwin, *Science and Humanism.*
Shaw, Bernard, *Prefaces.*
Snow, C. P., *The Masters.*
Spinoza, Benedict, *Correspondence.*
Whitehead, Alfred North, *Science and the Modern World.*
Wiener, Norbert, *God and Golem, Inc.*

INDEX OF NAMES

INDEX OF SUBJECTS

Boston Studies in the Philosophy of Science

Editor: Robert S. Cohen, *Boston University*

1. M.W. Wartofsky (ed.): *Proceedings of the Boston Colloquium for the Philosophy of Science, 1961/1962.* [Synthese Library 6] 1963 ISBN 90-277-0021-4
2. R.S. Cohen and M.W. Wartofsky (eds.): *Proceedings of the Boston Colloquium for the Philosophy of Science, 1962/1964.* In Honor of P. Frank. [Synthese Library 10] 1965 ISBN 90-277-9004-0
3. R.S. Cohen and M.W. Wartofsky (eds.): *Proceedings of the Boston Colloquium for the Philosophy of Science, 1964/1966.* In Memory of Norwood Russell Hanson. [Synthese Library 14] 1967 ISBN 90-277-0013-3
4. R.S. Cohen and M.W. Wartofsky (eds.): *Proceedings of the Boston Colloquium for the Philosophy of Science, 1966/1968.* [Synthese Library 18] 1969 ISBN 90-277-0014-1
5. R.S. Cohen and M.W. Wartofsky (eds.): *Proceedings of the Boston Colloquium for the Philosophy of Science, 1966/1968.* [Synthese Library 19] 1969 ISBN 90-277-0015-X
6. R.S. Cohen and R.J. Seeger (eds.): *Ernst Mach, Physicist and Philosopher.* [Synthese Library 27] 1970 ISBN 90-277-0016-8
7. M. Čapek: *Bergson and Modern Physics.* A Reinterpretation and Re-evaluation. [Synthese Library 37] 1971 ISBN 90-277-0186-5
8. R.C. Buck and R.S. Cohen (eds.): *PSA 1970.* Proceedings of the 2nd Biennial Meeting of the Philosophy and Science Association (Boston, Fall 1970). In Memory of Rudolf Carnap. [Synthese Library 39] 1971 ISBN 90-277-0187-3; Pb 90-277-0309-4
9. A.A. Zinov'ev: *Foundations of the Logical Theory of Scientific Knowledge (Complex Logic).* Translated from Russian. Revised and enlarged English Edition, with an Appendix by G.A. Smirnov, E.A. Sidorenko, A.M. Fedina and L.A. Bobrova. [Synthese Library 46] 1973 ISBN 90-277-0193-8; Pb 90-277-0324-8
10. L. Tondl: *Scientific Procedures.* A Contribution Concerning the Methodological Problems of Scientific Concepts and Scientific Explanation.Translated from Czech. [Synthese Library 47] 1973 ISBN 90-277-0147-4; Pb 90-277-0323-X
11. R.J. Seeger and R.S. Cohen (eds.): *Philosophical Foundations of Science.* Proceedings of Section L, 1969, American Association for the Advancement of Science. [Synthese Library 58] 1974 ISBN 90-277-0390-6; Pb 90-277-0376-0
12. A. Grünbaum: *Philosophical Problems of Space and Times.* 2nd enlarged ed. [Synthese Library 55] 1973 ISBN 90-277-0357-4; Pb 90-277-0358-2
13. R.S. Cohen and M.W. Wartofsky (eds.): *Logical and Epistemological Studies in Contemporary Physics.* Proceedings of the Boston Colloquium for the Philosophy of Science, 1969/72, Part I. [Synthese Library 59] 1974 ISBN 90-277-0391-4; Pb 90-277-0377-9
14. R.S. Cohen and M.W. Wartofsky (eds.): *Methodological and Historical Essays in the Natural and Social Sciences.* Proceedings of the Boston Colloquium for the Philosophy of Science, 1969/72, Part II. [Synthese Library 60] 1974 ISBN 90-277-0392-2; Pb 90-277-0378-7
15. R.S. Cohen, J.J. Stachel and M.W. Wartofsky (eds.): *For Dirk Struik.* Scientific, Historical and Political Essays in Honor of Dirk J. Struik. [Synthese Library 61] 1974 ISBN 90-277-0393-0; Pb 90-277-0379-5
16. N. Geschwind: *Selected Papers on Language and the Brains.* [Synthese Library 68] 1974 ISBN 90-277-0262-4; Pb 90-277-0263-2
17. B.G. Kuznetsov: *Reason and Being.* Translated from Russian. Edited by C.R. Fawcett and R.S. Cohen. 1987 ISBN 90-277-2181-5

18. P. Mittelstaedt: *Philosophical Problems of Modern Physics.* Translated from the revised 4th German edition by W. Riemer and edited by R.S. Cohen. [Synthese Library 95] 1976
ISBN 90-277-0285-3; Pb 90-277-0506-2
19. H. Mehlberg: *Time, Causality, and the Quantum Theory.* Studies in the Philosophy of Science. Vol. I: *Essay on the Causal Theory of Time.* Vol. II: *Time in a Quantized Universe.* Translated from French. Edited by R.S. Cohen. 1980 Vol. I: ISBN 90-277-0721-9; Pb 90-277-1074-0
Vol. II: ISBN 90-277-1075-9; Pb 90-277-1076-7
20. K.F. Schaffner and R.S. Cohen (eds.): *PSA 1972.* Proceedings of the 3rd Biennial Meeting of the Philosophy of Science Association (Lansing, Michigan, Fall 1972). [Synthese Library 64] 1974 ISBN 90-277-0408-2; Pb 90-277-0409-0
21. R.S. Cohen and J.J. Stachel (eds.): *Selected Papers of Léon Rosenfeld.* [Synthese Library 100] 1979 ISBN 90-277-0651-4; Pb 90-277-0652-2
22. M. Čapek (ed.): *The Concepts of Space and Time.* Their Structure and Their Development. [Synthese Library 74] 1976 ISBN 90-277-0355-8; Pb 90-277-0375-2
23. M. Grene: *The Understanding of Nature.* Essays in the Philosophy of Biology. [Synthese Library 66] 1974 ISBN 90-277-0462-7; Pb 90-277-0463-5
24. D. Ihde: *Technics and Praxis.* A Philosophy of Technology. [Synthese Library 130] 1979
ISBN 90-277-0953-X; Pb 90-277-0954-8
25. J. Hintikka and U. Remes: *The Method of Analysis.* Its Geometrical Origin and Its General Significance. [Synthese Library 75] 1974 ISBN 90-277-0532-1; Pb 90-277-0543-7
26. J.E. Murdoch and E.D. Sylla (eds.): *The Cultural Context of Medieval Learning.* Proceedings of the First International Colloquium on Philosophy, Science, and Theology in the Middle Ages, 1973. [Synthese Library 76] 1975 ISBN 90-277-0560-7; Pb 90-277-0587-9
27. M. Grene and E. Mendelsohn (eds.): *Topics in the Philosophy of Biology.* [Synthese Library 84] 1976 ISBN 90-277-0595-X; Pb 90-277-0596-8
28. J. Agassi: *Science in Flux.* [Synthese Library 80] 1975
ISBN 90-277-0584-4; Pb 90-277-0612-3
29. J.J. Wiatr (ed.): *Polish Essays in the Methodology of the Social Sciences.* [Synthese Library 131] 1979 ISBN 90-277-0723-5; Pb 90-277-0956-4
30. P. Janich: *Protophysics of Time.* Constructive Foundation and History of Time Measurement. Translated from German. 1985 ISBN 90-277-0724-3
31. R.S. Cohen and M.W. Wartofsky (eds.): *Language, Logic, and Method.* 1983
ISBN 90-277-0725-1
32. R.S. Cohen, C.A. Hooker, A.C. Michalos and J.W. van Evra (eds.): *PSA 1974.* Proceedings of the 4th Biennial Meeting of the Philosophy of Science Association. [Synthese Library 101] 1976 ISBN 90-277-0647-6; Pb 90-277-0648-4
33. G. Holton and W.A. Blanpied (eds.): *Science and Its Public.* The Changing Relationship. [Synthese Library 96] 1976 ISBN 90-277-0657-3; Pb 90-277-0658-1
34. M.D. Grmek, R.S. Cohen and G. Cimino (eds.): *On Scientific Discovery.* The 1977 Erice Lectures. 1981 ISBN 90-277-1122-4; Pb 90-277-1123-2
35. S. Amsterdamski: *Between Experience and Metaphysics.* Philosophical Problems of the Evolution of Science. Translated from Polish. [Synthese Library 77] 1975
ISBN 90-277-0568-2; Pb 90-277-0580-1
36. M. Marković and G. Petrović (eds.): *Praxis.* Yugoslav Essays in the Philosophy and Methodology of the Social Sciences. [Synthese Library 134] 1979
ISBN 90-277-0727-8; Pb 90-277-0968-8

37. H. von Helmholtz: *Epistemological Writings.* The Paul Hertz / Moritz Schlick Centenary Edition of 1921. Translated from German by M.F. Lowe. Edited with an Introduction and Bibliography by R.S. Cohen and Y. Elkana. [Synthese Library 79] 1977 ISBN 90-277-0290-X; Pb 90-277-0582-8
38. R.M. Martin: *Pragmatics, Truth and Language.* 1979 ISBN 90-277-0992-0; Pb 90-277-0993-9
39. R.S. Cohen, P.K. Feyerabend and M.W. Wartofsky (eds.): *Essays in Memory of Imre Lakatos.* [Synthese Library 99] 1976 ISBN 90-277-0654-9; Pb 90-277-0655-7
40. Not published.
41. Not published.
42. H.R. Maturana and F.J. Varela: *Autopoiesis and Cognition.* The Realization of the Living. With a Preface to "Autopoiesis' by S. Beer. 1980 ISBN 90-277-1015-5; Pb 90-277-1016-3
43. A. Kasher (ed.): *Language in Focus: Foundations, Methods and Systems.* Essays in Memory of Yehoshua Bar-Hillel. [Synthese Library 89] 1976 ISBN 90-277-0644-1; Pb 90-277-0645-X
44. T.D. Thao: *Investigations into the Origin of Language and Consciousness.* 1984 ISBN 90-277-0827-4
45. F.G.-I. Nagasaka (ed.): *Japanese Studies in the Philosophy of Science.* 1997 ISBN 0-7923-4781-1
46. P.L. Kapitza: *Experiment, Theory, Practice.* Articles and Addresses. Edited by R.S. Cohen. 1980 ISBN 90-277-1061-9; Pb 90-277-1062-7
47. M.L. Dalla Chiara (ed.): *Italian Studies in the Philosophy of Science.* 1981 ISBN 90-277-0735-9; Pb 90-277-1073-2
48. M.W. Wartofsky: *Models.* Representation and the Scientific Understanding. [Synthese Library 129] 1979 ISBN 90-277-0736-7; Pb 90-277-0947-5
49. T.D. Thao: *Phenomenology and Dialectical Materialism.* Edited by R.S. Cohen. 1986 ISBN 90-277-0737-5
50. Y. Fried and J. Agassi: *Paranoia.* A Study in Diagnosis. [Synthese Library 102] 1976 ISBN 90-277-0704-9; Pb 90-277-0705-7
51. K.H. Wolff: *Surrender and Cath.* Experience and Inquiry Today. [Synthese Library 105] 1976 ISBN 90-277-0758-8; Pb 90-277-0765-0
52. K. Kosík: *Dialectics of the Concrete.* A Study on Problems of Man and World. 1976 ISBN 90-277-0761-8; Pb 90-277-0764-2
53. N. Goodman: *The Structure of Appearance.* [Synthese Library 107] 1977 ISBN 90-277-0773-1; Pb 90-277-0774-X
54. H.A. Simon: *Models of Discovery* and Other Topics in the Methods of Science. [Synthese Library 114] 1977 ISBN 90-277-0812-6; Pb 90-277-0858-4
55. M. Lazerowitz: *The Language of Philosophy.* Freud and Wittgenstein. [Synthese Library 117] 1977 ISBN 90-277-0826-6; Pb 90-277-0862-2
56. T. Nickles (ed.): *Scientific Discovery, Logic, and Rationality.* 1980 ISBN 90-277-1069-4; Pb 90-277-1070-8
57. J. Margolis: *Persons and Mind.* The Prospects of Nonreductive Materialism. [Synthese Library 121] 1978 ISBN 90-277-0854-1; Pb 90-277-0863-0
58. G. Radnitzky and G. Andersson (eds.): *Progress and Rationality in Science.* [Synthese Library 125] 1978 ISBN 90-277-0921-1; Pb 90-277-0922-X
59. G. Radnitzky and G. Andersson (eds.): *The Structure and Development of Science.* [Synthese Library 136] 1979 ISBN 90-277-0994-7; Pb 90-277-0995-5

60. T. Nickles (ed.): *Scientific Discovery.* Case Studies. 1980 ISBN 90-277-1092-9; Pb 90-277-1093-7
61. M.A. Finocchiaro: *Galileo and the Art of Reasoning.* Rhetorical Foundation of Logic and Scientific Method. 1980 ISBN 90-277-1094-5; Pb 90-277-1095-3
62. W.A. Wallace: *Prelude to Galileo.* Essays on Medieval and 16th-Century Sources of Galileo's Thought. 1981 ISBN 90-277-1215-8; Pb 90-277-1216-6
63. F. Rapp: *Analytical Philosophy of Technology.* Translated from German. 1981 ISBN 90-277-1221-2; Pb 90-277-1222-0
64. R.S. Cohen and M.W. Wartofsky (eds.): *Hegel and the Sciences.* 1984 ISBN 90-277-0726-X
65. J. Agassi: *Science and Society.* Studies in the Sociology of Science. 1981 ISBN 90-277-1244-1; Pb 90-277-1245-X
66. L. Tondl: *Problems of Semantics.* A Contribution to the Analysis of the Language of Science. Translated from Czech. 1981 ISBN 90-277-0148-2; Pb 90-277-0316-7
67. J. Agassi and R.S. Cohen (eds.): *Scientific Philosophy Today.* Essays in Honor of Mario Bunge. 1982 ISBN 90-277-1262-X; Pb 90-277-1263-8
68. W. Krajewski (ed.): *Polish Essays in the Philosophy of the Natural Sciences.* Translated from Polish and edited by R.S. Cohen and C.R. Fawcett. 1982 ISBN 90-277-1286-7; Pb 90-277-1287-5
69. J.H. Fetzer: *Scientific Knowledge.* Causation, Explanation and Corroboration. 1981 ISBN 90-277-1335-9; Pb 90-277-1336-7
70. S. Grossberg: *Studies of Mind and Brain.* Neural Principles of Learning, Perception, Development, Cognition, and Motor Control. 1982 ISBN 90-277-1359-6; Pb 90-277-1360-X
71. R.S. Cohen and M.W. Wartofsky (eds.): *Epistemology, Methodology, and the Social Sciences.* 1983. ISBN 90-277-1454-1
72. K. Berka: *Measurement.* Its Concepts, Theories and Problems. Translated from Czech. 1983 ISBN 90-277-1416-9
73. G.L. Pandit: *The Structure and Growth of Scientific Knowledge.* A Study in the Methodology of Epistemic Appraisal. 1983 ISBN 90-277-1434-7
74. A.A. Zinov'ev: *Logical Physics.* Translated from Russian. Edited by R.S. Cohen. 1983 [*see also* Volume 9] ISBN 90-277-0734-0
75. G-G. Granger: *Formal Thought and the Sciences of Man.* Translated from French. With and Introduction by A. Rosenberg. 1983 ISBN 90-277-1524-6
76. R.S. Cohen and L. Laudan (eds.): *Physics, Philosophy and Psychoanalysis.* Essays in Honor of Adolf Grünbaum. 1983 ISBN 90-277-1533-5
77. G. Böhme, W. van den Daele, R. Hohlfeld, W. Krohn and W. Schäfer: *Finalization in Science.* The Social Orientation of Scientific Progress. Translated from German. Edited by W. Schäfer. 1983 ISBN 90-277-1549-1
78. D. Shapere: *Reason and the Search for Knowledge.* Investigations in the Philosophy of Science. 1984 ISBN 90-277-1551-3; Pb 90-277-1641-2
79. G. Andersson (ed.): *Rationality in Science and Politics.* Translated from German. 1984 ISBN 90-277-1575-0; Pb 90-277-1953-5
80. P.T. Durbin and F. Rapp (eds.): *Philosophy and Technology.* [*Also* Philosophy and Technology Series, Vol. 1] 1983 ISBN 90-277-1576-9
81. M. Marković: *Dialectical Theory of Meaning.* Translated from Serbo-Croat. 1984 ISBN 90-277-1596-3
82. R.S. Cohen and M.W. Wartofsky (eds.): *Physical Sciences and History of Physics.* 1984. ISBN 90-277-1615-3

83. É. Meyerson: *The Relativistic Deduction.* Epistemological Implications of the Theory of Relativity. Translated from French. With a Review by Albert Einstein and an Introduction by Milić Čapek. 1985 ISBN 90-277-1699-4
84. R.S. Cohen and M.W. Wartofsky (eds.): *Methodology, Metaphysics and the History of Science.* In Memory of Benjamin Nelson. 1984 ISBN 90-277-1711-7
85. G. Tamás: *The Logic of Categories.* Translated from Hungarian. Edited by R.S. Cohen. 1986 ISBN 90-277-1742-7
86. S.L. de C. Fernandes: *Foundations of Objective Knowledge.* The Relations of Popper's Theory of Knowledge to That of Kant. 1985 ISBN 90-277-1809-1
87. R.S. Cohen and T. Schnelle (eds.): *Cognition and Fact.* Materials on Ludwik Fleck. 1986 ISBN 90-277-1902-0
88. G. Freudenthal: *Atom and Individual in the Age of Newton.* On the Genesis of the Mechanistic World View. Translated from German. 1986 ISBN 90-277-1905-5
89. A. Donagan, A.N. Perovich Jr and M.V. Wedin (eds.): *Human Nature and Natural Knowledge.* Essays presented to Marjorie Grene on the Occasion of Her 75th Birthday. 1986 ISBN 90-277-1974-8
90. C. Mitcham and A. Hunning (eds.): *Philosophy and Technology II.* Information Technology and Computers in Theory and Practice. [*Also* Philosophy and Technology Series, Vol. 2] 1986 ISBN 90-277-1975-6
91. M. Grene and D. Nails (eds.): *Spinoza and the Sciences.* 1986 ISBN 90-277-1976-4
92. S.P. Turner: *The Search for a Methodology of Social Science.* Durkheim, Weber, and the 19th-Century Problem of Cause, Probability, and Action. 1986. ISBN 90-277-2067-3
93. I.C. Jarvie: *Thinking about Society.* Theory and Practice. 1986 ISBN 90-277-2068-1
94. E. Ullmann-Margalit (ed.): *The Kaleidoscope of Science.* The Israel Colloquium: Studies in History, Philosophy, and Sociology of Science, Vol. 1. 1986 ISBN 90-277-2158-0; Pb 90-277-2159-9
95. E. Ullmann-Margalit (ed.): *The Prism of Science.* The Israel Colloquium: Studies in History, Philosophy, and Sociology of Science, Vol. 2. 1986 ISBN 90-277-2160-2; Pb 90-277-2161-0
96. G. Márkus: *Language and Production.* A Critique of the Paradigms. Translated from French. 1986 ISBN 90-277-2169-6
97. F. Amrine, F.J. Zucker and H. Wheeler (eds.): *Goethe and the Sciences: A Reappraisal.* 1987 ISBN 90-277-2265-X; Pb 90-277-2400-8
98. J.C. Pitt and M. Pera (eds.): *Rational Changes in Science.* Essays on Scientific Reasoning. Translated from Italian. 1987 ISBN 90-277-2417-2
99. O. Costa de Beauregard: *Time, the Physical Magnitude.* 1987 ISBN 90-277-2444-X
100. A. Shimony and D. Nails (eds.): *Naturalistic Epistemology.* A Symposium of Two Decades. 1987 ISBN 90-277-2337-0
101. N. Rotenstreich: *Time and Meaning in History.* 1987 ISBN 90-277-2467-9
102. D.B. Zilberman: *The Birth of Meaning in Hindu Thought.* Edited by R.S. Cohen. 1988 ISBN 90-277-2497-0
103. T.F. Glick (ed.): *The Comparative Reception of Relativity.* 1987 ISBN 90-277-2498-9
104. Z. Harris, M. Gottfried, T. Ryckman, P. Mattick Jr, A. Daladier, T.N. Harris and S. Harris: *The Form of Information in Science.* Analysis of an Immunology Sublanguage. With a Preface by Hilary Putnam. 1989 ISBN 90-277-2516-0
105. F. Burwick (ed.): *Approaches to Organic Form.* Permutations in Science and Culture. 1987 ISBN 90-277-2541-1

Boston Studies in the Philosophy of Science

106. M. Almási: *The Philosophy of Appearances.* Translated from Hungarian. 1989
ISBN 90-277-2150-5
107. S. Hook, W.L. O'Neill and R. O'Toole (eds.): *Philosophy, History and Social Action.* Essays in Honor of Lewis Feuer. With an Autobiographical Essay by L. Feuer. 1988
ISBN 90-277-2644-2
108. I. Hronszky, M. Fehér and B. Dajka: *Scientific Knowledge Socialized.* Selected Proceedings of the 5th Joint International Conference on the History and Philosophy of Science organized by the IUHPS (Veszprém, Hungary, 1984). 1988 ISBN 90-277-2284-6
109. P. Tillers and E.D. Green (eds.): *Probability and Inference in the Law of Evidence.* The Uses and Limits of Bayesianism. 1988 ISBN 90-277-2689-2
110. E. Ullmann-Margalit (ed.): *Science in Reflection.* The Israel Colloquium: Studies in History, Philosophy, and Sociology of Science, Vol. 3. 1988
ISBN 90-277-2712-0; Pb 90-277-2713-9
111. K. Gavroglu, Y. Goudaroulis and P. Nicolacopoulos (eds.): *Imre Lakatos and Theories of Scientific Change.* 1989 ISBN 90-277-2766-X
112. B. Glassner and J.D. Moreno (eds.): *The Qualitative-Quantitative Distinction in the Social Sciences.* 1989 ISBN 90-277-2829-1
113. K. Arens: *Structures of Knowing.* Psychologies of the 19th Century. 1989
ISBN 0-7923-0009-2
114. A. Janik: *Style, Politics and the Future of Philosophy.* 1989 ISBN 0-7923-0056-4
115. F. Amrine (ed.): *Literature and Science as Modes of Expression.* With an Introduction by S. Weininger. 1989 ISBN 0-7923-0133-1
116. J.R. Brown and J. Mittelstrass (eds.): *An Intimate Relation.* Studies in the History and Philosophy of Science. Presented to Robert E. Butts on His 60th Birthday. 1989
ISBN 0-7923-0169-2
117. F. D'Agostino and I.C. Jarvie (eds.): *Freedom and Rationality.* Essays in Honor of John Watkins. 1989 ISBN 0-7923-0264-8
118. D. Zolo: *Reflexive Epistemology.* The Philosophical Legacy of Otto Neurath. 1989
ISBN 0-7923-0320-2
119. M. Kearn, B.S. Philips and R.S. Cohen (eds.): *Georg Simmel and Contemporary Sociology.* 1989 ISBN 0-7923-0407-1
120. T.H. Levere and W.R. Shea (eds.): *Nature, Experiment and the Science.* Essays on Galileo and the Nature of Science. In Honour of Stillman Drake. 1989 ISBN 0-7923-0420-9
121. P. Nicolacopoulos (ed.): *Greek Studies in the Philosophy and History of Science.* 1990
ISBN 0-7923-0717-8
122. R. Cooke and D. Costantini (eds.): *Statistics in Science.* The Foundations of Statistical Methods in Biology, Physics and Economics. 1990 ISBN 0-7923-0797-6
123. P. Duhem: *The Origins of Statics.* Translated from French by G.F. Leneaux, V.N. Vagliente and G.H. Wagner. With an Introduction by S.L. Jaki. 1991 ISBN 0-7923-0898-0
124. H. Kamerlingh Onnes: *Through Measurement to Knowledge.* The Selected Papers, 1853-1926. Edited and with an Introduction by K. Gavroglu and Y. Goudaroulis. 1991
ISBN 0-7923-0825-5
125. M. Čapek: *The New Aspects of Time: Its Continuity and Novelties.* Selected Papers in the Philosophy of Science. 1991 ISBN 0-7923-0911-1
126. S. Unguru (ed.): *Physics, Cosmology and Astronomy, 1300–1700.* Tension and Accommodation. 1991 ISBN 0-7923-1022-5

Boston Studies in the Philosophy of Science

127. Z. Bechler: *Newton's Physics on the Conceptual Structure of the Scientific Revolution.* 1991 ISBN 0-7923-1054-3
128. É. Meyerson: *Explanation in the Sciences.* Translated from French by M-A. Siple and D.A. Siple. 1991 ISBN 0-7923-1129-9
129. A.I. Tauber (ed.): *Organism and the Origins of Self.* 1991 ISBN 0-7923-1185-X
130. F.J. Varela and J-P. Dupuy (eds.): *Understanding Origins.* Contemporary Views on the Origin of Life, Mind and Society. 1992 ISBN 0-7923-1251-1
131. G.L. Pandit: *Methodological Variance.* Essays in Epistemological Ontology and the Methodology of Science. 1991 ISBN 0-7923-1263-5
132. G. Munévar (ed.): *Beyond Reason.* Essays on the Philosophy of Paul Feyerabend. 1991 ISBN 0-7923-1272-4
133. T.E. Uebel (ed.): *Rediscovering the Forgotten Vienna Circle.* Austrian Studies on Otto Neurath and the Vienna Circle. Partly translated from German. 1991 ISBN 0-7923-1276-7
134. W.R. Woodward and R.S. Cohen (eds.): *World Views and Scientific Discipline Formation.* Science Studies in the [former] German Democratic Republic. Partly translated from German by W.R. Woodward. 1991 ISBN 0-7923-1286-4
135. P. Zambelli: *The Speculum Astronomiae and Its Enigma.* Astrology, Theology and Science in Albertus Magnus and His Contemporaries. 1992 ISBN 0-7923-1380-1
136. P. Petitjean, C. Jami and A.M. Moulin (eds.): *Science and Empires.* Historical Studies about Scientific Development and European Expansion. ISBN 0-7923-1518-9
137. W.A. Wallace: *Galileo's Logic of Discovery and Proof.* The Background, Content, and Use of His Appropriated Treatises on Aristotle's *Posterior Analytics.* 1992 ISBN 0-7923-1577-4
138. W.A. Wallace: *Galileo's Logical Treatises.* A Translation, with Notes and Commentary, of His Appropriated Latin Questions on Aristotle's *Posterior Analytics.* 1992 ISBN 0-7923-1578-2
   Set (137 + 138) ISBN 0-7923-1579-0
139. M.J. Nye, J.L. Richards and R.H. Stuewer (eds.): *The Invention of Physical Science.* Intersections of Mathematics, Theology and Natural Philosophy since the Seventeenth Century. Essays in Honor of Erwin N. Hiebert. 1992 ISBN 0-7923-1753-X
140. G. Corsi, M.L. dalla Chiara and G.C. Ghirardi (eds.): *Bridging the Gap: Philosophy, Mathematics and Physics.* Lectures on the Foundations of Science. 1992 ISBN 0-7923-1761-0
141. C.-H. Lin and D. Fu (eds.): *Philosophy and Conceptual History of Science in Taiwan.* 1992 ISBN 0-7923-1766-1
142. S. Sarkar (ed.): *The Founders of Evolutionary Genetics.* A Centenary Reappraisal. 1992 ISBN 0-7923-1777-7
143. J. Blackmore (ed.): *Ernst Mach – A Deeper Look.* Documents and New Perspectives. 1992 ISBN 0-7923-1853-6
144. P. Kroes and M. Bakker (eds.): *Technological Development and Science in the Industrial Age.* New Perspectives on the Science–Technology Relationship. 1992 ISBN 0-7923-1898-6
145. S. Amsterdamski: *Between History and Method.* Disputes about the Rationality of Science. 1992 ISBN 0-7923-1941-9
146. E. Ullmann-Margalit (ed.): *The Scientific Enterprise.* The Bar-Hillel Colloquium: Studies in History, Philosophy, and Sociology of Science, Volume 4. 1992 ISBN 0-7923-1992-3
147. L. Embree (ed.): *Metaarchaeology.* Reflections by Archaeologists and Philosophers. 1992 ISBN 0-7923-2023-9
148. S. French and H. Kamminga (eds.): *Correspondence, Invariance and Heuristics.* Essays in Honour of Heinz Post. 1993 ISBN 0-7923-2085-9
149. M. Bunzl: *The Context of Explanation.* 1993 ISBN 0-7923-2153-7

Boston Studies in the Philosophy of Science

150. I.B. Cohen (ed.): *The Natural Sciences and the Social Sciences.* Some Critical and Historical Perspectives. 1994 ISBN 0-7923-2223-1
151. K. Gavroglu, Y. Christianidis and E. Nicolaidis (eds.): *Trends in the Historiography of Science.* 1994 ISBN 0-7923-2255-X
152. S. Poggi and M. Bossi (eds.): *Romanticism in Science.* Science in Europe, 1790–1840. 1994 ISBN 0-7923-2336-X
153. J. Faye and H.J. Folse (eds.): *Niels Bohr and Contemporary Philosophy.* 1994 ISBN 0-7923-2378-5
154. C.C. Gould and R.S. Cohen (eds.): *Artifacts, Representations, and Social Practice.* Essays for Marx W. Wartofsky. 1994 ISBN 0-7923-2481-1
155. R.E. Butts: *Historical Pragmatics.* Philosophical Essays. 1993 ISBN 0-7923-2498-6
156. R. Rashed: *The Development of Arabic Mathematics: Between Arithmetic and Algebra.* Translated from French by A.F.W. Armstrong. 1994 ISBN 0-7923-2565-6
157. I. Szumilewicz-Lachman (ed.): *Zygmunt Zawirski: His Life and Work.* With Selected Writings on Time, Logic and the Methodology of Science. Translations by Feliks Lachman. Ed. by R.S. Cohen, with the assistance of B. Bergo. 1994 ISBN 0-7923-2566-4
158. S.N. Haq: *Names, Natures and Things.* The Alchemist Jābir ibn Ḥayyān and His *Kitāb al-Aḥjār* (Book of Stones). 1994 ISBN 0-7923-2587-7
159. P. Plaass: *Kant's Theory of Natural Science.* Translation, Analytic Introduction and Commentary by Alfred E. and Maria G. Miller. 1994 ISBN 0-7923-2750-0
160. J. Misiek (ed.): *The Problem of Rationality in Science and its Philosophy.* On Popper vs. Polanyi. The Polish Conferences 1988–89. 1995 ISBN 0-7923-2925-2
161. I.C. Jarvie and N. Laor (eds.): *Critical Rationalism, Metaphysics and Science.* Essays for Joseph Agassi, Volume I. 1995 ISBN 0-7923-2960-0
162. I.C. Jarvie and N. Laor (eds.): *Critical Rationalism, the Social Sciences and the Humanities.* Essays for Joseph Agassi, Volume II. 1995 ISBN 0-7923-2961-9
Set (161–162) ISBN 0-7923-2962-7
163. K. Gavroglu, J. Stachel and M.W. Wartofsky (eds.): *Physics, Philosophy, and the Scientific Community.* Essays in the Philosophy and History of the Natural Sciences and Mathematics. In Honor of Robert S. Cohen. 1995 ISBN 0-7923-2988-0
164. K. Gavroglu, J. Stachel and M.W. Wartofsky (eds.): *Science, Politics and Social Practice.* Essays on Marxism and Science, Philosophy of Culture and the Social Sciences. In Honor of Robert S. Cohen. 1995 ISBN 0-7923-2989-9
165. K. Gavroglu, J. Stachel and M.W. Wartofsky (eds.): *Science, Mind and Art.* Essays on Science and the Humanistic Understanding in Art, Epistemology, Religion and Ethics. Essays in Honor of Robert S. Cohen. 1995 ISBN 0-7923-2990-2
Set (163–165) ISBN 0-7923-2991-0
166. K.H. Wolff: *Transformation in the Writing.* A Case of Surrender-and-Catch. 1995 ISBN 0-7923-3178-8
167. A.J. Kox and D.M. Siegel (eds.): *No Truth Except in the Details.* Essays in Honor of Martin J. Klein. 1995 ISBN 0-7923-3195-8
168. J. Blackmore: *Ludwig Boltzmann, His Later Life and Philosophy, 1900–1906.* Book One: A Documentary History. 1995 ISBN 0-7923-3231-8
169. R.S. Cohen, R. Hilpinen and R. Qiu (eds.): *Realism and Anti-Realism in the Philosophy of Science.* Beijing International Conference, 1992. 1996 ISBN 0-7923-3233-4
170. I. Kuçuradi and R.S. Cohen (eds.): *The Concept of Knowledge.* The Ankara Seminar. 1995 ISBN 0-7923-3241-5

Boston Studies in the Philosophy of Science

171. M.A. Grodin (ed.): *Meta Medical Ethics*: The Philosophical Foundations of Bioethics. 1995 ISBN 0-7923-3344-6
172. S. Ramirez and R.S. Cohen (eds.): *Mexican Studies in the History and Philosophy of Science.* 1995 ISBN 0-7923-3462-0
173. C. Dilworth: *The Metaphysics of Science*. An Account of Modern Science in Terms of Principles, Laws and Theories. 1995 ISBN 0-7923-3693-3
174. J. Blackmore: *Ludwig Boltzmann, His Later Life and Philosophy, 1900–1906* Book Two: The Philosopher. 1995 ISBN 0-7923-3464-7
175. P. Damerow: *Abstraction and Representation.* Essays on the Cultural Evolution of Thinking. 1996 ISBN 0-7923-3816-2
176. M.S. Macrakis: *Scarcity's Ways: The Origins of Capital.* A Critical Essay on Thermodynamics, Statistical Mechanics and Economics. 1997 ISBN 0-7923-4760-9
177. M. Marion and R.S. Cohen (eds.): *Québec Studies in the Philosophy of Science.* Part I: Logic, Mathematics, Physics and History of Science. Essays in Honor of Hugues Leblanc. 1995 ISBN 0-7923-3559-7
178. M. Marion and R.S. Cohen (eds.): *Québec Studies in the Philosophy of Science.* Part II: Biology, Psychology, Cognitive Science and Economics. Essays in Honor of Hugues Leblanc. 1996 ISBN 0-7923-3560-0
Set (177–178) ISBN 0-7923-3561-9
179. Fan Dainian and R.S. Cohen (eds.): *Chinese Studies in the History and Philosophy of Science and Technology.* 1996 ISBN 0-7923-3463-9
180. P. Forman and J.M. Sánchez-Ron (eds.): *National Military Establishments and the Advancement of Science and Technology.* Studies in 20th Century History. 1996 ISBN 0-7923-3541-4
181. E.J. Post: *Quantum Reprogramming*. Ensembles and Single Systems: A Two-Tier Approach to Quantum Mechanics. 1995 ISBN 0-7923-3565-1
182. A.I. Tauber (ed.): *The Elusive Synthesis: Aesthetics and Science.* 1996 ISBN 0-7923-3904-5
183. S. Sarkar (ed.): *The Philosophy and History of Molecular Biology: New Perspectives.* 1996 ISBN 0-7923-3947-9
184. J.T. Cushing, A. Fine and S. Goldstein (eds.): *Bohmian Mechanics and Quantum Theory: An Appraisal.* 1996 ISBN 0-7923-4028-0
185. K. Michalski: *Logic and Time.* An Essay on Husserl's Theory of Meaning. 1996 ISBN 0-7923-4082-5
186. G. Munévar (ed.): *Spanish Studies in the Philosophy of Science.* 1996 ISBN 0-7923-4147-3
187. G. Schubring (ed.): *Hermann Günther Graßmann (1809–1877): Visionary Mathematician, Scientist and Neohumanist Scholar.* Papers from a Sesquicentennial Conference. 1996 ISBN 0-7923-4261-5
188. M. Bitbol: *Schrödinger's Philosophy of Quantum Mechanics.* 1996 ISBN 0-7923-4266-6
189. J. Faye, U. Scheffler and M. Urchs (eds.): *Perspectives on Time.* 1997 ISBN 0-7923-4330-1
190. K. Lehrer and J.C. Marek (eds.): *Austrian Philosophy Past and Present.* Essays in Honor of Rudolf Haller. 1996 ISBN 0-7923-4347-6
191. J.L. Lagrange: *Analytical Mechanics*. Translated and edited by Auguste Boissonade and Victor N. Vagliente. Translated from the *Mécanique Analytique, novelle édition* of 1811. 1997 ISBN 0-7923-4349-2
192. D. Ginev and R.S. Cohen (eds.): *Issues and Images in the Philosophy of Science.* Scientific and Philosophical Essays in Honour of Azarya Polikarov. 1997 ISBN 0-7923-4444-8

Boston Studies in the Philosophy of Science

193. R.S. Cohen, M. Horne and J. Stachel (eds.): *Experimental Metaphysics*. Quantum Mechanical Studies for Abner Shimony, Volume One. 1997 ISBN 0-7923-4452-9
194. R.S. Cohen, M. Horne and J. Stachel (eds.): *Potentiality, Entanglement and Passion-at-a-Distance*. Quantum Mechanical Studies for Abner Shimony, Volume Two. 1997 ISBN 0-7923-4453-7; Set 0-7923-4454-5
195. R.S. Cohen and A.I. Tauber (eds.): *Philosophies of Nature: The Human Dimension*. 1997 ISBN 0-7923-4579-7
196. M. Otte and M. Panza (eds.): *Analysis and Synthesis in Mathematics*. History and Philosophy. 1997 ISBN 0-7923-4570-3
197. A. Denkel: *The Natural Background of Meaning*. 1999 ISBN 0-7923-5331-5
198. D. Baird, R.I.G. Hughes and A. Nordmann (eds.): *Heinrich Hertz: Classical Physicist, Modern Philosopher*. 1999 ISBN 0-7923-4653-X
199. A. Franklin: *Can That be Right?* Essays on Experiment, Evidence, and Science. 1999 ISBN 0-7923-5464-8
200. D. Raven, W. Krohn and R.S. Cohen (eds.): *The Social Origins of Modern Science*. 2000 ISBN 0-7923-6457-0
201. Reserved
202. Reserved
203. B. Babich and R.S. Cohen (eds.): *Nietzsche, Theories of Knowledge, and Critical Theory*. Nietzsche and the Sciences I. 1999 ISBN 0-7923-5742-6
204. B. Babich and R.S. Cohen (eds.): *Nietzsche, Epistemology, and Philosophy of Science*. Nietzsche and the Science II. 1999 ISBN 0-7923-5743-4
205. R. Hooykaas: *Fact, Faith and Fiction in the Development of Science*. The Gifford Lectures given in the University of St Andrews 1976. 1999 ISBN 0-7923-5774-4
206. M. Fehér, O. Kiss and L. Ropolyi (eds.): *Hermeneutics and Science*. 1999 ISBN 0-7923-5798-1
207. R.M. MacLeod (ed.): *Science and the Pacific War*. Science and Survival in the Pacific, 1939-1945. 1999 ISBN 0-7923-5851-1
208. I. Hanzel: *The Concept of Scientific Law in the Philosophy of Science and Epistemology*. A Study of Theoretical Reason. 1999 ISBN 0-7923-5852-X
209. G. Helm; R.J. Deltete (ed./transl.): *The Historical Development of Energetics*. 1999 ISBN 0-7923-5874-0
210. A. Orenstein and P. Kotatko (eds.): *Knowledge, Language and Logic*. Questions for Quine. 1999 ISBN 0-7923-5986-0
211. R.S. Cohen and H. Levine (eds.): *Maimonides and the Sciences*. 2000 ISBN 0-7923-6053-2
212. H. Gourko, D.I. Williamson and A.I. Tauber (eds.): *The Evolutionary Biology Papers of Elie Metchnikoff*. 2000 ISBN 0-7923-6067-2
213. S. D'Agostino: *A History of the Ideas of Theoretical Physics*. Essays on the Nineteenth and Twentieth Century Physics. 2000 ISBN 0-7923-6094-X
214. S. Lelas: *Science and Modernity*. Toward An Integral Theory of Science. 2000 ISBN 0-7923-6303-5
215. E. Agazzi and M. Pauri (eds.): *The Reality of the Unobservable*. Observability, Unobservability and Their Impact on the Issue of Scientific Realism. 2000 ISBN 0-7923-6311-6
216. P. Hoyningen-Huene and H. Sankey (eds.): *Incommensurability and Related Matters*. 2001 ISBN 0-7923-6989-0
217. A. Nieto-Galan: *Colouring Textiles*. A History of Natural Dyestuffs in Industrial Europe. 2001 ISBN 0-7923-7022-8

Boston Studies in the Philosophy of Science

218. J. Blackmore, R. Itagaki and S. Tanaka (eds.): *Ernst Mach's Vienna 1895–1930*. Or Phenomenalism as Philosophy of Science. 2001 ISBN 0-7923-7122-4
219. R. Vihalemm (ed.): *Estonian Studies in the History and Philosophy of Science*. 2001 ISBN 0-7923-7189-5
220. W. Lefèvre (ed.): *Between Leibniz, Newton, and Kant*. Philosophy and Science in the Eighteenth Century. 2001 ISBN 0-7923-7198-4
221. T.F. Glick, M.Á. Puig-Samper and R. Ruiz (eds.): *The Reception of Darwinism in the Iberian World*. Spain, Spanish America and Brazil. 2001 ISBN 1-4020-0082-0
222. U. Klein (ed.): *Tools and Modes of Representation in the Laboratory Sciences*. 2001 ISBN 1-4020-0100-2
223. P. Duhem: *Mixture and Chemical Combination*. And Related Essays. Edited and translated, with an introduction, by Paul Needham. 2002 ISBN 1-4020-0232-7
224. J.C. Boudri: *What was Mechanical about Mechanics*. The Concept of Force Betweem Metaphysics and Mechanics from Newton to Lagrange. 2002 ISBN 1-4020-0233-5
225. B.E. Babich (ed.): *Hermeneutic Philosophy of Science, Van Gogh's Eyes, and God*. Essays in Honor of Patrick A. Heelan, S.J. 2002 ISBN 1-4020-0234-3
226. D. Davies Villemaire: *E.A. Burtt, Historian and Philosopher*. A Study of the Author of The Metaphysical Foundations of Modern Physical Science. 2002 ISBN 1-4020-0428-1
227. L.J. Cohen: *Knowledge and Language*. Selected Essays of L. Jonathan Cohen. Edited and with an introduction by James Logue. 2002 ISBN 1-4020-0474-5
228. G.E. Allen and R.M. MacLeod (eds.): *Science, History and Social Activism: A Tribute to Everett Mendelsohn*. 2002 ISBN 1-4020-0495-0
229. O. Gal: *Meanest Foundations and Nobler Superstructures*. Hooke, Newton and the "Compounding of the Celestiall Motions of the Planetts". 2002 ISBN 1-4020-0732-9
230. R. Nola: *Rescuing Reason*. A Critique of Anti-Rationalist Views of Science and Knowledge. 2003 Hb: ISBN 1-4020-1042-7; Pb ISBN 1-4020-1043-5
231. J. Agassi: *Science and Culture*. 2003 ISBN 1-4020-1156-3
232. M.C. Galavotti (ed.): *Observation and Experiment in the Natural and Social Science*. 2003 ISBN 1-4020-1251-9

Also of interest:

R.S. Cohen and M.W. Wartofsky (eds.): *A Portrait of Twenty-Five Years Boston Colloquia for the Philosophy of Science, 1960-1985*. 1985 ISBN Pb 90-277-1971-3

Previous volumes are still available.

KLUWER ACADEMIC PUBLISHERS – DORDRECHT / BOSTON / LONDON